STUDENT SOLUTIONS MANUAL

NANCY S. BOUDREAU

BOWLING GREEN STATE UNIVERSITY

A First Course In
BUSINESS STATISTICS

EIGHTH EDITION

McCLAVE ▪ BENSON ▪ SINCICH

Prentice
Hall

Upper Saddle River, NJ 07458

Acquisitions Editor: Kathy Boothby Sestak
Supplement Editor: Joanne Wendelken
Special Projects Manager: Barbara A. Murray
Production Editor: Dawn Murrin
Supplement Cover Manager: Paul Gourhan
Supplement Cover Designer: PM Workshop Inc.
Manufacturing Buyer: Lisa McDowell

© 2001
by Prentice Hall
Upper Saddle River, NJ 07458

Printed in the United States of America

10 9 8 7 6 5 4 3 2

ISBN 0-13-018992-8

Prentice-Hall International (UK) Limited, London
Prentice-Hall of Australia Pty. Limited, Sydney
Prentice-Hall Canada, Inc., Toronto
Prentice-Hall Hispanoamericana, S.A., Mexico
Prentice-Hall of India Private Limited, New Delhi
Pearson Education Asia Pte. Ltd., Singapore
Prentice-Hall of Japan, Inc., Tokyo
Editora Prentice-Hall do Brazil, Ltda., Rio de Janeiro

Contents

Preface

This solutions manual is designed to accompany the text, *A First Course in Business Statistics*, Eighth Edition, by James T. McClave, P. George Benson, and Terry Sincich. It provides answers to most odd-numbered exercises for each chapter in the text. Other methods of solution may also be appropriate; however, the author has presented one that she believes to be most instructive to the beginning Statistics student. The student should first attempt to solve the assigned exercises without help from this manual. Then, if unsuccessful, the solution in the manual will clarify points necessary to the solution. The student who successfully solves an exercise should still refer to the manual's solution. Many points are clarified and expanded upon to provide maximum insight into and benefit from each exercise.

Instructors will also benefit from the use of this manual. It will save time in preparing presentations of the solutions and possibly provide another point of view regarding their meaning.

Some of the exercises are subjective in nature and thus omitted from the Answer Key at the end of *A First Course in Business Statistics,* Eighth Edition. The subjective decisions regarding these exercises have been made and are explained by the author. Solutions based on these decisions are presented; the solution to this type of exercise is often most instructive. When an alternative interpretation of an exercise may occur, the author has often addressed it and given justification for the approach taken.

I would like to thank Kelly Barber for typing and creating the art work and Brenda Dobson for her assistance and for typing this work.

Nancy S. Boudreau
Bowling Green State University
Bowling Green, Ohio

Statistics, Data, and Statistical Thinking
Chapter 1

1.1 Statistics is a science that deals with the collection, classification, analysis, and interpretation of information or data. It is a meaningful, useful science with a broad, almost limitless scope of applications to business, government, and the physical and social sciences.

1.3 The four elements of a descriptive statistics problem are:

1. The population or sample of interest. This is the collection of all the units upon which the variable is measured.
2. One or more variables that are to be investigated. These are the types of data that are to be collected.
3. Tables, graphs, or numerical summary tools. These are tools used to display the characteristic of the sample or population.
4. Conclusions about the data based on the patterns revealed. These are summaries of what the summary tools revealed about the population or sample.

1.5 The first major method of collecting data is from a published source. These data have already been collected by someone else and is available in a published source. The second method of collecting data is from a designed experiment. These data are collected by a researcher who exerts strict control over the experimental units in a study. These data are measured directly from the experimental units. The third method of collecting data is from a survey. These data are collected by a researcher asking a group of people one or more questions. Again, these data are collected directly from the experimental units or people. The final method of collecting data is observationally. These data are collected directly from experimental units by simply observing the experimental units in their natural environment and recording the values of the desired characteristics.

1.7 A population is a set of existing units such as people, objects, transactions, or events. A variable is a characteristic or property of an individual population unit such as height of a person, time of a reflex, amount of a transaction, etc.

1.9 A representative sample is a sample that exhibits characteristics similar to those possessed by the target population. A representative sample is essential if inferential statistics is to be applied. If a sample does not possess the same characteristics as the target population, then any inferences made using the sample will be unreliable.

1.11 A population is a set of existing units such as people, objects, transactions, or events. A process is a series of actions or operations that transform inputs to outputs. A process produces or generates output over time. Examples of processes are assembly lines, oil refineries, and stock prices.

1.13 The data consisting of the classifications A, B, C, and D are qualitative. These data are nominal and thus are qualitative. After the data are input as 1, 2, 3, and 4, they are still nominal and thus qualitative. The only differences between the two data sets are the names of the categories. The numbers associated with the four groups are meaningless.

1.15   a.   The population of interest is all citizens of the United States.

       b.   The variable of interest is the view of each citizen as to whether the president is doing a good or bad job. It is qualitative.

       c.   The sample is the 2000 individuals selected for the poll.

       d.   The inference of interest is to estimate the proportion of all citizens who believe the president is doing a good job.

       e.   The method of data collection is a survey.

       f.   It is not very likely that the sample will be representative of the population of all citizens of the United States. By selecting phone numbers at random, the sample will be limited to only those people who have telephones. Also, many people share the same phone number, so each person would not have an equal chance of being contacted. Another possible problem is the time of day the calls are made. If the calls are made in the evening, those people who work in the evening would not be represented.

1.17   a.   The population of interest is all employees in the U.S.

       b.   The variable of interest is whether an employee is likely to remain in his/her job in the next five years if he/she is provided with mentoring.

       c.   Since the answer to the question would be either 'yes' or 'no', the variable is qualitative.

       d.   The sample is the 1000 employees in the U.S. who were actually surveyed.

       e.   Since 62% of those surveyed indicated that they would remain in their jobs for the next five years if they received mentoring, we could infer that the majority of all workers would remain in their jobs for the next five years if they receive mentoring.

1.19   a.   Length of maximum span can take on values such as 15 feet, 50 feet, 75 feet, etc. Therefore, it is quantitative.

       b.   The number of vehicle lanes can take on values such as 2, 4, etc. Therefore, it is quantitative.

       c.   The answer to this item is "yes" or "no," which are not numeric. Therefore, it is qualitative.

       d.   Average daily traffic could take on values such as 150 vehicles, 3,579 vehicles, 53,295 vehicles, etc. Therefore, it is quantitative.

       e.   Condition can take on values "good," "fair," or "poor," which are not numeric. Therefore, it is qualitative.

       f.   The length of the bypass or detour could take on values such as 1 mile, 4 miles, etc. Therefore, it is quantitative.

       g.   Route type can take on values "interstate," U.S.," "state," "county," or "city," which are not numeric. Therefore, it is qualitative.

1.21 a. The population from which the sample was selected is the set of all department store executives.

b. There are two variables measured by the authors. They are job-satisfaction and Machiavellian rating for each of the executives.

c. The sample is the set of 218 department store executives who completed the questionnaire.

d. The method of data collection is a survey.

e. The inference made by the authors is that those executives with higher job-satisfaction scores are likely to have a lower 'mach' rating.

1.23 a. Some possible questions are:

1. In your opinion, why has the banking industry consolidated in the past few years? Check all that apply.

a. Too many small banks with not enough capital.
b. A result of the Savings and Loan scandals.
c. To eliminate duplicated resources in the upper management positions.
d. To provide more efficient service to the customers.
e. To provide a more complete list of financial opportunities for the customers.
f. Other. Please list.

2. Using a scale from 1 to 5, where 1 means strongly disagree and 5 means strongly agree, indicate your agreement to the following statement: "The trend of consolidation in the banking industry will continue in the next five years."

1 strongly disagree   2 disagree   3 no opinion   4 agree   5 strongly agree

b. The population of interest is the set of all bank presidents in the United States.

c. It would be extremely difficult and costly to obtain information from all 10,000 bank presidents. Thus, it would be more efficient to sample just 200 bank presidents. However, by sending the questionnaires to only 200 bank presidents, one risks getting the results from a sample which is not representative of the population. The sample must be chosen in such a way that the results will be representative of the entire population of bank presidents in order to be of any use.

1.25 a. The population of interest is the collection of all major U.S. firms.

b. The variable of interest is whether the firm offers job-sharing to its employees or not.

c. The sample is the set of 1,035 firms selected.

d. The government might want to estimate the proportion of all firms that offer job-sharing to their employees.

1.27 a. The population of interest is the set of all New York accounting firms employing two or more professionals. There are two variables of interest: Whether or not the firm uses audit sampling methods, and if so, whether or not it uses random sampling. The sample is the set of 163 firms whose responses were useable. The inference of interest to the New York Society of CPAs is the proportion of all New York accounting firms employing two or more professionals that use sampling methods in auditing their clients.

b.   The four responses that were unusable could have been returned blank or could have been filled out incorrectly.

c.   Any time a survey is mailed it is questionable whether the returned questionnaires represent a random sample. Often times, only those with very strong opinions return the surveys. In such a case, the returned surveys would not be representative of the entire population.

2.1   First, we find the frequency of the grade A. The sum of the frequencies for all five grades must be 200. Therefore, subtract the sum of the frequencies of the other four grades from 200. The frequency for grade A is:

$$200 - (36 + 90 + 30 + 28) = 200 - 184 = 16$$

To find the relative frequency for each grade, divide the frequency by the total sample size, 200. The relative frequency for the grade B is $36/200 = .18$. The rest of the relative frequencies are found in a similar manner and appear in the table:

Grade on Statistics Exam	Frequency	Relative Frequency
A: 90–100	16	.08
B: 80– 89	36	.18
C: 65– 79	90	.45
D: 50– 64	30	.15
F: Below 50	28	.14
Total	200	1.00

2.3   a.   To form a relative frequency bar chart, we must first convert the percents to relative frequencies by dividing the percents by 100%. The relative frequency bar chart for **Banks** is:

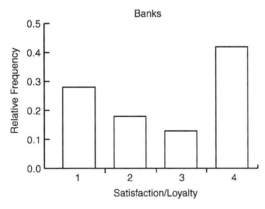

1: Totally Satisfied and Very Loyal
2: Totally Satisfied and Not Very Loyal
3: Not Totally Satisfied and Very Loyal
4: Not Totally Satisfied and Not Very Loyal

The relative frequency bar chart for Department Stores is:

1: Totally Satisfied and Very Loyal
2: Totally Satisfied and Not Very Loyal
3: Not Totally Satisfied and Very Loyal
4: Not Totally Satisfied and Not Very Loyal

b. Since the data are qualitative, we could have described them using pie charts.

c. For the banking industry, a little over a quarter of those who are totally satisfied are very loyal. This is a relatively small percentage. However, in the department stores, only 4% of those who are totally satisfied are very loyal. This indicates that very few department store customers are very loyal.

2.5 a. We must first compute the relative frequency for each response. To find the relative frequency, we divide the frequency by the total sample size, 240. For the first category, the relative frequency is 154/240 = .642. The rest of the relative frequencies are found in a similar manner and are shown in the table.

Response	Number of Investors	Relative Frequency
Seek formal explanation	154	.642
Seek CEO performance review	49	.204
Dismiss CEO	20	.083
Seek no action	17	.071
TOTAL	240	1.000

b. The relative frequency graph is:

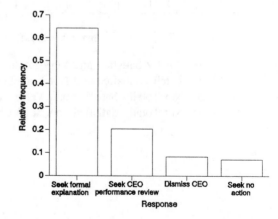

c.   If the chief executive officer and the board of directors differed on company strategy, almost 2/3 of the large investors would seek formal explanation (.642). Approximately 20% (.204) would seek CEO performance review. Very few would dismiss the CEO (.083) or would seek no action (.071).

2.7   a.   The variable measured by Performark is the length of time it took for each advertiser to respond back.

        b.   The pie chart is:

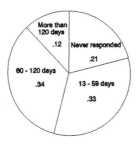

        c.   Twenty-one percent of .21 × 17,000 = 3,570 of the advertisers never respond to the sales lead.

        d.   The information from the pie chart does not indicate how effective the "bingo cards" are. It just indicates how long it takes advertisers to respond, if at all.

2.9   a.   The variable measured in the survey was the length of time small businesses used the Internet per week.

        b.   A bar graph of the data is:

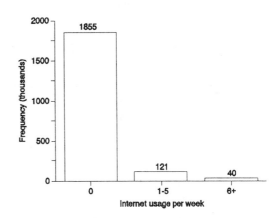

        c.   The proportion of the 2,016 small businesses that use the Internet on a weekly basis is (121 + 40)/2,016 = 161/2,016 = .08.

2.11   a.   The Pareto diagram is:

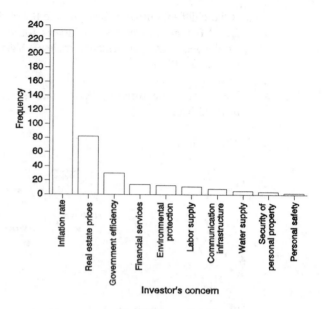

b.   The environmental factor of most concern is "Inflation rate" with 233/402 = .58 or 58% of the investors indicating this as their most serious concern. The second most serious concern was "Real Estate prices." Over 20% ((82/402) × 100% = 20.4%) of the investors chose this concern. Each of the other categories were chosen by less than 10% of the investors.

c.   Two factors out of 10 represent 20% of the factors. The two factors are "Inflation rate" and "Real estate prices." These two factors represent ((233 + 82)/402 = .78) 78% of the investors. This is very close to 80%.

2.13   To find the number of measurements for each measurement class, multiply the relative frequency by the total number of observations, $n = 500$. The frequency table is:

Measurement Class	Relative Frequency	Frequency
.5 − 2.5	.10	500(.10) = 50
2.5 − 4.5	.15	500(.15) = 75
4.5 − 6.5	.25	500(.25) = 125
6.5 − 8.5	.20	500(.20) = 100
8.5 − 10.5	.05	500(.05) = 25
10.5 − 12.5	.10	500(.10) = 50
12.5 − 14.5	.10	500(.10) = 50
14.5 − 16.5	.05	500(.05) = 25
		500

The frequency histogram is:

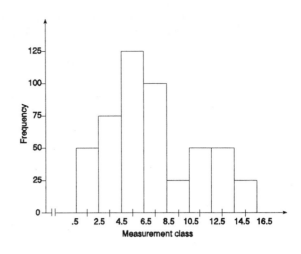

2.15   a.   This is a frequency histogram because the number of observations is graphed for each interval rather than the relative frequency.

        b.   There are 14 measurement classes.

        c.   There are 49 measurements in the data set.

2.17   a.   Using MINITAB, the stem-and-leaf displays for the two types of companies are:

            Character Stem-and-Leaf Display

```
Stem-and-Leaf of Technology Companies    N = 28
Leaf Unit = 10

    9    -0  444311000
   14     0  01334555689
    8     1  12345
    3     2
    3     3  119
```

            Character Stem-and-Leaf Display

```
Stem-and-Leaf of Industrial Companies    N = 13
Leaf Unit = 10

   (7)   -0  5333200
    6     0  004
    3     1  156
```

        b.   The stock price changes in the technology companies are much more variable than that of the industrial companies. The technology company stock price changes range from -44% to 392% while the industrial company stock price changes range from -59% to 161%. In addition, over half (7 out of 13) of the industrial companies had a negative stock price change while only 9 out of 28 of the technology companies had negative stock price changes.

        c.   In order for the stock prices to more than double, the percent change must be 100% or more. Of the 28 technology companies, 8 had stock price changes of 100% or higher. Thus, the percentage is (8/28)*100% = 28.6%.

2.19   a.   Almost half (14) of the bid prices were between $99.50 and $102.50. Seventy percent (21) of the bid prices were between $96.50 and $105.50. Only one bid price was greater than $105.50.

b.	The total number of bonds with bid prices greater than $96.50 is $3 + 14 + 4 + 1 = 22$. The proportion of the total is $22/33 = .733$.

2.21	a.	Using MINITAB, the stem-and-leaf display is:

```
Stem-and-Leaf of PENALTY         N = 38
Leaf Unit = 10000

(28)    0  (0)0(1)(1)111(2)(2)222222(3)(3)33334444899
 10     1  (0)0239
  5     2
  5     3 0
  4     4 0
  3     5
  3     6
  3     7
  3     8 5
  2     9 3
  1    10 0
```

b.	See the circled leaves in part **a**.

c.	Most of the penalties imposed for Clean Air Act violations are relatively small compared to the penalties imposed for other violations. All but one of the penalties for Clean Air Act violations are below the median penalty imposed.

2.23	a.	Using MINITAB, the three frequency histograms are as follows (the same starting point and class interval were used for each):

```
Histogram of C1        N = 25

Tenth Performance

Midpoint    Count
    4.00       0
    8.00       0
   12.00       1   *
   16.00       5   *****
   20.00      10   **********
   24.00       6   ******
   28.00       0
   32.00       2   **
   36.00       0
   40.00       1   *


Histogram of C2        N = 25

Thirtieth Performance

Midpoint    Count
    4.00       1   *
    8.00       9   *********
   12.00      12   ************
   16.00       2   **
   20.00       1   *


Histogram of C3        N = 25

Fiftieth Performance

Midpoint    Count
    4.00       3   ***
    8.00      15   ***************
   12.00       4   ****
   16.00       2   **
   20.00       1   *
```

b. The histogram for the tenth performance shows a much greater spread of the observations than the other two histograms. The thirtieth performance histogram shows a shift to the left—implying shorter completion times than for the tenth performance. In addition, the fiftieth performance histogram shows an additional shift to the left compared to that for the thirtieth performance. However, the last shift is not as great as the first shift. This agrees with statements made in the problem.

2.25 a. The percentage of realizations of V with values ranging from .425 to .675 is approximately $13.5 + 11.5 + 8 + 6 + 5.75 = 44.75\%$.

b. The norm constraint level V that has approximately 10% of the realizations less than it is approximately .325.

2.27 a. $\sum x = 3 + 8 + 4 + 5 + 3 + 4 + 6 = 33$

b. $\sum x^2 = 3^2 + 8^2 + 4^2 + 5^2 + 3^2 + 4^2 + 6^2 = 175$

c. $\sum (x - 5)^2 = (3 - 5)^2 + (8 - 5)^2 + (4 - 5)^2 + (5 - 5)^2 + (3 - 5)^2 + (4 - 5)^2$
$+ (6 - 5)^2 = 20$

d. $\sum (x - 2)^2 = (3 - 2)^2 + (8 - 2)^2 + (4 - 2)^2 + (5 - 2)^2 + (3 - 2)^2 + (4 - 2)^2$
$+ (6 - 2)^2 = 71$

e. $\left(\sum x\right)^2 = (3 + 8 + 4 + 5 + 3 + 4 + 6)^2 = 33^2 = 1089$

2.29 a. $\sum x = 6 + 0 + (-2) + (-1) + 3 = 6$

b. $\sum x^2 = 6^2 + 0^2 + (-2)^2 + (-1)^2 + 3^2 = 50$

c. $\sum x^2 - \dfrac{\left(\sum x\right)^2}{5} = 50 - \dfrac{6^2}{5} = 50 - 7.2 = 42.8$

2.31 Assume the data are a sample. The sample mean is:

$$\bar{x} = \frac{\sum x}{n} = \frac{3.2 + 2.5 + 2.1 + 3.7 + 2.8 + 2.0}{6} = \frac{16.3}{6} = 2.717$$

The median is the average of the middle two numbers when the data are arranged in order (since $n = 6$ is even). The data arranged in order are: 2.0, 2.1, 2.5, 2.8, 3.2, 3.7. The middle two numbers are 2.5 and 2.8. The median is:

$$\frac{2.5 + 2.8}{2} = \frac{5.3}{2} = 2.65$$

2.33 a. $\bar{x} = \dfrac{\sum x}{n} = \dfrac{7 + \cdots + 4}{6} + \dfrac{15}{6} = 2.5$

Median $= \dfrac{3 + 3}{2} = 3$ (mean of 3rd and 4th numbers, after ordering)

Mode $= 3$

b. $\overline{x} = \dfrac{\sum x}{n} = \dfrac{2 + \cdots + 4}{13} = \dfrac{40}{13} = 3.08$

Median = 3 (7th number, after ordering)

Mode = 3

c. $\overline{x} = \dfrac{\sum x}{n} = \dfrac{51 + \cdots + 37}{10} = \dfrac{496}{10} = 49.6$

Median $= \dfrac{48 + 50}{2} = 49$ (mean of 5th and 6th numbers, after ordering)

Mode = 50

2.35   The median is the average of the middle two numbers (since n is even) once the measurements have been arranged in order. The median is:

$$\text{Median} = \dfrac{106{,}161 + 152{,}240}{2} = 129{,}200.5$$

The median number of passengers per cruise line is 129,200.5. Half of the cruise lines had total number of passengers greater than 129,200.5 and half had less.

The mean is: $\overline{x} = \dfrac{\sum x}{n} = \dfrac{1{,}581{,}058}{8} = 197{,}632.25$

The average number of passengers per cruise line is 197,632.25. Since the mean is quite a bit larger than the median, the data are skewed to the right.

2.37   a. From the printout, the mean number of Superfund sites per county is 5.24, the median is 3 and the mode is 2. Since the mean is larger than the median and the median is larger than the mode, the data are skewed to the right. The average number of sites per county is 5.24. Half of the counties have three or fewer Superfund sites. More counties have two sites than any other number.

b. The county with the most sites has 48 sites. This number is much larger than any other in the data set.

c. We know that $\overline{x} = \dfrac{\sum x}{n}$. Thus, $\sum x = n\overline{x} = 75(5.24) = 393$. If we eliminate the value of 48 from the data set, then $\overline{x} = \dfrac{\sum x}{n} = \dfrac{(393 - 48)}{75 - 1} = \dfrac{345}{74} = 4.66$

If we eliminate 48 from the data set, the median is the average of the middle two numbers and is Median $= \dfrac{3 + 3}{2} = 3$

The mode is the number occurring most frequently and is still 2.

Thus, the median and the mode did not change, while the mean dropped from 5.24 to 4.66.

2.39   a. Due to the "elite" superstars, the salary distribution is skewed to the right. Since this implies that the median is less than the mean, the players' association would want to use the median.

b. The owners, by the logic of part **a**, would want to use the mean.

2.41   a.   For the "Joint exchange offer with prepack" firms, the mean time is 2.6545 months, and the median is 1.5 months. Thus, the average time spent in bankruptcy for "Joint" firms is 2.6545 months, while half of the firms spend 1.5 months or less in bankruptcy.

For the "No prefiling vote held" firms, the mean time is 4.2364 months, and the median is 3.2 months. Thus, the average time spent in bankruptcy for "No prefiling vote held" firms is 4.2364 months, while half of the firms spend 3.2 months or less in bankruptcy.

For the "Prepack solicitation only" firms, the mean time is 1.8185 months, and the median is 1.4 months. Thus, the average time spent in bankruptcy for "Prepack solicitation only" firms is 1.8185 months, while half of the firms spend 1.4 months or less in bankruptcy.

b.   Since the means and medians for the three groups of firms differ quite a bit, it would be unreasonable to use a single number to locate the center of the time in bankruptcy. Three different "centers" should be used.

2.43   a.   The primary disadvantage of using the range to compare variability of data sets is that the two data sets can have the same range and be vastly different with respect to data variation. Also, the range is greatly affected by extreme measures.

b.   The sample variance is the sum of the squared deviations from the sample mean divided by the sample size minus 1. The population variance is the sum of the squared deviations from the population mean divided by the population size.

c.   The variance of a data set can never be negative. The variance of a sample is the sum of the *squared* deviations from the mean divided by $n - 1$. The square of any number, positive or negative, is always positive. Thus, the variance will be positive.

The variance is usually greater than the standard deviation. However, it is possible for the variance to be smaller than the standard deviation. If the data are between 0 and 1, the variance will be smaller than the standard deviation. For example, suppose the data set is .8, .7, .9, .5, and .3. The sample mean is:

$$\bar{x} = \frac{\sum x}{n} = \frac{.8 + .7 + .9 + .5 + .3}{.5} = \frac{3.2}{5} = .64$$

The sample variance is:

$$s^2 = \frac{\sum x^2 - \frac{(\sum x)^2}{n}}{n - 1} = \frac{2.28 - \frac{3.2^2}{5}}{5 - 1} = \frac{2.28 - 2.048}{4} = \frac{.325}{4} = .058$$

The standard deviation is $s = \sqrt{.058} = .241$

2.45 a. Range $= 4 - 0 = 4$

$$s^2 = \frac{\sum x^2 - \frac{\left(\sum x\right)^2}{n}}{n-1} = \frac{22 - \frac{8^2}{5}}{4-1} = 2.3 \qquad s = \sqrt{2.3} = 1.52$$

b. Range $= 6 - 0 = 6$

$$s^2 = \frac{\sum x^2 - \frac{\left(\sum x\right)^2}{n}}{n-1} = \frac{63 - \frac{17^2}{7}}{7-1} = 3.619 \qquad s = \sqrt{3.619} = 1.90$$

c. Range $= 8 - (-2) = 10$

$$s^2 = \frac{\sum x^2 - \frac{\left(\sum x\right)^2}{n}}{n-1} = \frac{154 - \frac{30^2}{10}}{10-1} = 7.111 \qquad s = \sqrt{7.111} = 2.67$$

d. Range $= 2 - (-3) = 5$

$$s^2 = \frac{\sum x^2 - \frac{\left(\sum x\right)^2}{n}}{n-1} = \frac{29 - \frac{(-5)^2}{18}}{18-1} = 1.624 \qquad s = \sqrt{1.624} = 1.274$$

2.47 a. $\sum x = 3 + 1 + 10 + 10 + 4 = 28$

$\sum x^2 = 3^2 + 1^2 + 10^2 + 10^2 + 4^2 = 226$

$\bar{x} = \dfrac{\sum x}{n} = \dfrac{28}{5} = 5.6$

$$s^2 = \frac{\sum x^2 - \frac{\left(\sum x\right)^2}{n}}{n-1} = \frac{226 - \frac{28^2}{5}}{5-1} = \frac{69.2}{4} = 17.3 \qquad s = \sqrt{17.3} = 4.1593$$

b. $\sum x = 8 + 10 + 32 + 5 = 55$

$\sum x^2 = 8^2 + 10^2 + 32^2 + 5^2 = 1213$

$\bar{x} = \dfrac{\sum x}{n} = \dfrac{55}{4} = 13.75$ feet

$$s^2 = \frac{\sum x^2 - \frac{\left(\sum x\right)^2}{n}}{n-1} = \frac{1213 - \frac{55^2}{4}}{4-1} = \frac{456.75}{3} = 152.25 \text{ square feet}$$

$s = \sqrt{152.25} = 12.339$ feet

c. $\sum x = -1 + (-4) + (-3) + 1 + (-4) + (-4) = -15$

$\sum x^2 = (-1)^2 + (-4)^2 + (-3)^2 + 1^2 + (-4)^2 + (-4)^2 = 59$

$\bar{x} = \dfrac{\sum x}{n} = \dfrac{-15}{6} = -2.5$

$$s^2 = \frac{\sum x^2 - \frac{\left(\sum x\right)^2}{n}}{n-1} = \frac{59 - \frac{(-15)^2}{6}}{6-1} = \frac{21.5}{5} = 4.3 \qquad s = \sqrt{4.3} = 2.0736$$

d. $\sum x = \dfrac{1}{5} + \dfrac{1}{5} + \dfrac{1}{5} + \dfrac{2}{5} + \dfrac{1}{5} + \dfrac{4}{5} = \dfrac{10}{5} = 2$

$$\sum x^2 = \left[\dfrac{1}{5}\right]^2 + \left[\dfrac{1}{5}\right]^2 + \left[\dfrac{1}{5}\right]^2 + \left[\dfrac{2}{5}\right]^2 + \left[\dfrac{1}{5}\right]^2 + \left[\dfrac{4}{5}\right]^2 = \dfrac{24}{25} = .96$$

$$\bar{x} = \dfrac{\sum x}{n} = \dfrac{2}{6} = \dfrac{1}{3} = .33 \text{ ounce}$$

$$s^2 = \dfrac{\sum x^2 - \dfrac{(\sum x)^2}{n}}{n-1} = \dfrac{\dfrac{24}{25} - \dfrac{2^2}{6}}{6-1} = \dfrac{.2933}{5} = .0587 \text{ square ounce}$$

$$s = \sqrt{.0587} = .2422 \text{ ounce}$$

2.49   a.   For Buick, the range is:

         Range = Largest − Smallest = \$36,695 − \$19,335 = \$17,360

         For Cadillac, the range is:

         Range = Largest − Smallest = \$48,520 − \$34,820 = \$13,700

     b.   For Chevrolet, the range is:

         Range = Largest − Smallest = \$45,575 − \$9,373 = \$36,202

     c.   No. With only the values of the ranges, we have no idea which manufacturer produces only luxury cars. The prices of luxury cars can vary over a large range of values, so the range for the prices could be very large, or if the manufacturer makes only a few luxury models which are comparably priced, the range could be small.

2.51   a.   The mean value for the U.S. City Average Index for the data in the table is:

$$\bar{x} = \dfrac{\sum x}{n} = \dfrac{3607.3}{24} = 150.3042$$

         The mean value for the Chicago Index for the data in the table is:

$$\bar{x} = \dfrac{\sum x}{n} = \dfrac{3622.9}{24} = 150.9542$$

     b.   For the U.S. City Average Index, the range = largest measurement − smallest measurement = 153.7 − 146.2 = 7.5

         For the Chicago Index, the range = largest measurement − smallest measurement = 154.3 − 146.5 = 7.8

c. The standard deviation for the U.S. City Average Index is:

$$s = \sqrt{\frac{\sum x^2 - \frac{(\sum x)^2}{n}}{n - 1}} = \sqrt{\frac{542,325.89 - \frac{3607.3^2}{24}}{24 - 1}} = \sqrt{5.8117} = 2.4108$$

The standard deviation for the Chicago Index for the data in the table is:

$$s = \sqrt{\frac{\sum x^2 - \frac{(\sum x)^2}{n}}{n - 1}} = \sqrt{\frac{547,054.25 - \frac{3622.9^2}{24}}{24 - 1}} = \sqrt{7.0609} = 2.6572$$

d. The Chicago Index displays the greater variation about its mean for this time period. This is evident by the larger standard deviation and range for the Chicago Index.

2.53 a. The range is the largest measurement − the smallest measurement = 510.0 − 54.8 = 455.2

$$\sum x = 182.6 + 226.0 + 342.1 + 510.0 + 119.3 + 378.0 + 54.8 = 1812.8$$
$$\sum x^2 = 182.6^2 + 226.0^2 + 342.1^2 + 510.0^2 + 119.3^2 + 378.0^2 + 54.8^2 = 621,670.7$$
$$s^2 = \frac{\sum x^2 - \frac{(\sum x)^2}{n}}{n - 1} = \frac{621,670.7 - \frac{1812.8^2}{7}}{7 - 1} = \frac{152,207.2943}{6} = 25,367.88238$$
$$s = \sqrt{25,367.88235} = 159.2730$$

b. The range is $455.2 million.

The variance is 25,367.88238 million dollars squared.

The standard deviation is $159.2730 million.

c. If America West had a loss of $50 million, the range would increase since the smallest measurement decreased. The data are more spread out now.

If America West had a loss of $50 million, the standard deviation would increase since the data set is more spread out. $(s = \sqrt{34,069.27667} = 184.5786 > 159.2730)$

2.55 According to the Empirical Rule:

a. Approximately 68% of the measurements will be contained in the interval $\bar{x} - s$ to $\bar{x} + s$.

b. Approximately 95% of the measurements will be contained in the interval $\bar{x} - 2s$ to $\bar{x} + 2s$.

c. Essentially all the measurements will be contained in the interval $\bar{x} - 3s$ to $\bar{x} + 3s$.

2.57 a. $\bar{x} = \frac{\sum x}{n} = \frac{206}{25} = 8.24$

$$s^2 = \frac{\sum x^2 - \frac{(\sum x)^2}{n}}{n - 1} = \frac{1778 - \frac{206^2}{25}}{25 - 1} = 3.357 \qquad s = \sqrt{s^2} = 1.83$$

b.

| | Number of Measurements | |
Interval	in Interval	Percentage
$\bar{x} \pm s$, or (6.41, 10.07)	18	18/25 = .72 or 72%
$\bar{x} \pm 2s$, or (4.58, 11.90)	24	24/25 = .96 or 96%
$\bar{x} \pm 3s$, or (2.75, 13.73)	25	25/25 = 1 or 100%

c.   The percentages in part **b** are in agreement with Chebyshev's Rule and agree fairly well with the percentages given by the Empirical Rule.

d.   Range $= 12 - 5 = 7$

$s \approx$ range/4 $= 7/4 = 1.75$

The range approximation provides a satisfactory estimate of $s = 1.83$ from part **a**.

2.59   a.   Since the sample mean (18.2) is larger than the sample median (15), it indicates that the distribution of years is skewed to the right. In addition, the maximum number of years is 50 and the minimum is 2. If the distribution were symmetric, the mean and median should be about halfway between these two numbers. Halfway between the maximum and minimum values is 26, which is much larger than either the mean or the median.

b.   The standard deviation can be estimated by the range divided by either 4 or 6. For this distribution, the range is:

Range = Largest $-$ smallest $= 50 - 2 = 48$.

Dividing the range by 4, we get an estimate of the standard deviation to be $48/4 = 12$.

Dividing the range by 6, we get an estimate of the standard deviation to be $48/6 = 8$.

Thus, the standard deviation should be somewhere between 8 and 12. For this problem, the standard deviation is $s = 10.64$. This value falls in the estimated range of 8 to 12.

c.   First, we calculate the number of standard deviations from the mean the value of 40 years is. To do this, we first subtract the mean and then divide by the value of the standard deviation.

Number of standard deviations is $\dfrac{40 - \bar{x}}{s} = \dfrac{40 - 18.2}{10.64} = 2.05 \approx 2$

Using Chebyshev's Rule, we know that at most $1/k^2$ or $1/2^2 = 1/4$ of the data will be more than 2 standard deviations from the mean. Thus, this would indicate that at most 25% of the Generation Xers responded with 40 years or more.

Next, we calculate the number of standard deviations from the mean the value of 8 years is.

Number of standard deviations is $\dfrac{8 - \bar{x}}{s} = \dfrac{8 - 18.2}{10.64} = -.96 \approx -1$

Using Chebyshev's Rule, we get no information about the data within 1 standard deviation of the mean. However, we know the median (15) is more than 8. By definition, 50% of the data are larger than the median. Thus, at least 50% of the Generation Xers responded with 8 years or more. No additional information can be obtained with the information given.

2.61 a. More than half of the spillage amounts are less than or equal to 48 metric tons and almost all (44 out of 50) are below 104 metric tons. There appear to be three outliers, values which are much different than the others. These three values are larger than 216 metric tons.

b. From the graph in part **a**, the data are not mound shaped. Thus, we must use Chebyshev's Rule. This says that at least 8/9 of the measurements will fall within 3 standard deviations of the mean. Since most of the observations will be within 3 standard deviations of the mean, we could use this interval to predict the spillage amount of the next major oil spill. From the printout, the mean is 59.820 and the standard deviation is 53.362. The interval would be:

$$\bar{x} \pm 3s \Rightarrow 59.82 \pm 3(53.362) \Rightarrow 59.82 \pm 160.086 \Rightarrow (-100.266, 219.906)$$

Since an oil spillage amount cannot be negative, we would predict that the spillage amount of the next major oil spill will be between 0 and 219.906 metric tons.

2.63 a. Since the data are not mound-shaped, the Empirical Rule would not be appropriate for describing bankruptcy times.

b. Chebyshev's Rule says that at least 75% of the data will fall within 2 standard deviations of the mean. From the printout, the mean is 2.549 and the standard deviation is 1.828. The interval $\bar{x} \pm 2s$ is $2.549 \pm 2(1.828)$ or $(-1.107, 6.205)$. Thus, at least 75% of the bankruptcy times should fall within -1.107 and 6.205 months. Since the number of months cannot be negative, at least 95% of the bankruptcy times are less than 6.205 months.

c. From the data in Exercise 2.24, 47 of the 49 observations fall in this interval or $47/49 = .959$ or 95.9%. This is at least 75%. It is also very close to the 95% used with the Empirical Rule.

d. Because the data set is skewed to the right, the median is a better estimate of the center of the distribution than the mean. Thus, we would estimate that a firm would be in bankruptcy approximately 1.7 months. From the interval in part **b**, we would be very confident that the firm would be in bankruptcy no more than 6.2 months.

2.65 Since we do not know if the distribution of the heights of the trees is mound-shaped, we need to apply Chebyshev's Rule. We know $\mu = 30$ and $\sigma = 3$. Therefore,

$$\mu \pm 3\sigma \Rightarrow 30 \pm 3(3) \Rightarrow 30 \pm 9 \Rightarrow (21, 39)$$

According to Chebyshev's Rule, at least 8/9 or .89 of the tree heights on this piece of land fall within this interval and at most $\frac{1}{9}$ or .11 of the tree heights will fall above the interval. However, the buyer will only purchase the land if at least $\frac{1000}{5000}$ or .20 of the tree heights are at least 40 feet tall. Therefore, the buyer should not buy the piece of land.

2.67 We know $\mu = 25$ and $\sigma = .1$. Therefore,

$$\mu \pm 2\sigma \Rightarrow 25 \pm 2(.1) \Rightarrow 25 \pm .2 \Rightarrow (24.8, 25.2)$$

The machine is shut down for adjustment if the contents of two consecutive bags fall more than 2 standard deviations from the mean (i.e., outside the interval (24.8, 25.2)). Therefore, the machine was shut down yesterday at 11:30 (25.23 and 25.25 are outside the interval) and again at 4:00 (24.71 and 25.31 are outside the interval).

2.69   Using the definition of a percentile:

	Percentile	Percentage Above	Percentage Below
a.	75th	25%	75%
b.	50th	50%	50%
c.	20th	80%	20%
d.	84th	16%	84%

2.71   We first compute z-scores for each x value.

a. $z = \dfrac{x - \mu}{\sigma} = \dfrac{100 - 50}{25} = 2$

b. $z = \dfrac{x - \mu}{\sigma} = \dfrac{1 - 4}{1} = -3$

c. $z = \dfrac{x - \mu}{\sigma} = \dfrac{0 - 200}{100} = -2$

d. $z = \dfrac{x - \mu}{\sigma} = \dfrac{10 - 5}{3} = 1.67$

The above z-scores indicate that the x value in part **a** lies the greatest distance above the mean and the x value of part **b** lies the greatest distance below the mean.

2.73   Since the 90th percentile of the study sample in the subdivision was .00372 mg/L, which is less than the USEPA level of .015 mg/L, the water customers in the subdivision are not at risk of drinking water with unhealthy lead levels.

2.75   a.   To calculate the U.S. merchandise trade balance for each of the ten countries, take the exports minus the imports.

Country	U.S. Merchandise Trade Balance (in billions)
Brazil	6,289.2
China	−49,695.3
Egypt	3,177.9
France	−4,671.5
Italy	−10,412.8
Japan	−56,114.7
Mexico	−14,549.1
Panama	1,168.9
Sweden	−3,984.8
Singapore	−2,378.4

b. To find z-scores, we must first calculate the sample mean and standard deviation.

$$\bar{x} = \frac{\sum x}{n} = \frac{-131,170.6}{10} = -13,117.06$$

$$s^2 = \frac{\sum x^2 - \frac{(\sum x)^2}{n}}{n-1} = \frac{6,032,962,855 - \frac{(131,170.6)^2}{10}}{10-1} = \frac{4,312,390,225}{9}$$
$$= 479,154,469.4$$

$$s = \sqrt{479,154,469.4} = 21,889.6$$

Japan: $z = \dfrac{x - \bar{x}}{s} = \dfrac{-56,114.7 - (-13,117.06)}{21,889.6} = -1.96$

The relative position of the U.S. trade balance with Japan is 1.96 standard deviations below the mean. This indicates that this observation is small compared to the other U.S. trade balances.

Egypt: $z = \dfrac{x - \bar{x}}{s} = \dfrac{3,177.9 - (-13,117.06)}{21,889.6} = .74$

The relative position of the U.S. trade balance with Egypt is .74 standard deviations above the mean. This indicates that this observation is larger than the average of the U.S. trade balances.

2.77   a.   From the printout, the 10th percentile is 0. Ten percent of the observations are less than or equal to 0.

b.   From the printout, the 95% percentile is 21. Ninety-five percent of the observations are less than or equal to 21.

c.   The z-score for the county with 48 Superfund sites is:

$$z = \frac{x - \bar{x}}{s} = \frac{48 - 5.24}{7.224} = 5.90$$

d.   Yes. A score of 48 is almost 6 standard deviations from the mean. We know that for any data set almost all (at least 8/9 using Chebyshev's Rule) of the observations are within 3 standard deviations of the mean. To be almost 6 standard deviations from the mean is very unusual.

2.79   To determine if the measurements are outliers, compute the z-score.

a. $z = \dfrac{x - \bar{x}}{s} = \dfrac{65 - 57}{11} = .727$ Since this z-score is less than 3 in magnitude, 65 is not an outlier.

b. $z = \dfrac{x - \bar{x}}{s} = \dfrac{21 - 57}{11} = -3.273$ Since this z-score is more than 3 in magnitude, 21 is an outlier.

c. $z = \dfrac{x - \bar{x}}{s} = \dfrac{72 - 57}{11} = 1.364$ Since this z-score is less than 3 in magnitude, 72 is not an outlier.

d. $z = \dfrac{x - \bar{x}}{s} = \dfrac{98 - 57}{11} = 3.727$ Since this z-score is more than 3 in magnitude, 98 is an outlier.

2.81   The interquartile range is IQR $= Q_U - Q_L = 85 - 60 = 25$.

The lower inner fence $= Q_L - 1.5(\text{IQR}) = 60 - 1.5(25) = 22.5$.

The upper inner fence $= Q_U + 1.5(\text{IQR}) = 85 + 1.5(25) = 122.5$.

The lower outer fence $= Q_L - 3(\text{IQR}) = 60 - 3(25) = -15$.

The upper outer fence $= Q_U + 3(\text{IQR}) = 85 + 3(25) = 160$.

With only this information, the box plot would look something like the following:

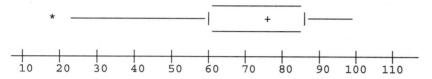

The whiskers extend to the inner fences unless no data points are that small or that large. The upper inner fence is 122.5. However, the largest data point is 100, so the whisker stops at 100. The lower inner fence is 22.5. The smallest data point is 18, so the whisker extends to 22.5. Since 18 is between the inner and outer fences, it is designated with a *. We do not know if there is any more than one data point below 22.5, so we cannot be sure that the box plot is entirely correct.

2.83   a.   Using MINITAB, the box plot for sample A is given below.

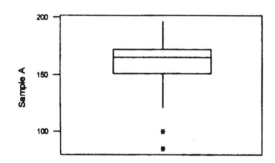

Using MINITAB, the box plot for sample B is given below.

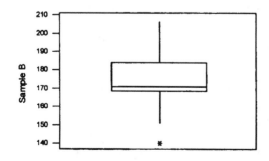

Methods for Describing Sets of Data   21

b. The range for sample A is larger than the range for sample B. The descriptive measures for sample A (Q_L, median, and Q_U) are all smaller than those of sample B. Both samples look somewhat symmetric (excluding outliers) since the whiskers on each box plot are the same length on both sides.

In sample A, the measurement 85 is an outlier. It lies outside the outer fence.

$$\begin{aligned}
\text{Lower outer fence} &= \text{Lower hinge} - 3(\text{IQR})\\
&\approx 151 - 3(172 - 151)\\
&= 151 - 3(21)\\
&= 88
\end{aligned}$$

In addition, the observation of 100 may be an outlier. It lies outside the inner fence, but inside the outer fence.

In Sample B, the observation of 140 may be an outlier. It lies outside the inner fence, but inside the outer fence.

2.85 a. The median bankruptcy times for "Prepack" and "Joint" firms are almost the same. They are both less than the median bankruptcy time of the "None" firms.

b. The range of the "Prepack" firms is less than the other two, while the range of the "None" firms is the largest. The interquartile range of the "Prepack" firms is less than the other two, while the interquartile range of the "Joint" firms is larger than the other two.

c. No. The interquartile range for the "Prepack" firms is the smallest which corresponds to the smallest standard deviation. However, the second smallest interquartile range corresponds to the "none" firms. The second smallest standard deviation corresponds to the "Joint" firms.

d. Yes. There is evidence of two outliers in the "Prepack" firms. These are indicated by the two *'s. There is also evidence of two outliers in the "None" firms. These are indicated by the two *'s.

2.87 a. Using MINITAB, the box plot is:

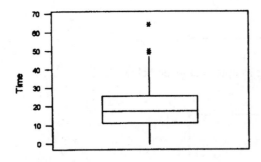

The median is about 18. The data appear to be skewed to the right since there are 3 suspect outliers to the right and none to the left. The variability of the data is fairly small because the IQR is fairly small, approximately $26 - 10 = 16$.

b. The customers associated with the suspected outliers are customers 268, 269, and 264.

c. In order to find the z-scores, we must first find the mean and standard deviation.

$$\bar{x} = \frac{\sum x}{n} = \frac{815}{40} = 20.375$$

$$s^2 = \frac{\sum x^2 - \frac{(\sum x)^2}{n}}{n - 1} = \frac{24129 - \frac{815^2}{40}}{40 - 1} = 192.90705$$

$$s = \sqrt{192.90705} = 13.89$$

The z-scores associated with the suspected outliers are:

Customer 268 $z = \dfrac{49 - 20.375}{13.89} = 2.06$

Customer 269 $z = \dfrac{50 - 20.375}{13.89} = 2.13$

Customer 264 $z = \dfrac{64 - 20.375}{13.89} = 3.14$

All the z-scores are greater than 2. These are very unusual values.

2.89   The relative frequency histogram is:

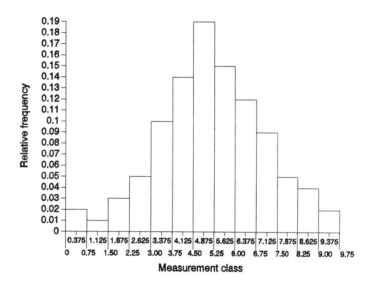

2.91   a. $z = \dfrac{x - \mu}{\sigma} = \dfrac{50 - 60}{10} = -1$

$z = \dfrac{70 - 60}{10} = 1$

$z = \dfrac{80 - 60}{10} = 2$

b. $z = \dfrac{x - \mu}{\sigma} = \dfrac{50 - 60}{5} = -2$

$z = \dfrac{70 - 60}{5} = 2$

$z = \dfrac{80 - 60}{5} = 4$

c. $z = \dfrac{x - \mu}{\sigma} = \dfrac{50 - 40}{10} = 1$

$z = \dfrac{70 - 40}{10} = 3$

$z = \dfrac{80 - 40}{10} = 4$

d. $z = \dfrac{x - \mu}{\sigma} = \dfrac{50 - 40}{100} = .1$

$z = \dfrac{70 - 40}{100} = .3$

$z = \dfrac{80 - 40}{100} = .4$

2.93 a. $\sum x = 13 + 1 + 10 + 3 + 3 = 30$

$\sum x^2 = 13^2 + 1^2 + 10^2 + 3^2 + 3^2 = 288$

$\bar{x} = \sum x = \dfrac{30}{5} = 6$

$s^2 = \dfrac{\sum x^2 - \dfrac{(\sum x)^2}{n}}{n - 1} = \dfrac{288 - \dfrac{30^2}{5}}{5 - 1} = \dfrac{108}{4} = 27$ $\qquad s = \sqrt{27} = 5.20$

b. $\sum x = 13 + 6 + 6 + 0 = 25$

$\sum x^2 = 13^2 + 6^2 + 6^2 + 0^2 = 241$

$\bar{x} = \dfrac{\sum x}{n} = \dfrac{25}{4} = 6.25$

$s^2 = \dfrac{\sum x^2 - \dfrac{(\sum x)^2}{n}}{n - 1} = \dfrac{241 - \dfrac{25^2}{4}}{4 - 1} = \dfrac{84.75}{3} = 28.25$ $\qquad s = \sqrt{28.25} = 5.32$

c. $\sum x = 1 + 0 + 1 + 10 + 11 + 11 + 15 = 49$

$\sum x^2 = 1^2 + 0^2 + 1^2 + 10^2 + 11^2 + 11^2 + 15^2 = 569$

$\bar{x} = \dfrac{\sum x}{n} = \dfrac{49}{7} = 7$

$s^2 = \dfrac{\sum x^2 - \dfrac{(\sum x)^2}{n}}{n - 1} = \dfrac{569 - \dfrac{49^2}{7}}{7 - 1} = \dfrac{226}{6} = 37.67$ $\qquad s = \sqrt{37.67} = 6.14$

d. $\sum x = 3 + 3 + 3 + 3 = 12$

$\sum x^2 = 3^2 + 3^2 + 3^2 + 3^2 = 36$

$\bar{x} = \dfrac{\sum x}{n} = \dfrac{12}{4} = 3$

$s^2 = \dfrac{\sum x^2 - \dfrac{(\sum x)^2}{n}}{n-1} = \dfrac{36 - \dfrac{12^2}{4}}{4-1} = \dfrac{0}{3} = 0 \qquad\qquad s = \sqrt{0} = 0$

2.95   The range is found by taking the largest measurement in the data set and subtracting the smallest measurement. Therefore, it only uses two measurements from the whole data set. The standard deviation uses every measurement in the data set. Therefore, it takes every measurement into account—not just two. The range is affected by extreme values more than the standard deviation.

2.97   a.   To display the status, we use a pie chart. From the pie chart, we see that 58% of the Beanie babies are retired and 42% are current.

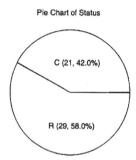
Pie Chart of Status

C (21, 42.0%)

R (29, 58.0%)

b.   Using Minitab, a histogram of the values is:

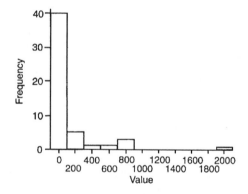

Most (40 of 50) Beanie babies have values less than $100. Of the remaining 10, 5 have values between $100 and $300, 1 has a value between $300 and $500, 1 has a value between $500 and $700, 2 have values between $700 and $900, and 1 has a value between $1900 and $2100.

c.   A plot of the value versus the age of the Beanie Baby is as follows:

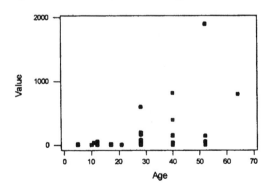

From the plot, it appears that as the age increases, the value tends to increase.

2.99   a.   First, we must compute the total processing times by adding the processing times of the three departments. The total processing times are as follows:

Request	Total Processing Time	Request	Total Processing Time	Request	Total Processing Time
1	13.3	17	19.4*	33	23.4*
2	5.7	18	4.7	34	14.2
3	7.6	19	9.4	35	14.3
4	20.0*	20	30.2	36	24.0*
5	6.1	21	14.9	37	6.1
6	1.8	22	10.7	38	7.4
7	13.5	23	36.2*	39	17.7*
8	13.0	24	6.5	40	15.4
9	15.6	25	10.4	41	16.4
10	10.9	26	3.3	42	9.5
11	8.7	27	8.0	43	8.1
12	14.9	28	6.9	44	18.2*
13	3.4	29	17.2*	45	15.3
14	13.6	30	10.2	46	13.9
15	14.6	31	16.0	47	19.9*
16	14.4	32	11.5	48	15.4
				49	14.3*
				50	19.0

The stem-and-leaf displays with the appropriate leaves circled are as follows:

Stem-and-leaf of Mkt
Leaf Unit = 0.10

```
   6     0   112446
   7     1   3
  14     2   (0)024699
  16     3   2(5)
  22     4   (0)(0)(1)577
 (10)    5   0344556889
  18     6   000(2)2247(9)(9)
   8     7   003(8)
   4     8   (0)7
   2     9
   2    10   0
   1    11   0
```

Stem-and-leaf of Engr
Leaf Unit = 0.10

```
   7     0   4466699
  14     1   333378(8)
  19     2   (1)22(4)6
  23     3   1568
  (5)    4   24688
  22     5   233
  19     6   (0)12(3)9
  14     7   (2)(2)379
   9     8
   9     9   66
   7    10   0
   6    11   3
   5    12   02(3)
   2    13   (0)
   1    14   (4)
```

	Stem-and-leaf of Accnt Leaf Unit = 0.10			Stem-and-leaf of Total Leaf Unit = 1.00

```
   Stem-and-leaf of Accnt          Stem-and-leaf of Total
       Leaf Unit = 0.10                Leaf Unit = 1.00

   19    0   11111111111(2)2333444    1    0   1
   (8)   0   55556(8)88                3    0   33
   23    1   00                        5    0   45
   21    1   7(9)                      11    0   666677
   19    2   00(2)3                    17    0   888999
   15    2                            21    1   0000
   15    3   23                        (5)   1   33333
   13    3   78                       24    1   4(4)44445555
   11    4                            14    1   66(7)(7)
   11    4                            10    1   (8)9(9)(9)
   11    5                             6    2   (0)
   11    5   8                         5    2   (3)
   10    6   2                         4    2   (4)4
    9    6                            HI   30, (36)
    9    7   (0)
    8    7
    8    8   4
   HI   (99), (105), (135), 144,
         (182), 220, (300)
```

Of the 50 requests, 10 were lost. For each of the three departments, the processing times for the lost requests are scattered throughout the distributions. The processing times for the departments do not appear to be related to whether the request was lost or not. However, the total processing times for the lost requests appear to be clustered towards the high side of the distribution. It appears that if the total processing time could be kept under 17 days, 76% of the data could be maintained, while reducing the number of lost requests to 1.

b. For the Marketing department, if the maximum processing time was set at 6.5 days, 78% of the requests would be processed, while reducing the number of lost requests by 4. For the Engineering department, if the maximum processing time was set at 7.0 days, 72% of the requests would be processed, while reducing the number of lost requests by 5. For the Accounting department, if the maximum processing time was set at 8.5 days, 86% of the requests would be processed, while reducing the number of lost requests by 5.

2.101 a. One reason the plot may be interpreted differently is that no scale is given on the vertical axis. Also, since the plot almost reaches the horizontal axis at 3 years, it is obvious that the bottom of the plot has been cut off. Another important factor omitted is who responded to the survey.

b. A scale should be added to the vertical axis. Also, that scale should start at 0.

2.103 a. Since the mean is greater than the median, the distribution of the radiation levels is skewed to the right.

b. $\bar{x} \pm s \Rightarrow 10 \pm 3 \Rightarrow (7, 13)$; $\bar{x} \pm 2s \Rightarrow 10 \pm 2(3) \Rightarrow (4, 16)$; $\bar{x} \pm 3s \Rightarrow 10 \pm 3(3) \Rightarrow (1, 19)$

Interval	Chebyshev's	Empirical
(7, 13)	At least 0	$\approx 68\%$
(4, 16)	At least 75%	$\approx 95\%$
(1, 19)	At least 88.9%	$\approx 100\%$

Since the data are skewed to the right, Chebyshev's Rule is probably more appropriate in this case.

c.   The background level is 4. Using Chebyshev's Rule, at least 75% or .75(50) $\approx$ 38 homes are above the background level. Using the Empirical Rule, $\approx$ 97.5% or .975(50) $\approx$ 49 homes are above the background level.

d. $z = \dfrac{x - \bar{x}}{s} = \dfrac{20 - 10}{3} = 3.333$

It is unlikely that this new measurement came from the same distribution as the other 50. Using either Chebyshev's Rule or the Empirical Rule, it is very unlikely to see any observations more than 3 standard deviations from the mean.

2.105   a.   Using MINITAB, the stem-and-leaf plot for an NFL team's current value is:

Character Stem-and-Leaf Display

```
Stem-and-leaf of Value    N = 30
Leaf Unit = 10

    2    2  99
   12    3  0000111222
   (9)   3  566779999
    9    4  0124
    5    4  68
    3    5  0
    2    5
    2    6  0
    1    6  6
```

b.   Yes. Most of the values are in the neighborhood of $300 to $400 million. However, there are a few teams with very large values. The distribution is skewed to the right.

c.   From the stem-and-leaf display above, the median is the average of the 15th and 16th values. The values from the plot are 360 and 370, giving a median of 365. The actual values are 369 and 371. The average of 369 and 371 is 370. Thus, the median is 370.

d.   To calculate the z-scores for the Denver Broncos, we must first compute the means and standard deviations for the two variables.

Current Value:

$$\bar{x} = \frac{\sum x}{n} = \frac{11,560}{30} = 385.33$$

$$s^2 = \frac{\sum x^2 - \dfrac{(\sum x)^2}{n}}{n - 1} = \frac{4,688,054 - \dfrac{(11,560)^2}{30}}{30 - 1} = \frac{233,600.667}{29} = 8,055.1954$$

$$s = \sqrt{8,055.1954} = 89.751$$

The z-score for the Denver Broncos current value is:

$$z = \frac{x - \overline{x}}{s} = \frac{427 - 385.33}{89.751} = 0.46$$

Operating Income:

$$\overline{x} = \frac{\sum x}{n} = \frac{590.40}{30} = 19.68$$

$$s^2 = \frac{\sum x^2 - \frac{(\sum x)^2}{n}}{n - 1} = \frac{16,782.94 - \frac{(590.4)^2}{30}}{30 - 1} = \frac{5,163.868}{29} = 178.0644$$

$$s = \sqrt{178.0644} = 13.344$$

The z-score for the Denver Broncos operating income is:

$$z = \frac{x - \overline{x}}{s} = \frac{5.0 - 19.68}{13.344} = -1.10$$

e. The z-score for the current value is 0.46. The Denver Broncos' current value is .46 standard deviations above the mean current value of all NFL teams. The z-score for operating income is -1.10. The Denver Broncos' operating income is 1.10 standard deviations below the mean operating income of all NFL teams.

f. There are several teams that have a positive current value z-score (value above the mean of 385.33) and a negative operating income z-score (value below the mean of 19.68). These teams are:

Carolina Panthers
New England Patriots
Denver Broncos
Seattle Seahawks
Pittsburgh Steelers
Cincinnati Bengals

g. Using MINITAB, the box plot is:

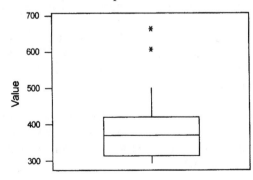

From the box plot, there are 2 potential outliers. These 2 points lie outside the inner fences but inside the outer fences. These potential outliers are associated with the Dallas Cowboys (663) and the Washington Redskins (607). The z-scores associated with these potential outliers are:

The z-score for the Dallas Cowboys' current value is:

$$z = \frac{x - \bar{x}}{s} = \frac{663 - 385.33}{89.751} = 3.09$$

The z-score for the Washington Redskins' current value is:

$$z = \frac{x - \bar{x}}{s} = \frac{607 - 385.33}{89.751} = 2.47$$

Using the z-score value, the current value associated with the Dallas Cowboys is an outlier.

h. To investigate the trend between an NFL team's current value and its operating income, we will construct a plot of the current value against the operating income.

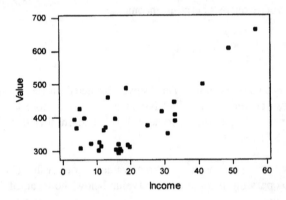

From the plot, it appears that as a team's operating income tends to increase, the current value also tends to increase.

2.109 a. The histograms for both baseball teams are skewed to the right. Most players have relatively low salaries while only a few have relatively high salaries. However, the shapes of the two histograms are quite different. For the Tampa Bay Devil Rays, 22 of the 33 players (66.7%) make $250,000 or less. For the Baltimore Orioles, only 5 of the 28 players (17.9%) make $500,000 or less. For the Tampa Bay Devil Rays, only 3 of 33 players (9.1%) make more than $2,500,000. However, for the Baltimore Orioles, 13 of the 28 players (46.4%) make more than $2,500,000. These percents are quite different for the two teams.

b. For the Baltimore Orioles, the average salary is $2,685,200 and the standard deviation is $2,222,790. Two standard deviations is 2($2,222,790) = $4,445,580. By Chebyshev's Rule, at least 75% of the salaries of the Baltimore Orioles will fall within 2 standard deviations of the mean or between $0 and $7,130,780.

For the Tampa Bay Devil Rays, the average salary is $830,700 and the standard deviation is $1,349,450. Two standard deviations is 2($1,349,450) = $2,698,900. By Chebyshev's Rule, at least 75% of the salaries of the Tampa Bay Devil Rays will fall within 2 standard deviations of the mean or between $0 and $3,529,600.

2.111   The time series plot for the data is:

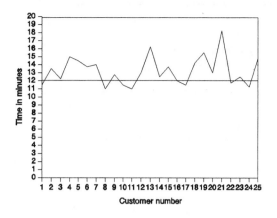

Of the 25 observations, only 7 are less than the claimed number of 12 minutes. Thus, the claim that "your hood will be open less than 12 minutes when we service your car" is probably not true.

2.113   a.   A relative frequency histogram is:

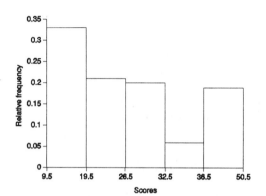

From the histogram, about a third of the sample scored in the "Very relaxed/confident" category. About an equal amount scored in categories "Very anxious," "Some mild anxiety," and "Generally relaxed/comfortable." Very few scored in the "Anxious/tense" category.

b.   From the tables, the means and standard deviations for the male and female teachers are very similar. The mean for the females is slightly lower than the mean for the males, but the standard deviation is slightly larger. Thus, the average female has a little less anxiety towards computers, but the distribution of female scores is slightly wider than that of the males. But again, these differences are so small that there probably is no difference between male and female teachers concerning their anxiety toward computers.

3.1   a.   Since the probabilities must sum to 1,

$$P(E_3) = 1 - P(E_1) - P(E_2) - P(E_4) - P(E_5) = 1 - .1 - .2 - .1 - .1 = .5$$

b. $P(E_3) = 1 - P(E_3) - P(E_2) - P(E_4) - P(E_5)$
     $\Rightarrow 2P(E_3) = 1 - .1 - .2 - .1 \Rightarrow 2P(E_3) = .6 \Rightarrow P(E_3) = .3$

c. $P(E_3) = 1 - P(E_1) - P(E_2) - P(E_4) - P(E_5) = 1 - .1 - .1 - .1 - .1 = .6$

3.3 $P(A) = P(1) + P(2) + P(3) = .05 + .20 + .30 = .55$
      $P(B) = P(1) + P(3) + P(5) = .05 + .30 + .15 = .50$
      $P(C) = P(1) + P(2) + P(3) + P(5) = .05 + .20 + .30 + .15 = .70$

3.5   If we denote the marbles as B_1, B_2, R_1, R_2, R_3, then the ten equally likely sample points in the sample space would be:

$$S: \begin{bmatrix} (B_1, B_2), (B_1, R_1), (B_1, R_2), (B_1, R_3), (B_2, R_1) \\ (B_2, R_2), (B_2, R_3), (R_1, R_2), (R_1, R_3), (R_2, R_3) \end{bmatrix}$$

Notice that order is ignored, as the only concern is whether or not a marble is selected. Each of these 10 would be equally likely, implying that each occurs with a probability 1/10.

$$P(A) = \frac{1}{10} \qquad P(B) = 6\left[\frac{1}{10}\right] = \frac{6}{10} = \frac{3}{5} \qquad P(C) = 3\left[\frac{1}{10}\right] = \frac{3}{10}$$

3.7   a.   The experiment consists of selecting 159 employees and asking each to indicate how strongly he/she agreed or disagreed with the statement "I believe that management is committed to CQI." There are five sample points: "Strongly agree," "Agree," "Neither agree nor disagree," "Disagree," and "Strongly disagree."

b.   Since we have frequencies for each of the sample points, good estimates of the probabilities are the relative frequencies. To find the relative frequencies, divide all of the frequencies by the sample size of 159. The estimates of the probabilities are:

Strongly Agree	Agree	Neither Agree Nor Disagree	Disagree	Strongly Disagree
.189	.403	.258	.113	.038

c.   The probability that an employee agrees or strongly agrees with the statement is .189 + .403 = .592.

d.   The probability that an employee does not strongly agree with the statement is equal to the sum of all the probabilities except that for "strongly agree" = .403 + .258 + .113 + .038 = .812.

3.9   There are $\begin{pmatrix} 6 \\ 3 \end{pmatrix} = \dfrac{6!}{3!3!} = \dfrac{6 \cdot 5 \cdot 4 \cdot 3 \cdot 2 \cdot 1}{3 \cdot 2 \cdot 1 \cdot 3 \cdot 2 \cdot 1} = 20$ possible ways to select 3 cars

from 6. Only one of these combinations includes all three lemons, so the probability that dealer A receives all three lemons is 1/20.

3.11   a.   Define the following events:

H: {news story was related to Hispanics}

N: {news story was not related to Hispanics}

Of the 11,855 news stories, 118 were related to Hispanics and 11,855 - 118 = 11,737 were not related to Hispanics. Thus,

$$P(N) = 11{,}737/11{,}855 = .990$$

$$P(H) = 118/11{,}855 = .010$$

b.   Define the following events:

C: {Hispanic news story focused on crime}

D: {Hispanic news story focused on drugs}

P: {Hispanic news story focused on politics}

Thus, $P(C) = 23/118 = .195$.

The probability that the news story focused on crime or drugs is

$$P(C) + P(D) = 23/118 + 1/118 = 24/118 = .203.$$

The probability that the news story did not focus on politics or crime is

$$10/118 + 10/118 + 1/118 + 5/118 + 1/118 + 3/118 + 3/118 + 3/118 + 11/118 +$$
$$5/118 + 5/118 + 10/118 + 1/118 = 68/118 = .576$$

3.13   a.   The odds in favor of an Oxford Shoes win are $\dfrac{1}{3}$ to $1 - \dfrac{1}{3} = \dfrac{2}{3}$ or 1 to 2.

b.   If the odds in favor of Oxford Shoes are 1 to 1, then the probability that Oxford Shoes wins is $\dfrac{1}{1 + 1} = \dfrac{1}{2}$.

c.   If the odds against Oxford Shoes are 3 to 2, then the odds in favor of Oxford Shoes are 2 to 3. Therefore, the probability that Oxford Shoes wins is $\dfrac{2}{2 + 3} = \dfrac{2}{5}$.

3.15   a.   The four classifications are:

           (1)   Raise a broad mix of crops
           (2)   Raise livestock
           (3)   Use chemicals sparingly
           (4)   Use techniques for regenerating the soil

Let us define the following events:

      A_1:   Raise a broad mix of crops
      A_2:   Do not raise a broad mix of crops
      B_1:   Raise livestock
      B_2:   Do not raise livestock
      C_1:   Use chemical sparingly
      C_2:   Do not use chemical sparingly
      D_1:   Use techniques for regenerating the soil
      D_2:   Do not use techniques for regenerating the soil

Each farmer is classified as using or not using each of the four techniques. Thus, the sample points are:

$$A_1B_1C_1D_1, A_1B_1C_1D_2, A_1B_1C_2D_1, A_1B_2C_1D_1, A_2B_1C_1D_1, A_1B_1C_2D_2,$$
$$A_1B_2C_1D_2, A_2B_1C_1D_2, A_1B_2C_2D_1, A_2B_1C_2D_1, A_2B_2C_1D_1, A_1B_2C_2D_2,$$
$$A_2B_1C_2D_2, A_2B_2C_1D_2, A_2B_2C_2D_1, A_2B_2C_2D_2$$

     b.   Since there are 16 classification sets or 16 sample points, the probability of any one sample point is 1/16. The probability that a farmer will be classified as unlikely on all four criteria is:

$$P(A_2B_2C_2D_2) = 1/16$$

     c.   The probability that a farmer will be classified as likely on at least three of the criteria is:

$$P(A_1B_1C_1D_1) + P(A_1B_1C_1D_2) + P(A_1B_1C_2D_1) + P(A_1B_2C_1D_1) + P(A_2B_1C_1D_1)$$
$$= 1/16 + 1/16 + 1/16 + 1/16 + 1/16 = 5/16$$

3.17   Two events are mutually exclusive if they have no sample points in common. A possible Venn diagram of two mutually exclusive events is:

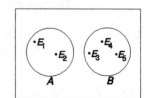

3.19   a. $P(A) = P(E_1) + P(E_2) + P(E_3) + P(E_5) + P(E_6) = \dfrac{1}{5} + \dfrac{1}{5} + \dfrac{1}{5} + \dfrac{1}{20} + \dfrac{1}{10} = \dfrac{15}{20} = \dfrac{3}{4}$

     b. $P(B) = P(E_2) + P(E_3) + P(E_4) + P(E_7) = \dfrac{1}{5} + \dfrac{1}{5} + \dfrac{1}{20} + \dfrac{1}{5} = \dfrac{13}{20}$

     c. $P(A \cup B) = P(E_1) + P(E_2) + P(E_3) + P(E_4) + P(E_5) + P(E_6) + P(E_7)$
$$= \dfrac{1}{5} + \dfrac{1}{5} + \dfrac{1}{5} + \dfrac{1}{20} + \dfrac{1}{20} + \dfrac{1}{10} + \dfrac{1}{5} = 1$$

     d. $P(A \cap B) = P(E_2) + P(E_3) = \dfrac{1}{5} + \dfrac{1}{5} = \dfrac{2}{5}$

e. $P(A^c) = 1 - P(A) = 1 - \dfrac{3}{4} = \dfrac{1}{4}$

f. $P(B^c) = 1 - P(B) = 1 - \dfrac{13}{20} = \dfrac{7}{20}$

g. $P(A \cup A^c) = P(E_1) + P(E_2) + P(E_3) + P(E_4) + P(E_5) + P(E_6) + P(E_7)$
$$= \dfrac{1}{5} + \dfrac{1}{5} + \dfrac{1}{5} + \dfrac{1}{20} + \dfrac{1}{20} + \dfrac{1}{10} + \dfrac{1}{5} = 1$$

h. $P(A^c \cap B) = P(E_4) + P(E_7) = \dfrac{1}{20} + \dfrac{1}{5} = \dfrac{5}{20} = \dfrac{1}{4}$

3.21   a. $P(A) = .50 + .10 + .05 = .65$

b. $P(B) = .10 + .07 + .50 + .05 = .72$

c. $P(C) = .25$

d. $P(D) = .05 + .03 = .08$

e. $P(A^c) = .25 + .07 + .03 = .35$ (Note: $P(A^c) = 1 - P(A) = 1 - .65 = .35$)

f. $P(A \cup B) = P(B) = .10 + .07 + .50 + .05 = .72$

g. $P(A \cap C) = 0$

h. Two events are mutually exclusive if they have no sample points in common or if the probability of their intersection is 0.

$P(A \cap B) = P(A) = .50 + .10 + .05 = .65$. Since this is not 0, A and B are not mutually exclusive.

$P(A \cap C) = 0$. Since this is 0, A and C are mutually exclusive.

$P(A \cap D) = .05$. Since this is not 0, A and D are not mutually exclusive.

$P(B \cap C) = 0$. Since this is 0, B and C are mutually exclusive.

$P(B \cap D) = .05$. Since this is not 0, B and D are not mutually exclusive.

$P(C \cap D) = 0$. Since this is 0, C and D are mutually exclusive.

3.23   a. $B \cap C$

b. A^c

c. $C \cup B$

d. $A \cap C^c$

3.25   a. $P(A) = 8/28.44 = .281$

$P(B) = 7.84/28.44 = .276$

$P(C) = 1.24/28.44 = .044$

$P(D) = (1.0 + 1.24)/28.44 = 2.24/28.44 = .079$

$P(E) = 1.24/28.44 = .044$

b. $P(A \cap B) = 0/28.44 = 0$

c. $P(A \cup B) = (7.84 + 8)/28.44 = 15.84/28.44 = .557$

d. $P(B^c \cap E) = 0$

e. $P(A \cup E) = 9.24/28.44 = .325$

f.   Two events are mutually exclusive if they have no sample points in common or if the probability of their intersection is 0.

$P(A \cap B) = 0$. Since this is 0, A and B are mutually exclusive.

$P(A \cap C) = 0$. Since this is 0, A and C are mutually exclusive.

$P(A \cap D) = 0$. Since this is 0, A and D are mutually exclusive.

$P(A \cap E) = 0$. Since this is 0, A and E are mutually exclusive.

$P(B \cap C) = 1.24/28.44 = .044$. Since this is not 0, B and C are not mutually exclusive.

$P(B \cap D) = 2.24/28.44 = .079$. Since this is not 0, B and D are not mutually exclusive.

$P(B \cap E) = 1.24/28.44 = .044$. Since this is not 0, B and E are not mutually exclusive.

$P(C \cap D) = 1.24/28.44 = .044$. Since this is not 0, C and D are not mutually exclusive.

$P(C \cap E) = 1.24/28.44 = .044$. Since this is not 0, C and E are not mutually exclusive.

$P(D \cap E) = 1.24/28.44 = .044$. Since this is not 0, D and E are not mutually exclusive.

3.27   a.   A sample point is an event that cannot be decomposed into two or more other events. In this example, there are nine sample points. Let 1 be Warehouse 1, 2 be Warehouse 2, 3 be Warehouse 3, and let R be Regular, S be Stiff, and E be Extra stiff. The sample points are:

(1, R)   (1, S)   (1, E)
(2, R)   (2, S)   (2, E)
(3, R)   (3, S)   (3, E)

b.   The *sample space* of an experiment is the collection of all its sample points.

c. $P(C) = P(3, R) + P(3, S) + P(3, E)$
$= \quad .11 \quad + \quad .07 \quad + \quad .06$
$= .24$

d. $P(F) = P(1, E) + P(2, E) + P(3, E)$
    $\quad\quad = \quad 0 \quad + \quad .04 \quad + \quad .06$
    $\quad\quad = .10$

e. $P(A) = P(1, R) + P(1, S) + P(1, E)$
    $\quad\quad = \quad .41 \quad + \quad .06 \quad + \quad 0$
    $\quad\quad = .47$

f. $P(D) = P(1, R) + P(2, R) + P(3, R)$
    $\quad\quad = \quad .41 \quad + \quad .10 \quad + \quad .11$
    $\quad\quad = .62$

g. $P(E) = P(1, S) + P(2, S) + P(3, S)$
    $\quad\quad = \quad .06 \quad + \quad .15 \quad + \quad .07$
    $\quad\quad = .28$

3.29 a. Define the following events:

   D: {manufacturer is DaimlerChrysler}

   F: {manufacturer is Ford}

   G: {manufacturer is General Motors}

   C: {type is car}

   T: {type is truck}

   There are 6 sample points: $D,C \quad D,T \quad F,C \quad F,T \quad G,C \quad G,T$

   b. $P(C) = 1,050,100/2,333,200 = .450$

   $P(T) = 1,283,100/2,333,200 = .550$

   $P(F) = 788,600/2,333,200 = .338$

   c. $P(D \cup F) = P(D) + P(F) = (560,200 + 788,600)/2,333,200 = 1,348,800/2,333,200$
   $\quad\quad\quad\quad\quad = .578$

   $P(D \cap F) = 0$

   d. $P(C \cap G) = 550,500/2,333,200 = .236$

   $P(T \cap G) = 433,900/2,333,200 = .186$

3.31 a. Yes, the probabilities in the table sum to 1.

   $.05 + .16 + .05 + .19 + .32 + .05 + .11 + .05 + .02 = 1$

   b. $P(A) = .05 + .16 + .05 = .26$
   $P(B) = .05 + .19 + .11 = .35$
   $P(C) = .05 + .16 + .19 + .32 = .72$
   $P(D) = .05 + .05 + .11 + .05 + .02 = .28$
   $P(E) = .05$

c. $P(A \cup B) = .05 + .16 + .05 + .19 + .11 = .56$
$P(A \cap B) = .05$
$P(A \cup C) = .05 + .16 + .05 + .19 + .32 = .77$

d. $P(A^c) = 1 - P(A) = 1 - .26 = .74$

The probability that a managerial prospect is not highly motivated is .74. Only about 1/4 of the prospects are highly motivated.

e. The pairs of events that are mutually exclusive have no sample points in common.

$A \cap B$ contains the event "Prospect places in the high motivation category and in the high talent category." Therefore, A and B are not mutually exclusive.

$A \cap C$ contains the event "Prospect places in the high motivation category and in the medium or high talent category." Therefore, A and C are not mutually exclusive.

$A \cap D$ contains the event "Prospect places in the high motivation category and in the low talent category." Therefore, A and D are not mutually exclusive.

$A \cap E$ contains the event "Prospect places in the high motivation category and in the high talent category." Therefore, A and E are not mutually exclusive.

$B \cap C$ contains the event "Prospect places in the high talent category and in the medium or high motivation category." Therefore, B and C are not mutually exclusive.

$B \cap D$ contains the event "Prospect places in the high talent category and in the low motivation category." Therefore, B and D are not mutually exclusive.

$B \cap E$ contains the event "Prospect places in the high talent category and in the high motivation category." Therefore, B and E are not mutually exclusive.

$C \cap D$ contains no events. Therefore, C and D are mutually exclusive.

$C \cap E$ contains the event "Prospect places in the high talent category and in the high motivation category." Therefore, C and E are not mutually exclusive.

$D \cap E$ contains no events. Therefore, D and E are mutually exclusive.

3.33 a. $P(A) = P(E_1) + P(E_2) + P(E_3)$
$= .2 + .3 + .3$
$= .8$

$P(B) = P(E_2) + P(E_3) + P(E_5)$
$= .3 + .3 + .1$
$= .7$

$P(A \cap B) = P(E_2) + P(E_3)$
$= .3 + .3$
$= .6$

b. $P(E_1 \mid A) = \dfrac{P(E_1 \cap A)}{P(A)} = \dfrac{P(E_1)}{P(A)} = \dfrac{.2}{.8} = .25$

$P(E_2 \mid A) = \dfrac{P(E_2 \cap A)}{P(A)} = \dfrac{P(E_2)}{P(A)} = \dfrac{.3}{.8} = .375$

$P(E_3 \mid A) = \dfrac{P(E_3 \cap A)}{P(A)} = \dfrac{P(E_3)}{P(A)} = \dfrac{.3}{.8} = .375$

The original sample point probabilities are in the proportion .2 to .3 to .3 or 2 to 3 to 3.

The conditional probabilities for these sample points are in the proportion .25 to .375 to .375 or 2 to 3 to 3.

c.   (1) $P(B \mid A) = P(E_2 \mid A) + P(E_3 \mid A)$
$$= \quad .375 \quad + \quad .375 \quad \text{(from part } \mathbf{b})$$
$$= .75$$

     (2) $P(B \mid A) = \dfrac{P(A \cap B)}{P(A)} = \dfrac{.6}{.8} = .75$ (from part **a**)

The two methods do yield the same result.

d.   If A and B are independent events, $P(B \mid A) = P(B)$.

From part **c**, $P(B \mid A) = .75$. From part **a**, $P(B) = .7$.

Since $.75 \neq .7$, A and B are not independent events.

3.35   a. $P(A) = P(E_1) + P(E_3) = .22 + .15 = .37$

       b. $P(B) = P(E_2) + P(E_3) + P(E_4) = .31 + .15 + .22 = .68$

       c. $P(A \cap B) = P(E_3) = .15$

       d. $P(A \mid B) = \dfrac{P(A \cap B)}{P(B)} = \dfrac{.15}{.68} = .2206$

       e. $P(B \cap C) = 0$

       f. $P(C \mid B) = \dfrac{P(C \cap B)}{P(B)} = \dfrac{0}{.68} = 0$

       g.   For pair A and B: A and B are not independent because $P(A \mid B) \neq P(A)$ or $.2206 \neq .37$.

For pair A and C:

$$P(A \cap C) = P(E_1) = .22$$
$$P(C) = P(E_1) + P(E_5) = .22 + .1 = .32$$
$$P(A \mid C) = \dfrac{P(A \cap C)}{P(C)} = \dfrac{.22}{.32} = .6875$$

A and C are not independent because $P(A \mid C) \neq P(A)$ or $.6875 \neq .37$.

For pair B and C: B and C are not independent because $P(C \mid B) \neq P(C)$ or $0 \neq .32$.

3.37   a. $P(A \cap C) = 0 \Rightarrow A$ and C are mutually exclusive.
           $P(B \cap C) = 0 \Rightarrow B$ and C are mutually exclusive.

       b. $P(A) = P(1) + P(2) + P(3) = .20 + .05 + .30 = .55$
           $P(B) = P(3) + P(4) = .30 + .10 = .40$
           $P(C) = P(5) + P(6) = .10 + .25 = .35$

$$P(A \cap B) = P(3) = .30$$

$$P(A \mid B) = \frac{P(A \cap B)}{P(B)} = \frac{.30}{.40} = .75$$

A and B are independent if $P(A \mid B) = P(A)$. Since $P(A \mid B) = .75$ and $P(A) = .55$, A and B are not independent.

Since A and C are mutually exclusive, they are not independent. Similarly, since B and C are mutually exclusive, they are not independent.

c.   Using the probabilities of sample points,
$$P(A \cup B) = P(1) + P(2) + P(3) + P(4) = .20 + .05 + .30 + .10 = .65$$

Using the additive rule,
$$P(A \cup B) = P(A) + P(B) - P(A \cap B) = .55 + .40 - .30 = .65$$

Using the probabilities of sample points,
$$P(A \cup C) = P(1) + P(2) + P(3) + P(5) + P(6)$$
$$= .20 + .05 + .30 + .10 + .25 = .90$$

Using the additive rule,
$$P(A \cup C) = P(A) + P(C) - P(A \cap C) = .55 + .35 - 0 = .90$$

3.39   a. $P(A \cap B) = P(A) \cdot P(B) = (.4)(.2) = .08$
$P(A \mid B) = P(A) = .4$
$P(A \cup B) = P(A) + P(B) - P(A \cap B) = .4 + .2 - .08 = .52$

b. $P(A \cap B) = P(A \mid B) \cdot P(B) = (.6)(.2) = .12$

$$P(B \mid A) = \frac{P(A \cap B)}{P(A)} = \frac{.12}{.40} = .30$$

3.41   a.   Define the following events:

E: {winner was from Eastern Division}

C: {winner was from Central Division}

W: {winner was from Western Division}

N: {winner was from National League}

A: {winner was from American League}

$$P(E \mid A) = \frac{P(E \cap A)}{P(A)} = \frac{5/9}{6/9} = \frac{5}{6} = .833$$

b. $$P(N \mid C) = \frac{P(N \cap C)}{P(C)} = \frac{1/9}{2/9} = \frac{1}{2} = .5$$

c. $$P(C \cup W \mid N) = \frac{P([C \cup W] \cap N)}{P(N)} = \frac{1/9}{3/9} = \frac{1}{3} = .333$$

3.43   First, define the following event:

        A: {CVSA correctly determines the veracity of a suspect} $P(A) = .98$ (from claim)

    a.   The event that the CVSA is correct for all four suspects is the event $A \cap A \cap A \cap A$.
       $P(A \cap A \cap A \cap A) = .98(.98)(.98)(.98) = .9224$

    b.   The event that the CVSA is incorrect for at least one of the four suspects is the event
       $(A \cap A \cap A \cap A)^c$. $P(A \cap A \cap A \cap A)^c = 1 - P(A \cap A \cap A \cap A)$
       $= 1 - .9224 = .0776$

3.45   a.   We will define the events the same as in Exercise 3.10.

       There are a total of $9 \times 2 = 18$ sample points for this experiment. There are 9 sources of
       CO poisoning, and each source of poisoning has 2 possible outcomes, fatal or nonfatal.
       Suppose we introduce some notation to make it easier to write down the sample points. Let
       FI = Fire, AU = Auto exhaust, FU = Furnace, K = Kerosene or spaceheater,
       AP = Appliance, OG = Other gas-powered motors, FP = Fireplace, O = Other, and
       U = Unknown. Also, let F = Fatal and N = Nonfatal. The 18 sample points are:

           FI, F   AU, F   FU, F   K, F   AP, F   OG, F   FP, F   O, F   U, F
           FI, N   AU, N   FU, N   K, N   AP, N   OG, N   FP, N   O, N   U, N

       $P(F \mid FI) = 63/116 = .543$

    b. $P(AU \mid N) = 178/807 = .221$

    c. $P(U \mid F) = 9/174 = .052$

    d.   The event "not fire or fireplace" would include $AU, FU, K, AP, OG, O,$ and U.

       $P(AU \cup FU \cup K \cup AP \cup OG \cup O \cup U \mid N)$
          $= (178 + 345 + 18 + 63 + 73 + 19 + 42)/807 = 738/807 = .914$

3.47   Let us define the following events:

        S: {School is a subscriber}
        N: {School never uses the CCN broadcast}
        F: {School uses the CCN broadcast more than five times per week}

    From the problem, $P(S) = .40$, $P(N \mid S) = .05$, and $P(F \mid S) = .20$.

    a. $P(S \cap N) = P(N \mid S) \, P(S) = .05(.40) = .02$

    b. $P(S \cap F) = P(F \mid S) \, P(S) = .20(.40) = .08$

3.49   Define the following events:

        A: {Seed carries single spikelets}
        B: {Seed carries paired spikelets}
        C: {Seed produces ears with single spikelets}
        D: {Seed produces ears with paired spikelets}

    From the problem, $P(A) = .4$, $P(B) = .6$, $P(C \mid A) = .29$, $P(D \mid A) = .71$, $P(C \mid B) = .26$, and
    $P(D \mid B) = .74$.

a. $P(A \cap C) = P(C \mid A)P(A) = .29(.4) = .116$

b. $P(D) = P(A \cap D) + P(B \cap D) = P(D \mid A)P(A) + P(D \mid B)P(B) = .71(.4) + .74(.6)$
    $= .284 + .444 = .728$

3.51 Define the following events:

   A: {Selected firm implemented TQM}
   B: {Selected firm's sales increased}

From the information given, $P(A) = 30/100 = .3$, $P(B) = 60/100 = .6$, and $P(A \mid B) = 20/60$ $= 1/3$.

a. $P(A) = 30/100 = .3$
    $P(B) = 60/100 = .6$

b. If A and B are independent, $P(A \mid B) = P(A)$. However, $P(A \mid B) = 1/3 \neq P(A) = .3$. Thus, A and B are not independent.

c. Now, $P(A \mid B) = 18/60 = .3$. Since $P(A \mid B) = .3 = P(A) = .3$, A and B are independent.

3.53 a. The number of samples of size $n = 3$ elements that can be selected from a population of $N = 600$ is:

$$\begin{bmatrix} N \\ n \end{bmatrix} = \begin{bmatrix} 600 \\ 3 \end{bmatrix} = \frac{600!}{3!597!} = \frac{600(599)(598)}{3(2)(1)} = 35,820,200$$

b. If random sampling is employed, then each sample is equally likely. The probability that any sample is selected is $1/35,820,200$.

c. To draw a random sample of three elements from 600, we will number the elements from 1 to 600. Then, starting in an arbitrary position in Table I, Appendix B, we will select three numbers by going either down a column or across a row. Suppose that we start in the first three positions of column 8 and row 17. We will proceed down the column until we select three different numbers, skipping 000 and any numbers between 601 and 999. The first sample drawn will be 448, 298, and 136 (skip 987). The second sample drawn will be 47, 263, and 287. The 20 samples selected are:

Sample Number	Items Selected	Sample Number	Items Selected
1	448, 298, 136	11	345, 420, 152
2	47, 263, 287	12	144, 68, 485
3	153, 147, 222	13	490, 54, 178
4	360, 86, 357	14	428, 297, 549
5	205, 587, 254	15	186, 256, 261
6	563, 408, 258	16	90, 383, 232
7	428, 356, 543	17	438, 430, 352
8	248, 410, 197	18	129, 493, 496
9	542, 355, 208	19	440, 253, 81
10	399, 313, 563	20	521, 300, 15

None of the samples contain the same three elements. Because the probability in part **b** was so small, it would be very unlikely to have any two samples with the same elements.

3.55 a. If we randomly select one account from the 5,382 accounts, the probability of selecting account 3,241 is $1/5,382 = .000186$.

b. To draw a random sample of 10 accounts from 5,382, we will number the accounts from 1 to 5,382. Then, starting in an arbitrary position in Table I, Appendix B, we will select 10 numbers by going either down a column or across a row. Suppose that we start in the first four positions of column 10 and row 5. We will proceed down the column until we select 10 different numbers, skipping 0000 and any numbers between 5,382 and 9,999. The sample drawn will be:

1505, 4884, 1256, 1798, 3159, 2084, 0827, 2635, 4610, 2217

c. No. If the samples are randomly selected, any sample of size 10 is equally likely. The total number of ways to select 10 accounts from 5,382 is:

$$\begin{pmatrix} N \\ n \end{pmatrix} = \begin{pmatrix} 5,382 \\ 10 \end{pmatrix} = \frac{5,382!}{10!5,372!} = \frac{5,382(5381)(5380) \dots (5373)}{10(9)(8) \dots (1)}$$
$$= 5.572377607 \times 10^{30}$$

The probability that any one sample is selected is $1/5.572377607 \times 10^{30}$. Each of the two samples shown have the same probability of occurring.

3.57 a. Give each stock in the NYSE-Composite Transactions table of the Wall Street Journal a number (1 to m). Using Table I of Appendix B, pick a starting point and read down using the same number of digits as in m until you have n different numbers between 1 and m, inclusive.

3.59 (1) The probabilities of all sample points must lie between 0 and 1, inclusive.
      (2) The probabilities of all the sample points in the sample space must sum to 1.

3.61 $P(A \cap B) = .4$, $P(A \mid B) = .8$

Since the $P(A \mid B) = \dfrac{P(A \cap B)}{P(B)}$, substitute the given probabilities into the formula and solve for $P(B)$.

$$.8 = \frac{.4}{P(B)} \Rightarrow P(B) = \frac{.4}{.8} = .5$$

3.63 a. $P(A \cap B) = 0$

$P(B \cap C) = P(2) = .2$

$P(A \cup C) = P(1) + P(2) + P(3) + P(5) + P(6) = .3 + .2 + .1 + .1 + .2 = .9$

$P(A \cup B \cup C) = P(1) + P(2) + P(3) + P(4) + P(5) + P(6)$
$= .3 + .2 + .1 + .1 + .1 + .2 = 1$

$P(B^c) = P(1) + P(3) + P(5) + P(6) = .3 + .1 + .1 + .2 = .7$

$P(A^c \cap B) = P(2) + P(4) = .2 + .1 = .3$

$$P(B \mid C) = \frac{P(B \cap C)}{P(C)} = \frac{P(2)}{P(2) + P(5) + P(6)} = \frac{.2}{.2 + .1 + .2} = \frac{.2}{.5} = .4$$

$$P(B \mid A) = \frac{P(B \cap A)}{P(A)} = \frac{0}{P(A)} = 0$$

b. Since $P(A \cap B) = 0$, and $P(A) \cdot P(B) > 0$, these two would not be equal, implying A and B are not independent. However, A and B are mutually exclusive, since $P(A \cap B) = 0$.

c. $P(B) = P(2) + P(4) = .2 + .1 = .3$. But $P(B \mid C)$, calculated above, is .4. Since these are not equal, B and C are not independent. Since $P(B \cap C) = .2$, B and C are not mutually exclusive.

3.65 a. $6! = 6 \cdot 5 \cdot 4 \cdot 3 \cdot 2 \cdot 1 = 720$

b. $\begin{pmatrix} 10 \\ 9 \end{pmatrix} = \dfrac{10!}{9!(10-9)!} = \dfrac{10 \cdot 9 \cdot 8 \cdot \ \cdots \ \cdot 1}{9 \cdot 8 \cdot 7 \cdot \ \cdots \ \cdot 1 \cdot 1} = 10$

c. $\begin{pmatrix} 10 \\ 1 \end{pmatrix} = \dfrac{10!}{1!(10-1)!} = \dfrac{10 \cdot 9 \cdot 8 \cdot \ \cdots \ \cdot 1}{1 \cdot 9 \cdot 8 \cdot \ \cdots \ \cdot 1} = 10$

d. $\begin{pmatrix} 6 \\ 3 \end{pmatrix} = \dfrac{6!}{3!(6-3)!} = \dfrac{6 \cdot 5 \cdot 4 \cdot 3 \cdot 2 \cdot 1}{3 \cdot 2 \cdot 1 \cdot 3 \cdot 2 \cdot 1} = 20$

e. $0! = 1$

3.67 a. From the problem, it states the 25% of American adults smoke cigarettes. Thus, $P(A) = .25$.

b. Again, from the problem, it says that of the smokers, 13% attempted to quit smoking. Thus, $P(B \mid A) = .13$.

c. $P(A^c) = 1 - P(A) = 1 - .25 = .75$. The probability that an American adult does not smoke is .75.

d. $P(A \cap B) = P(B \mid A)\, P(A) = .13(.25) = .0325$.

3.69 Define the following events:

  T: {Technical staff}
  N: {Nontechnical staff}
  U: {Under 20 years with company}
  O: {Over 20 years with company}
  R_1: {Retire at age 65}
  R_2: {Retire at age 68}

The probabilities for each sample point are given in table form.

	U		O	
	T	N	T	N
R_1	$\dfrac{31}{200}$	$\dfrac{5}{200}$	$\dfrac{45}{200}$	$\dfrac{12}{200}$
R_2	$\dfrac{59}{200}$	$\dfrac{25}{200}$	$\dfrac{15}{200}$	$\dfrac{8}{200}$

Each sample point consists of three characteristics: type of staff (T or N), years with the company, (U or O), and age plan to retire (R_1 or R_2).

a. $P(T) = P(T \cap U \cap R_1) + P(T \cap U \cap R_2) + P(T \cap O \cap R_1) + P(T \cap O \cap R_2)$

$$= \frac{31}{200} + \frac{59}{200} + \frac{45}{200} + \frac{15}{200} = \frac{150}{200} = .75$$

b. $P(O) = P(O \cap T \cap R_1) + P(O \cap T \cap R_2) + P(O \cap N \cap R_1) + P(O \cap N \cap R_2)$

$$= \frac{45}{200} + \frac{15}{200} + \frac{12}{200} + \frac{8}{200} = \frac{80}{200} = .4$$

$$P(R_2 \cap O) = P(R_2 \cap O \cap T) + P(R_2 \cap O \cap N) = \frac{15}{200} + \frac{8}{200} = \frac{23}{200} = .115$$

Thus, $P(R_2 \mid O) = \dfrac{P(R_2 \cap O)}{P(O)} = \dfrac{.115}{.4} = .2875$

c. $P(T) = .75$ from **a**.

$$P(U \cap T) = P(U \cap T \cap R_1) + P(U \cap T \cap R_2) = \frac{31}{200} + \frac{59}{200} = \frac{90}{200} = .45$$

Thus, $P(U \mid T) = \dfrac{P(U \cap T)}{P(T)} = \dfrac{.45}{.75} = .6$

d. $P(O \cap N \cap R_1) = \dfrac{12}{200} = .06$

e. If A and B are independent, then $P(A \mid B) = P(A)$ or $P(R_2 \mid T) = P(R_2)$.

$$P(A \mid B) = \frac{P(A \cap B)}{P(B)} = \frac{\dfrac{59}{200} + \dfrac{15}{200}}{\dfrac{31}{200} + \dfrac{59}{200} + \dfrac{45}{200} + \dfrac{15}{200}} = \frac{\dfrac{74}{200}}{\dfrac{150}{200}} = \frac{74}{150} = .4933$$

$$P(A) = \frac{59}{200} + \frac{25}{200} + \frac{15}{200} + \frac{8}{200} = \frac{107}{200} = .535$$

$.4933 \neq .535$; therefore, A and B are not independent events.

f. The employee does not plan to retire at age 68 or the employee is not on the technical staff.

g. Yes, B and E are mutually exclusive events. An employee cannot be on the technical staff and on the nontechnical staff at the same time.

3.71 Suppose we define the following events:

   A: {Southwest Airline is selected}
   B: {Continental Airline is selected}
   C: {Flight arrived on time}
   D: {Flight was late}

a. Since one airline is to be selected at random, each airline is equally likely. Thus, the probability of selecting any one airline is 1/10.

$$P(A) = 1/10$$
$$P(B) = 1/10$$

b. $P(C \mid B) = 64.1/100 = .641$
$P(D \mid B) = (100 - 64.1)/100 = 35.9/100 = .359$

c. Since these figures are reported by the airline, these are probably upper bounds. The airlines would want to have a high "on-time" percentage. Thus, they would probably report a percentage that is higher than the actual percentage.

3.73 Define the following events:

A: {The watch is accurate}
N: {The watch is not accurate}

Assuming the manufacturer's claim is correct,

$P(N) = .05$ and $P(A) = 1 - P(N) = 1 - .05 = .95$

The sample space for the purchase of four of the manufacturer's watches is listed below.

(A, A, A, A)	(N, A, A, A)	(A, N, N, A)	(N, A, N, N)
(A, A, A, N)	(A, A, N, N)	(N, A, N, A)	(N, N, A, N)
(A, A, N, A)	(A, N, A, N)	(N, N, A, A)	(N, N, N, A)
(A, N, A, A)	(N, A, A, N)	(A, N, N, N)	(N, N, N, N)

a. All four watches not being accurate as claimed is the sample point (N, N, N, N).

Assuming the watches purchased operate independently and the manufacturer's claim is correct,

$P(N, N, N, N) = P(N)P(N)P(N)P(N) = .05^4 = .00000625$

b. The sample points in the sample space that consist of exactly two watches failing to meet the claim are listed below.

(A, A, N, N)	(N, A, A, N)
(A, N, A, N)	(N, A, N, A)
(A, N, N, A)	(N, N, A, A)

The probability that exactly two of the four watches fail to meet the claim is the sum of the probabilities of these six sample points.

Assuming the watches purchased operate independently and the manufacturer's claim is correct,

$P(A, A, N, N) = P(A)P(A)P(N)P(N) = (.95)(.95)(.05)(.05) = .00225625$

All six of the sample points will have the same probability. Therefore, the probability that exactly two of the four watches fail to meet the claim when the manufacturer's claim is correct is

$6(0.00225625) = .0135$

c. The sample points in the sample space that consist of three of the four watches failing to meet the claim are listed below.

$$(A, N, N, N) \quad (N, N, A, N)$$
$$(N, A, N, N) \quad (N, N, N, A)$$

The probability that three of the four watches fail to meet the claim is the sum of the probabilities of the four sample points.

Assuming the watches purchased operate independently and the manufacturer's claim is correct,

$$P(A, N, N, N) = P(A)P(N)P(N)P(N) = (.95)(.05)(.05)(.05) = .00011875$$

All four of the sample points will have the same probability. Therefore, the probability that three of the four watches fail to meet the claim when the manufacturer's claim is correct is

$$4(.00011875) = .000475$$

If this event occurred, we would tend to doubt the validity of the manufacturer's claim since its probability of occurring is so small.

d. All four watches tested failing to meet the claim is the sample point (N, N, N, N).

Assuming the watches purchased operate independently and the manufacturer's claim is correct,

$$P(N, N, N, N) = P(N)P(N)P(N)P(N) = (.05)^4 = .00000625$$

Since the probability of observing this event is so small if the claim is true, we have strong evidence against the validity of the claim. However, we do not have conclusive proof that the claim is false. There is still a chance the event can occur (with probability .00000625) although it is extremely small.

3.75   Define the following events:

A: {Acupoll predicts the success of a particular product}
B: {Product is successful}

From the problem, we know

$$P(A \mid B) = .89 \text{ and } P(B) = .90$$

Thus, $P(A \cap B) = P(A \mid B)P(B) = .89(.90) = .801$

3.77   Define the following events:

G: {regularly use the golf course}
T: {regularly use the tennis courts}

Given: $P(G) = .7$ and $P(T) = .5$

The event "uses neither facility" can be written as $G^c \cap T^c$ or $(G \cup T)^c$. We are given $P(G^c \cap T^c) = P[(G \cup T)^c] = .05$. The complement of the event "uses neither facility" is the event "uses at least one of the two facilities" which can be written as $G \cup T$.

$$P(G \cup T) = 1 - P[(G \cup T)^c] = 1 - .05 = .95$$

From the additive rule, $P(G \cup T) = P(G) + P(T) - P(G \cap T)$

$$\Rightarrow .95 = .7 + .5 - P(G \cap T)$$
$$\Rightarrow P(G \cap T) = .25$$

a.   The Venn Diagram is:

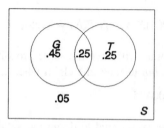

b. $P(G \cup T) = .95$ from above.

c. $P(G \cap T) = .25$ from above.

d. $P(G \mid T) = \dfrac{P(G \cap T)}{P(T)} = \dfrac{.25}{.5} = .5$

3.79   Define the following events:

   A: {Family with young children has income above the poverty line, but less than \$25,000}
   B: {Family with young children has unemployed parents or no parents}
   C: {Family with young children has income below the poverty line}

a. $P(A) = (2\% + 22\%)/100\% = 24\%/100\% = .24$

b. $P(B) = (1\% + 7\% + 2\%)/100\% = 10\%/100\% = .1$

c. $P(C) = (7\% + 7\%)/100\% = 14\%/100\% = .14$

3.81   Define the following events:

   S_1:   {Salesman makes sale on the first visit}
   S_2:   {Salesman makes a sale on the second visit}

$P(S_1) = .4 \qquad P(S_2 \mid S_1^c) = .65$

The sample points of the experiment are:

$S_1 \cap S_2^c$
$S_1^c \cap S_2$
$S_1^c \cap S_2^c$

The probability the salesman will make a sale is:

$$P(S_1 \cap S_2^c) + P(S_1^c \cap S_2) = P(S_1) + P(S_2 \mid S_1^c)P(S_1^c) = .4 + .65(1 - .4) = .4 + .39 = .79$$

3.83   Define the following events:

O_1: {Component #1 operates properly}
O_2: {Component #2 operates properly}
O_3: {Component #3 operates properly}

$P(O_1) = 1 - P(O_1^c) = 1 - .12 = .88$
$P(O_2) = 1 - P(O_2^c) = 1 - .09 = .91$
$P(O_3) = 1 - P(O_3^c) = 1 - .11 = .89$

a. $P(\text{System operates properly}) = P(O_1 \cap O_2 \cap O_3)$
$= P(O_1)P(O_2)P(O_3)$
(since the three components operate independently)
$= (.88)(.91)(.89) = .7127$

b. $P(\text{System fails}) = 1 - P(\text{system operates properly})$
$= 1 - .7127 \text{ (see part a)}$
$= .2873$

3.85   a.   The possible pairs of accounts that could be obtained are:

(0001, 0002) (0001, 0003) (0001, 0004) (0001, 0005)
(0002, 0003) (0002, 0004) (0002, 0005)
(0003, 0004) (0003, 0005)
(0004, 0005)

b.   There are 10 possible pairs of accounts that could be obtained. In a random sample, all 10 pairs of accounts have an equal chance of being selected. The probability of selecting any one of the 10 pairs is 1/10. Therefore, the probability of selecting accounts 0001 and 0004 is 1/10.

c.   Since only two accounts have a balance of $1,000 (0001 and 0004), the probability of selecting two accounts that each have a balance of $1,000 is 1/10.

Since there are only three accounts that do not have a balance of $1,000 (0002, 0003, and 0005), there are three possible pairs of accounts in which each has a balance other than $1,000 (0002, 0003), (0002, 0005), and (0003, 0005)). Therefore, the probability of selecting a pair of accounts in which each has a balance other than $1,000 is 3/10.

3.87   Define the following events:

A: {Take tough action early}
B: {Take tough action late}
C: {Never take tough action}
D: {Wisconsin}
E: {Illinois}
F: {Arkansas}
G: {Louisiana}

a. $P(D \cup G) = P(D) + P(G)$ (since D and G are mutually exclusive)
$$= \frac{0}{151} + \frac{37}{151} + \frac{9}{151} + \frac{1}{151} + \frac{21}{151} + \frac{15}{151} = \frac{83}{151} = .550$$

b. $P((D \cup G)^c) = 1 - P(D \cup G)$

$$= 1 - \frac{83}{151} = \frac{68}{151} = .450$$

c. $P(C) = \frac{9}{151} + \frac{11}{151} + \frac{6}{151} + \frac{15}{151} = \frac{41}{151} = .272$

d. $P(F \cap C) = \frac{6}{151} = .040$

e. $P(C \mid F) = \frac{P(F \cap C)}{P(F)} = \frac{\frac{6}{151}}{\frac{33}{151}} = \frac{6}{33} = .182$

f. $P((F \cup G) \mid A) = \frac{P((F \cup G) \cap A)}{P(A)} = \frac{\frac{5}{151} + \frac{1}{151}}{\frac{7}{151}} = \frac{6}{7} = .857$

g. $P(C \mid F) = \frac{P(F \cap C)}{P(F)} = \frac{\frac{6}{151}}{\frac{33}{151}} = \frac{6}{33} = .182$

4.1   A random variable is a rule that assigns one and only one value to each sample point of an experiment.

4.3   a.   The closing price of a particular stock on the New York Stock Exchange is discrete. It can take on only a countable number of values.

        b.   The number of shares of a particular stock that are traded on a particular day is discrete. It can take on only a countable number of values.

        c.   The quarterly earnings of a particular firm is discrete. It can take on only a countable number of values.

        d.   The percentage change in yearly earnings between 1996 and 1997 for a particular firm is continuous. It can take on any value in an interval.

        e.   The number of new products introduced per year by a firm is discrete. It can take on only a countable number of values.

        f.   The time until a pharmaceutical company gains approval from the U.S. Food and Drug Administration to market a new drug is continuous. It can take on any value in an interval of time.

4.5   The number of occupied units in an apartment complex at any time is a discrete random variable, as is the number of shares of stock traded on the New York Stock Exchange on a particular day. Two examples of continuous random variables are the length of time to complete a building project and the weight of a truckload of oranges.

4.7   An economist might be interested in the percentage of the work force that is unemployed, or the current inflation rate, both of which are continuous random variables.

4.9   The manager of a clothing store might be concerned with the number of employees on duty at a specific time of day, or the number of articles of a particular type of clothing that are on hand.

4.11   a.   We know $\sum p(x) = 1$. Thus, $p(2) + p(3) + p(5) + p(8) + p(10) = 1$

$$\Rightarrow p(5) = 1 - p(2) - p(3) - p(8) - p(10) = 1 - .15 - .10 - .25 - .25 = .25$$

        b. $P(x = 2 \text{ or } x = 10) = P(x = 2) + P(x = 10) = .15 + .25 = .40$

        c. $P(x \le 8) = P(x = 2) + P(x = 3) + P(x = 5) + P(x = 8) = .15 + .10 + .25 + .25 = .75$

4.13   a. $P(x \le 3) = p(1) + p(3) = .1 + .2 = .3$

        b. $P(x < 3) = p(1) = .1$

c. $P(x = 7) = .2$

d. $P(x \geq 5) = p(5) + p(7) + p(9) = .4 + .2 + .1 = .7$

e. $P(x > 2) = p(3) + p(5) + p(7) + p(9) = .2 + .4 + .2 + .1 = .9$

f. $P(3 \leq x \leq 9) = p(3) + p(5) + p(7) + p(9) = .2 + .4 + .2 + .1 = .9$

4.15   a. $\mu = E(x) = \sum xp(x)$
$$= 10(.05) + 20(.20) + 30(.30) + 40(.25) + 50(.10) + 60(.10)$$
$$= .5 + 4 + 9 + 10 + 5 + 6 = 34.5$$

$\sigma^2 = E(x - \mu)^2 = \sum (x - \mu)^2 p(x)$
$$= (10 - 34.5)^2(.05) + (20 - 34.5)^2(.20) + (30 - 34.5)^2(.30)$$
$$+ (40 - 34.5)^2(.25) + (50 - 34.5)^2(.10) + (60 - 34.5)^2(.10)$$
$$= 30.0125 + 42.05 + 6.075 + 7.5625 + 24.025 + 65.025 = 174.75$$

$\sigma = \sqrt{174.75} = 13.219$

b.

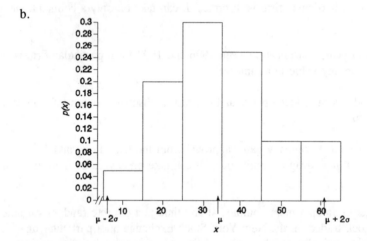

c. $\mu \pm 2\sigma \Rightarrow 34.5 \pm 2(13.219) \Rightarrow 34.5 \pm 26.438 \Rightarrow (8.062, 60.938)$

$P(8.062 < x < 60.938) = p(10) + p(20) + p(30) + p(40) + p(50) + p(60)$
$$= .05 + .20 + .30 + .25 + .10 + .10 = 1.00$$

4.17   a. $P(x = 4) = .2592$

b. $P(x < 2) = P(x = 0) + P(x = 1) = .0102 + .0768 = .0870$

c. $P(x \geq 3) = P(x = 3) + P(x = 4) + P(x = 5) = .3456 + .2592 + .0778 = .6826$

4.19   a.   Since the number of observations is very large, the relative frequencies or proportions should reflect the probabilities very well.

b.   Let x = household income. Then $P(x > \$200,000) = 1.1/100 = .011$
$P(x > \$100,000) = 4.1/100 + 1.1/100 = 5.2/100 = .052$
$P(x < \$100,000) = 18.5/100 + 19.0/100 + 15.9/100 + 12.8/100 + 9.1/100 + 13.8/100$
$$+ 5.7/100 = 94.8/100 = .948$$
$P(\$30,000 < x < \$49,999) = 12.8/100 + 9.1/100 = 21.9/100 = .219$

c.

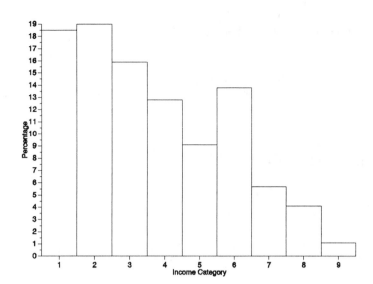

d. $P(\text{category } 6) = P(\$50,000 < x < \$74,999) = 13.8/100 = .138$
$P(\text{category 1 or 9}) = P(x < \$10,000) + P(x \geq \$200,000)$
$$= 18.5/100 + 1.1/100 = 19.6/100 = .196$$

4.21 $\mu_1 = E(a_1) = \sum xp(x) = 3(.60) + 2(.25) + 1(.10) + 0(.05) = 1.80 + .50 + .10 + 0 = 2.40$
$\mu_2 = E(a_2) = \sum xp(x) = 2(.60) + 1(.30) + 0(.10) = 1.20 + .30 + 0 = 1.50$
$\mu_3 = E(a_3) = \sum xp(x) = 1(.90) + 0(.10) = .90 + 0 = .90$
$\mu_4 = E(a_4) = \sum xp(x) = 1(.90) + 0(.10) = .90 + 0 = .90$
$\mu_5 = E(a_5) = \sum xp(x) = 1(.90) + 0(.10) = .90 + 0 = .90$
$\mu_6 = E(a_6) = \sum xp(x) = 2(.70) + 1(.25) + 0(.05) = 1.40 + .25 + 0 = 1.65$

4.23 a. If a large number of measurements are observed, then the relative frequencies should be very good estimators of the probabilities.

b. $E(x) = \sum xp(x) = 1(.01) + 2(.04) + 3(.04) + 4(.08) + 5(.10) + 6(.15) + 7(.25) + 8(.20)$
$\qquad + 9(.08) + 10(.05)$
$$= .01 + .08 + .12 + .32 + .50 + .90 + 1.75 + 1.60 + .72 + .50$$
$$= 6.50$$

The average number of checkout lanes per store is 6.5.

c. $\sigma^2 = \sum (x - \mu)^2 p(x) = (1 - 6.5)^2(.01) + (2 - 6.5)^2(.04) + (3 - 6.5)^2(.04)$
$\qquad + (4 - 6.5)^2(.08) + (5 - 6.5)^2(.10) + (6 - 6.5)^2(.15)$
$\qquad + (7 - 6.5)^2(.25) + (8 - 6.5)^2(.20) + (9 - 6.5)^2(.08)$
$\qquad + (10 - 6.5)^2(.05)$
$$= .3025 + .8100 + .4900 + .5000 + .2250 + .0375 + .0625$$
$$+ .4500 + .5000 + .6125$$
$$= 3.99$$

$\sigma = \sqrt{3.99} = 1.9975$

d. Chebyshev's Rule says that at least 0 of the observations should fall in the interval $\mu \pm \sigma$.

Chebyshev's Rule says that at least 75% of the observations should fall in the interval $\mu \pm 2\sigma$.

Random Variables and Probability Distributions

e. $\mu \pm \sigma \Rightarrow 6.5 \pm 1.9975 \Rightarrow (4.5025, 8.4975)$
$P(4.5025 \le x \le 8.4975) = .10 + .15 + .25 + .20 = .70$
This is at least 0.

$\mu \pm 2\sigma \Rightarrow 6.5 \pm 2(1.9975) \Rightarrow 6.5 \pm 3.995 \Rightarrow (2.505, 10,495)$
$P(2.505 \le x \le 10.495) = .04 + .08 + .10 + .15 + .25 + .20 + .08 + .05 = .95$
This is at least .75 or 75%.

4.25 a. $\dfrac{6!}{2!(6-2)!} = \dfrac{6!}{2!4!} = \dfrac{6 \cdot 5 \cdot 4 \cdot 3 \cdot 2 \cdot 1}{(2 \cdot 1)(4 \cdot 3 \cdot 2 \cdot 1)} = 15$

b. $\begin{pmatrix} 5 \\ 2 \end{pmatrix} = \dfrac{5!}{2!(5-2)!} = \dfrac{5!}{2!3!} = \dfrac{5 \cdot 4 \cdot 3 \cdot 2 \cdot 1}{(2 \cdot 1)(3 \cdot 2 \cdot 1)} = 10$

c. $\begin{pmatrix} 7 \\ 0 \end{pmatrix} = \dfrac{7!}{0!(7-0)!} = \dfrac{7!}{0!7!} = \dfrac{7 \cdot 6 \cdot 5 \cdot 4 \cdot 3 \cdot 2 \cdot 1}{(1)(7 \cdot 6 \cdot 5 \cdot 4 \cdot 3 \cdot 2 \cdot 1)} = 1$
(Note: $0! = 1$)

d. $\begin{pmatrix} 6 \\ 6 \end{pmatrix} = \dfrac{6!}{6!(6-6)!} = \dfrac{6!}{6!0!} = \dfrac{6 \cdot 5 \cdot 4 \cdot 3 \cdot 2 \cdot 1}{(6 \cdot 5 \cdot 4 \cdot 3 \cdot 2 \cdot 1)(1)} = 1$

e. $\begin{pmatrix} 4 \\ 3 \end{pmatrix} = \dfrac{4!}{3!(4-3)!} = \dfrac{4!}{3!1!} = \dfrac{4 \cdot 3 \cdot 2 \cdot 1}{(3 \cdot 2 \cdot 1)(1)} = 4$

4.27 a. $P(x = 1) = \dfrac{5!}{1!4!}(.2)^1(.8)^4 = \dfrac{5 \cdot 4 \cdot 3 \cdot 2 \cdot 1}{(1)(4 \cdot 3 \cdot 2 \cdot 1)}(.2)^1(.8)^4 = 5(.2)^1(.8)^4 = .4096$

b. $P(x = 2) = \dfrac{4!}{2!2!}(.6)^2(.4)^2 = \dfrac{4 \cdot 3 \cdot 2 \cdot 1}{(2 \cdot 1)(2 \cdot 1)}(.6)^2(.4)^2 = 6(.6)^2(.4)^2 = .3456$

c. $P(x = 0) = \dfrac{3!}{0!3!}(.7)^0(.3)^3 = \dfrac{3 \cdot 2 \cdot 1}{(1)(3 \cdot 2 \cdot 1)}(.7)^0(.3)^3 = 1(.7)^0(.3)^3 = .027$

d. $P(x = 3) = \dfrac{5!}{3!2!}(.1)^3(.9)^2 = \dfrac{5 \cdot 4 \cdot 3 \cdot 2 \cdot 1}{(3 \cdot 2 \cdot 1)(2 \cdot 1)}(.1)^3(.9)^2 = 10(.1)^3(.9)^2 = .0081$

e. $P(x = 2) = \dfrac{4!}{2!2!}(.4)^2(.6)^2 = \dfrac{4 \cdot 3 \cdot 2 \cdot 1}{(2 \cdot 1)(2 \cdot 1)}(.4)^2(.6)^2 = 6(.4)^2(.6)^2 = .3456$

f. $P(x = 1) = \dfrac{3!}{1!2!}(.9)^1(.1)^2 = \dfrac{3 \cdot 2 \cdot 1}{(1)(2 \cdot 1)}(.9)^1(.1)^2 = 3(.9)^1(.1)^2 = .027$

4.29 a. $\mu = np = 25(.5) = 12.5$

$\sigma^2 = np(1-p) = 25(.5)(.5) = 6.25$

$\sigma = \sqrt{\sigma^2} = \sqrt{6.25} = 2.5$

b. $\mu = np = 80(.2) = 16$

$\sigma^2 = np(1-p) = 80(.2)(.8) = 12.8$

$\sigma = \sqrt{\sigma^2} = \sqrt{12.8} = 3.578$

c. $\mu = np = 100(.6) = 60$

$\sigma^2 = np(1 - p) = 100(.6)(.4) = 24$

$\sigma = \sqrt{\sigma^2} = \sqrt{24} = 4.899$

d. $\mu = np = 70(.9) = 63$

$\sigma^2 = np(1 - p) = 70(.9)(.1) = 6.3$

$\sigma = \sqrt{\sigma^2} = \sqrt{6.3} = 2.510$

e. $\mu = np = 60(.8) = 48$

$\sigma^2 = np(1 - p) = 60(.8)(.2) = 9.6$

$\sigma = \sqrt{\sigma^2} = \sqrt{9.6} = 3.098$

f. $\mu = np = 1,000(.04) = 40$

$\sigma^2 = np(1 - p) = 1,000(.04)(.96) = 38.4$

$\sigma = \sqrt{\sigma^2} = \sqrt{38.4} = 6.197$

4.31   Define the following events:

   A: {Taxpayer is audited}
   B: {Taxpayer has income less than $100,000)
   C: {Taxpayer has income of $100,000 or higher}

a.   From the information given in the problem,

   $P(A \mid B) = 15/1000 = .015$
   $P(A \mid C) = 30/1000 = .030$

b.   Let x = number of taxpayers with incomes under $100,000 who are audited. Then x is a binomial random variable with $n = 5$ and $p = .015$.

$$P(x = 1) = \binom{5}{1}.015^1.985^{(5-1)}\frac{5!}{1!4!}.015^1.985^{(4)} = .0706$$

$$P(x > 1) = 1 - [P(x = 0) + P(x = 1)]$$

$$= 1 - \left[\binom{5}{0}.015^0.985^{(5-0)} + .0706\right]$$

$$= 1 - \left[\frac{5!}{0!5!}.015^0.985^5 + .0706\right]$$

$$= 1 - [.9272 + .0706] = 1 - .9978 = .0022$$

c. Let x = number of taxpayers with incomes of \$100,000 or more who are audited. Then x is a binomial random variable with $n = 5$ and $p = .030$.

$$P(x = 1) = \begin{pmatrix} 5 \\ 1 \end{pmatrix} .03^1 .97^{(5-1)} \frac{5!}{1!4!} .03^1 .97^4 = .1328$$

$$P(x > 1) = 1 - [P(x = 0) + P(x = 1)]$$

$$= 1 - \left[\begin{pmatrix} 5 \\ 0 \end{pmatrix} .03^0 .97^{(5-0)} + .1328 \right]$$

$$= 1 - \left[\frac{5!}{0!5!} .03^0 .97^5 + .1328 \right]$$

$$= 1 - [.8587 + .1328] = 1 - .9915 = .0085$$

d. Let x = number of taxpayers with incomes under \$100,000 who are audited. Then x is a binomial random variable with $n = 2$ and $p = .015$.

Let y = number of taxpayers with incomes \$100,000 or more who are audited. Then y is a binomial random variable with $n = 2$ and $p = .030$.

$$P(x = 0) = \begin{pmatrix} 2 \\ 0 \end{pmatrix} .015^0 .985^{(2-0)} = \frac{2!}{0!2!} .015^0 .985^2 = .9702$$

$$P(y = 0) = \begin{pmatrix} 2 \\ 0 \end{pmatrix} .03^0 .97^{(2-0)} = \frac{2!}{0!2!} .03^0 .97^2 = .9409$$

$$P(x = 0)P(y = 0) = .9702(.9409) = .9129$$

e. We must assume that the variables defined as x and y are binomial random variables. We must assume that the trials are identical, the probability of success is the same from trial to trial, and that the trials are independent.

4.33 a. In order to be a binomial random variable, the five characteristics must hold.

1. For this problem, there are 5 American adults surveyed. We will assume that these 5 trials are identical.

2. For each person surveyed, there are 2 possible outcomes: adult believes children will have a higher standard of living (S) or adult does not believe children will have a higher standard of living (F).

3. The probability that an adult believes his/her children will have a higher standard of living remains constant from trial to trial. For this problem, we will assume that this probability is $P(S) = .60$ for each trial.

4. We will assume that the opinion of one adult is independent of any other.

5. The random variable x is the number of adults who believe their children will have a higher standard of living in 5 trials.

Thus, x is a binomial random variable.

b. The value of p, the probability an adult believes his/her children will have a higher standard of living is .60.

c. Using Table II, Appendix B, with $n = 5$ and $p = .6$,

$$P(x = 3) = P(x \leq 3) - P(x \leq 2) = .663 - .317 = .346$$

d. Using Table II, Appendix B, with $n = 5$ and $p = .6$,

$$P(x \leq 2) = .317$$

4.35 a. Let x = the number of disasters in 25 shuttle missions. Then x is a binomial random variable with $n = 25$ and $p = 1/60{,}000$ (NASA's failure-rate estimate).

$$P(x = 0) = \binom{25}{0} (1/60{,}000)^0 (59{,}999/60{,}000)^{25-0} = \frac{25!}{0!\,25!} (1/60{,}000)^0 (59{,}999/60{,}000)^{25}$$
$$= .9996$$

b. Let x = the number of disasters in 25 shuttle missions. Then x is a binomial random variable with $n = 25$ and $p = 1/35$ (Air Force's failure-rate estimate).

$$P(x = 0) = \binom{25}{0} (1/35)^0 (34/35)^{25-0} = \frac{25!}{0!\,25!} (1/35)^0 (34/35)^{25}$$
$$= .4845$$

c. In both parts **a** and **b** we must assume that the flights are independent from trial to trial, and that the probability of disaster is the same from trial to trial.

d. The probability of at least one disaster in 25 trials would be $P(x \geq 1) = 1 - P(x = 0)$.

For NASA's failure-rate, $P(x \geq 1) = 1 - P(x = 0) = 1 - .9996 = .0004$.

For the Air Force's failure-rate, $P(x \geq 1) = 1 - P(x = 0) = 1 - .4845 = .5155$.

Since a disaster did happen, the Air Force's estimate is much more plausible because the probability is much higher than that using NASA'a failure-rate estimate.

e. Let x = the number of catastrophic failures in 10 shuttle missions. Then x is a binomial random variable with $n = 10$ and $p = 1/120$ (NASA's new failure-rate estimate).

$$P(x \geq 1) = 1 - P(x = 0) = 1 - \binom{10}{0} (1/120)^0 (119/120)^{10-0}$$
$$= 1 - \frac{10!}{0!\,10!} (1/120)^0 (119/120)^{10} = 1 - .9197 = .0803$$

4.37 a. $\mu = E(x) = np = 800(.65) = 520$
$\sigma = \sqrt{npq} = \sqrt{800(.65)(.35)} = 13.491$

b. Half of the 800 food items is 400. A value of $x = 400$ would have a z-score of:

$$z = \frac{x - \mu}{\sigma} = \frac{400 - 520}{13.491} = -8.895$$

Random Variables and Probability Distributions

Since the z-score associated with 400 items is so small (-8.895), it would be virtually impossible to observe less than half without any traces of pesticides if the 65% value was correct.

4.39   a.   The random variable x is discrete since it can assume a countable number of values $(0, 1, 2, \ldots)$.

b.   This is a Poisson probability distribution with $\lambda = 3$.

c.   In order to graph the probability distribution, we need to know the probabilities for the possible values of x. Using Table III of Appendix B with $\lambda = 3$:

$p(0) = .050$
$p(1) = P(x \le 1) - P(x = 0) = .199 - .050 = .149$
$p(2) = P(x \le 2) - P(x \le 1) = .423 - .199 = .224$
$p(3) = P(x \le 3) - P(x \le 2) = .647 - .423 = .224$
$p(4) = P(x \le 4) - P(x \le 3) = .815 - .647 = .168$
$p(5) = P(x \le 5) - P(x \le 4) = .916 - .815 = .101$
$p(6) = P(x \le 6) - P(x \le 5) = .966 - .916 = .050$
$p(7) = P(x \le 7) - P(x \le 6) = .988 - .966 = .022$
$p(8) = P(x \le 8) - P(x \le 7) = .996 - .988 = .008$
$p(9) = P(x \le 9) - P(x \le 8) = .999 - .996 = .003$
$p(10) \approx .001$

The probability distribution of x in graphical form is:

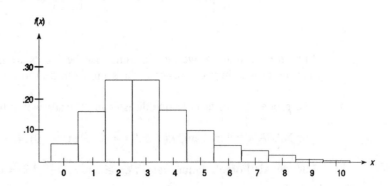

d. $\mu = \lambda = 3$
$\sigma^2 = \lambda = 3$

$\sigma = \sqrt{3} = 1.7321$

e.   The mean of x is the same as the mean of the probability distribution, $\mu = \lambda = 3$.

The standard deviation of x is the same as the standard deviation of the probability distribution, $\sigma = 1.7321$.

4.41 $\mu = \lambda = 1.5$

Using Table III of Appendix B:

a. $P(x \le 3) = .934$

b. $P(x \ge 3) = 1 - P(x \le 2) = 1 - .809 = .191$

c. $P(x = 3) = P(x \le 3) - P(x \le 2) = .934 - .809 = .125$

d. $P(x = 0) = .223$

e. $P(x > 0) = 1 - P(x = 0) = 1 - .223 = .777$

f. $P(x > 6) = 1 - P(x \le 6) = 1 - .999 = .001$

4.43   a. $E(x) = \mu = \lambda = 4$

$\sigma = \sqrt{\lambda} = \sqrt{4} = 2$

b. $z = \dfrac{x - \mu}{\sigma} = \dfrac{1 - 4}{2} = -1.5$

c.   Using Table III, Appendix B, with $\lambda = 4$,

$P(x \le 6) = .889$

d.   The experiment consists of counting the number of bank failures per year. We assume the probability a bank fails in a year is the same for each year. We must also assume that the number of bank failures in one year is independent of the number in any other year.

4.45   a. $\sigma = \sqrt{\sigma^2} = \sqrt{\lambda} = \sqrt{4} = 2$

b. $P(x > 10) = 1 - P(x \le 10)$
$= 1 - .977$ (Table III, Appendix B)
$= .003$

No. The probability that a sample of air from the plant exceeds the EPA limit is only .003. Since this value is very small, it is not very likely that this will occur.

c.   The experiment consists of counting the number of parts per million of vinyl chloride in air samples. We must assume the probability of a part of vinyl chloride appearing in a million parts of air is the same for each million parts of air. We must also assume the number of parts of vinyl chloride in one million parts of air is independent of the number in any other one million parts of air.

4.47   a.   In the problem, it is stated that $E(x) = .03$. This is also the value of λ.

$\sigma^2 = \lambda = .03$

b.   The experiment consists of counting the number of deaths or missing persons in a three- year interval. We must assume that the probability of a death or missing person in a three-year period is the same for any three-year period. We must also assume that the number of deaths or missing persons in any three-year period is independent of the number of deaths or missing persons in any other three-year period.

c. $P(x = 1) = \dfrac{\lambda^1 e^{-\lambda}}{1!} = \dfrac{.03^1 e^{-.03}}{1!} = .0291$

$P(x = 0) = \dfrac{\lambda^0 e^{-\lambda}}{0!} = \dfrac{.03^0 e^{-.03}}{0!} = .9704$

4.49   a.   From the problem, $\lambda = .37$. Thus, $\sigma = \sqrt{\lambda} = \sqrt{.37} = .6083$

b.   In order to plot the distribution of x, we must first calculate the probabilities.

$$P(x = 0) = \frac{\lambda^0 e^{-\lambda}}{0!} = \frac{.37^0 e^{-.37}}{0!} = .6907$$

$$P(x = 1) = \frac{.37^1 e^{-.37}}{1!} = .2556$$

$$P(x = 2) = \frac{.37^2 e^{-.37}}{2!} = .0473$$

$$P(x = 3) = \frac{.37^3 e^{-.37}}{3!} = .0058$$

$$P(x = 4) = \frac{.37^4 e^{-.37}}{4!} = .0005$$

$$P(x = 5) = \frac{.37^5 e^{-.37}}{5!} = .00004$$

The plot of the distribution is:

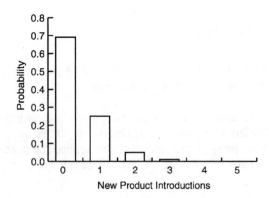

c. $P(x > 2) = 1 - P(x \leq 2) = 1 - .6907 - .2556 - .0473 = .0064$

Since this probability is so small, it would be very unlikely that a mainframe manufacturer would introduce more than 2 new products per year.

$P(x < 1) = P(x = 0) = .6907$

Since this probability is not small, it would not be unusual for a mainframe manufacturer to introduce less than 1 new product per year.

4.51 a. $f(x) = \dfrac{1}{d - c}$ $(c \le x \le d)$

$$\dfrac{1}{d - c} = \dfrac{1}{7 - 3} = \dfrac{1}{4}$$

$$f(x) = \begin{cases} \dfrac{1}{4} & (3 \le x \le 7) \\ 0 & \text{otherwise} \end{cases}$$

   b. $\mu = \dfrac{c + d}{2} = \dfrac{3 + 7}{2} = \dfrac{10}{2} = 5$

$$\sigma = \dfrac{d - c}{\sqrt{12}} = \dfrac{7 - 3}{\sqrt{12}} = \dfrac{4}{\sqrt{12}} = 1.155$$

   c. $\mu \pm \sigma \Rightarrow 5 \pm 1.155 \Rightarrow (3.845, 6.155)$

$$P(\mu - \sigma \le x \le \mu + \sigma) = P(3.845 \le x \le 6.155) = \dfrac{b - a}{d - c} = \dfrac{6.155 - 3.845}{7 - 3} = \dfrac{2.31}{4}$$
$$= .5775$$

4.53 $f(x) = \dfrac{1}{d - c} = \dfrac{1}{200 - 100} = \dfrac{1}{100} = .01$

$$f(x) = \begin{cases} .01 & (100 \le x \le 200) \\ 0 & \text{otherwise} \end{cases}$$

$$\mu = \dfrac{c + d}{2} = \dfrac{100 + 200}{2} = \dfrac{300}{2} = 150$$

$$\sigma = \dfrac{d - c}{\sqrt{12}} = \dfrac{200 - 100}{\sqrt{12}} = \dfrac{100}{\sqrt{12}} = 28.8675$$

   a. $\mu \pm 2\sigma \Rightarrow 150 \pm 2(28.8675) \Rightarrow 150 \pm 57.735 \Rightarrow (92.265, 207.735)$

$$P(x < 92.265) + P(x > 207.735) = P(x < 100) + P(x > 200)$$
$$= \qquad 0 \qquad + \qquad 0$$
$$= 0$$

   b. $\mu \pm 3\sigma \Rightarrow 150 \pm 3(28.8675) \Rightarrow 150 \pm 86.6025 \Rightarrow (63.3975, 236.6025)$

$$P(63.3975 < x < 236.6025) = P(100 < x < 200) = (200 - 100)(.01) = 1$$

   c. From **a**, $\mu \pm 2\sigma \Rightarrow (92.265, 207.735)$.

$$P(92.265 < x < 207.735) = P(100 < x < 200) = (200 - 100)(.01) = 1$$

4.55 a. For layer 2, let x = amount loss. Since the amount of loss is random between .01 and .05 million dollars, the uniform distribution for x is:

$$f(x) = \dfrac{1}{d - c} \qquad (c \le x \le d)$$

$$\dfrac{1}{d - c} = \dfrac{1}{.05 - .01} = \dfrac{1}{.04} = 25$$

Therefore, $f(x) = \begin{cases} 25 & (.01 \le x \le .05) \\ 0 & \text{otherwise} \end{cases}$

A graph of the distribution looks like the following:

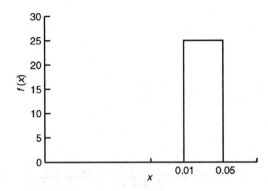

$$\mu = \frac{c + d}{2} = \frac{.01 + .05}{2} = .03$$

$$\sigma = \frac{d - c}{\sqrt{12}} = \frac{.05 - .01}{\sqrt{12}} = .0115, \ \sigma^2 = (.0115)^2 = .00013$$

The mean loss for layer 2 is .03 million dollars and the variance of the loss for layer 2 is .00013 million dollars squared.

b. For layer 6, let $x = $ amount loss. Since the amount of loss is random between .50 and 1.00 million dollars, the uniform distribution for x is:

$$f(x) = \frac{1}{d - c} \quad (c \le x \le d)$$

$$\frac{1}{d - c} = \frac{1}{1.00 - .50} = \frac{1}{.50} = 2$$

Therefore, $f(x) = \begin{cases} 2 & (.50 \le x \le 1.00) \\ 0 & \text{otherwise} \end{cases}$

A graph of the distribution looks like the following:

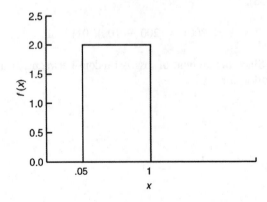

$$\mu = \frac{c + d}{2} = \frac{.50 + 1.00}{2} = .75$$

$$\sigma = \frac{d - c}{\sqrt{12}} = \frac{1.00 - .50}{\sqrt{12}} = .1443, \ \sigma^2 = (.1443)^2 = .0208$$

The mean loss for layer 6 is .75 million dollars and the variance of the loss for layer 6 is .0208 million dollars squared.

c. A loss of \$10,000 corresponds to $x = .01$. $P(x > .01) = 1$

A loss of \$25,000 corresponds to $x = .025$.

$$P(x < .025) = (\text{Base})(\text{Height}) = (x - c)\left[\frac{1}{d - c}\right] = (.025 - .01)\left[\frac{1}{.05 - .01}\right]$$
$$= .015(25) = .375$$

d. A loss of \$750,000 corresponds to $x = .75$. A loss of \$1,000,000 corresponds to $x = 1$.

$$P(.75 < x < 1) = (\text{Base})(\text{Height}) = (d - x)\left[\frac{1}{d - c}\right] = (1.00 - .75)\left[\frac{1}{1.00 - .50}\right]$$
$$= .25(2) = .5$$

A loss of \$900,000 corresponds to $x = .90$.

$$P(x > .9) = (\text{Base})(\text{Height}) = (d - x)\left[\frac{1}{d - c}\right] = (1.00 - .90)\left[\frac{1}{1.00 - .50}\right]$$
$$= .10(2) = .20$$

$$P(x = .9) = 0$$

4.57   To construct a relative frequency histogram for the data, we can use 7 measurement classes.

$$\text{Interval width} = \frac{\text{Largest number} - \text{smallest number}}{\text{Number of classes}} = \frac{98.0716 - .7434}{7} = 13.9$$

We will use an interval width of 14 and a starting value of .74335.

The measurement classes, frequencies, and relative frequencies are given in the table below.

Class	Measurement Class	Class Frequency	Class Relative Frequency
1	.74335 − 14.74335	6	6/40 = .15
2	14.74335 − 28.74335	4	.10
3	28.74335 − 42.74335	6	.15
4	42.74335 − 56.74335	6	.15
5	56.74335 − 70.74335	5	.125
6	70.74335 − 84.74335	4	.10
7	84.74335 − 98.74335	9	.225
		40	1.000

The histogram looks like the data could be from a uniform distribution. The last class (84.74335 − 98.74335) has a few more observations in it than we would expect. However, we cannot expect a perfect graph from a sample of only 40 observations.

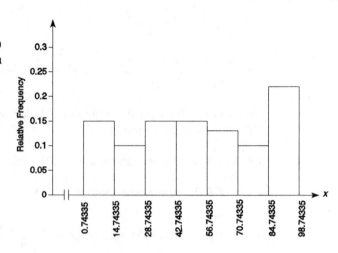

4.59   a.   The amount dispensed by the beverage machine is a continuous random variable since it can take on any value between 6.5 and 7.5 ounces.

      b.   Since the amount dispensed is random between 6.5 and 7.5 ounces, x is a uniform random variable.

$$f(x) = \frac{1}{d - c} \quad (c \le x \le d)$$

$$\frac{1}{d - c} = \frac{1}{7.5 - 6.5} = \frac{1}{1} = 1$$

Therefore, $f(x) = \begin{cases} 1 & (6.5 \le x \le 7.5) \\ 0 & \text{otherwise} \end{cases}$

The graph is as follows:

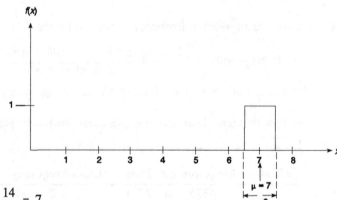

      c. $\mu = \dfrac{c + d}{2} = \dfrac{6.5 + 7.5}{2} = \dfrac{14}{2} = 7$

$$\sigma = \frac{d - c}{\sqrt{12}} = \frac{7.5 - 6.5}{\sqrt{12}} = .2887$$

$\mu \pm 2\sigma \Rightarrow 7 \pm 2(.2887) \Rightarrow 7 \pm .5774 \Rightarrow (6.422, 7.577)$

      d. $P(x \ge 7) = (7.5 - 7)(1) = .5$

      e. $P(x < 6) = 0$

f. $P(6.5 \leq x \leq 7.25) = (7.25 - 6.5)(1) = .75$

g. The probability that the next bottle filled will contain more than 7.25 ounces is:

$$P(x > 7.25) = (7.5 - 7.25)(1) = .25$$

The probability that the next 6 bottles filled will contain more than 7.25 ounces is:

$$P[(x > 7.25) \cap (x > 7.25) \cap (x > 7.25) \cap (x > 7.25) \cap (x > 7.25) \cap (x > 7.25)]$$
$$= [P(x > 7.25)]^6 = .25^6 = .0002$$

4.61   Table IV in the text gives the area between $z = 0$ and $z = z_0$. In this exercise, the answers may thus be read directly from the table by looking up the appropriate z.

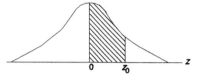

a. $P(0 < z < 2.0) = .4772$

b. $P(0 < z < 3.0) = .4987$

c. $P(0 < z < 1.5) = .4332$

d. $P(0 < z < .80) = .2881$

4.63   Using Table IV of Appendix B:

a. $P(z \leq z_0) = .2090$

$A = .5000 - .2090 = .2910$

Look up the area .2910 in the body of Table IV; $z_0 = -.81$.

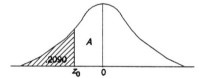

(z_0 is negative since the graph shows z_0 is on the left side of 0.)

b. $P(z \leq z_0) = .7090$

$P(z \leq z_0) = P(z \leq 0) + P(0 \leq z \leq z_0)$
$= .5 + P(0 \leq z \leq z) = .7090$

Therefore, $P(0 \leq z \leq z_0) = .7090 - .5 = .2090$

Look up the area .2090 in the body of Table IV; $z_0 \approx .55$.

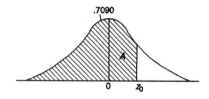

c. $P(-z_0 \leq z < z_0) = .8472$

$P(-z_0 \leq z < z_0) = 2P(0 \leq z \leq z_0)$
$2P(0 \leq z \leq z_0) = .8472$

Therefore, $P(0 \leq z \leq z_0) = .4236$.

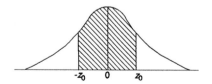

Look up the area .4236 in the body of Table IV; $z_0 = 1.43$.

d. $P(-z_0 \leq z < z_0) = .1664$

$P(-z_0 \leq z \leq z_0) = 2P(0 \leq z \leq z_0)$
$2P(0 \leq z \leq z_0) = .1664$

Therefore, $P(0 \leq z \leq z_0) = .0832$.

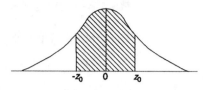

Look up the area .0832 in the body of Table IV; $z_0 = .21$.

e. $P(z_0 \leq z \leq 0) = .4798$

$P(z_0 \leq z \leq 0) = P(0 \leq z \leq -z_0)$

Look up the area .4798 in the body of Table IV; $z_0 = -2.05$.

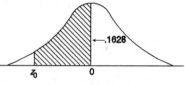

f. $P(-1 < z < z_0) = .5328$

$P(-1 < z < z_0)$
$= P(-1 < z < 0) + P(0 < z < z_0)$
$= .5328$

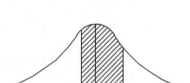

$P(0 < z < 1) + P(0 < z < z_0) = .5328$

Thus, $P(0 < z < z_0) = .5328 - .3413 = .1915$

Look up the area .1915 in the body of Table IV; $z_0 = .50$.

4.65 a. $P(10 \leq x \leq 12) = P\left[\dfrac{10 - 11}{2} \leq z \leq \dfrac{12 - 11}{2}\right]$

$= P(-0.50 \leq z \leq 0.50)$
$= A_1 + A_2$
$= .1915 + .1915 = .3830$

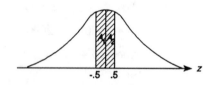

b. $P(6 \leq x \leq 10) = P\left[\dfrac{6 - 11}{2} \leq z \leq \dfrac{10 - 11}{2}\right]$

$= P(-2.50 \leq z \leq -0.50)$
$= P(-2.50 \leq z \leq 0)$
$\quad - P(-0.50 \leq z \leq 0)$
$= .4938 - .1915 = .3023$

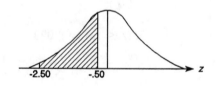

c. $P(13 \leq x \leq 16) = P\left[\dfrac{13 - 11}{2} \leq z \leq \dfrac{16 - 11}{2}\right]$

$= P(1.00 \leq z \leq 2.50)$
$= P(0 \leq z \leq 2.50)$
$\quad - P(0 \leq z \leq 1.00)$
$= .4938 - .3413 = .1525$

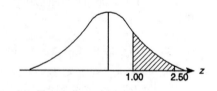

d. $P(7.8 \le x \le 12.6)$

$$= P\left[\frac{7.8 - 11}{2} \le z \le \frac{12.6 - 11}{2}\right]$$

$= P(-1.60 \le z \le 0.80)$
$= A_1 + A_2$
$= .4452 + .2881 = .7333$

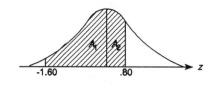

e. $P(x \ge 13.24) = P\left[z \ge \frac{13.24 - 11}{2}\right]$

$= P(z \ge 1.12)$
$= A_2 = .5 - A_1$
$= .5000 - .3686 = .1314$

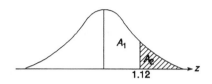

f. $P(x \ge 7.62) = P\left[z \ge \frac{7.62 - 11}{2}\right]$

$= P(z \ge -1.69)$
$= A_1 + A_2$
$= .4545 + .5000 = .9545$

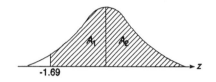

4.67 a. Let x = crop yield. The random variable x has a normal distribution with $\mu = 1,500$ and $\sigma = 250$.

$$P(x < 1,600) = P\left[z < \frac{1,600 - 1,500}{250}\right] = P(z < .4) = .5 + .1554 = .6554$$
(Using Table IV)

b. Let x_1 = crop yield in first year and x_2 = crop yield in second year. If x_1 and x_2 are independent, then the probability that the farm will lose money for two straight years is:

$$P(x_1 < 1,600)\, P(x_2 < 1,600) = P\left[z_1 < \frac{1,600 - 1,500}{250}\right] P\left[z_2 < \frac{1,600 - 1,500}{250}\right]$$

$= P(z_1 < .4)\, P(z_2 < .4) = (.5 + .1554)(.5 + .1554) = .6554(.6554) = .4295$
(Using Table IV)

c. $P(1,500 - 2\sigma \le x \le 1,500 + 2\sigma) =$

$$P\left[\frac{[1,500 - 2\sigma] - 1,500}{\sigma} \le z \le \frac{[1,500 + 2\sigma] - 1,500}{\sigma}\right]$$
$= P(-2 \le z \le 2) = 2P(0 \le z \le 2) = 2(.4772) = .9544$ \hspace{1cm} (Using Table IV)

4.69 a. Let x = passenger demand. The random variable x has a normal distribution with $\mu = 125$ and $\sigma = 45$.

For the Boeing 727, the probability that the passenger demand will exceed the capacity is:

$$P(x > 148) = P\left[z > \frac{148 - 125}{45}\right] = P(z > .51) = .5 - .1950 = .3050$$
(using Table IV)

For the Boeing 757, the probability that the passenger demand will exceed the capacity is:

$$P(x > 182) = P\left[z > \frac{182 - 125}{45}\right] = P(z > 1.27) = .5 - .3890 = .1020$$

b.   For the Boeing 727, the probability that the flight will depart with one or more empty seats is:

$$P(x \leq 147) = P\left[z \leq \frac{147 - 125}{45}\right] = P(z \leq .49) = .5 + .1879 = .6879$$

For the Boeing 757, the probability that the flight will depart with one or more empty seats is:

$$P(x \leq 181) = P\left[z \leq \frac{181 - 125}{45}\right] = P(z \leq 1.24) = .5 + .3925 = .8925$$

c.   For the Boeing 727, the probability that the spill is more than 100 passengers is:

$$P(x > 248) = P\left[z > \frac{248 - 125}{45}\right] = P(z > 2.73) = .5 - .4968 = .0032$$

4.71   a.   Using Table IV, Appendix B, with $\mu = 20.2$ and $\sigma = .65$,

$$P(20 < x < 21) = P\left[\frac{20 - 20.2}{.65} < z < \frac{21 - 20.2}{.65}\right] = P(-.31 < z < 1.23)$$
$$= P(-.31 < z < 0) + P(0 \leq z < 1.23) = .1217 + .3907 = .5124$$

b. $P(x \leq 19.84) = P\left[z \leq \frac{19.84 - 20.2}{.65}\right] = P(z \leq -.55) = .5 - .2088 = .2912$

Since the probability of observing a sardine with a length of 19.84 cm or smaller is not small ($p = .2912$), this is not an unusual event if the sardine was, in fact, two years old. Thus, it is likely that the sardine is two years old.

c. $P(x \geq 22.01) = P\left[z \geq \frac{22.01 - 20.2}{.65}\right] = P(z \geq 2.78) = .5 - .4973 = .0027$

Since the probability of observing a sardine with a length of 22.01 cm or larger is so small ($p = .0027$), this would be a very unusual event if the sardine was, in fact, two years old. Thus, it is unlikely that the sardine is two years old.

4.73   a.   Using Table IV, Appendix B, with $\mu = 8.72$ and $\sigma = 1.10$,

$$P(x < 6) = P\left[z < \frac{6 - 8.72}{1.10}\right] = P(z < -2.47) = .5 - .4932 = .0068$$

Thus, approximately .68% of the games would result in fewer than 6 hits.

b.   The probability of observing fewer than 6 hits in a game is $p = .0068$. The probability of observing 0 hits would be even smaller. Thus, it would be extremely unusual to observe a no hitter.

4.75   Let x = the amount of dye discharged. The random variable x
        is normally distributed with σ = 4.

        We want P(shade is unacceptable) $\leq$ .01

        $\Rightarrow P(x > 6) \leq$ .01

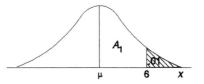

        Then A_1 = .50 − .01 = .49. Look up the area .49 in the body of Table IV, Appendix B; (take the
        closest value) z_0 = 2.33.

        To find μ, substitute into the z-score formula:

        $$z = \frac{x - \mu}{\sigma}$$

        $$2.33 = \frac{6 - \mu}{.4}$$

        $$\mu = 6 - .4(2.33) = 5.068$$

4.77   a.   IQR $= Q_U - Q_L$ = 195 − 72 = 123

        b.   IQR/s = 123/95 = 1.295

        c.   Yes. Since IQR is approximately 1.3, this implies that the data are approximately normal.

4.79   a.   Using MINITAB, the stem-and-leaf display is:

```
Stem-and-leaf of X          N = 28
Leaf Unit = 0.10

     5      11266
     6   2  1
     8   3  35
    11   4  035
    14   5  039
    14   6  3457
    10   7  346
     7   8  24469
     2      47
```

        Since the data do not form a mound-shape, it indicates that the data may not be normally
        distributed.

        b.   Using MINITAB, the descriptive statistics are:

Variable	N	Mean	Median	TrMean	StDev	SE Mean
X	28	5.511	6.100	5.519	2.765	0.5230

Variable	Minimum	Maximum	Q1	Q3
X	1.100	9.700	3.350	8.050

        The standard deviation is 2.765.

        c.   Using the printout from MINITAB in part **b**, Q_L = 3.35, and Q_U = 8.05. The IQR
              $= Q_U - Q_L$ = 8.05 − 3.35 = 4.7. If the data are normally distributed, then IQR/s $\approx$ 1.3.

        For this data, IQR/s = 4.7/2.765 = 1.70. This is a fair amount larger than 1.3, which
        indicates that the data may not be normally distributed.

Random Variables and Probability Distributions   69

d.   Using MINITAB, the normal probability plot is:

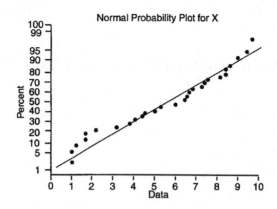

The data at the extremes are not particularly on a straight line. This indicates that the data are not normally distributed.

4.81   Using MINITAB, the stem-and-leaf display for the data is:

```
Stem-and-leaf of Score    N = 121
Leaf Unit = 1.0

    1    3  6
    1    4
    1    4
    1    5
    1    5
    1    6
    1    6
    3    7  04
    7    7  5889
    8    8  1
   38    8  666666677777788888888899999999
  (56)   9  00000001111111111111222222222233333333333333333333444444444
   27    9  5555555555556666666667777889
```

From the plot, the data appear to be very skewed to the left. This implies that the data are not normal.
Using MINITAB, the descriptive statistics are:

Variable	N	Mean	Median	TrMean	StDev	SE Mean
Score	121	90.504	92.000	91.321	6.961	0.633

Variable	Minimum	Maximum	Q1	Q3
Score	36.000	99.000	88.500	94.000

$\bar{x} \pm s \Rightarrow 90.504 \pm 6.961 \Rightarrow (83.543, 97.465)$

$\bar{x} \pm 2s \Rightarrow 90.504 \pm 2(6.961) \Rightarrow 90.504 \pm 13.922 \Rightarrow (76.582, 104.426)$

$\bar{x} \pm 3s \Rightarrow 90.504 \pm 3(6.961) \Rightarrow 90.504 \pm 20.883 \Rightarrow (69.621, 111.387)$

Of the 121 measurements, 110 are in the interval $83.543 - 97.465$. The proportion is $110/121 = .909$. This is much larger than the proportion stated by the Empirical Rule.

Of the 121 measurements, 117 are in the interval 76.582 − 104.426. The proportion is 117/121 = .967. This is close to the proportion stated by the Empirical Rule.

Of the 121 measurements, 120 are in the interval 69.621 − 111.387. The proportion is 120/121 = .992. This is close to the proportion stated by the Empirical Rule.

Since the proportion in the first interval is so large compared to what is stated by the Empirical Rule, this would imply that the data are not normal.

IQR = Q_U − Q_L = 94 − 88.5 = 5.5. IQR/s = 5.5/6.961 = .790. If the data are normally distributed, this ratio should be close to 1.3. Since .790 is not fairly close to 1.3, this indicates that the data are not normal.

Using MINITAB, the normal probability plot is:

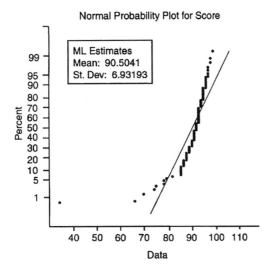

Normal Probability Plot for Score

The data do not form a straight line. This indicates that the data are not normally distributed.

4.83   Let x = maximum number of years one expects to spend with any one employer. For this problem, μ = 18.2 and σ = 10.64. The minimum value for x is 0, which has a z-score of:

$$z = \frac{x - \mu}{\sigma} = \frac{0 - 18.2}{10.64} = -1.71$$

Since the smallest possible value of x is only 1.71 standard deviations from the mean, it is very unlikely that the data are normal. A normal distribution will have about .0436 or 4.36% of the observations more than 1.71 standard deviations below the mean.

$$(P(z < -1.71) = .5 - .4564 = .0436)$$

For this data set, there are no observations more than 1.71 standard deviations below the mean.

4.85 a. Using Table II, $P(x \leq 11) = .345$

$$\mu = np = 25(.5) = 12.5, \sigma = \sqrt{npq} = \sqrt{25(.5)(.5)} = 2.5$$

Using the normal approximation,

$$P(x \leq 11) \approx P\left[z \leq \frac{(11 + .5) - 12.5}{2.5}\right] = P(z \leq -.40) = .5 - .1554 = .3446$$

b. Using Table II, $P(x \geq 16) = 1 - P(x \leq 15) = 1 - .885 = .115$

Using the normal approximation,

$$P(x \geq 16) \approx P\left[z \geq \frac{(16 - .5) - 12.5}{2.5}\right] = P(z \geq 1.2) = .5 - .3849 = .1151$$
$$\text{(from Table IV, Appendix B)}$$

c. Using Table II, $P(8 \leq x \leq 16) = P(x \leq 16) - P(x \leq 7) = .946 - .022 = .924$

Using the normal approximation,

$$P(8 \leq x \leq 16) \approx P\left[\frac{(8 - .5) - 12.5}{2.5} \leq z \leq \frac{(16 + .5) - 12.5}{2.5}\right]$$
$$= P(-2.0 \leq z \leq 1.6) = .4772 + .4452 = .9224$$
$$\text{(from Table IV, Appendix B)}$$

4.87 Let x = number of items with incorrect prices in 10,000 trials. Thus, x is a binomial random variable with $n = 10,000$ and $p = 1/30 = .033$.

$$\mu \pm 3\sigma \Rightarrow np \pm 3\sqrt{npq} \Rightarrow 10,000(.033) \pm 3\sqrt{10,000(.033)(.967)}$$
$$\Rightarrow 330 \pm 3\sqrt{319.11} \Rightarrow 330 \pm 3(17.864) \Rightarrow 330 \pm 53.591 \Rightarrow (276.409, 383.591)$$

Since the interval lies in the range 0 to 10,000, we can use the normal approximation to approximate the probabilities.

a. $$P(x \geq 100) \approx P\left[z \geq \frac{(100 - .5) - 330}{17.864}\right] = P(z \geq -12.90)$$
$$= P(-12.90 \leq z < 0) + .5 \approx .5 + .5 = 1$$

b. Let x = number of items with incorrect prices in 100 trials. Thus, x is a binomial random variable with $n = 100$ and $p = 1/30 = .033$.

$$\mu \pm 3\sigma \Rightarrow np \pm 3\sqrt{npq} \Rightarrow 100(.033) \pm 3\sqrt{100(.033)(.967)}$$
$$\Rightarrow 3.3 \pm 3\sqrt{3.191} \Rightarrow 3.3 \pm 3(1.786) \Rightarrow 3.3 \pm 5.358 \Rightarrow (-2.058, 8.658)$$

Since the interval does not lie in the range 0 to 100, the normal approximation will not be appropriate.

4.89　a.　Let x_1 = number of patients out of 500,000 who experience serious post-laser vision problems after being operated on by corneal specialists. Then x_1 is a binomial random variable with n = 500,000 and p_1 = .01

$$E(x_1) = np_1 = 500,000(.01) = 5,000$$

Let x_2 = number of patients out of 500,000 who experience serious post-laser vision problems after being operated on by opthalmologists. Then x_2 is a binomial random variable with n = 500,000 and p_2 = .05

$$E(x_2) = np_2 = 500,000(.05) = 25,000$$

b.　Let x_2 = number of patients out of 400 who experience serious post-laser vision problems after being operated on by opthalmologists. Then x_2 is a binomial random variable with n = 400 and p_2 = .05.

$$\mu \pm 3\sigma \Rightarrow np_2 \pm 3\sqrt{np_2q_2} \Rightarrow 400(.05) \pm 3\sqrt{400(.05)(.95)} \Rightarrow 20 \pm 3(4.3589)$$
$$\Rightarrow 20 \pm 13.0767 \Rightarrow (6.9233, 33.0767)$$

Since the interval lies in the range 0 to 400, we can use the normal approximation to approximate the probability.

$$P(x \geq 20) \approx P\left[z \geq \frac{(20 - .5) - 20}{4.3589}\right] = P(z \geq -.11) = .5 + .0438 = .5438$$

c.　Let x_1 = number of patients out of 400 who experience serious post-laser vision problems after being operated on by corneal specialists. Then x_1 is a binomial random variable with n = 400 and p_1 = .01.

$$\mu \pm 3\sigma \Rightarrow np_1 \pm 3\sqrt{np_1q_1} \Rightarrow 400(.01) \pm 3\sqrt{400(.01)(.99)} \Rightarrow 4 \pm 3(1.99)$$
$$\Rightarrow 4 \pm 5.97 \Rightarrow (-1.97, 9.97)$$

Since the interval does not lie in the range 0 to 400, we cannot use the normal approximation to approximate the probability.

4.91　Let x equal the number of catastrophes due to booster failure.

a.　In order to approximate the binomial distribution with the normal distribution, the interval $\mu \pm 3\sigma$ should lie in the range 0 to n.

$$\mu \pm 3\sigma \Rightarrow np \pm 3\sqrt{npq} \Rightarrow 25\left[\frac{1}{35}\right] \pm 3\sqrt{25\left[\frac{1}{35}\right]\left[1 - \frac{1}{35}\right]}$$
$$\Rightarrow .714 \pm 3(.833) \Rightarrow (-1.785, 3.213)$$

Since the interval calculated does not lie in the range 0 to 25, we should not use the normal approximation.

b. $P(x \geq 1) \approx P\left[z \geq \dfrac{(1 - .5) - .714}{.833}\right]$

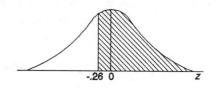

$= P(z \geq -.26)$

$= .5000 + .1026 = .6026$

(Using Table IV in Appendix B.)

The exact probability is .5155 and the approximate probability is .6026. The approximation is quite a bit off, but this is not surprising since in part **a** we decided that we should not use the normal approximation.

c. Referring to part **a**, recalculate the interval $\mu \pm 3\sigma$ using $n = 100$ instead of $n = 25$.

$$\mu \pm 3\sigma \Rightarrow np \pm 3\sqrt{npq} \Rightarrow 100\left[\dfrac{1}{35}\right] \pm 3\sqrt{100\left[\dfrac{1}{35}\right]\left[1 - \dfrac{1}{35}\right]}$$

$$\Rightarrow 2.857 \pm 3(1.666) \Rightarrow (-2.141, 7.855)$$

Since the interval calculated does not lie in the range 0 to 100 we should not use the normal approximation when $n = 100$.

Recalculate the interval $\mu \pm 3\sigma$ using $n = 500$.

$$\mu \pm 3\sigma \Rightarrow np \pm 3\sqrt{npq} \Rightarrow 500\left[\dfrac{1}{35}\right] \pm 3\sqrt{500\left[\dfrac{1}{35}\right]\left[1 - \dfrac{1}{35}\right]}$$

$$\Rightarrow 14.286 \pm 3(3.725) \Rightarrow (3.111, 25.461)$$

Since the interval calculated does lie in the range 0 to 500, we can use the normal approximation when $n = 500$.

Since we can use the normal approximation when $n = 500$, we can use it when $n = 1,000$.

$$\mu \pm 3\sigma \Rightarrow np \pm 3\sqrt{npq} \Rightarrow 1000\left[\dfrac{1}{35}\right] \pm 3\sqrt{1000\left[\dfrac{1}{35}\right]\left[1 - \dfrac{1}{35}\right]}$$

$$\Rightarrow 28.571 \pm 3(5.268) \Rightarrow (12.767, 44.375)$$

Since the interval calculated does lie in the range 0 to 1,000, we can use the normal approximation when $n = 1,000$.

d. x is a binomial random variable with $n = 1000$ and $p = \dfrac{1}{35}$.

$P(x > 25) \approx P\left[z > \dfrac{(25 + .5) - 28.571}{5.268}\right]$

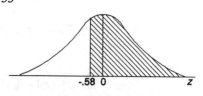

$= P(z > -.58)$

$= .5000 + .2190 = .7190$

(Using Table IV in Appendix B.)

4.93 a. If 80% of the passengers pass through without their luggage being inspected, then 20% will be detained for luggage inspection. The expected number of passengers detained will be:

$$E(x) = np = 1,500(.2) = 300$$

b. For $n = 4,000$, $E(x) = np = 4,000(.2) = 800$

c. $P(x > 600) \approx P\left[z > \dfrac{(600 + .5) - 800}{\sqrt{4000(.2)(.8)}}\right] = P(z > -7.89) = .5 + .5 = 1.0$

4.95 a. If $\lambda = 1$, $a = 1$, then $e^{-\lambda a} = e^{-1} = .367879$

b. If $\lambda = 1$, $a = 2.5$, then $e^{-\lambda a} = e^{-2.5} = .082085$

c. If $\lambda = 2.5$, $a = 3$, then $e^{-\lambda a} = e^{-7.5} = .000553$

d. If $\lambda = 5$, $a = .3$, then $e^{-\lambda a} = e^{-1.5} = .223130$

4.97 Using Table V in Appendix B:

a. $P(x \le 3) = 1 - P(x > 3) = 1 - e^{-2.5(3)} = 1 - e^{-7.5} = 1 - .000553 = .999447$

b. $P(x \le 4) = 1 - P(x > 4) = 1 - e^{-2.5(4)} = 1 - e^{-10} = 1 - .000045 = .999955$

c. $P(x \le 1.6) = 1 - P(x > 1.6) = 1 - e^{-2.5(1.6)} = 1 - e^{-4} = 1 - .018316 = .981684$

d. $P(x \le .4) = 1 - P(x > .4) = 1 - e^{-2.5(.4)} = 1 - e^{-1} = 1 - .367879 = .632121$

4.99 a. $\lambda = 1/\mu = 1/10.54 = .0949$

b. $\mu = 10.54$; $\sigma = 1/\lambda = 1/.0949 = 10.54$

On average, the University of Michigan hockey team scored every 10.54 minutes.

Since we know the distribution of time-between-goals is not symmetric, we use Chebyshev's Rule to describe the data. We know at least 75% of all times-between-goals will fall within 2 standard deviations of the mean.

$\mu \pm 2\sigma \Rightarrow 10.54 \pm 2(10.54) \Rightarrow 10.54 \pm 21.08 \Rightarrow (-10.54, 31.62)$

Since we know that no time can be negative, we know that at least 75% of all times-between-goals are less than 31.62 minutes.

c. To graph the exponential distribution of x when $\lambda = .0949$ we need to calculate $f(x)$ for certain values of x. Using a calculator:

$f(x) = \lambda e^{-\lambda x} = .0949\, e^{-.0949x}$

$f(1) = .0949\, e^{-.0949(1)} = .0863$

$f(5) = .0949\, e^{-.0949(5)} = .0590$

$f(10) = .0949\, e^{-.0949(10)} = .0367$

$f(15) = .0949\, e^{-.0949(15)} = .0229$

$f(20) = .0949\, e^{-.0949(20)} = .0142$

$f(25) = .0949\, e^{-.0949(25)} = .0088$

$$f(30) = .0949\, e^{-.0949(30)} = .0055$$

The graph of the exponential distribution is:

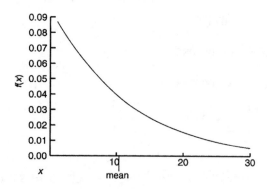

d. $P(x < 2) = 1 - e^{-.0949(2)} = 1 - e^{-.1898} = 1 - .8271 = .1729$

4.101 a. $P(x) = e^{-\lambda x} = e^{-.5x}$

b. $P(x \geq 4) = e^{-.5(4)} = e^{-2} = .135335$ (Table V, Appendix B)

c. $\mu = \dfrac{1}{\lambda} = \dfrac{1}{.5} = 2$

$P(x > \mu) = P(x > 2) = e^{-.5(2)} = e^{-1} = .367879$ (Table V, Appendix B)

d. For all exponential distributions, $\mu = \dfrac{1}{\lambda}$

$P(x > \mu) = P\left[x > \dfrac{1}{\lambda}\right] = e^{-\lambda(1/\lambda)} = e^{-1} = .367879.$ Thus,

regardless of the value of λ, the probability that x is larger than the mean is always .367879.

e. $P(x > 5) = e^{-.5(5)} = e^{-2.5} = .082085$ (Table V, Appendix B)

If 10,000 units are sold, approximately $10,000(.082085) = 820.85$ will perform satisfactorily for more than 5 years.

$P(x \leq 1) = 1 - P(x > 1) = 1 - e^{-.5(1)} = 1 - e^{-.5} = 1 - .606531 = .39469$

If 10,000 units are sold, approximately $10,000(.393469) = 3934.69$ will fail within 1 year.

f. $P(x < a) \leq .05$
$\Rightarrow 1 - P(x \geq a) \leq .05$
$\Rightarrow P(x \geq a) \geq .95$
$\Rightarrow e^{-.5a} \geq .95$

Using Table V, Appendix B, $e^{-.05}$ is closest to .95 (yet larger).

Thus, $.05 = .5a \Rightarrow a = .1$

The warranty should be for approximately .1 year or $.1(365) = 36.5$ or 37 days.

4.103   a.   Let x = interarrival time of jobs. Then x has an exponential distribution with a mean of $\mu = 1.25$ minutes and $\lambda = 1/1.25$.

$$P(x \leq 1) = 1 - P(x > 1) = 1 - e^{-1/1.25} = 1 - e^{-.8} = 1 - .449329 = .550671$$

b.   Let y = amount of time the machine operates before breaking down. Then y has an exponential distribution with a mean of $\mu = 540$ minutes and $\lambda = 1/540$.

$$P(y > 720) = e^{-720/540} = e^{-1.333333} = .263597$$

4.105   a–b.   The different samples of $n = 2$ with replacement and their means are:

Possible Samples	$\bar{x}$	Possible Samples	$\bar{x}$
0, 0	0	4, 0	2
0, 2	1	4, 2	3
0, 4	2	4, 4	4
0, 6	3	4, 6	5
2, 0	1	6, 0	3
2, 2	2	6, 2	4
2, 4	3	6, 4	5
2, 6	4	6, 6	6

c.   Since each sample is equally likely, the probability of any 1 being selected is $\dfrac{1}{4}\left(\dfrac{1}{4}\right) = \dfrac{1}{16}$

d. $P(\bar{x} = 0) = \dfrac{1}{16}$

$P(\bar{x} = 1) = \dfrac{1}{16} + \dfrac{1}{16} = \dfrac{2}{16}$

$P(\bar{x} = 2) = \dfrac{1}{16} + \dfrac{1}{16} + \dfrac{1}{16} = \dfrac{3}{16}$

$P(\bar{x} = 3) = \dfrac{1}{16} + \dfrac{1}{16} + \dfrac{1}{16} + \dfrac{1}{16} = \dfrac{4}{16}$

$P(\bar{x} = 4) = \dfrac{1}{16} + \dfrac{1}{16} + \dfrac{1}{16} = \dfrac{3}{16}$

$P(\bar{x} = 5) = \dfrac{1}{16} + \dfrac{1}{16} = \dfrac{2}{16}$

$P(\bar{x} = 6) = \dfrac{1}{16}$

$\bar{x}$	$p(\bar{x})$
0	1/16
1	2/16
2	3/16
3	4/16
4	3/16
5	2/16
6	1/16

e.

4.107   If the observations are independent of each other, then

$$P(1, 1) = p(1)p(1) = .2(.2) = .04$$
$$P(1, 2) = p(1)p(2) = .2(.3) = .06$$
$$P(1, 3) = p(1)p(3) = .2(.2) = .04$$

etc.

a.

Possible Samples	$\bar{x}$	$p(\bar{x})$	Possible Samples	$\bar{x}$	$p(\bar{x})$
1, 1	1	.04	3, 4	3.5	.04
1, 2	1.5	.06	3, 5	4	.02
1, 3	2	.04	4, 1	2.5	.04
1, 4	2.5	.04	4, 2	3	.06
1, 5	3	.02	4, 3	3.5	.04
2, 1	1.5	.06	4, 4	4	.04
2, 2	2	.09	4, 5	4.5	.02
2, 3	2.5	.06	5, 1	3	.02
2, 4	3	.06	5, 2	3.5	.03
2, 5	3.5	.03	5, 3	4	.02
3, 1	2	.04	5, 4	4.5	.02
3, 2	2.5	.06	5, 5	5	.01
3, 3	3	.04			

Summing the probabilities, the probability distribution of $\bar{x}$ is:

$\bar{x}$	$p(\bar{x})$
1	.04
1.5	.12
2	.17
2.5	.20
3	.20
3.5	.14
4	.08
4.5	.04
5	.01

b.

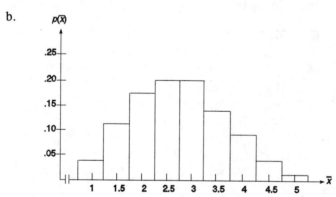

c. $P(\bar{x} \geq 4.5) = .04 + .01 = .05$

d.   No. The probability of observing $\bar{x} = 4.5$ or larger is small (.05).

4.109 a. For a sample of size $n = 2$, the sample mean and sample median are exactly the same. Thus, the sampling distribution of the sample median is the same as that for the sample mean (see Exercise 4.107**a**).

b. The probability histogram for the sample median is identical to that for the sample mean (see Exercise 4.107**b**).

4.111 a. $\mu_{\bar{x}} = \mu = 100, \ \sigma_{\bar{x}} = \dfrac{\sigma}{\sqrt{n}} = \dfrac{\sqrt{100}}{\sqrt{4}} = 5$

b. $\mu_{\bar{x}} = \mu = 100, \ \sigma_{\bar{x}} = \dfrac{\sigma}{\sqrt{n}} = \dfrac{\sqrt{100}}{\sqrt{25}} = 2$

c. $\mu_{\bar{x}} = \mu = 100, \ \sigma_{\bar{x}} = \dfrac{\sigma}{\sqrt{n}} = \dfrac{\sqrt{100}}{\sqrt{100}} = 1$

d. $\mu_{\bar{x}} = \mu = 100, \ \sigma_{\bar{x}} = \dfrac{\sigma}{\sqrt{n}} = \dfrac{\sqrt{100}}{\sqrt{50}} = 1.414$

e. $\mu_{\bar{x}} = \mu = 100, \ \sigma_{\bar{x}} = \dfrac{\sigma}{\sqrt{n}} = \dfrac{\sqrt{100}}{\sqrt{500}} = .447$

f. $\mu_{\bar{x}} = \mu = 100, \ \sigma_{\bar{x}} = \dfrac{\sigma}{\sqrt{n}} = \dfrac{\sqrt{100}}{\sqrt{1000}} = .316$

4.113 a. $\mu = \sum xp(x) = 1(.1) + 2(.4) + 3(.4) + 8(.1) = 2.9$

$\sigma^2 = \sum (x - \mu)^2 p(x) = (1 - 2.9)^2(.1) + (2 - 2.9)^2(.4) + (3 - 2.9)^2(.4) + (8 - 2.9)^2(.1)$
$= .361 + .324 + .004 + 2.601 = 3.29$

$\sigma = \sqrt{3.29} = 1.814$

b. The possible samples, values of $\bar{x}$, and associated probabilities are listed:

Possible Samples	x	$p(x)$	Possible Samples	x	$p(x)$
1, 1	1	.01	3, 1	2	.04
1, 2	1.5	.04	3, 2	2.5	.16
1, 3	2	.04	3, 3	3	.16
1, 8	4.5	.01	3, 8	5.5	.04
2, 1	1.5	.04	8, 1	4.5	.01
2, 2	2	.16	8, 2	5	.04
2, 3	2.5	.16	8, 3	5.5	.04
2, 8	5	.04	8, 8	8	.01

$P(1, 1) = p(1)p(1) = .1(.1) = .01$
$P(1, 2) = p(1)p(2) = .1(.4) = .04$
$P(1, 3) = p(1)p(3) = .1(.4) = .04$
  etc.

The sampling distribution of $\bar{x}$ is:

$\bar{x}$	$p(\bar{x})$
1	.01
1.5	.08
2	.24
2.5	.32
3	.16
4.5	.02
5	.08
5.5	.08
8	.01
	1.00

c. $\mu_{\bar{x}} = E(\bar{x}) = \sum \bar{x}p(\bar{x}) = 1(.01) + 1.5(.08) + 2(.24) + 2.5(.32) + 3(.16) + 4.5(.02)$
$$+ 5(.08) + 5.5(.08) + 8(.01)$$
$$= 2.9 = \mu$$

$\sigma_{\bar{x}}^2 = \sum (\bar{x} - \mu_{\bar{x}})^2 p(\bar{x}) = (1 - 2.9)^2(.01) + (1.5 - 2.9)^2(.08) + (2 - 2.9)^2(.24)$
$$+ (2.5 - 2.9)^2(.32) + (3 - 2.9)^2(.16) + (4.5 - 2.9)^2(.02)$$
$$+ (5 - 2.9)^2(.08) + (5.5 - 2.9)^2(.08) + (8 - 2.9)^2(.01)$$
$$= .0361 + .1568 + .1944 + .0512 + .0016 + .0512 + .3528$$
$$+ .5408 + .2601$$
$$= 1.645$$

$$\sigma_{\bar{x}} = \sqrt{1.645} = 1.283$$

$$\sigma_{\bar{x}} = \sigma/\sqrt{n} = 1.814/\sqrt{2} = 1.283$$

4.115 In Exercise 4.114, it was determined that the mean and standard deviation of the sampling distribution of the sample mean are 20 and 2 respectively. Using Table IV, Appendix B:

a. $P(\bar{x} < 16) = P\left(z < \dfrac{16 - 20}{2}\right) = P(z < -2) = .5 - .4772 = .0228$

b. $P(\bar{x} > 23) = P\left(z > \dfrac{23 - 20}{2}\right) = P(z > 1.50) = .5 - .4332 = .0668$

c. $P(\bar{x} > 25) = P\left(z > \dfrac{25 - 20}{2}\right) = P(z > 2.5) = .5 - .4938 = .0062$

d. $P(16 < \bar{x} < 22) = P\left(\dfrac{16 - 20}{2} < z < \dfrac{22 - 20}{2}\right) = P(-2 < z < 1)$
$$= .4772 + .3413 = .8185$$

e. $P(\bar{x} < 14) = P\left(z < \dfrac{14 - 20}{2}\right) = P(z < -3) = .5 - .4987 = .0013$

4.117   a.   By the Central Limit Theorem, the sampling distribution of $\bar{x}$ is approximately normal with

$$\mu_{\bar{x}} = \mu \text{ and } \sigma_{\bar{x}} = \sigma/\sqrt{n}.$$

   b.   Let $\mu = 18.5$. Since we do not know σ we will estimate it with $s = 6$.

$$P(\bar{x} \geq 19.1) \approx P\left(z \geq \frac{19.1 - 18.5}{6/\sqrt{344}}\right) = P(z \geq 1.85) = .5 - .4678 = .0322$$

   c.   Let $\mu = 19.5$. Since we do not know σ we will estimate it with $s = 6$.

$$P(\bar{x} \geq 19.1) \approx P\left(z \geq \frac{19.1 - 19.5}{6/\sqrt{344}}\right) = P(z \geq -1.24) = .5 + .3925 = .8925$$

   d.   If $P(\bar{x} \geq 19.1) = .5$, then the population mean must be equal to 19.1. (For a normal distribution, half of the distribution is above the mean and half is below the mean.)

   e.   If $P(\bar{x} \geq 19.1) = .2$, then the population mean is less than 19.1. We know the probability that $\bar{x}$ is greater than the mean is .5. Since $P(x \geq 19.1) = .2$ which is less than .5, we know that 19.1 must be to the right of the mean. Thus, the population mean must be less than 19.1.

4.119   a.   For $n = 36$, $\mu_{\bar{x}} = \mu = 406$ and $\sigma_{\bar{x}} = \sigma/\sqrt{n} = 10.1/\sqrt{36} = 1.6833$. By the Central Limit Theorem, the sampling distribution is approximately normal (n is large).

   b. $P(\bar{x} \leq 400.8) = P\left(z \leq \dfrac{400.8 - 406}{1.6833}\right) = P(z \leq -3.09) = .5 - .4990 = .0010$

   (using Table IV, Appendix B)

   c.   The first. If the true value of μ is 406, it would be extremely unlikely to observe an $\bar{x}$ as small as 400.8 or smaller (probability .0010). Thus, we would infer that the true value of μ is less than 406.

4.121   a.   If the melatonin is not effective, then the mean time to fall asleep will be the same as for the placebo as will the standard deviation. Thus,

$$\mu_{\bar{x}} = \mu = 15; \quad \sigma_{\bar{x}} = \frac{\sigma}{\sqrt{n}} = \frac{5}{\sqrt{40}} = .7906$$

$$P(\bar{x} < 6) = P\left(z < \frac{6 - 15}{.7906}\right) = P(z < -11.38) \approx .5 - .5 = 0.$$

b. First, calculate $\bar{x}$ and s.

$$\bar{x} = \frac{\sum x}{n} = \frac{222.1}{40} = 5.5525$$

$$s^2 = \frac{\sum x^2 - \frac{(\sum x)^2}{n}}{n-1} = \frac{1570.4 - \frac{(222.1)^2}{40}}{40-1} = \frac{337.18975}{39} = 8.6459$$

$$s = \sqrt{8.6459} = 2.9404$$

A point estimate for the true value of μ is $\bar{x} = 5.5525$.

The estimate of the standard deviation of the $\bar{x}$ distribution is

$$s_{\bar{x}} = \frac{s}{\sqrt{n}} = \frac{2.9404}{\sqrt{40}} = .4649$$

We would expect most values $\bar{x}$ to fall within 2 standard deviations of the mean of μ. Thus, we would estimate that the true value of μ will fall between

$$\bar{x} - 2s_{\bar{x}} \text{ and } \bar{x} + 2s_{\bar{x}}$$

$\Rightarrow 5.5525 - 2(.4649)$ and $5.5525 + 2(.4649)$
$\Rightarrow 5.5525 - .9298$ and $5.5525 + .9298$
$\Rightarrow 4.6227$ and 6.4823

Thus, we would be pretty sure that the true value of μ will be between 4.6227 and 6.4823 minutes. Since these values are much less than the 15 minutes needed with the placebo, it appears that melatonin is effective against insomnia.

4.123 a.   Discrete — The number of damaged inventory items is countable.

b.   Continuous — The average monthly sales can take on any value within an acceptable limit.

c.   Continuous — The number of square feet can take on any positive value.

d.   Continuous — The length of time we must wait can take on any positive value.

4.125 $p(x) = \begin{bmatrix} n \\ x \end{bmatrix} p^x q^{n-x}$ $x = 0, 1, 2, \ldots , n$

a. $P(x = 3) = p(3) = \begin{bmatrix} 7 \\ 3 \end{bmatrix} .5^3 .5^4 = \frac{7!}{3!4!} .5^3 .5^4 = 35(.125)(.0625) = .2734$

b. $P(x = 3) = p(3) = \begin{bmatrix} 4 \\ 3 \end{bmatrix} .8^3 .2^1 = \frac{4!}{3!1!} .8^3 .2^1 = 4(.512)(.2) = .4096$

c. $P(x = 1) = p(1) = \begin{bmatrix} 15 \\ 1 \end{bmatrix} .1^1 .9^{14} = \frac{15!}{1!14!} .1^1 .9^{14} = 15(.1)(.228768) = .3432$

4.127 From Table II, Appendix B:

a. $P(x = 14) = P(x \le 14) - P(x \le 13) = .584 - .392 = .192$

b. $P(x \le 12) = .228$

c. $P(x > 12) = 1 - P(x \le 12) = 1 - .228 = .772$

d. $P(9 \le x \le 18) = P(x \le 18) - P(x \le 8) = .992 - .005 = .987$

e. $P(8 < x < 18) = P(x \le 17) - P(x \le 8) = .965 - .005 = .960$

f. $\mu = np = 20(.7) = 14$

$\sigma^2 = npq = 20(.7)(.3) = 4.2, \sigma = \sqrt{4.2} = 2.049$

g. $\mu \pm 2\sigma \Rightarrow 14 \pm 2(2.049) \Rightarrow 14 \pm 4.098 \Rightarrow (9.902, 18.098)$

$$P(9.902 < x < 18.098) = P(10 \le x \le 18) = P(x \le 18) - P(x \le 9)$$
$$= .992 - .017 = .975$$

4.129 a. $f(x) = \begin{cases} \dfrac{1}{d-c} = \dfrac{1}{90-10} = \dfrac{1}{80}, & 10 \le x \le 90 \\ 0 & \text{otherwise} \end{cases}$

b. $\mu = \dfrac{c+d}{2} = \dfrac{10+90}{2} = 50$

$\sigma = \dfrac{d-c}{\sqrt{12}} = \dfrac{90-10}{\sqrt{12}} = 23.094011$

c. The interval $\mu \pm 2\sigma \Rightarrow 50 \pm 2(23.094) \Rightarrow 50 \pm 46.188 \Rightarrow (3.812, 96.188)$ is indicated on the graph.

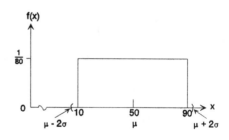

d. $P(x \le 60) = \text{Base(height)} = (60-10)\dfrac{1}{80} = \dfrac{5}{8} = .625$

e. $P(x \ge 90) = 0$

f. $P(x \le 80) = \text{Base(height)} = (80-10)\dfrac{1}{80} = \dfrac{7}{8} = .875$

g. $P(\mu - \sigma \le x \le \mu + \sigma) = P(50 - 23.094 \le x \le 50 + 23.094)$

$$= P(26.906 \le x \le 73.094)$$
$$= \text{Base(height)}$$
$$= (73.094 - 26.906)\left[\frac{1}{80}\right] = \frac{46.188}{80} = .577$$

h. $P(x > 75) = \text{Base(height)} = (90 - 75)\dfrac{1}{80} = \dfrac{15}{80} = .1875$

4.131 $\mu = np = 100(.5) = 50, \ \sigma = \sqrt{npq} = \sqrt{100(.5)(.5)} = 5$

a. $P(x \le 48) = P\left[z \le \dfrac{(48 + .5) - 50}{5}\right]$

$$= P(z \le -.30)$$
$$= .5 - .1179 = .3821$$

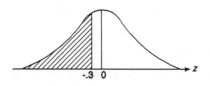

b. $P(50 \le x \le 65)$

$$= P\left[\dfrac{(50 - .5) - 50}{5} \le z \le \dfrac{(65 + .5) - 50}{5}\right]$$
$$= P(-.10 \le z \le 3.10)$$
$$= .0398 + .5000 = .5398$$

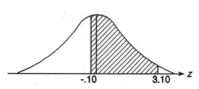

c. $P(x \ge 70) = P\left[z \ge \dfrac{(70 - .5) - 50}{5}\right]$

$$= (z \ge 3.90)$$
$$= .5 - .5 = 0$$

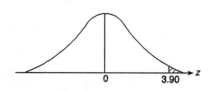

d. $P(55 \le x \le 58)$

$$= P\left[\dfrac{(55 - .5) - 50}{5} \le z \le \dfrac{(58 + .5) - 50}{5}\right]$$
$$= P(.90 \le z \le 1.70)$$
$$= P(0 \le z \le 1.70) - P(0 \le z \le .90)$$
$$= .4554 - .3159 = .1395$$

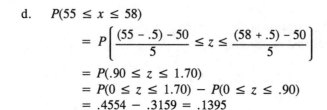

e. $P(x = 62)$

$$= P\left[\dfrac{(62 - .5) - 50}{5} \le z \le \dfrac{(62 + .5) - 50}{5}\right]$$
$$= P(2.30 \le z \le 2.50)$$
$$= P(0 \le z \le 2.50) - (0 \le z \le 2.30)$$
$$= .4938 - .4893 = .0045$$

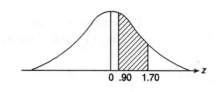
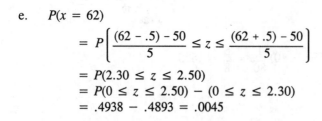

f. $P(x \le 49 \text{ or } x \ge 72)$

$$= P\left[z \le \dfrac{(49 + .5) - 50}{5}\right] + P\left[z \ge \dfrac{(72 - .5) - 50}{5}\right]$$
$$= P(z \le -.10) + P(z \ge 4.30)$$
$$= (.5 - .0398) + (.5 - .5) = .4602$$

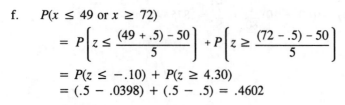
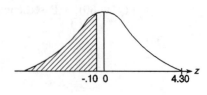
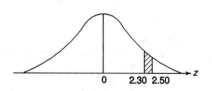

4.133 a. $P(x \le 2) = 1 - P(x > 2) = 1 - e^{-3(2)} = 1 - e^{-6} = 1 - .002479 = .997521$
(Using Table V, Appendix B)

b. $P(x > 3) = e^{-3(3)} = e^{-9} = .000123$ (Using Table V, Appendix B)

c. $P(x = 1) = 0$

d. $P(x \le 7) = 1 - P(x > 7) = 1 - e^{-3(7)} = 1 - e^{-21} = 1 - .000000 = 1$

e. $P(4 \le x \le 12) = P(x > 4) - P(x > 12) = e^{-3(4)} - e^{-3(12)} = e^{-12} - e^{-36}$
$= .000006 - .000000 = .000006$

4.135 Given: $\mu = 100$ and $\sigma = 10$

n	1	5	10	20	30	40	50
$\dfrac{\sigma}{\sqrt{n}}$	10	4.472	3.162	2.236	1.826	1.581	1.414

The graph of $\sigma/\sqrt{n}$ against n is given here:

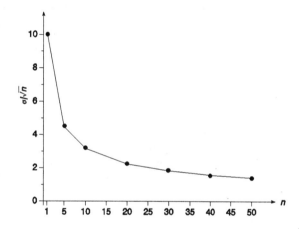

4.137 a. In order for the number of deaths to follow a Poisson distribution, we must assume that the probability of a death is the same for any week. We must also assume that the number of deaths in any week is independent of any other week.

The first assumption may not be valid. The probability of a death may not be the same for every week. The number of passengers varies from week to week, so the probability of a death may change. Also, things such as weather, which varies from week to week may increase or decrease the chance of derailment.

b. $E(x) = \lambda = 20$
$\sigma = \sqrt{\lambda} = \sqrt{20} = 4.4721$

c. The z-score corresponding to $x = 4$ is:
$$z = \frac{4 - 20}{4.4721} = -3.55$$

Since this z-score is more than 3 standard deviations from the mean, it would be very unlikely that only 4 or fewer deaths occur next week.

d. Using Table III, Appendix B with $\lambda = 20$,

$$P(x \le 4) = 0.000$$

This probability is consistent with the answer in part **c**. The probability of 4 or fewer deaths is essentially zero, which is very unlikely.

4.139 Let y be the profit on a metal part that is produced. Then y is $\$10$, $\$-2$, or $\$-1$, depending where it falls with respect to the tolerance limits.

Let x be the tensile strength of a particular metal part. The random variable x is normally distributed with $\mu = 25$ and $\sigma = 2$.

$$z = \frac{x - \mu}{\sigma} = \frac{21 - 25}{2} = -2$$

$$z = \frac{x - \mu}{\sigma} = \frac{30 - 25}{2} = 2.5$$

$$
\begin{aligned}
P(y = 10) &= P(x \text{ falls within the tolerance limits}) \\
&= P(21 < x < 30) = P(-2 < z < 2.5) \\
&= P(-2 < z < 0) + P(0 < z < 2.5) \\
&= P(0 < z < 2) + P(0 < z < 2.5) \\
&= .4772 + .4938 \\
&= .9710
\end{aligned}
$$

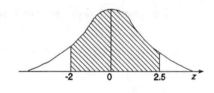

$$
\begin{aligned}
P(y = -2) &= P(x \text{ falls below the lower tolerance limit}) \\
&= P(x < 21) = P(z < -2) \\
&= .5000 - P(-2 < z < 0) \\
&= .5000 - P(0 < z < 2) \\
&= .5000 - .4772 \\
&= .0228
\end{aligned}
$$

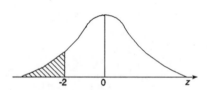

$$
\begin{aligned}
P(y = -1) &= P(x \text{ falls above the upper tolerance limit}) \\
&= P(x > 30) = P(z > 2.5) \\
&= .5000 - P(0 < z < 2.5) \\
&= .5000 - .4938 \\
&= .0062
\end{aligned}
$$

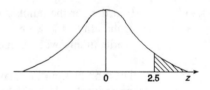

The probability distribution of y is given below:

y	10	-2	-1
$p(y)$	.9710	.0228	.0062

$$
\begin{aligned}
E(y) = \sum yp(y) &= 10(.9710) + -2(.0228) + -1(.0062) \\
&= 9.71 - .0456 - .0062 \\
&= \$9.6582
\end{aligned}
$$

4.141   Let x be the noise level per jet takeoff in a neighborhood near the airport. The random variable x is approximately normally distributed with $\mu = 100$ and $\sigma = 6$.

a. $P(x > 108) = P\left(z > \dfrac{108 - 100}{6}\right)$

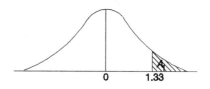

   $= P(z > 1.33)$
   $= .5000 - P(0 \leq z \leq 1.33)$
   $= .5000 - .4082$
   $= .0918$

b. $P(x = 100) = 0$

c.   Given $P(x < 105) = .95$ and $\sigma = 6$,

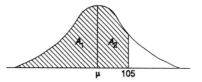

   $P(x < 105) = P(x < \mu) + P(\mu < x < 105)$
   $\qquad\qquad = .5 + .45 = A_1 + A_2$
   Looking up the area $A_2 = .45$ in Table IV, $z_0 = 1.645$.

   Since $z = 1.645$, $x = 105$ and $\sigma = 6$,

   $$z_0 = \frac{x_0 - \mu}{\sigma} \Rightarrow 1.645 = \frac{105 - \mu}{6} \Rightarrow 9.87 = 105 - \mu$$

   Hence, $\mu = 95.13$

   Since $\mu = 100$, the mean level of noise must be lowered $100 - 95.13 = 4.87$ decibels.

4.143   Let x = interarrival time between patients.   Then x is an exponential random variable with a mean of 4 minutes and $\lambda = 1/4$.

a. $P(x < 1) = 1 - P(x \geq 1)$
   $\qquad\qquad = 1 - e^{-1/4}$
   $\qquad\qquad = 1 - e^{-.25}$
   $\qquad\qquad = 1 - .778801$
   $\qquad\qquad = .221199$

b.   Assuming that the interarrival times are independent,
   P(next 4 interarrival times are all less than 1 minute)
   $\qquad\qquad = \{P(x < 1)\}^4$
   $\qquad\qquad = .221199^4$
   $\qquad\qquad = .002394$

c. $P(x > 10) = e^{-10(1/4)}$
   $\qquad\qquad = e^{-2.5}$
   $\qquad\qquad = .082085$

4.145　Using MINITAB, the stem-and-leaf display is:

```
Stem-and-leaf of Time      N  = 49
Leaf Unit = 0.10

 (26)    1  00001122222344444445555679
  23     2  11446799
  15     3  002899
   9     4  11125
   4     5  24
   2     6
   2     7  8
   1     8
   1     9
   1    10  1
```

The data are skewed to the right, and do not appear to be normally distributed.

Using MINITAB, the descriptive statistics are:

Variable	N	Mean	Median	TrMean	StDev	SE Mean
Time	49	2.549	1.700	2.333	1.828	0.261

Variable	Minimum	Maximum	Q1	Q3
Time	1.000	10.100	1.350	3.500

$\bar{x} \pm s \Rightarrow 2.549 \pm 1.828 \Rightarrow (0.721, 4.377)$

$\bar{x} \pm 2s \Rightarrow 2.549 \pm 2(1.828) \Rightarrow 2.549 \pm 3.656 \Rightarrow (-1.107, 6.205)$

$\bar{x} \pm 3s \Rightarrow 2.549 \pm 3(1.828) \Rightarrow 2.549 \pm 5.484 \Rightarrow (-2.935, 8.033)$

Of the 49 measurements, 44 are in the interval (0.721, 4.377). The proportion is 44/49 = .898. This is much larger than the proportion (.68) stated by the Empirical Rule.

Of the 49 measurements, 47 are in the interval $(-1.107, 6.205)$. The proportion is 47/49 = .959. This is close to the proportion (.95) stated by the Empirical Rule.

Of the 49 measurements, 48 are in the interval $(-2.935, 8.033)$. The proportion is 48/49 = .980. This is smaller than the proportion (1.00) stated by the Empirical Rule.

This would imply that the data are not normal.

$IQR = Q_U - Q_L = 3.500 - 1.350 = 2.15$.　$IQR/s = 2.15/1.828 = 1.176$. If the data are normally distributed, this ratio should be close to 1.3. Since 1.176 is smaller than 1.3, this indicates that the data may not be normal.

Using MINITAB, the normal probability plot is:

Normal Probability Plot for Time

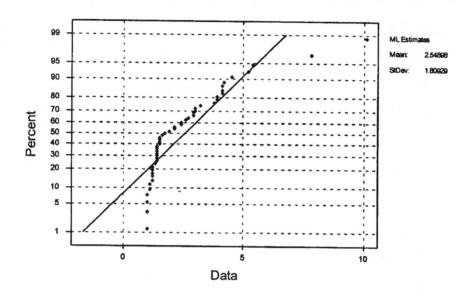

Since this plot is not a straight line, the data are not normal.

All four checks indicate that the data are not normal.

4.147   a.   The distribution of x has a mean of $\mu = 26$ and a standard deviation of σ. There is no information given to indicate the shape of the distribution.

b.   The distribution of $\bar{x}$ has a mean of $\mu_{\bar{x}} = \mu = 26$ and a standard deviation of $\sigma_{\bar{x}} = \sigma/\sqrt{n}$. Since $n = 200$ is sufficiently large, the Central Limit Theorem says that the sampling distribution of $\bar{x}$ is approximately normal.

c.   If $\sigma = 20$, then $\sigma_{\bar{x}} = \sigma/\sqrt{n} = 20/\sqrt{200} = 1.4142$.

$$P(\bar{x} > 26.8) = P\left(z > \frac{26.8 - 26}{1.4142}\right) = P(z > .57) = .5 - .2157 = .2843$$

d.   If $\sigma = 10$, then $\sigma_{\bar{x}} = \sigma/\sqrt{n} = 10/\sqrt{200} = .7071$

$$P(\bar{x} > 26.8) = P\left(z > \frac{26.8 - 26}{.7071}\right) = P(z > 1.13) = .5 - .3708 = .1292$$

5.1   a.   For $\alpha = .10$, $\alpha/2 = .10/2 = .05$. $z_{\alpha/2} = z_{.05}$ is the z-score with .05 of the area to the right of it. The area between 0 and $z_{.05}$ is $.5 - .05 = .4500$. Using Table IV, Appendix B, $z_{.05} = 1.645$.

       b.   For $\alpha = .01$, $\alpha/2 = .01/2 = .005$. $z_{\alpha/2} = z_{.005}$ is the z-score with .005 of the area to the right of it. The area between 0 and $z_{.005}$ is $.5 - .005 = .4950$. Using Table IV, Appendix B, $z_{.005} = 2.58$.

       c.   For $\alpha = .05$, $\alpha/2 = .05/2 = .025$. $z_{\alpha/2} = z_{.025}$ is the z-score with .025 of the area to the right of it. The area between 0 and $z_{.025}$ is $.5 - .025 = .4750$. Using Table IV, Appendix B, $z_{.025} = 1.96$.

       d.   For $\alpha = .20$, $\alpha/2 = .20/2 = .10$. $z_{\alpha/2} = z_{.10}$ is the z-score with .10 of the area to the right of it. The area between 0 and $z_{.10}$ is $.5 - .10 = .4000$. Using Table IV, Appendix B, $z_{.10} = 1.28$.

5.3   a.   For confidence coefficient .95, $\alpha = .05$ and $\alpha/2 = .05/2 = .025$. From Table IV, Appendix B, $z_{.025} = 1.96$. The confidence interval is:

$$\bar{x} \pm z_{.025}\frac{s}{\sqrt{n}} \Rightarrow 28 \pm 1.96\frac{\sqrt{12}}{\sqrt{75}} \Rightarrow 28 \pm .784 \Rightarrow (27.216, 28.784)$$

       b. $\bar{x} \pm z_{.025}\frac{s}{\sqrt{n}} \Rightarrow 102 \pm 1.96\frac{\sqrt{22}}{\sqrt{200}} \Rightarrow 102 \pm .65 \Rightarrow (101.35, 102.65)$

       c. $\bar{x} \pm z_{.025}\frac{s}{\sqrt{n}} \Rightarrow 15 \pm 1.96\frac{.3}{\sqrt{100}} \Rightarrow 15 \pm .0588 \Rightarrow (14.9412, 15.0588)$

       d. $\bar{x} \pm z_{.025}\frac{s}{\sqrt{n}} \Rightarrow 4.05 \pm 1.96\frac{.83}{\sqrt{100}} \Rightarrow 4.05 \pm .163 \Rightarrow (3.887, 4.213)$

       e.   No. Since the sample size in each part was large (n ranged from 75 to 200), the Central Limit Theorem indicates that the sampling distribution of $\bar{x}$ is approximately normal.

5.5   a.   For confidence coefficient .95, $\alpha = .05$ and $\alpha/2 = .05/2 = .025$. From Table IV, Appendix B, $z_{.025} = 1.96$. The confidence interval is:

$$\bar{x} \pm z_{.025}\frac{s}{\sqrt{n}} \Rightarrow 33.9 \pm 1.96\frac{3.3}{\sqrt{100}} \Rightarrow 33.9 \pm .647 \Rightarrow (33.253, 34.547)$$

       b. $\bar{x} \pm z_{.025}\frac{s}{\sqrt{n}} \Rightarrow 33.9 \pm 1.96\frac{3.3}{\sqrt{400}} \Rightarrow 33.9 \pm .323 \Rightarrow (33.577, 34.223)$

c. For part **a**, the width of the interval is 2(.647) = 1.294. For part **b**, the width of the interval is 2(.323) = .646. When the sample size is quadrupled, the width of the confidence interval is halved.

5.7 a. The population from which the sample was drawn is the set of all adult smokers in the U.S.

b. The 95% confidence interval is (19.7, 20.3). We are 95% confident that the mean number of cigarettes smoked per day by all smokers is between 19.7 and 20.3.

c. Since the sample size is so large ($n = 11,000$), no assumptions are necessary. The Central Limit Theorem indicates that the sampling distribution of $\bar{x}$ is approximately normal.

d. Since the entire 95% confidence interval is above the value of 15, the claim made by the tobacco industry researcher is probably not true.

5.9 From the printout, the 95% confidence interval is (.3526, .4921). We are 95% confident that the true mean correlation coefficient between appraisal participation and a subordinate's satisfaction with the appraisal is between .3526 and .4921.

5.11 a. Some preliminary calculations are:

$$\bar{x} = \frac{\sum x}{n} = \frac{2406}{36} = 66.83$$

$$s^2 = \frac{\sum x^2 - \frac{(\sum x)^2}{n}}{n-1} = \frac{168,016 - \frac{(2406)^2}{36}}{36-1} = 206.143$$

$$s = \sqrt{206.143} = 14.3577$$

For confidence coefficient .95, $\alpha = 1 - .95 = .05$ and $\alpha/2 = .05/2 = .025$. From Table IV, Appendix B, $z_{.025} = 1.96$. The confidence interval is:

$$\bar{x} \pm z_{.025}\frac{s}{\sqrt{n}} \Rightarrow 66.83 \pm 1.96\frac{14.3577}{\sqrt{36}} \Rightarrow 66.83 \pm 4.69 \Rightarrow (62.14, 71.52)$$

We are 95% confident that the mean raw test score for all twenty-five year olds is between 62.14 and 71.52.

b. We must assume that the sample is random and that the observations are independent.

c. From the printout, the 95% confidence interval is (41.009, 49.602). We are 95% confident that the mean raw test score for all sixty year olds is between 41.009 and 49.602.

5.13 a. $P(-t_0 < t < t_0) = .95$ where df = 10

Because of symmetry, the statement can be written

$$P(0 < t < t_0) = .475 \quad \text{where df} = 10$$
$$\Rightarrow P(t \geq t_0) = .025$$
$$t_0 = 2.228$$

Inferences Based on a Single Sample: Estimation with Confidence Intervals

b. $P(t \le -t_0 \text{ or } t \ge t_0) = .05$ where df $= 10$

$$\Rightarrow 2P(t \ge t_0) = .05$$
$$\Rightarrow P(t \ge t_0) = .025 \text{ where df } = 10$$
$$t_0 = 2.228$$

c. $P(t \le t_0) = .05$ where df $= 10$

Because of symmetry, the statement can be written

$$\Rightarrow P(t \ge -t_0) = .05 \text{ where df } = 10$$
$$t_0 = -1.812$$

d. $P(t < -t_0 \text{ or } t > t_0) = .10$ where df $= 20$
$$\Rightarrow 2P(t > t_0) = .10$$
$$\Rightarrow P(t > t_0) = .05 \text{ where df } = 20$$
$$t_0 = 1.725$$

e. $P(t \le -t_0 \text{ or } t \ge t_0) = .01$ where df $= 5$
$$\Rightarrow 2P(t \ge t_0) = .01$$
$$\Rightarrow P(t \ge t_0) = .005 \text{ where df } = 5$$
$$t_0 = 4.032$$

5.15 For this sample,

$$\bar{x} = \frac{\sum x}{n} = \frac{1567}{16} = 97.9375$$

$$s^2 = \frac{\sum x^2 - \dfrac{(\sum x)^2}{n}}{n - 1} = \frac{155,867 - \dfrac{1567^2}{16}}{16 - 1} = 159.9292$$

$$s = \sqrt{s^2} = 12.6463$$

a. For confidence coefficient, .80, $\alpha = 1 - .80 = .20$ and $\alpha/2 = .20/2 = .10$. From Table VI, Appendix B, with df $= n - 1 = 16 - 1 = 15$, $t_{.10} = 1.341$. The 80% confidence interval for μ is:

$$\bar{x} \pm t_{.10}\frac{s}{\sqrt{n}} \Rightarrow 97.94 \pm 1.341\frac{12.6463}{\sqrt{16}} \Rightarrow 97.94 \pm 4.240 \Rightarrow (93.700, 102.180)$$

b. For confidence coefficient, .95, $\alpha = 1 - .95 = .05$ and $\alpha/2 = .05/2 = .025$. From Table VI, Appendix B, with df $= n - 1 = 24 - 1 = 23$, $t_{.025} = 2.131$. The 95% confidence interval for μ is:

$$\bar{x} \pm t_{.025}\frac{s}{\sqrt{n}} \Rightarrow 97.94 \pm 2.131\frac{12.6463}{\sqrt{16}} \Rightarrow 97.94 \pm 6.737 \Rightarrow (91.203, 104.677)$$

The 95% confidence interval for μ is wider than the 80% confidence interval for μ found in part **a**.

c. For part **a**:

We are 80% confident that the true population mean lies in the interval 93.700 to 102.180.

For part **b**:

We are 95% confident that the true population mean lies in the interval 91.203 to 104.677.

The 95% confidence interval is wider than the 80% confidence interval because the more confident you want to be that μ lies in an interval, the wider the range of possible values.

5.17   a.   First, we must compute some preliminary satistics:

$$\bar{x} = \frac{\sum x}{n} = \frac{28.856}{10} = 2.8856$$

$$s^2 = \frac{\sum x^2 - \frac{(\sum x)^2}{n}}{n - 1} = \frac{221.90161 - \frac{(28.856)^2}{10}}{10 - 1} = 15.4039$$

$$s = \sqrt{s^2} = \sqrt{15.4039} = 3.925$$

For confidence coefficient .99, $\alpha = .01$ and $\alpha/2 = .01/2 = .005$. From Table VI, Appendix B, with df $= n - 1 = 10 - 1 = 9$, $t_{.005} = 3.250$. The confidence interval is:

$$\bar{x} \pm t_{.005} \frac{s}{\sqrt{n}} \Rightarrow 2.8856 \pm 3.250 \frac{3.925}{\sqrt{10}} \Rightarrow 2.8856 \pm 4.034 \Rightarrow (-1.148, 6.919)$$

b.   First, we must compute some preliminary statistics:

$$\bar{x} = \frac{\sum x}{n} = \frac{4.083}{10} = .4083$$

$$s^2 = \frac{\sum x^2 - \frac{(\sum x)^2}{n}}{n - 1} = \frac{2.227425 - \frac{(4.083)^2}{10}}{10 - 1} = .06226$$

$$s = \sqrt{s^2} = \sqrt{.06226} = .2495$$

For confidence coefficient .99, $\alpha = .01$ and $\alpha/2 = .01/2 = .005$. From Table VI, Appendix B, with df $= n - 1 = 10 - 1 = 9$, $t_{.005} = 3.250$. The confidence interval is:

$$\bar{x} \pm t_{.005} \frac{s}{\sqrt{n}} \Rightarrow .4083 \pm 3.250 \frac{.2495}{\sqrt{10}} \Rightarrow .4083 \pm .2564 \Rightarrow (.1519, .6647)$$

c.   We are 99% confident that the mean lead level in water specimens from Crystal Lake Manors is between -1.148 and 6.919 or 0 and 6.919 μ/L since no value can be less than 0.

We are 99% confident that the mean copper level in water specimens from Crystal Lake Manors is between .1519 and .6647 mg/L.

d. The phrase "99% confident" means that if repeated samples of size n were selected and 99% confidence intervals constructed for the mean, 99% of all intervals constructed would contain the mean.

5.19 a. For confidence coefficient .99, $\alpha = .01$ and $\alpha/2 = .01/2 = .005$. From Table VI, Appendix B, with df $= n - 1 = 3 - 1 = 2$, $t_{.005} = 9.925$. The confidence interval is:

$$\bar{x} \pm t_{.005}\frac{s}{\sqrt{n}} \Rightarrow 49.3 \pm 9.925\frac{1.5}{\sqrt{3}} \Rightarrow 49.3 \pm 8.60 \Rightarrow (40.70, 57.90)$$

b. We are 99% confident that the mean percentage of B(a)p removed from all soil specimens using the poison is between 40.70% and 57.90%.

c. We must assume that the distribution of the percentages of B(a)p removed from all soil specimens using the poison is normal.

5.21 First we make some preliminary calculations:

$$\bar{x} = \frac{\sum x}{n} = \frac{1479.9}{8} = 184.9875$$

$$s^2 = \frac{\sum x^2 - \frac{\left(\sum x\right)^2}{n}}{n - 1} = \frac{453,375.17 - \frac{1479.9^2}{8}}{8 - 1} = 25,658.88124$$

$$s = \sqrt{25,658.88124} = 160.1839$$

For confidence coefficient .95, $\alpha = .05$ and $\alpha/2 = .025$. From Table VI, Appendix B, with df $= n - 1 = 8 - 1 = 7$, $t_{.025} = 2.365$. The 95% confidence interval is:

$$\bar{x} \pm t_{.05}\frac{s}{\sqrt{n}} \Rightarrow 184.9875 \pm 2.365\frac{160.1839}{\sqrt{8}} \Rightarrow 184.9875 \pm 133.9384 \Rightarrow (51.0491, 318.9259)$$

We must assume that the population of private colleges' and universities' endowments are normally distributed.

5.23 a. For confidence coefficient .99, $\alpha = .01$ and $\alpha/2 = .01/2 = .005$. From Table VI, Appendix B, with df $= n - 1 = 22 - 1 = 21$, $t_{.005} = 2.831$. The confidence interval is:

$$\bar{x} \pm t_{\alpha/2}\frac{s}{\sqrt{n}} \Rightarrow 22.455 \pm 2.831\frac{18.518}{\sqrt{22}} \Rightarrow 22.455 \pm 11.177 \Rightarrow (11.278, 33.632)$$

b. We are 99% confident that the mean number of full-time employees at office furniture dealers in Tampa is between 11.278 and 33.632.

c. In order for the confidence interval to be valid, we must assume that the distribution of the number of full-time employees at all office furniture dealers in Tampa is normal and that the sample was a random sample.

d. If the 22 observations in the sample were the top-ranked furniture dealers in Tampa, then the sample was not a random sample. Thus, the validity of the interval is suspect.

5.25   The sample size is large enough if $\hat{p} \pm 3\sigma_{\hat{p}}$ lies within the interval $(0, 1)$.

$$\hat{p} \pm 3\sigma_{\hat{p}} \Rightarrow \hat{p} \pm 3\sqrt{\frac{pq}{n}} \Rightarrow \hat{p} \pm 3\sqrt{\frac{\hat{p}\hat{q}}{n}}$$

a.   When $n = 400$, $\hat{p} = .10$:

$$.10 \pm 3\sqrt{\frac{.10(1 - .10)}{400}} \Rightarrow .10 \pm .045 \Rightarrow (.055, .145)$$

Since the interval lies completely in the interval $(0, 1)$, the normal approximation will be adequate.

b.   When $n = 50$, $\hat{p} = .10$:

$$.10 \pm 3\sqrt{\frac{.10(1 - .10)}{50}} \Rightarrow .10 \pm .127 \Rightarrow (-.027, .227)$$

Since the interval does not lie completely in the interval $(0, 1)$, the normal approximation will not be adequate.

c.   When $n = 20$, $\hat{p} = .5$:

$$.5 \pm 3\sqrt{\frac{.5(1 - .5)}{20}} \Rightarrow .5 \pm .335 \Rightarrow (.165, .835)$$

Since the interval lies completely in the interval $(0, 1)$, the normal approximation will be adequate.

d.   When $n = 20$, $\hat{p} = .3$:

$$.3 \pm 3\sqrt{\frac{.3(1 - .3)}{20}} \Rightarrow .3 \pm .307 \Rightarrow (-.007, .607)$$

Since the interval does not lie completely in the interval $(0, 1)$, the normal approximation will not be adequate.

5.27   a.   The sample size is large enough if the interval $\hat{p} \pm 3\sigma_{\hat{p}}$ does not include 0 or 1.

$$\hat{p} \pm 3\sigma_{\hat{p}} \Rightarrow \hat{p} \pm 3\sqrt{\frac{pq}{n}} \Rightarrow \hat{p} \pm 3\sqrt{\frac{\hat{p}\hat{q}}{n}} \Rightarrow .46 \pm 3\sqrt{\frac{.46(1 - .46)}{225}} \Rightarrow .46 \pm .0997$$
$$\Rightarrow (.3603, .5597)$$

Since the interval lies within the interval $(0, 1)$, the normal approximation will be adequate.

b.   For confidence coefficient .95, $\alpha = .05$ and $\alpha/2 = .025$. From Table IV, Appendix B, $z_{.025} = 1.96$. The 95% confidence interval is:

$$\hat{p} \pm z_{.025}\sqrt{\frac{pq}{n}} \Rightarrow \hat{p} \pm 1.96\sqrt{\frac{\hat{p}\hat{q}}{n}} \Rightarrow .46 \pm 1.96\sqrt{\frac{.46(1 - .46)}{225}} \Rightarrow .46 \pm .065$$
$$\Rightarrow (.395, .525)$$

c.   We are 95% confident the true value of p will fall between .395 and .525.

d. "95% confidence interval" means that if repeated samples of size 225 were selected from the population and 95% confidence intervals formed, 95% of all confidence intervals will contain the true value of p.

5.29   a.   Of the 1000 observations, 29% said they would never give personal information to a company $\Rightarrow \hat{p} = .29$

To see if the sample size is sufficiently large:

$$\hat{p} \pm 3\sigma_{\hat{p}} \approx \hat{p} \pm 3\sqrt{\frac{\hat{p}\hat{q}}{n}} \Rightarrow .29 \pm 3\sqrt{\frac{.29(.71)}{1000}} \Rightarrow .29 \pm .043 \Rightarrow (.247, .333)$$

Since this interval is wholly contained in the interval $(0, 1)$, we may conclude that the normal approximation is reasonable.

b.   For confidence coefficient .95, $\alpha = 1 - .95 = .05$ and $\alpha/2 = .05/2 = .025$. From Table IV, Appendix B, $z_{.025} = 1.96$. The 95% confidence interval is:

$$\hat{p} \pm z_{.025}\sqrt{\frac{\hat{p}\hat{q}}{n}} \Rightarrow .29 \pm 1.96\sqrt{\frac{.29(.71)}{1000}} \Rightarrow .29 \pm .028 \Rightarrow (.262, .318)$$

We are 95% confident that the proportion of Internet users who would never give personal information to a company is between .262 and .318.

c.   We must assume that the sample is a random sample from the population.

5.31   a.   The point estimate for the proportion of major oil spills that are caused by hull failure is:

$$\hat{p} = \frac{x}{n} = \frac{12}{50} = .24$$

b.   To see if the sample size is sufficiently large:

$$\hat{p} \pm 3\sigma_{\hat{p}} \approx \hat{p} \pm 3\sqrt{\frac{\hat{p}\hat{q}}{n}} \Rightarrow .24 \pm 3\sqrt{\frac{.24(.76)}{50}} \Rightarrow .24 \pm .181 \Rightarrow (.059, .421)$$

Since this interval is wholly contained in the interval $(0, 1)$, we may conclude that the normal approximation is reasonable.

For confidence coefficient .95, $\alpha = .05$ and $\alpha/2 = .05/2 = .025$. From Table IV, Appendix B, $z_{.025} = 1.96$. The confidence interval is:

$$\hat{p} \pm z_{.025}\sqrt{\frac{pq}{n}} \approx \hat{p} \pm 1.96\sqrt{\frac{\hat{p}\hat{q}}{n}} \Rightarrow .24 \pm 1.96\sqrt{\frac{.24(.76)}{50}} \Rightarrow .24 \pm .118$$
$$\Rightarrow (.122, .358)$$

We are 95% confident that the true percentage of major oil spills that are caused by hull failure is between .122 and .358.

5.33   First, we must compute $\hat{p}$: $\hat{p} = \dfrac{x}{n} = \dfrac{282,200}{332,000} = .85$

To see if the sample size is sufficiently large:

$$\hat{p} \pm 3\sigma_{\hat{p}} \approx \hat{p} \pm 3\sqrt{\dfrac{\hat{p}\hat{q}}{n}} \Rightarrow .85 \pm 3\sqrt{\dfrac{.85(.15)}{332,000}} \Rightarrow .85 \pm .002 \Rightarrow (.848, .852)$$

Since this interval is wholly contained in the interval (0, 1), we may conclude that the normal approximation is reasonable.

For confidence coefficient .99, $\alpha = .01$ and $\alpha/2 = .01/2 = .005$. From Table IV, Appendix B, $z_{.005} = 2.58$. The confidence interval is:

$$\hat{p} \pm z_{.005}\sqrt{\dfrac{pq}{n}} \approx \hat{p} \pm 2.58\sqrt{\dfrac{\hat{p}\hat{q}}{n}} \Rightarrow .85 \pm 2.58\sqrt{\dfrac{.85(.15)}{332,000}} \Rightarrow .85 \pm .002 \Rightarrow (.848, .852)$$

We are 99% confident that the true percentage of items delivered on time by the U.S. Postal Service is between 84.8% and 85.2%.

5.35   To compute the necessary sample size, use

$$n = \dfrac{(z_{\alpha/2})^2\sigma^2}{B^2} \text{ where } \alpha = 1 - .95 = .05 \text{ and } \alpha/2 = .05/2 = .025.$$

From Table IV, Appendix B, $z_{.025} = 1.96$. Thus,

$$n = \dfrac{(1.96)^2(7.2)}{.3^2} = 307.328 \approx 308$$

You would need to take 308 samples.

5.37   a.   An estimate of σ is obtained from:

$$\text{range} \approx 4s$$
$$s \approx \dfrac{\text{range}}{4} = \dfrac{34 - 30}{4} = 1$$

To compute the necessary sample size, use

$$n = \dfrac{(z_{\alpha/2})^2\sigma^2}{B^2} \text{ where } \alpha = 1 - .90 = .10 \text{ and } \alpha/2 = .05.$$

From Table IV, Appendix B, $z_{.05} = 1.645$. Thus,

$$n = \dfrac{(1.645)^2(1)^2}{.2^2} = 67.65 \approx 68$$

Inferences Based on a Single Sample: Estimation with Confidence Intervals   97

b. A less conservative estimate of σ is obtained from:

$$\text{range} \approx 6s$$

$$s \approx \frac{\text{range}}{6} = \frac{34 - 30}{6} = .6667$$

Thus, $n = \frac{(z_{\alpha/2})^2\sigma^2}{B^2} = \frac{(1.645)^2(.6667)^2}{.2^2} = 30.07 \approx 31$

5.39   For confidence coefficient .90, $\alpha = .10$ and $\alpha/2 = .05$. From Table IV, Appendix B, $z_{.05} = 1.645$.

We know $\hat{p}$ is in the middle of the interval, so $\hat{p} = \frac{.54 + .26}{2} = .4$

The confidence interval is $\hat{p} \pm z_{.05}\sqrt{\frac{\hat{p}\hat{q}}{n}} \Rightarrow .4 \pm 1.645\sqrt{\frac{.4(.6)}{n}}$

We know $.4 - 1.645\sqrt{\frac{.4(.6)}{n}} = .26$

$$\Rightarrow .4 - \frac{.8059}{\sqrt{n}} = .26$$

$$\Rightarrow .4 - .26 = \frac{.8059}{\sqrt{n}} \Rightarrow \sqrt{n} = \frac{.8059}{.14} = 5.756$$

$$\Rightarrow n = 5.756^2 = 33.1 \approx 34$$

5.41   a. Of the 13,000 observations, 2,938 indicated that they were definitely not willing to pay such fees, $\Rightarrow \hat{p} = 2,938/13,000 = .226$.

To see if the sample size is sufficiently large:

$$\hat{p} \pm 3\sigma_{\hat{p}} \Rightarrow \hat{p} \pm 3\sqrt{\frac{pq}{n}} \Rightarrow \hat{p} \pm 3\sqrt{\frac{\hat{p}\hat{q}}{n}} \Rightarrow .226 \pm 3\sqrt{\frac{.226(.774)}{13,000}} \Rightarrow .226 \pm .011$$
$$\Rightarrow (.215, .237)$$

Since the interval lies within the interval (0, 1), the normal approximation will be adequate.

For confidence coefficient .95, $\alpha = .05$ and $\alpha/2 = .05/2 = .025$. From Table IV, Appendix B, $z_{.025} = 1.96$. The confidence interval is:

$$\hat{p} \pm z_{.05}\sqrt{\frac{pq}{n}} \Rightarrow \hat{p} \pm 1.96\sqrt{\frac{\hat{p}\hat{q}}{n}} \Rightarrow .226 \pm 1.96\sqrt{\frac{.226(.774)}{13,000}} \Rightarrow .226 \pm .007$$
$$\Rightarrow (.219, .233)$$

We are 95% confident that the proportion definitely unwilling to pay fees is between .219 and .233.

b. The width of the interval is $.233 - .219 = .014$. Since the interval is unnecessarily small, this indicates that the sample size was extremely large.

c. The bound is $B = .02$. For confidence coefficient .95, $\alpha = .05$ and $\alpha/2 = .05/2 = .025$. From Table IV, Appendix B, $z_{.025} = 1.96$. Thus,

$$n = \frac{(z_{\alpha/2})^2 pq}{B^2} = \frac{1.96^2 \, .226(.774)}{.02^2} = 1{,}679.97 \approx 1{,}680.$$

Thus, we would need a sample size of 1,680.

5.43 For confidence coefficient .90, $\alpha = .10$ and $\alpha/2 = .10/2 = .05$. From Table IV, Appendix B, $z_{.05} = 1.645$. Since we have no estimate given for the value of p, we will use .5. The sample size is:

$$n = \frac{z_{\alpha/2}^2 pq}{B^2} = \frac{1.645^2 \, (.5)(.5)}{.02^2} = 1{,}691.3 \approx 1{,}692$$

5.45 To compute the needed sample size, use

$$n = \frac{(z_{\alpha/2})^2 \sigma^2}{B^2} \quad \text{where } \alpha = 1 - .95 = .05 \text{ and } \alpha/2 = .05/2 = .025.$$

From Table IV, Appendix B, $z_{.025} = 1.96$.

Thus, for $s = 10$, $n = \dfrac{(1.96)^2(10)^2}{3^2} = 42.68 \approx 43$

For $s = 20$, $n = \dfrac{(1.96)^2(20)^2}{3^2} = 170.74 \approx 171$

For $s = 30$, $n = \dfrac{(1.96)^2(30)^2}{3^2} = 384.16 \approx 385$

5.47 The bound is $B = .05$. For confidence coefficient .99, $\alpha = 1 - .99 = .01$ and $\alpha/2 = .01/2 = .005$. From Table IV, Appendix B, $z_{.005} = 2.575$.

We estimate p with $\hat{p} = 11/27 = .407$. Thus,

$$n = \frac{(z_{\alpha/2})^2 pq}{B^2} = \frac{2.575^2(.407)(.593)}{.05^2} \approx 640.1 \Rightarrow 641$$

The necessary sample size would be 641. The sample was not large enough.

5.49 a. $P(t \le t_0) = .05$ where df $= 20$
$t_0 = -1.725$

b. $P(t \ge t_0) = .005$ where df $= 9$
$t_0 = 3.250$

c. $P(t \le -t_0 \text{ or } t \ge t_0) = .10$ where df $= 8$ is equivalent to
$P(t \ge t_0) = .10/2 = .05$ where df $= 8$
$t_0 = 1.860$

Inferences Based on a Single Sample: Estimation with Confidence Intervals

d. $P(t \le -t_0 \text{ or } t \ge t_0) = .01$ where df $= 17$ is equivalent to
$$P(t \ge t_0) = .01/2 = .005 \text{ where df} = 17$$
$$t_0 = 2.898$$

5.51 a. For confidence coefficient .99, $\alpha = .01$ and $\alpha/2 = .005$. From Table IV, Appendix B, $z_{.005} = 2.58$. The confidence interval is:

$$\bar{x} \pm z_{\alpha/2} \frac{s}{\sqrt{n}} \Rightarrow 32.5 \pm 2.58 \frac{30}{\sqrt{225}} \Rightarrow 32.5 \pm 5.16 \Rightarrow (27.34, 37.66)$$

b. The sample size is $n = \dfrac{(z_{\alpha/2})^2 \sigma^2}{B^2} = \dfrac{2.58^2 (30)^2}{.5^2} = 23{,}963.04 \approx 23{,}964$

c. "99% confidence" means that if repeated samples of size 225 were selected from the population and 99% confidence intervals constructed for the population mean, then 99% of all the intervals constructed will contain the population mean.

5.53 a. The 95% confidence interval is (298.6, 582.3).

b. We are 95% confident that the mean sales price is between $298,600 and $582,300.

c. "95% confidence" means that in repeated sampling, 95% of all confidence intervals constructed will contain the true mean salary and 5% will not.

d. Since the sample size is small ($n = 20$), we must assume that the distribution of sales prices is normal. From the stem-and-leaf display, it does not appear that the data come from a normal distribution. Thus, this confidence interval is probably not valid.

5.55 a. First we must compute $\hat{p}$: $\hat{p} = \dfrac{x}{n} = \dfrac{89{,}582}{102{,}263} = .876$

To see if the sample size is sufficiently large:

$$\hat{p} \pm 3\sigma_{\hat{p}} \approx \hat{p} \pm 3\sqrt{\frac{\hat{p}\hat{q}}{n}} \Rightarrow .876 \pm 3\sqrt{\frac{.876(.124)}{102{,}263}} \Rightarrow .876 \pm .003 \Rightarrow (.873, .879)$$

Since this interval is wholly contained in the interval (0, 1), we may conclude that the normal approximation is reasonable.

For confidence coefficient .99, $\alpha = .01$ and $\alpha/2 = .01/2 = .005$. From Table IV, Appendix B, $z_{.005} = 2.58$. The confidence interval is:

$$\hat{p} \pm z_{.005}\sqrt{\frac{pq}{n}} \approx \hat{p} \pm 2.58\sqrt{\frac{\hat{p}\hat{q}}{n}} \Rightarrow .876 \pm 2.58\sqrt{\frac{.876(.124)}{102{,}263}} \Rightarrow .876 \pm .003$$
$$\Rightarrow (.873, .879)$$

We are 99% confident that the true proportion of American adults who believe their health to be good to excellent is between .873 and .879.

5.57 a. For confidence coefficient .95, $\alpha = .05$ and $\alpha/2 = .025$. From Table IV, Appendix B, $z_{.025} = 1.96$. The confidence interval is:

$$\bar{x} \pm z_{\alpha/2}\frac{s}{\sqrt{n}}$$

Men: $7.4 \pm 1.96\frac{6.3}{\sqrt{159}} \Rightarrow 7.4 \pm .979 \Rightarrow (6.421, 8.379)$

We are 95% confident that the average distance to work for men in the central city is between 6.421 and 8.379 miles.

Women: $4.5 \pm 1.96\frac{4.2}{\sqrt{119}} \Rightarrow 4.5 \pm .755 \Rightarrow (3.745, 5.255)$

We are 95% confident that the average distance to work for women in the central city is between 3.745 and 5.255 miles.

b. Men: $9.3 \pm 1.96\frac{7.1}{\sqrt{138}} \Rightarrow 9.3 \pm 1.185 \Rightarrow (8.115, 10.485)$

We are 95% confident that the average distance to work for men in the suburbs is between 8.115 and 10.485 miles.

Women: $6.6 \pm 1.96\frac{5.6}{\sqrt{93}} \Rightarrow 6.6 \pm 1.138 \Rightarrow (5.462, 7.738)$

We are 95% confident that the average distance to work for women in the suburbs is between 5.462 and 7.738 miles.

5.59 a. For confidence coefficient .90, $\alpha = .10$ and $\alpha/2 = .05$. From Table IV, Appendix B, $z_{.05} = 1.645$. The 90% confidence interval is:

$$\bar{x} \pm z_{.05}\frac{\sigma}{\sqrt{n}} \Rightarrow \bar{x} \pm 1.645\frac{s}{\sqrt{n}} \Rightarrow 12.2 \pm 1.645\frac{10}{\sqrt{100}} \Rightarrow 12.2 \pm 1.645$$

$$\Rightarrow (10.555, 13.845)$$

b. For confidence coefficient .99, $\alpha = .01$ and $\alpha/2 = .005$. From Table IV, Appendix B, $z_{.005} = 2.58$.

The sample size is $n = \dfrac{\left(z_{\alpha/2}\right)^2\sigma^2}{B^2} = \dfrac{(2.58)^2(10)^2}{2^2} = 166.4 \approx 167$

You would need to take $n = 167$ samples.

5.61 a. First, we must estimate the standard deviation. The only information that we have is the values of the 20th, 50th, and 80th percentiles. Since the 20th percentile $35,100 is closer to the median, $50,000, than the 80th percentile, $73,000, the data are skewed. From Chebyshev's Rule, we know that at least $1 - 1/k^2$ of the observations are within k standard deviations of the mean. Thus, we want to find k such that $1 - 1/k^2 = .8 - .2 = .6$.

$1 - 1/k^2 = .6 \Rightarrow k^2 = 1/.4 = 2.5 \Rightarrow k \approx 1.6$

Thus, there are $2(1.6) = 3.2$ standard deviations in the interval from the 20th percentile to the 80th percentile. The standard deviation can be estimated by:

$$s \approx \frac{80\text{th} - 20\text{th}}{3.2} = \frac{73,000 - 35,100}{3.2} = 11,843.75$$

For confidence coefficient .98, $\alpha = .02$ and $\alpha/2 = .02/2 = .01$. From Table IV, Appendix B, $z_{.01} = 2.33$. Thus,

$$n = \frac{(z_{\alpha/2})^2 \sigma^2}{B^2} = \frac{2.33^2 (11,843.75)^2}{2,000^2} = 190.4 \approx 191$$

Thus, we would need a sample size of 191.

b. See part **a**.

c. We must assume that the distribution of salaries next year has a similar shape to the distribution of salaries in the sixth annual salary survey.

5.63   a. We would have to assume that the sample was a random sample. Since n is large, the Central Limit Theorem applies.

b. $\bar{x} = \dfrac{\sum x}{n} = \dfrac{586}{180} = 3.256$

$$s^2 = \frac{\sum x^2 - \dfrac{(\sum x)^2}{n}}{n-1} = \frac{2,640 - \dfrac{586^2}{180}}{180 - 1} = 4.0908; \quad s = \sqrt{4.0908} = 2.0226$$

For confidence coefficient .98, $\alpha = .02$ and $\alpha/2 = .02/2 = .01$. From Table IV, Appendix B, $z_{.01} = 2.33$. The 98% confidence interval is:

$$\bar{x} \pm 2.33\hat{\sigma}_{\bar{x}} \Rightarrow \bar{x} \pm 2.33 \left[\frac{s}{\sqrt{n}} \right] \sqrt{\frac{N-n}{N}} \Rightarrow 3.256 \pm 2.33 \left[\frac{2.0226}{\sqrt{180}} \right] \sqrt{\frac{8,521 - 180}{8,521}}$$

$$\Rightarrow 3.256 \pm .348 \Rightarrow (2.908, 3.604)$$

We are 98% confident that the mean subscription length is between 2.908 and 3.604 years.

5.65   The bound is $B = .1$. For confidence coefficient .99, $\alpha = 1 - .99 = .01$ and $\alpha/2 = .01/2 = .005$.

From Table IV, Appendix B, $z_{.005} = 2.575$.

We estimate p with $\hat{p}$ from Exercise 7.40 which is $\hat{p} = .636$. Thus,

$$n = \frac{(z_{\alpha/2})^2 pq}{B^2} \approx \frac{2.575^2 (.636)(.364)}{.1^2} = 153.5 \Rightarrow 154$$

The necessary sample size would be 154.

6.1   The null hypothesis is the "status quo" hypothesis, while the alternative hypothesis is the research hypothesis.

6.3   The "level of significance" of a test is α. This is the probability that the test statistic will fall in the rejection region when the null hypothesis is true.

6.5   The four possible results are:

1.   Rejecting the null hypothesis when it is true. This would be a Type I error.
2.   Accepting the null hypothesis when it is true. This would be a correct decision.
3.   Rejecting the null hypothesis when it is false. This would be a correct decision.
4.   Accepting the null hypothesis when it is false. This would be a Type II error.

6.7   When you reject the null hypothesis in favor of the alternative hypothesis, this does not prove the alternative hypothesis is correct. We are $100(1 - \alpha)\%$ confident that there is sufficient evidence to conclude that the alternative hypothesis is correct.

        If we were to repeatedly draw samples from the population and perform the test each time, approximately $100(1 - \alpha)\%$ of the tests performed would yield the correct decision.

6.9   Let p = student loan default rate in 2000. To see if the student loan default rate is less than .10, we test:

$$H_0: \ p = .10$$

$$H_a: \ p < .10$$

6.11   a.   A Type I error is rejecting the null hypothesis when it is true. In a murder trial, we would be concluding that the accused is guilty when, in fact, he/she is innocent.

            A Type II error is accepting the null hypothesis when it is false. In this case, we would be concluding that the accused is innocent when, in fact, he/she is guilty.

       b.   Both errors are bad. However, if an innocent person is found guilty of murder and is put to death, there is no way to correct the error. On the other hand, if a guilty person is set free, he/she could murder again.

       c.   In a jury trial, α is assumed to be smaller than β. The only way to convict the accused is for a unanimous decision of guilt. Thus, the probability of convicting an innocent person is set to be small.

       d.   In order to get a unanimous vote to convict, there has to be overwhelming evidence of guilt. The probability of getting a unanimous vote of guilt if the person is really innocent will be very small.

e. If a jury is predjuced against a guilty verdict, the value of α will decrease. The probability of convicting an innocent person will be even smaller if the jury if predjudiced against a guilty verdict.

f. If a jury is predjudiced against a guilty verdict, the value of β will increase. The probability of declaring a guilty person innocent will be larger if the jury is prejudiced against a guilty verdict.

6.13   a. Since the company must give proof the drug is safe, the null hypothesis would be the drug is unsafe. The alternative hypothesis would be the drug is safe.

b. A Type I error would be concluding the drug is safe when it is not safe. A Type II error would be concluding the drug is not safe when it is. α is the probability of concluding the drug is safe when it is not. β is the probability of concluding the drug is not safe when it is.

c. In this problem, it would be more important for α to be small. We would want the probability of concluding the drug is safe when it is not to be as small as possible.

6.15   a.
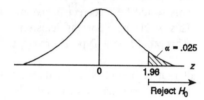

b.
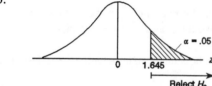

c.
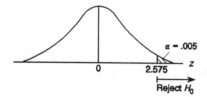

d.
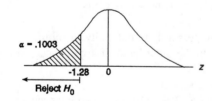

e.

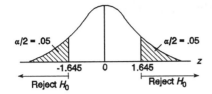

f.

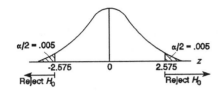

g. $P(z > 1.96) = .025$
$P(z > 1.645) = .05$
$P(z > 2.575) = .005$
$P(z < -1.28) = .1003$
$P(z < -1.645 \text{ or } z > 1.645) = .10$
$P(z < -2.575 \text{ or } z > 2.575) = .01$

6.17 a. $H_0: \mu = 100$
$H_a: \mu > 100$

The test statistic is $z = \dfrac{\bar{x} - \mu_0}{\sigma_{\bar{x}}} = \dfrac{\bar{x} - \mu_0}{\sigma/\sqrt{n}} = \dfrac{110 - 100}{60/\sqrt{100}} = 1.67$

The rejection region requires $\alpha = .05$ in the upper tail of the z-distribution. From Table IV, Appendix B, $z_{.05} = 1.645$. The rejection region is $z > 1.645$.

Since the observed value of the test statistic falls in the rejection region, ($z = 1.67 > 1.645$), H_0 is rejected. There is sufficient evidence to indicate the true population mean is greater than 100 at $\alpha = .05$.

b. $H_0: \mu = 100$
$H_a: \mu \neq 100$

The test statistic is $z = \dfrac{\bar{x} - \mu_0}{\sigma_{\bar{x}}} = \dfrac{110 - 100}{60/\sqrt{100}} = 1.67$

The rejection region requires $\alpha/2 = .05/2 = .025$ in each tail of the z-distribution. From Table IV, Appendix B, $z_{.025} = 1.96$. The rejection region is $z < -1.96$ or $z > 1.96$.

Since the observed value of the test statistic does not fall in the rejection region, ($z = 1.67 \not> 1.96$), H_0 is not rejected. There is insufficient evidence to indicate μ does not equal 100 at $\alpha = .05$.

c. In part a, we rejected H_0 and concluded the mean was greater than 100. In part b, we did not reject H_0. There was insufficient evidence to conclude the mean was different from 100. Because the alternative hypothesis in part a is more specific than the one in b, it is easier to reject H_0.

6.19   To determine if the mean point-spread error is different from 0, we test:

H_0: $\mu = 0$
H_a: $\mu \neq 0$

The test statistic is $z = \dfrac{\bar{x} - \mu_0}{\sigma_{\bar{x}}} = \dfrac{-1.6 - 0}{13.3/\sqrt{240}} = -1.86$

The rejection region requires $\alpha/2 = .01/2 = .005$ in each tail of the z distribution. From Table IV, Appendix B, $z_{.005} = 2.575$. The rejection region is $z > 2.575$ or $z < -2.575$.

Since the observed value of the test statistic does not fall in the rejection region ($z = -1.86 \not<$ -2.575), H_0 is not rejected. There is insufficient evidence to indicate that the true mean point-spread error is different from 0 at $\alpha = .01$.

6.21   a.   To determine whether the true mean PTSD score of all World War II aviator POWs is less than 16, we test:

H_0: $\mu = 16$
H_a: $\mu < 16$

b.   The test statistic is $z = \dfrac{\bar{x} - \mu_0}{\sigma_{\bar{x}}} = \dfrac{9 - 16}{9.32/\sqrt{33}} = -4.31$

The rejection region requires $\alpha = .10$ in the lower tail of the z-distribution. From Table IV, Appendix B, $z_{.10} = 1.28$. The rejection region is $z < -1.28$.

Since the observed value of the test statistic falls in the rejection region ($z = -4.31 <$ -1.28), H_0 is rejected. There is sufficient evidence to indicate that the true mean PTSD score of all World War II aviator POWs is less than 16 at $\alpha = .10$.

The practical implications of the test are that the World War II aviator POWs have a lower level PTSD level on the average than the POWs from Vietnam.

c.   The sample used in this study was a self-selected sample—only 33 of the 239 located survivors responded. Very often, self-selected respondents are not representative of the population. Here, those former POWs who are more comfortable with their lives may be more willing to respond than those who are less comfortable. Those who are less comfortable may be suffering more from PTSD than those who are more comfortable. Also, it may not be fair to compare the survivors from World War II to the survivors of Vietnam. The World War II survivors are more removed from their imprisonment than those from the Vietnam war. Also, many of the World War II POWs probably are no longer living. Again, those still alive may be the ones who are more comfortable with their lives.

6.23   a.   To determine if the process is not operating satisfactorily, we test:

H_0: $\mu = .250$
H_a: $\mu \neq .250$

The test statistic is $z = \dfrac{\bar{x} - \mu_0}{\sigma_{\bar{x}}} = \dfrac{.252475 - .250}{.00223/\sqrt{40}} = 7.02$

The rejection region requires $\alpha/2 = .01/2 = .005$ in each tail of the z-distribution. From Table IV, Appendix B, $z_{.005} = 2.58$. The rejection region is $z < -2.58$ or $z > 2.58$.

Since the observed value of the test statistic falls in the rejection region ($z = 7.02 > 2.58$), H_0 is rejected. There is sufficient information to indicate the process is performing in an unsatisfactory manner at $\alpha = .01$.

b. α is the probability of a Type I error. A Type I error, in this case, is to say the process is unsatisfactory when, in fact, it is satisfactory. The risk, then, is to the producer since he will be spending time and money to repair a process that is not in error.

β is the probability of a Type II error. A Type II error, in this case, is to say the process is satisfactory when it, in fact, is not. This is the consumer's risk since he could unknowingly purchase a defective product.

6.25 a. To determine if the sample data refute the manufacturer's claim, we test:

H_0: $\mu = 10$
H_a: $\mu < 10$

b. A Type I error is concluding the mean number of solder joints inspected per second is less than 10 when, in fact, it is 10 or more.

A Type II error is concluding the mean number of solder joints inspected per second is at least 10 when, in fact, it is less than 10.

c. H_0: $\mu = 10$
H_a: $\mu < 10$

The test statistic is $z = \dfrac{\bar{x} - \mu_0}{\sigma_{\bar{x}}} = \dfrac{9.29 - 10}{2.10/\sqrt{48}} = -2.34$

The rejection region requires $\alpha = .05$ in the lower tail of the z-distribution. From Table IV, Appendix B, $z_{.05} = 1.645$. The rejection region is $z < -1.645$.

Since the observed value of the test statistic falls in the rejection region ($z = -2.34 < -1.645$), H_0 is rejected. There is sufficient evidence to indicate the mean number of inspections per second is less than 10 at $\alpha = .05$.

6.27 a. Since the standard deviation is almost the same as the mean, and we know that fat intake cannot be negative, the distribution of fat intake per day is skewed to the right.

b. To determine if the mean fat intake for middle-age men on weight-control programs exceeds 30 grams, we test:

H_0: $\mu = 30$
H_a: $\mu > 30$

The test statistic is $z = \dfrac{\bar{x} - \mu_0}{\sigma_{\bar{x}}} = \dfrac{37 - 30}{32/\sqrt{64}} = 1.75$

The rejection statistics requires $\alpha = .10$ in the upper tail of the z-distribution. From Table IV, Appendix B, $z_{.10} = 1.28$. The rejection region is $z > 1.28$

Inferences Based on a Single Sample: Tests of Hypothesis   **107**

Since the observed value of the test statistic falls in the rejection region ($z = 1.75 > 1.28$), H_0 is rejected. There is sufficient evidence to indicate the mean fat intake for middle-age men on weight-control programs exceeds 30 grams at $\alpha = .10$.

c. For $\alpha = .05$, the rejection region requires $\alpha = .05$ in the upper tail of the z-distribution. From Table IV, Appendix B, $z_{.05} = 1.645$. The rejection region is $z > 1.645$. Since the observed value of the test statistic falls in the rejection region ($z = 1.75 > 1.645$), H_0 is rejected. The conclusion is the same.

For $\alpha = .01$, the rejection region requires $\alpha = .01$ in the upper tail of the z-distribution. From Table IV, Appendix B, $z_{.01} = 2.33$. The rejection region is $z > 2.33$. Since the observed value of the test statistic does not fall in the rejection region ($z = 1.75 \not> 2.33$), H_0 is not rejected. The conclusion is now different.

6.29 a. Since the p-value $= .10$ is greater than $\alpha = .05$, H_0 is not rejected.

b. Since the p-value $= .05$ is less than $\alpha = .10$, H_0 is rejected.

c. Since the p-value $= .001$ is less than $\alpha = .01$, H_0 is rejected.

d. Since the p-value $= .05$ is greater than $\alpha = .025$, H_0 is not rejected.

e. Since the p-value $= .45$ is greater than $\alpha = .10$, H_0 is not rejected.

6.31 p-value $= P(z \geq 2.17) = .5 - P(0 < z < 2.17) = .5 - .4850 = .0150$

(using Table IV, Appendix B)

6.33 a. The p-value reported by SAS is for a two-tailed test. Thus, $P(z \leq -1.63) + P(z \geq 1.63) = .1032$. For this one-tailed test, the p-value $= P(z \leq -1.63) = .1032/2 = .0516$.

Since the p-value $= .0516 > \alpha = .05$, H_0 is not rejected. There is insufficient evidence to indicate $\mu < 75$ at $\alpha = .05$.

b. For this one-tailed test, the p-value $= P(z \leq 1.63)$. Since $P(z \leq -1.63) = .1032/2 = .0516$, $P(z \leq 1.63) = 1 - .0516 = .9484$.

Since the p-value $= .9484 > \alpha = .10$, H_0 is not rejected. There is insufficient evidence to indicate $\mu < 75$ at $\alpha = .10$.

c. For this one-tailed test, the p-value $= P(z \geq 1.63) = .1032/2 = .0516$.

Since the p-value $= .0516 < \alpha = .10$, H_0 is rejected. There is sufficient evidence to indicate $\mu > 75$ at $\alpha = .10$.

d. For this two-tailed test, the p-value $= .1032$.

Since the p-value $= .1032 > \alpha = .01$, H_0 is not rejected. There is insufficient evidence to indicate $\mu \neq 75$ at $\alpha = .01$.

6.35 a. $z = \dfrac{\bar{x} - \mu_0}{\sigma_{\bar{x}}} = \dfrac{10.2 - 0}{31.3/\sqrt{50}} = 2.30$

b. For this two-sided test, the p-value $= P(z \geq 2.30) + P(z \leq -2.30) = (.5 - .4893) + (.5 - .4893) = .0214$. Since this value is so small, there is evidence to reject H_0. There is sufficient evidence to indicate the mean level of feminization is different from 0% for any value of $\alpha > .0214$.

c. $$z = \frac{\bar{x} - \mu_0}{\sigma_{\bar{x}}} = \frac{15.0 - 0}{25.1/\sqrt{50}} = 4.23$$

For this two-sided test, the p-value $= P(z \geq 4.23) + P(z \leq -4.23) \approx (.5 - .5) + (.5 - .5) = 0$. Since this value is so small, there is evidence to reject H_0. There is sufficient evidence to indicate the mean level of feminization is different from 0% for any value of $\alpha > 0.0$.

6.37 a. To determine if children in this age group perceive a risk associated with failure to wear helmets, we test:

H_0: $\mu = 2.5$
H_a: $\mu > 2.5$

b. The test statistic is $z = \dfrac{\bar{x} - \mu_0}{\sigma_{\bar{x}}} = \dfrac{3.39 - 2.5}{.80/\sqrt{797}} = 31.41$

p-value $= P(z \geq 31.41) \approx .5 - .5 = 0$

c. There is strong evidence to reject H_0 for any reasonable value of α. There is strong evidence to indicate the mean perceived risk associated with failure to wear helmets is greater than 2.5 for any reasonable value of α.

6.39 a. To determine whether Chinese smokers smoke, on average, more cigarettes a day in 1997 than in 1995, we test:

H_0: $\mu = 16.5$
H_a: $\mu > 16.5$

b. The test statistic is $z = \dfrac{\bar{x} - \mu_0}{\sigma_{\bar{x}}} = \dfrac{17.05 - 16.5}{5.21/\sqrt{200}} = 1.49$

The observed significance level is $p = P(z \geq 1.49) = .5 - .4319 = .0681$ (using Table IV, Appendix B).

Since the observed significance level (.0681) is not less than $\alpha = .05$, H_0 is not rejected. There is insufficient evidence to indicate that Chinese smokers smoke, on average, more cigarettes a day in 1997 than in 1995 at $\alpha = .05$.

If we used $\alpha = .10$, we would reject H_0. There is sufficient evidence to indicate that Chinese smokers smoke, on average, more cigarettes a day in 1997 than in 1995 at $\alpha = .10$.

c. The two-tailed test is inappropriate because we are interested in whether Chinese smokers, on average, smoke more cigarettes now than in 1995. This specifies only one-tail for the test.

6.41   a. $P(t > 1.440) = .10$
(Using Table VI, Appendix B, with df = 6)

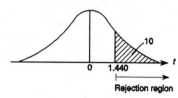

b. $P(t < -1.782) = .05$
(Using Table VI, Appendix B, with df = 12)

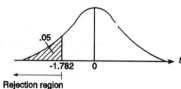

c. $P(t < -2.060) = P(t > 2.060) = .025$
(Using Table VI, Appendix B, with df = 25)

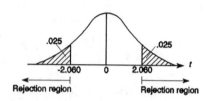

6.43   a.   The rejection region requires $\alpha/2 = .05/2 = .025$ in each tail of the t-distribution with df = $n - 1 = 14 - 1 = 13$. From Table VI, Appendix B, $t_{.025} = 2.160$. The rejection region is $t < -2.160$ or $t > 2.160$.

b.   The rejection region requires $\alpha = .01$ in the upper tail of the t-distribution with df = $n - 1 = 24 - 1 = 23$. From Table VI, Appendix B, $t_{.01} = 2.500$. The rejection region is $t > 2.500$.

c.   The rejection region requires $\alpha = .10$ in the upper tail of the t-distribution with df = $n - 1 = 9 - 1 = 8$. From Table VI, Appendix B, $t_{.10} = 1.397$. The rejection region is $t > 1.397$.

d.   The rejection region requires $\alpha = .01$ in the lower tail of the t-distribution with df = $n - 1 = 12 - 1 = 11$. From Table VI, Appendix B, $t_{.01} = 2.718$. The rejection region is $t < -2.718$.

e.   The rejection region requires $\alpha/2 = .10/2 = .05$ in each tail of the t-distribution with df = $n - 1 = 20 - 1 = 19$. From Table VI, Appendix B, $t_{.05} = 1.729$. The rejection region is $t < -1.729$ or $t > 1.729$.

f.   The rejection region requires $\alpha = .05$ in the lower tail of the t-distribution with df = $n - 1 = 4 - 1 = 3$. From Table VI, Appendix B, $t_{.05} = 2.353$. The rejection region is $t < -2.353$.

6.45   a.   We must assume that a random sample was drawn from a normal population.

   b.   The hypotheses are:

$$H_0: \ \mu = 1000$$
$$H_a: \ \mu > 1000$$

The test statistic is $t = 1.894$.

The p-value is .0382.

There is evidence to reject H_0 for $\alpha > .0382$. There is evidence to indicate the mean is greater than 1000 for $\alpha > .0382$.

   c.   The hypotheses are:

$$H_0: \ \mu = 1000$$
$$H_a: \ \mu \neq 1000$$

The test statistic is $t = 1.894$.

The p-value is $2(.0382) = .0764$.

There is no evidence to reject H_0 for $\alpha = .05$. There is insufficient evidence to indicate the mean is different than 1000 for $\alpha = .05$.

There is evidence to reject H_0 for $\alpha > .0764$. There is evidence to indicate the mean is different than 1000 for $\alpha > .0764$.

6.47   a.   To determine if the mean repellency percentage of the new mosquito repellent is less than 95, we test:

$$H_0: \mu = 95$$
$$H_a: \mu < 95$$

The test statistic is $t = \dfrac{\bar{x} - \mu_0}{s/\sqrt{n}} = \dfrac{83 - 95}{15/\sqrt{5}} = -1.79$

The rejection region requires $\alpha = .10$ in the lower tail of the t distribution. From Table VI, Appendix B, with df $= n - 1 = 5 - 1 = 4$, $t_{.10} = 1.533$. The rejection region is $t < -1.533$.

Since the observed value of the test statistic falls in the rejection region ($t = -1.79 < -1.533$), H_0 is rejected. There is sufficient evidence to indicate that the true mean repellency percentage of the new mosquito repellent is less than 95 at $\alpha = .10$.

   b.   We must assume that the population of percent repellencies is normally distributed.

6.49   Some preliminary calculations:

$$\bar{x} = \frac{\sum x}{n} = \frac{489}{5} = 97.8 \qquad s^2 = \frac{\sum x^2 - \dfrac{(\sum x)^2}{n}}{n - 1} = \frac{47{,}867 - \dfrac{489^2}{5}}{5 - 1} = 10.7$$

$$s = \sqrt{10.7} = 3.271$$

To determine if the mean recovery percentage of Aldrin exceeds 85% using the new MSPD method, we test:

H_0: $\mu = 85$
H_a: $\mu > 85$

The test statistic is $t = \dfrac{\bar{x} - \mu_0}{s/\sqrt{n}} = \dfrac{97.8 - 85}{3.271/\sqrt{5}} = 8.75$

The rejection region requires $\alpha = .05$ in the upper tail of the t-distribution with df $= n - 1 = 5 - 1 = 4$. From Table VI, Appendix B, $t_{.05} = 2.132$. The rejection region is $t > 2.132$.

Since the observed value of the test statistic falls in the rejection region ($t = 8.75 > 2.132$), H_0 is rejected. There is sufficient evidence to indicate that the true mean recovery percentage of Aldrin exceeds 85% using the new MSPD method at $\alpha = .05$.

6.51   a.   To determine if the plants meet the current OSHA standard, we test:

H_0: $\mu = .004$
H_a: $\mu > .004$

b.   First, compute the sample mean and standard deviation for plant 1's arsenic level:

$$\bar{x} = \frac{\sum x}{n} = \frac{.015}{2} = .0075$$

$$s^2 = \frac{\sum x^2 - \dfrac{(\sum x)^2}{n}}{n - 1} = \frac{.000125 - \dfrac{.015^2}{2}}{2 - 1} = .0000125$$

$$s = \sqrt{s^2} = .003536$$

The test statistic is $t = \dfrac{\bar{x} - \mu_0}{s/\sqrt{n}} = \dfrac{.0075 - .004}{.003536/\sqrt{2}} = 1.40$

The p-value $= P(t \geq 1.40)$. From Table VI, Appendix B, with df $= n - 1 = 2 - 1 = 1$, the p-value $> .10$.

Next, compute the sample mean and standard deviation for plant 2's arsenic level:

$$\bar{x} = \frac{\sum x}{n} = \frac{.14}{2} = .07$$

$$s^2 = \frac{\sum x^2 - \dfrac{(\sum x)^2}{n}}{n - 1} = \frac{.0106 - \dfrac{.14^2}{2}}{2 - 1} = .0008$$

$$s = \sqrt{s^2} = .0283$$

The test statistic is $t = \dfrac{\bar{x} - \mu_0}{s/\sqrt{n}} = \dfrac{.07 - .004}{.0283/\sqrt{2}} = 3.3$

The p-value $= P(t \geq 3.30)$. From Table VI, Appendix B, with df $= n - 1 = 2 - 1 = 1$, the $.05 < p$-value $< .10$.

c. For plant 1, the test statistic is $t = 1.40$ and the p-value $= .200$. There is no evidence to reject H_0 for $\alpha \leq .10$. There is insufficient evidence to indicate the mean level is greater than .004 for $\alpha \leq .10$.

For plant 2, the test statistic is $t = 3.30$ and the p-value $= .094$. There is no evidence to reject H_0 for $\alpha = .05$. There is insufficient evidence to indicate the mean level is greater than .004 for $\alpha = .05$. There is evidence to reject H_0 for $\alpha = .10$. There is sufficient evidence to indicate the mean level is greater than .004 for $\alpha = .10$.

6.53 The sample size is large enough if the interval $p_0 \pm 3\sigma_{\hat{p}}$ is contained in the interval $(0, 1)$.

a. $p_0 \pm 3\sqrt{\dfrac{p_0 q_0}{n}} \Rightarrow .975 \pm 3\sqrt{\dfrac{(.975)(.025)}{900}} \Rightarrow .975 \pm .016 \Rightarrow (.959, .991)$

Since the interval is contained in the interval $(0, 1)$, the sample size is large enough.

b. $p_0 \pm 3\sqrt{\dfrac{p_0 q_0}{n}} \Rightarrow .01 \pm 3\sqrt{\dfrac{(.01)(.99)}{125}} \Rightarrow .01 \pm .027 \Rightarrow (-.017, .037)$

Since the interval is not contained in the interval $(0, 1)$, the sample size is not large enough.

c. $p_0 \pm 3\sqrt{\dfrac{p_0 q_0}{n}} \Rightarrow .75 \pm 3\sqrt{\dfrac{(.75)(.25)}{40}} \Rightarrow .75 \pm .205 \Rightarrow (.545, .955)$

Since the interval is contained in the interval $(0, 1)$, the sample size is large enough.

d. $p_0 \pm 3\sqrt{\dfrac{p_0 q_0}{n}} \Rightarrow .75 \pm 3\sqrt{\dfrac{(.75)(.25)}{15}} \Rightarrow .75 \pm .335 \Rightarrow (.415, 1.085)$

Since the interval is not contained in the interval $(0, 1)$, the sample size is not large enough.

e. $p_0 \pm 3\sqrt{\dfrac{p_0 q_0}{n}} \Rightarrow .62 \pm 3\sqrt{\dfrac{(.62)(.38)}{12}} \Rightarrow .62 \pm .420 \Rightarrow (.120, 1.040)$

Since the interval is not contained in the interval $(0, 1)$, the sample size is not large enough.

6.55 From Exercise 5.26, $n = 50$ and since p is the proportion of consumers who do not like the snack food, $\hat{p}$ will be:

$$\hat{p} = \frac{\text{Number of 0's in sample}}{n} = \frac{29}{50} = .58$$

First, check to see if the normal approximation will be adequate:

$$p_0 \pm 3\sigma_{\hat{p}} \Rightarrow p_0 \pm 3\sqrt{\frac{pq}{n}} \approx p_0 \pm 3\sqrt{\frac{p_0 q_0}{n}} \Rightarrow .5 \pm 3\sqrt{\frac{.5(1 - .5)}{50}} \Rightarrow .5 \pm .2121$$
$$\Rightarrow (.2879, .7121)$$

Since the interval lies completely in the interval $(0, 1)$, the normal approximation will be adequate.

Inferences Based on a Single Sample: Tests of Hypothesis

a. H_0: $p = .5$
H_a: $p > .5$

The test statistic is $z = \dfrac{\hat{p} - p_0}{\sigma_{\hat{p}}} = \dfrac{\hat{p} - p_0}{\sqrt{\dfrac{p_0 q_0}{n}}} = \dfrac{.58 - .5}{\sqrt{\dfrac{.5(1 - .5)}{50}}} = 1.13$

The rejection region requires $\alpha = .10$ in the upper tail of the z-distribution. From Table IV, Appendix B, $z_{.10} = 1.28$. The rejection region is $z > 1.28$.

Since the observed value of the test statistic does not fall in the rejection region ($z = 1.13 \ngtr 1.28$), H_0 is not rejected. There is insufficient evidence to indicate the proportion of customers who do not like the snack food is greater than .5 at $\alpha = .10$.

b. p-value $= P(z \geq 1.13) = .5 - .3708 = .1292$

6.57 a. Some preliminary calculations are:

$\hat{p} = x/n = 23/33 = .697$

First we check to see if the normal approximation is adequate:

$p_0 \pm 3\sigma_{\hat{p}} \Rightarrow p_0 \pm 3\sqrt{\dfrac{p_0 q_0}{n}} \Rightarrow .6 \pm 3\sqrt{\dfrac{.6(.4)}{33}} \Rightarrow .6 \pm .256 \Rightarrow (.344, .856)$

Since the interval falls completely in the interval $(0, 1)$, the normal distribution will be adequate.

To determine if the cream will improve the skin of more than 60% of women over age 40, we test:

H_0: $p = .60$
H_a: $p > .60$

The test statistic is $z = \dfrac{\hat{p} - p_0}{\sqrt{\dfrac{p_0 q_0}{n}}} = \dfrac{.697 - .60}{\sqrt{\dfrac{.60(.40)}{33}}} = 1.14$

The rejection region requires $\alpha = .05$ in the upper tail of the z distribution. From Table IV, Appendix B, $z_{.05} = 1.645$. The rejection region is $z > 1.645$.

Since the observed value of the test statistic does not fall in the region ($z = 1.14 \ngtr 1.645$), H_0 is not rejected. There is insufficient evidence to indicate that the cream will improve the skin of more than 60% of women over age 40 at $\alpha = .05$.

b. The p-value is $p = P(z \geq 1.14) = .5 - P(0 < z < 1.14) = .5 - .3729 = .1271$. (using Table IV) The probability of observing our test statistic or anything more unusual, given H_0 is true, is .1271. Since this p-value is not very small, there is no evidence to indicate that H_0 is false.

6.59   Let p = proportion of patients taking the pill who reported an improved condition.

First we check to see if the normal approximation is adequate:

$$p_0 \pm 3\sigma_{\hat{p}} \Rightarrow p_0 \pm 3\sqrt{\frac{p_0 q_0}{n}} \Rightarrow .5 \pm 3\sqrt{\frac{.5(.5)}{7000}} \Rightarrow .5 \pm .018 \Rightarrow (.482, .518)$$

Since the interval falls completely in the interval (0, 1), the normal distribution will be adequate.

To determine if there really is a placebo effect at the clinic, we test:

H_0: $p = .5$
H_a: $p > .5$

The test statistic is $z = \dfrac{\hat{p} - p_0}{\sqrt{\dfrac{p_0 q_0}{n}}} = \dfrac{.7 - .5}{\sqrt{\dfrac{.5(.5)}{7000}}} = 33.47$

The rejection region requires $\alpha = .05$ in the upper tail of the z distribution. From Table IV, Appendix B, $z_{.05} = 1.645$. The rejection region is $z > 1.645$.

Since the observed value of the test statistic falls in the rejection region ($z = 33.47 > 1.645$), H_0 is rejected. There is sufficient evidence to indicate that there really is a placebo effect at the clinic at $\alpha = .05$.

6.61   a. $\hat{p} = 15/60 = .25$

To determine if the proportion of shoppers who fail in their attempts to purchase merchandise online is less than .39, we test:

H_0: $p = .39$
H_a: $p < .39$

The test statistic is $z = \dfrac{\hat{p} - p_0}{\sqrt{\dfrac{p_0 q_0}{n}}} = \dfrac{.25 - .39}{\sqrt{\dfrac{.39(.61)}{60}}} = -2.22$

The rejection region requires $\alpha = .01$ in the lower tail of the z-distribution. From Table IV, Appendix B, $z_{.05} = 2.33$. The rejection region is $z < -2.33$.

Since the observed value of the test statistic does not fall in the rejection region ($z = -2.22 \not< -2.33$), H_0 is not rejected. There is insufficient evidence to indicate the proportion of shoppers who fail in their attempts to purchase merchandise online is less than .39 at $\alpha = .01$.

b.   The observed significance level of the test is p-value = $P(z \le -2.22) = .5 - .4868 = .0132$. Since the p-value is greater than $\alpha = .01$, H_0 is not rejected.

6.63   a.   To determine whether the true proportion of toothpaste brands with the ADA seal of verifying effective decay prevention is less than .5, we test:

$$H_0: \ p = .5$$
$$H_a: \ p < .5$$

       b.   The p-value of the test is $.4610/2 = .2305$. Since the p-value is so large, there is no evidence to reject H_0 for any reasonable value of α. There is insufficient evidence to indicate the true proportion of toothpaste brands with the ADA seal of verifying effective decay prevention is less than .5 for $\alpha \leq .10$.

6.65   a. $P(x \geq 7) = 1 - P(x \leq 6) = 1 - .965 = .035$

       b. $P(x \geq 5) = 1 - P(x \leq 4) = 1 - .637 = .363$

       c. $P(x \geq 8) = 1 - P(x \leq 7) = 1 - .996 = .004$

       d. $P(x \geq 10) = 1 - P(x \leq 9) = 1 - .849 = .151$

$$\mu = np = 15(.5) = 7.5 \text{ and } \sigma = \sqrt{npq} = \sqrt{15(.5)(.5)} = 1.9365$$

$$P(x \geq 10) \approx P\left[z \geq \frac{(10 - .5) - 7.5}{1.9365}\right] = P(z \geq 1.03) = .5 - .3485 = .1515$$

       e. $P(x \geq 15) = 1 - P(x \leq 14) = 1 - .788 = .212$

$$\mu = np = 25(.5) = 12.5 \text{ and } \sigma = \sqrt{npq} = \sqrt{25(.5)(.5)} = 2.5$$

$$P(x \geq 15) \approx P\left[z \geq \frac{(15 - .5) - 12.5}{2.5}\right] = P(z \geq .80) = .5 - .2881 = .2119$$

6.67   To determine if the median is greater than 75, we test:

$$H_0: \ \eta = 75$$
$$H_a: \ \eta > 75$$

The test statistic is $S = $ number of measurements greater than $75 = 17$.

The p-value $= P(x \geq 17)$ where x is a binomial random variable with $n = 25$ and $p = .5$. From Table II,

$$p\text{-value} = P(x \geq 17) = 1 - P(x \leq 16) = 1 - .946 = .054$$

Since the p-value $= .054 < \alpha = .10$, H_0 is rejected. There is sufficient evidence to indicate the median is greater than 75 at $\alpha = .10$.

We must assume the sample was randomly selected from a continuous probability distribution.

Note: Since $n \geq 10$, we could use the large-sample approximation.

6.69   a.   To determine whether the median biting rate is higher in bright, sunny weather, we test:

$$H_0: \ \eta = 5$$
$$H_a: \ \eta > 5$$

b. The test statistic is $z = \dfrac{(S - .5) - .5n}{.5\sqrt{n}} = \dfrac{(95 - .5) - .5(122)}{.5\sqrt{122}} = 6.07$

(where S = number of observations greater than 5)

The p-value is $p = P(z \geq 6.07)$. From Table IV, Appendix B, $p = P(z \geq 6.07) \approx 0.0000$.

c. Since the observed p-value is less than α ($p = 0.0000 < .01$), H_0 is rejected. There is sufficient evidence to indicate that the median biting rate in bright, sunny weather is greater than 5 at $\alpha = .01$.

6.71 a. I would recommend the sign test because five of the sample measurements are of similar magnitude, but the 6th is about three times as large as the others. It would be very unlikely to observe this sample if the population were normal.

b. To determine if the airline is meeting the requirement, we test:

$H_0\!: \eta = 30$
$H_a\!: \eta < 30$

c. The test statistic is S = number of measurements less than $30 = 5$.

H_0 will be rejected if the p-value $< \alpha = .01$.

d. The test statistic is $S = 5$.

The p-value $= P(x \geq 5)$ where x is a binomial random variable with $n = 6$ and $p = .5$. From Table II,

p-value $= P(x \geq 5) = 1 - P(x \leq 4) = 1 - .891 = .109$

Since the p-value $= .109$ is not less than $\alpha = .01$, H_0 is not rejected. There is insufficient evidence to indicate the airline is meeting the maintenance requirement at $\alpha = .01$.

6.73 The smaller the p-value associated with a test of hypothesis, the stronger the support for the **alternative** hypothesis. The p-value is the probability of observing your test statistic or anything more unusual, given the null hypothesis is true. If this value is small, it would be very unusual to observe this test statistic if the null hypothesis were true. Thus, it would indicate the alternative hypothesis is true.

6.75 a. $H_0\!: \mu = 80$
$H_a\!: \mu < 80$

The test statistic is $t = \dfrac{\bar{x} - \mu_0}{s/\sqrt{n}} = \dfrac{72.6 - 80}{\sqrt{19.4}/\sqrt{20}} = -7.51$

The rejection region requires $\alpha = .05$ in the lower tail of the t-distribution with df $= n - 1 = 20 - 1 = 19$. From Table VI, Appendix B, $t_{.05} = 1.729$. The rejection region is $t < -1.729$.

Since the observed value of the test statistic falls in the rejection region ($-7.51 < -1.729$), H_0 is rejected. There is sufficient evidence to indicate that the mean is less than 80 at $\alpha = .05$.

b. H_0: $\mu = 80$
H_a: $\mu \neq 80$

The test statistic is $t = \dfrac{\bar{x} - \mu_0}{s/\sqrt{n}} = \dfrac{72.6 - 80}{\sqrt{19.4}/\sqrt{20}} = -7.51$

The rejection region requires $\alpha/2 = .01/2 = .005$ in each tail of the t-distribution with df $= n - 1 = 20 - 1 = 19$. From Table VI, Appendix B, $t_{.005} = 2.861$. The rejection region is $t < -2.861$ or $t > 2.861$.

Since the observed value of the test statistic falls in the rejection region ($-7.51 < -2.861$), H_0 is rejected. There is sufficient evidence to indicate that the mean is different from 80 at $\alpha = .01$.

6.77   a. H_0: $\mu = 8.3$
H_a: $\mu \neq 8.3$

The test statistic is $z = \dfrac{\bar{x} - \mu_0}{\sigma_{\bar{x}}} = \dfrac{8.2 - 8.3}{.79/\sqrt{175}} = -1.67$

The rejection region requires $\alpha/2 = .05/2 = .025$ in each tail of the z-distribution. From Table IV, Appendix B, $z_{.025} = 1.96$. The rejection region is $z < -1.96$ or $z > 1.96$.

Since the observed value of the test statistic does not fall in the rejection region ($-1.67 \not< -1.96$), H_0 is not rejected. There is insufficient evidence to indicate that the mean is different from 8.3 at $\alpha = .05$.

b. H_0: $\mu = 8.4$
H_a: $\mu \neq 8.4$

The test statistic is $z = \dfrac{\bar{x} - \mu_0}{\sigma_{\bar{x}}} = \dfrac{8.2 - 8.4}{.79/\sqrt{175}} = -3.35$

The rejection region is the same as part **b**, $z < -1.96$ or $z > 1.96$.

Since the observed value of the test statistic falls in the rejection region ($-3.35 < -1.96$), H_0 is rejected. There is sufficient evidence to indicate that the mean is different from 8.4 at $\alpha = .05$.

6.79   First, check to see if the normal approximation is adequate:

$$p_0 \pm 3\sigma_{\hat{p}} \Rightarrow p_0 \pm 3\sqrt{\dfrac{p_0 q_0}{n}} \Rightarrow .5 \pm 3\sqrt{\dfrac{(.5)(.5)}{100}} \Rightarrow .5 \pm .15 \Rightarrow (.35, .65)$$

Since the interval falls completely in the interval $(0,1)$, the normal distribution will be adequate.

$$\hat{p} = \dfrac{x}{n} = \dfrac{56}{100} = .56$$

To determine if more than half of all Diet Coke drinkers prefer Diet Pepsi, we test:

H_0: $p = .5$
H_a: $p > .5$

The test statistic is $z = \dfrac{\hat{p} - p_0}{\sqrt{\dfrac{p_0 q_0}{n}}} = \dfrac{.56 - .5}{\sqrt{\dfrac{.5(.5)}{100}}} = 1.20$

The rejection region requires $\alpha = .05$ in the upper tail of the z-distribution. From Table IV, Appendix B, $z_{.05} = 1.645$. The rejection region is $z > 1.645$.

Since the observed value of the test statistic does not fall in the rejection region ($z = 1.20 \not> 1.645$), H_0 is not rejected. There is insufficient evidence to indicate that more than half of all Diet Coke drinkers prefer Diet Pepsi at $\alpha = .05$.

Since H_0 was not rejected, there is no evidence that Diet Coke drinkers prefer Diet Pepsi.

6.81 a. To determine if the claim can be rejected, we test:

H_0: $\mu = .25$
H_a: $\mu < .25$

The test statistic is $z = \dfrac{\hat{p} - p_0}{\sqrt{\dfrac{p_0 q_0}{n}}} = \dfrac{.190 - .25}{\sqrt{\dfrac{.25(.75)}{195}}} = -1.93$

Since no α was given, we will use $\alpha = .05$. The rejection region requires $\alpha = .05$ in the lower tail of the z-distribution. From Table IV, Appendix B, $z_{.05} = 1.645$. The rejection region is $z < -1.645$.

Since the observed value of the test statistic falls in the rejection region ($z = -1.93 < -1.645$), H_0 is rejected. There is sufficient evidence to reject the claim that the "more than 25% of all U.S. businesses will have Web sites by the middle of 1995" at $\alpha = .05$.

b. This sample was self-selected and may not be representative of the population. The sample of readers who received the questionnaires was randomly selected. However, only 195 out of 1,500 returned the questionnaires. Usually those who return questionnaires have strong opinions one way or another, and thus, those responding to the questionnaire may not be representative.

6.83 a. The test statistic is $t = \dfrac{\bar{x} - \mu_0}{s/\sqrt{n}} = \dfrac{1173.6 - 1100}{36.3/\sqrt{3}} = 3.512$

The p-value $= P(t \geq 3.512)$. From Table VI with df $= n - 1 = 3 - 1 = 2$, $.025 < p\text{-value} < .05$.

b. The p-value $= .0362 = P(t \geq 3.512)$. Since this p-value is fairly small, there is evidence to reject H_0 for $\alpha > .0362$. There is evidence to indicate the mean length of life of a certain mechanical component is longer than 1100 hours.

c. A Type I error would be of most concern for this test. A Type I error would be concluding the mean lifetime is greater than 1100 hours when in fact the mean lifetime is not greater than 1100.

d. It is rather questionable whether a sample of 3 is representative of the population. If the sample is representative, then the conclusion is warranted.

6.85 a. First, check to see if n is large enough:

$$p_0 \pm 3\sigma_{\hat{p}} \Rightarrow p_0 \pm 3\sqrt{\frac{p_0 q_0}{n}} \Rightarrow .5 \pm 3\sqrt{\frac{.5(.5)}{250}} \Rightarrow .5 \pm .095 \Rightarrow (.405, .595)$$

Since the interval lies within the interval (0, 1), the normal approximation will be adequate.

To determine if there is evidence to reject the claim that no more than half of all manufacturers are dissatisfied with their trade promotion spending, we test:

H_0: $p = .5$
H_a: $p > .5$

The test statistic is $z = \dfrac{\hat{p} - p_0}{\sqrt{\dfrac{p_0 q_0}{n}}} = \dfrac{.91 - .5}{\sqrt{\dfrac{.5(.5)}{250}}} = 12.97$

The rejection region requires $\alpha = .02$ in the upper tail of the z-distribution. From Table IV, Appendix B, $z_{.02} = 2.05$. The rejection region is $z > 2.05$.

Since the observed value of the test statistic falls in the rejection region ($z = 12.97 > 2.05$), H_0 is rejected. There is sufficient evidence to reject the claim that no more than half of all manufacturers are dissatisfied with their trade promotion spending at $\alpha = .02$.

b. The observed significance level is p-value $= P(z \geq 12.97) \approx .5 - .5 = 0$. Since this p-value is so small, H_0 will be rejected for any reasonable value of α.

c. First, we must define the rejection region in terms of $\hat{p}$.

$$\hat{p} = p_0 + z_\alpha \sigma_{\hat{p}} = .5 + 2.05\sqrt{\frac{.5(.5)}{250}} = .565$$

$$\beta = P(\hat{p} < .565 \mid p = .55) = P\left(z < \frac{.565 - .55}{\sqrt{\frac{.55(.45)}{250}}}\right) = P(z < .48) = .5 + .1844 = .6844$$

6.87 a. To determine if the production process should be halted, we test:

H_0: $\mu = 3$
H_a: $\mu > 3$

where μ = mean amount of PCB in the effluent.

The test statistic is $z = \dfrac{\bar{x} - \mu_0}{\sigma_{\bar{x}}} = \dfrac{3.1 - 3}{.5/\sqrt{50}} = 1.41$

The rejection region requires $\alpha = .01$ in the upper tail of the z-distribution. From Table IV, Appendix B, $z_{.01} = 2.33$. The rejection region is $z > 2.33$.

Since the observed value of the test statistic does not fall in the rejection region, ($z = 1.41 \not> 2.33$), H_0 is not rejected. There is insufficient evidence to indicate the mean amount of PCB in the effluent is more than 3 parts per million at $\alpha = .01$. Do not halt the manufacturing process.

b. As plant manager, I do not want to shut down the plant unnecessarily. Therefore, I want $\alpha = P(\text{shut down plant when } \mu = 3)$ to be small.

c. The p-value is $p = P(z \geq 1.41) = .5 - .4207 = .0793$. Since the p-value is not less than $\alpha = .01$, H_0 is not rejected.

6.89   a. Since only 70 of the 80 customers responded to the question, only the 70 will be included.

To determine if the median amount spent on hamburgers at lunch at McDonald's is less than $2.25, we test:

H_0: $\eta = 2.25$
H_a: $\eta < 2.25$

S = number of measurements less than $2.25 = 20$.

The test statistic is $z = \dfrac{(S - .5) - .5n}{.5\sqrt{n}} = \dfrac{(20 - .5) - .5(70)}{.5\sqrt{70}} = -3.71$

No α was given in the exercise. We will use $\alpha = .05$. The rejection region requires $\alpha = .05$ in the lower tail of the z-distribution. From Table IV, Appendix B, $z_{.05} = 1.645$. The rejection region is $z > 1.645$.

Since the observed value of the test statistic does not fall in the rejection region ($z = -3.71 \not> 1.645$), H_0 is not rejected. There is insufficient evidence to indicate that the median amount spent on hamburgers at lunch at McDonald's is less than $2.25 at $\alpha = .05$.

b. No. The survey was done in Boston only. The eating habits of those living in Boston are probably not representative of all Americans.

c. We must assume that the sample is randomly selected from a continuous probability distribution.

6.91   a. To determine if the average high technology stock is riskier than the market as a whole, we test:

H_0: $\mu = 1$
H_a: $\mu > 1$

b. The test statistic is $t = \dfrac{\bar{x} - \mu_0}{s/\sqrt{n}}$

The rejection region requires $\alpha = .10$ in the upper tail of the t-distribution with df $= n - 1 = 15 - 1 = 14$. From Table VI, Appendix B, $t_{.10} = 1.345$. The rejection region is $t > 1.345$.

c. We must assume the population of beta coefficients of technology stocks is normally distributed.

d. The test statistic is $t = \dfrac{\bar{x} - \mu_0}{s/\sqrt{n}} = \dfrac{1.23 - 1}{.37/\sqrt{15}} = 2.41$

Since the observed value of the test statistic falls in the rejection region ($t = 2.41 > 1.345$), H_0 is rejected. There is sufficient evidence to indicate the mean high technology stock is riskier than the market as a whole at $\alpha = .10$.

e. From Table VI, Appendix B, with df $= n - 1 = 15 - 1 = 14$, $.01 < P(t \geq 2.41)$ $< .025$. Thus, $.01 < p$-value $< .025$. The probability of observing this test statistic, $t = 2.41$, or anything more unusual is between .01 and .025. Since this probability is small, there is evidence to indicate the null hypothesis is false for $\alpha = .05$.

6.93   a. To determine if the mean price of a new home in November 2000 exceeds \$209,700, we test:

H_0: $\mu = 209,700$

H_a: $\mu > 209,700$

b. The test statistic is $z = \dfrac{\bar{x} - \mu_0}{\sigma_{\bar{x}}} = \dfrac{216,981 - 209,700}{19,805/\sqrt{32}} = 2.08$

The p-value $= P(z \geq 2.08) = .5 - .4812 = .0188$.

Since the p-value is fairly small, there is evidence to reject H_0. There is sufficient evidence to indicate the mean price of a new home in November 2000 exceeds \$209,700 for any value of $\alpha > .0188$.

7.1   a.   For confidence coefficient .95, $\alpha = .05$ and $\alpha/2 = .025$. From Table IV, Appendix B, $z_{.025} = 1.96$. The confidence interval is:

$$(\bar{x}_1 - \bar{x}_2) \pm z_{.025} \sqrt{\frac{\sigma_1^2}{n_1} + \frac{\sigma_2^2}{n_2}} \Rightarrow (5{,}275 - 5{,}240) \pm 1.96 \sqrt{\frac{150^2}{400} + \frac{200^2}{400}}$$

$$\Rightarrow 35 \pm 24.5 \Rightarrow (10.5,\ 59.5)$$

We are 95% confident that the difference between the population means is between 10.5 and 59.5.

   b.   The test statistic is $z = \dfrac{(\bar{x}_1 - \bar{x}_2) - (\mu_1 - \mu_2)}{\sqrt{\dfrac{\sigma_1^2}{n_1} + \dfrac{\sigma_2^2}{n_2}}} = \dfrac{(5275 - 5240) - 0}{\sqrt{\dfrac{150^2}{400} + \dfrac{200^2}{400}}} = 2.8$

The p-value of the test is $P(z \le -2.8) + P(z \ge 2.8) = 2P(z \ge 2.8) = 2(.5 - .4974)$
$$= 2(.0026) = .0052$$

Since the p-value is so small, there is evidence to reject H_0. There is evidence to indicate the two population means are different for $\alpha > .0052$.

   c.   The p-value would be half of the p-value in part **b**. The p-value $= P(z \ge 2.8) = .5 - .4974$ $= .0026$. Since the p-value is so small, there is evidence to reject H_0. There is evidence to indicate the mean for population 1 is larger than the mean for population 2 for $\alpha > .0026$.

   d.   The test statistic is $z = \dfrac{(\bar{x}_1 - \bar{x}_2) - (\mu_1 - \mu_2)}{\sqrt{\dfrac{\sigma_1^2}{n_1} + \dfrac{\sigma_2^2}{n_2}}} = \dfrac{(5275 - 5240) - 25}{\sqrt{\dfrac{150^2}{400} + \dfrac{200^2}{400}}} = .8$

The p-value of the test is $P(z \le -.8) + P(z \ge .8) = 2P(z \ge .8) = 2(.5 - .2881)$
$$= 2(.2119) = .4238$$

Since the p-value is so large, there is no evidence to reject H_0. There is no evidence to indicate that the difference in the 2 population means is different from 25 for $\alpha \le .10$.

   e.   We must assume that we have two independent random samples.

7.3   a.   No. Both populations must be normal.

   b.   No. Both populations variances must be equal.

c.   No. Both populations must be normal.

d.   Yes.

e.   No. Both populations must be normal.

7.5   Some preliminary calculations are:

$$\bar{x}_1 = \frac{\sum x_1}{n_1} = \frac{11.8}{5} = 2.36 \qquad s_1^2 = \frac{\sum x_1^2 - \dfrac{\left(\sum x_1\right)^2}{n_1}}{n_1 - 1} = \frac{30.78 - \dfrac{(11.8)^2}{5}}{5 - 1} = .733$$

$$\bar{x}_2 = \frac{\sum x_2}{n_2} = \frac{14.4}{4} = 3.6 \qquad s_2^2 = \frac{\sum x_2^2 - \dfrac{\left(\sum x_2\right)^2}{n_2}}{n_2 - 1} = \frac{53.1 - \dfrac{(14.4)^2}{4}}{4 - 1} = .42$$

a. $$s_p^2 = \frac{(n_1 - 1)s_1^2 + (n_2 - 1)s_2^2}{n_1 + n_2 - 2} = \frac{(5 - 1).773 + (4 - 1).42}{5 + 4 - 2} = \frac{4.192}{7} = .5989$$

b. $H_0: \mu_1 - \mu_2 = 0$
       $H_a: \mu_1 - \mu_2 < 0$

The test statistic is $t = \dfrac{(\bar{x}_1 - \bar{x}_2) - D_0}{\sqrt{s_p^2\left[\dfrac{1}{n_1} + \dfrac{1}{n_2}\right]}} = \dfrac{(2.36 - 3.6) - 0}{\sqrt{.5989\left[\dfrac{1}{5} + \dfrac{1}{4}\right]}} = \dfrac{-1.24}{.5191} = -2.39$

The rejection region requires $\alpha = .10$ in the lower tail of the t-distribution with df $= n_1 + n_2 - 2 = 5 + 4 - 2 = 7$. From Table VI, Appendix B, $t_{.10} = 1.415$. The rejection region is $t < -1.415$.

Since the test statistic falls in the rejection region ($t = -2.39 < -1.415$), H_0 is rejected. There is sufficient evidence to indicate that $\mu_2 > \mu_1$ at $\alpha = .10$.

c.   A small sample confidence interval is needed because $n_1 = 5 < 30$ and $n_2 = 4 < 30$.

For confidence coefficient .90, $\alpha = .10$ and $\alpha/2 = .05$. From Table VI, Appendix B, with df $= n_1 + n_2 - 2 = 5 + 4 - 2 = 7$, $t_{.05} = 1.895$. The 90% confidence interval for $(\mu_1 - \mu_2)$ is:

$$(\bar{x}_1 - \bar{x}_2) \pm t_{.05}\sqrt{s_p^2\left[\dfrac{1}{n_1} + \dfrac{1}{n_2}\right]} \Rightarrow (2.36 - 3.6) \pm 1.895\sqrt{.5989\left[\dfrac{1}{5} + \dfrac{1}{4}\right]}$$

$$\Rightarrow -1.24 \pm .98 \Rightarrow (-2.22, -0.26)$$

d.   The confidence interval in part **c** provides more information about $(\mu_1 - \mu_2)$ than the test of hypothesis in part **b**. The test in part **b** only tells us that μ_2 is greater than μ_1. However, the confidence interval estimates what the difference is between μ_1 and μ_2.

7.7   a.   The test statistic is $z = -1.576$ and the p-value $= .1150$. Since the p-value is not small, there is no evidence to reject H_0 for $\alpha \leq .10$. There is insufficient evidence to indicate the two population means differ for $\alpha \leq .10$.

b. If the alternative hypothesis had been one-tailed, the p-value would be half of the value for the two-tailed test. Here, p-value $= .1150/2 = .0575$.

There is no evidence to reject H_0 for $\alpha = .05$. There is insufficient evidence to indicate the mean for population 1 is less than the mean for population 2 at $\alpha = .05$.

There is evidence to reject H_0 for $\alpha > .0575$. There is sufficient evidence to indicate the mean for population 1 is less than the mean for population 2 at $\alpha > .0575$.

7.9 a. $$s_p^2 = \frac{(n_1 - 1)s_1^2 + (n_2 - 1)s_2^2}{n_1 + n_2 - 2} = \frac{(17 - 1)3.4^2 + (12 - 1)4.8^2}{17 + 12 - 2} = 16.237$$

The test statistic is $$t = \frac{(\bar{x}_1 - \bar{x}_2) - 0}{\sqrt{s_p^2\left[\dfrac{1}{n_1} + \dfrac{1}{n_2}\right]}} = \frac{(5.4 - 7.9) - 0}{\sqrt{16.237\left[\dfrac{1}{17} + \dfrac{1}{12}\right]}} = -1.646$$

The p-value $= P(t \le -1.646) + P(t \ge 1.646) = 2P(t \ge 1.646)$.

Using Table VI with df $= n_1 + n_2 = 17 + 12 - 2 = 27$, $P(t \ge 1.646)$ is between .05 and .10. Thus, $2(.05) < p$-value $< 2(.10)$ or $.10 < p$-value $< .20$.

These values correspond to those found in the printout.

Since the p-value is not small, there is no evidence to reject H_0. There is no evidence to indicate the means are different for $\alpha \le .10$.

b. For confidence coefficient .95, $\alpha = .05$ and $\alpha/2 = .025$. From Table VI, Appendix B, with df $= n_1 + n_2 - 2 = 17 + 12 - 2 = 27$, $t_{.025} = 2.052$. The confidence interval is:

$$(\bar{x}_1 - \bar{x}_2) \pm t_{.025}\sqrt{s_p^2\left[\frac{1}{n_1} + \frac{1}{n_2}\right]} \quad \text{where } t \text{ has 27 df}$$

$$\Rightarrow (5.4 - 7.9) \pm 2.052\sqrt{16.237\left[\frac{1}{17} + \frac{1}{12}\right]} \Rightarrow -2.50 \pm 3.12 \Rightarrow (-5.62, 0.62)$$

7.11 a. Let $\mu_1 = $ mean ingratiatory score for managers and $\mu_2 = $ mean ingratiatory score for clerical personnel. To determine if there is a difference in ingratiatory behavior between managers and clerical personnel, we test:

$H_0: \mu_1 = \mu_2$

$H_a: \mu_1 \ne \mu_2$

b. The test statistic is $$z = \frac{(\bar{x}_1 - \bar{x}_2) - D_o}{\sqrt{\dfrac{s_1^2}{n_1} + \dfrac{s_2^2}{n_2}}} = \frac{(2.41 - 1.90) - 0}{\sqrt{\dfrac{(.74)^2}{288} + \dfrac{(.59)^2}{110}}} = 7.17$$

The rejection region requires $\alpha/2 = .05/2 = .025$ in each tail of the z-distribution. From Table IV, Appendix B, $z_{.025} = 1.96$. The rejection region is $z < -1.96$ or $z > 1.96$.

Since the observed value of the test statistic falls in the rejection region ($z = 7.17 > 1.96$), H_0 is rejected. There is sufficient evidence to indicate a difference in ingratiatory behavior between managers and clerical personnel at $\alpha = .05$.

c.   For confidence coefficient .95, $\alpha = .05$ and $\alpha/2 = .05/2 = .025$. From Table IV, Appendix B, $z_{.025} = 1.96$. The 95% confidence interval is:

$$(\bar{x}_1 - \bar{x}_2) \pm z_{.025}\sqrt{\frac{s_1^2}{n_1} + \frac{s_2^2}{n_2}} \Rightarrow (2.41 - 1.90) \pm 1.96\sqrt{\frac{.74^2}{288} + \frac{.59^2}{110}}$$

$$\Rightarrow .51 \pm .14 \Rightarrow (.37, .65)$$

We are 95% confident that the difference in mean ingratiatory scores between managers and clerical personnel is between .37 and .65. Since this interval does not contain 0, it is consistent with the test of hypothesis which rejected the hypothesis that there was no difference in mean scores for the two groups.

7.13   a.   Let μ_1 = mean age of nonpurchasers and μ_2 = mean age of purchasers.

To determine if there is a difference in the mean age of purchasers and nonpurchasers, we test:

$$H_0: \mu_1 - \mu_2 = 0$$
$$H_a: \mu_1 - \mu_2 \neq 0$$

The test statistic is $t = 1.9557$ (from printout).

The rejection region requires $\alpha/2 = .10/2 = .05$ in each tail of the t-distribution with df $= n_1 + n_2 - 2 = 20 + 20 - 2 = 38$. From Table VI, Appendix B, $t_{.05} \approx 1.684$. The rejection region is $t < -1.684$ or $t > 1.684$.

Since the observed value of the test statistic falls in the rejection region ($t = 1.9557 > 1.684$), H_0 is rejected. There is sufficient evidence to indicate the mean age of purchasers and nonpurchasers differ at $\alpha = .10$.

b.   The necessary assumptions are:

1.   Both sampled populations are approximately normal.
2.   The population variances are equal.
3.   The samples are randomly and independently sampled.

c.   The observed significance level is $p = .0579$. Since the p-value is less than α ($.0579 < .10$), H_0 is rejected. This is the same result as in part **a**.

d. For confidence coefficient .90, $\alpha = 1 - .90 = .10$ and $\alpha/2 = .10/2 = .05$. From Table VI, Appendix B, with df $= 38$, $t_{.05} \approx 1.684$. The confidence interval is:

$$(\bar{x}_2 - \bar{x}_1) \pm t_{.05}\sqrt{s_p^2\left[\frac{1}{n_2} + \frac{1}{n_1}\right]} \Rightarrow (39.8 - 47.2) \pm 1.684\sqrt{143.1684\left[\frac{1}{20} + \frac{1}{20}\right]}$$

$$\Rightarrow -7.4 \pm 6.382 \Rightarrow (-13.772, -1.028)$$

We are 90% confident that the difference in mean ages between purchasers and nonpurchasers is between -13.772 and -1.028.

7.15 a. Yes. The mean wastes for cities of industrialized countries are all greater than 2 while the mean wastes for cities of middle-income countries are all less than 1.0.

b. Let $\mu_1 =$ mean waste for cities in industrialized countries and $\mu_2 =$ mean waste for cities in middle-income countries. To determine if the mean waste generation rates of cities in industrialized and middle-income countries differ, we test:

H_0: $\mu_1 - \mu_2 = 0$
H_a: $\mu_1 - \mu_2 \neq 0$

The test statistic is $t = \dfrac{(\bar{x}_1 - \bar{x}_2) - D_0}{\sqrt{s_p^2\left[\frac{1}{n_1} + \frac{1}{n_2}\right]}} = 19.73$ (from printout)

The rejection region is $t < -2.228$ or $t > 2.228$ (from printout).

Since the observed value of the test statistic falls in the rejection region ($t = 19.73 > 2.228$), H_0 is rejected. There is sufficient evidence to indicate that the mean waste generation rates of cities in industrialized and middle-income countries differ at $\alpha = .05$.

7.17 For each of the characteristics, we wish to determine if there is a difference in mean score between the two groups. Thus, we test:

H_0: $\mu_1 = \mu_2$
H_a: $\mu_1 \neq \mu_2$

For the characteristic "Use of creative ideas", the p-value $< .001$. Since the p-value is so small, we would reject H_0. There is sufficient evidence to indicate a difference in the mean "use of creative ideas" between those who reported positive spillover to family life and those who did not report positive work spillover. Since the sample mean score for those who reported a positive spillover is greater than that for those who did not report a positive spillover, we conclude that those who reported a positive spillover had a higher use of creative ideas.

For the characteristic "Communication", the p-value is between .001 and .01. Since the p-value is so small, we would reject H_0. There is sufficient evidence to indicate a difference in the mean "communication" between those who reported positive spillover to family life and those who did not report positive work spillover. Since the sample mean score for those who reported a positive spillover is greater than that for those who did not report a positive spillover, we conclude that those who reported a positive spillover had a higher use of communication.

For the characteristic "Utilization of information", the p-value $> .05$. Since the p-value is not small, we would not reject H_0. There is insufficient evidence to indicate a difference in the mean "utilization of information" between those who reported positive spillover to family life and those who did not report positive work spillover.

For the characteristic "Participation in decisions", the p-value is between .01 and .05. Since the p-value is so small, we would reject H_0. There is sufficient evidence to indicate a difference in the mean "participation in decisions" between those who reported positive spillover to family life and those who did not report positive work spillover. Since the sample mean score for those who reported a positive spillover is greater than that for those who did not report a positive spillover, we conclude that those who reported a positive spillover had a higher participation in decisions.

For the characteristic "Good use of skills", the p-value $< .001$. Since the p-value is so small, we would reject H_0. There is sufficient evidence to indicate a difference in the mean "good use of skills" between those who reported positive spillover to family life and those who did not report positive work spillover. Since the sample mean score for those who reported a positive spillover is greater than that for those who did not report a positive spillover, we conclude that those who reported a positive spillover had a higher use of skills.

For the characteristic "Age", the p-value $> .05$. Since the p-value is not small, we would not reject H_0. There is insufficient evidence to indicate a difference in the mean "age" between those who reported positive spillover to family life and those who did not report positive work spillover.

For the characteristic "Education", the p-value $> .05$. Since the p-value is not small, we would not reject H_0. There is insufficient evidence to indicate a difference in the mean "education" between those who reported positive spillover to family life and those who did not report positive work spillover.

7.19   a.   Let μ_1 = mean number of cigarettes per week for the treatment group and μ_2 = mean number of cigarettes per week for the control group.

For confidence coefficient .95, $\alpha = .05$ and $\alpha/2 = .025$. From Table VI, Appendix B, with df $= n_1 + n_2 - 2 = 35 + 17 - 2 = 50$, $t_{.025} \approx 2.021$. The confidence interval is:

$$(\bar{x}_1 - \bar{x}_2) \pm t_{.025}\sqrt{s_p^2\left[\frac{1}{n_1} + \frac{1}{n_2}\right]}$$

For Beginning time period:

$$s_p^2 = \frac{(n_1 - 1)s_1^2 + (n_2 - 1)s_2^2}{n_1 + n_2 - 2} = \frac{(35 - 1)71.20^2 + (17 - 1)67.45^2}{35 + 17 - 2} = 4903.06$$

$$\Rightarrow (165.09 - 159.00) \pm 2.021\sqrt{4903.06\left[\frac{1}{35} + \frac{1}{17}\right]}$$

$$\Rightarrow 6.09 \pm 41.835 \Rightarrow (-35.745, 47.925)$$

We are 95% confident that the difference in the mean number of cigarettes smoked per week for the two groups is between -35.745 and 47.925.

For First follow-up period:

$$s_p^2 = \frac{(n_1 - 1)s_1^2 + (n_2 - 1)s_2^2}{n_1 + n_2 - 2} = \frac{(35 - 1)69.08^2 + (17 - 1)66.80^2}{35 + 17 - 2} = 4672.91$$

$$\Rightarrow (105.00 - 157.24) \pm 2.021 \sqrt{4672.91\left[\frac{1}{35} + \frac{1}{17}\right]}$$

$$\Rightarrow -52.24 \pm 40.842 \Rightarrow (-93.082, -11.398)$$

We are 95% confident that the difference in the mean number of cigarettes smoked per week for the two groups is between -93.082 and -11.398.

For Second follow-up period:

$$s_p^2 = \frac{(n_1 - 1)s_1^2 + (n_2 - 1)s_2^2}{n_1 + n_2 - 2} = \frac{(35 - 1)69.08^2 + (17 - 1)65.73^2}{35 + 17 - 2} = 4627.53$$

$$\Rightarrow (111.11 - 159.52) \pm 2.021 \sqrt{4627.53\left[\frac{1}{35} + \frac{1}{17}\right]}$$

$$\Rightarrow -48.41 \pm 40.643 \Rightarrow (-89.053, -7.767)$$

We are 95% confident that the difference in the mean number of cigarettes smoked per week for the two groups is between -89.053 and -7.767.

For Third follow-up period:

$$s_p^2 = \frac{(n_1 - 1)s_1^2 + (n_2 - 1)s_2^2}{n_1 + n_2 - 2} = \frac{(35 - 1)67.59^2 + (17 - 1)64.41^2}{35 + 17 - 2} = 4434.08$$

$$\Rightarrow (120.20 - 157.88) \pm 2.021 \sqrt{4434.08\left[\frac{1}{35} + \frac{1}{17}\right]}$$

$$\Rightarrow -37.68 \pm 39.784 \Rightarrow (-77.464, 2.104)$$

We are 95% confident that the difference in the mean number of cigarettes smoked per week for the two groups is between -77.464 and 2.104.

For Fourth follow-up period:

$$s_p^2 = \frac{(n_1 - 1)s_1^2 + (n_2 - 1)s_2^2}{n_1 + n_2 - 2} = \frac{(35 - 1)74.09^2 + (17 - 1)67.01^2}{35 + 17 - 2} = 5169.65$$

$$\Rightarrow (123.63 - 162.17) \pm 2.021 \sqrt{5169.65\left[\frac{1}{35} + \frac{1}{17}\right]}$$

$$\Rightarrow -38.54 \pm 42.958 \Rightarrow (-81.498, 4.418)$$

We are 95% confident that the difference in the mean number of cigarettes smoked per week for the two groups is between -81.498 and 4.418.

b. For each time period, we must make the following assumptions:

    1. Both populations being sampled from are normal
    2. The two population variances are equal.
    3. Independent random samples are selected from each population.

7.21 a. The rejection region requires $\alpha = .05$ in the upper tail of the t-distribution with $df = n_D - 1 = 12 - 1 = 11$. From Table VI, Appendix B, $t_{.05} = 1.796$. The rejection region is $t > 1.796$.

    b. From Table VI, with $df = n_D - 1 = 24 - 1 = 23$, $t_{.10} = 1.319$. The rejection region is $t > 1.319$.

    c. From Table VI, with $df = n_D - 1 = 4 - 1 = 3$, $t_{.025} = 3.182$. The rejection region is $t > 3.182$.

    d. From Table VI, with $df = n_D - 1 = 8 - 1 = 7$, $t_{.01} = 2.998$. The rejection region is $t > 2.998$.

7.23 Let μ_1 = mean of population 1 and μ_2 = mean of population 2.

    a. H_0: $\mu_D = 0$
        H_a: $\mu_D < 0$ where $\mu_D = \mu_1 - \mu_2$

    b. The test statistic is $t = -5.29$ and the p-value $= .0002$.

        Since the p-value is so small, there is evidence to reject H_0. There is evidence to indicate the mean for population 2 is larger than the mean for population 1 for $\alpha > .0002$.

    c. The confidence interval is $(-5.284, -2.116)$. We are 95% confident the difference in the 2 population means is between -5.284 and -2.116.

    d. We must assume that the population of differences is normal, and the sample of differences is randomly selected.

7.25 Some preliminary calculations:

Pair	Difference $x - y$
1	$55 - 44 = 11$
2	$68 - 55 = 13$
3	$40 - 25 = 15$
4	$55 - 56 = -1$
5	$75 - 62 = 13$
6	$52 - 38 = 14$
7	$49 - 31 = 18$

$$\bar{x}_D = \frac{\sum x_D}{n_D} = \frac{83}{7} = 11.86$$

$$s_D^2 = \frac{\sum x_D^2 - \frac{\left(\sum x_D\right)^2}{n_D}}{n_D - 1} = \frac{1205 - \frac{83^2}{7}}{7 - 1} = 36.8095$$

$$s_D = \sqrt{s_D^2} = \sqrt{36.8095} = 6.0671$$

a. H_0: $\mu_D = 10$
    H_a: $\mu_D \neq 10$ where $\mu_D = (\mu_1 - \mu_2)$

The test statistic is $t = \dfrac{\bar{x}_D - D_0}{s_D/\sqrt{n_D}} = \dfrac{11.86 - 10}{6.0671/\sqrt{7}} = \dfrac{1.86}{2.2931} = .81$

The rejection region requires $\alpha/2 = .05/2 = .025$ in each tail of the t-distribution with df $= n_D - 1 = 7 - 1 = 6$. From Table VI, Appendix B, $t_{.025} = 2.447$. The rejection region is $t < -2.447$ or $t > 2.447$.

Since the observed value of the test statistic does not fall in the rejection region ($t = .81 \not> 2.447$), H_0 is not rejected. There is insufficient evidence to conclude $\mu_D \neq 10$ at $\alpha = .05$.

b. p-value $= P(t \leq -.81) + P(t \geq .81) = 2P(t \geq .81)$

Using Table VI, Appendix B, with df $= 6$, $P(t \geq .81)$ is greater than .10.

Thus, $2P(t \geq .81)$ is greater than .20.

The probability of observing a value of t as large as .81 or as small as $-.81$ if, in fact, $\mu_D = 10$ is greater than .20. We would conclude that there is insufficient evidence to suggest $\mu_D \neq 10$.

7.27   Some preliminary calculations are:

Operator	Difference (Before - After)
1	5
2	3
3	9
4	7
5	2
6	-2
7	-1
8	11
9	0
10	5

$$\bar{x}_D = \frac{\sum x_D}{n_D} = \frac{39}{10} = 3.9$$

$$s_{D^2} = \frac{\sum x_{D^2} - \frac{\left(\sum x_D\right)^2}{n_D}}{n_D - 1} = \frac{319 - \frac{39^2}{10}}{10 - 1} = 18.5444$$

$$s_D = \sqrt{18.5444} = 4.3063$$

a. To determine if the new napping policy reduced the mean number of customer complaints, we test:

$H_0: \mu_D = 0$
$H_a: \mu_D > 0$

The test statistic is $t = \dfrac{\bar{x}_D - 0}{\dfrac{s_D}{\sqrt{n_D}}} = \dfrac{3.9 - 0}{\dfrac{4.3063}{\sqrt{10}}} = 2.864$

The rejection region requires $\alpha = .05$ in the upper tail of the t-distribution with df $= n_D - 1 = 10 - 1 = 9$. From Table VI, Appendix B, $t_{.05} = 1.833$. The rejection region is $t > 1.833$.

Since the observed value of the test statistic falls in the rejection region ($t = 2.864 > 1.833$), H_0 is rejected. There is sufficient evidence to indicate the new napping policy reduced the mean number of customer complaints at $\alpha = .05$.

b. In order for the above test to be valid, we must assume that

1. The population of differences is normal
2. The differences are randomly selected

7.29 a. To determine if on average, the economists were more optimistic about the prospects for low inflation in late 1999 than they were for Spring 2000, we test:

$H_0: \mu_D = 0$
$H_a: \mu_D < 0$

b. Some preliminary calculations are:

Economist	Difference (1999 - 2000)
1	$-.4$
2	0
3	0
4	$-.5$
5	$-.1$
6	$-.5$
7	0
8	$-.3$
9	$-.1$

$$\bar{x}_D = \frac{\sum x_D}{n_D} = \frac{-1.9}{9} = -.211$$

$$s_{D^2} = \frac{\sum x_{D^2} - \frac{\left(\sum x_D\right)^2}{n_D}}{n_D - 1} = \frac{.77 - \frac{(-1.9)^2}{9}}{9 - 1} = .0461$$

$$s_D = \sqrt{.0461} = .2147$$

The test statistic is $t = \dfrac{\bar{x}_D - 0}{\dfrac{s_D}{\sqrt{n_D}}} = \dfrac{-.211 - 0}{\dfrac{.2147}{\sqrt{9}}} = -2.948$

The rejection region requires $\alpha = .05$ in the lower tail of the t-distribution with df $= n_D - 1$ $= 9 - 1 = 8$. From Table VI, Appendix B, $t_{.05} = 1.860$. The rejection region is $t < -1.860$.

Since the observed value of the test statistic falls in the rejection region ($t = -2.948 < -1.860$), H_0 is rejected. There is sufficient evidence to indicate on average, the economists were more optimistic about the prospects for low inflation in late 1999 than they were for Spring 2000 at $\alpha = .05$.

7.31   a.   Let μ_D = mean difference in pupil dilation between pattern 1 and pattern 2.

To determine if the pupil dilation differs for the two patterns, we test:

H_0: $\mu_D = 0$
H_a: $\mu_D \neq 0$

b.   The test statistic is $t = 5.76$ and the p-value $= .000$. Since the p-value is so small, there is strong evidence to reject H_0. There is evidence to indicate that the pupil dilation differs for the two patterns for $\alpha > .0000$.

The p-value is not exactly 0. The p-value $= P(t \leq -5.76) + P(t \geq 5.76)$. Rounded off to 4 decimal places, the p-value is .0000.

The 95% confidence interval is $(.150, .328)$. We are 95% confident that the mean difference in pupil dilation is between .150 and .328. Since both values are greater than zero, there is evidence to indicate the mean pupil dilation for pattern 1 is greater than the mean dilation for pattern 2.

c.   The paired difference design is better. There is much variation in pupil dilation from person to person. By using the paired difference design, we can eliminate the person to person differences.

7.33   Let μ_1 = mean number of swims by male rat pups and μ_2 = mean number of swims by female rat pups. Then $\mu_D = \mu_1 - \mu_2$. To determine if there is a difference in the mean number of swims required by male and female rat pups, we test:

H_0: $\mu_D = 0$
H_a: $\mu_D \neq 0$

The test statistic is $t = 0.46$ (from printout)

The p-value is $p = 0.65$.

Since the p-value is greater than α $(p = .65 > .10)$, H_0 is not rejected. There is insufficient evidence to indicate there is a difference in the mean number of swims required by male and female rat pups at $\alpha = .10$.

7.35 $n_1 = n_2 = \dfrac{(z_{\alpha/2})^2(\sigma_1^2 + \sigma_2^2)}{B^2}$

For confidence coefficient .95, $\alpha = 1 - .95 = .05$ and $\alpha/2 = .05/2 = .025$. From Table IV, Appendix B, $z_{.025} = 1.96$.

$$n_1 = n_2 = \frac{1.96^2(14 + 14)}{1.8^2} = 33.2 \approx 34$$

7.37   For confidence coefficient .95, $\alpha = .05$ and $\alpha/2 = .05/2 = .025$. From Table IV, Appendix B, $z_{.025} = 1.96$.

$$n_1 = n_2 = \frac{(z_{\alpha/2})^2(\sigma_1^2 + \sigma_2^2)}{B^2} = \frac{1.96^2(3.189^2 + 2.355^2)}{1.5^2} = 26.8 \approx 27$$

We would need to sample 27 specimens from each location.

7.39   For confidence coefficient .95, $\alpha = 1 - .95 = .05$ and $\alpha/2 = .05/2 = .025$. From Table IV, Appendix B, $z_{.025} = 1.96$. From Exercise 7.33, $s_D = 3.515$.

$$n_D = \frac{(z_{\alpha/2})^2(\sigma_D^2)}{B^2} = \frac{(1.96)^2(3.515^2)}{1.5^2} = 21.1 \approx 22$$

7.41   For confidence coefficient .95, $\alpha = 1 - .95 = .05$ and $\alpha/2 = .025$. From Table IV, Appendix B, $z_{.025} = 1.96$.

$$n_1 = n_2 = \frac{(z_{\alpha/2})^2(\sigma_1^2 + \sigma_2^2)}{B^2} = \frac{(1.96)^2(35^2 + 80^2)}{10^2} = 292.9 \approx 293$$

7.43   To test H_0: $\sigma_1^2 = \sigma_2^2$ against H_a: $\sigma_1^2 \neq \sigma_2^2$, the rejection region is $F > F_{\alpha/2}$ with $\nu_1 = 10$ and $\nu_2 = 12$.

a. $\alpha = .20$, $\alpha/2 = .10$
      Reject H_0 if $F > F_{.10} = 2.19$ (Table VIII, Appendix B)

b. $\alpha = .10$, $\alpha/2 = .05$
      Reject H_0 if $F > F_{.05} = 2.75$ (Table IX, Appendix B)

c. $\alpha = .05$, $\alpha/2 = .025$
     Reject H_0 if $F > F_{.025} = 3.37$ (Table X, Appendix B)

d. $\alpha = .02$, $\alpha/2 = .01$
     Reject H_0 if $F > F_{.01} = 4.30$ (Table XI, Appendix B)

7.45   a.   To determine if a difference exists between the population variances, we test:

$$H_0: \sigma_1^2 = \sigma_2^2$$
$$H_a: \sigma_1^2 \neq \sigma_2^2$$

The test statistic is $F = \dfrac{s_2^2}{s_1^2} = \dfrac{8.75}{3.87} = 2.26$

The rejection region requires $\alpha/2 = .10/2 = .05$ in the upper tail of the F-distribution with $v_1 = n_2 - 1 = 27 - 1 = 26$ and $v_2 = n_1 - 1 = 12 - 1 = 11$. From Table IX, Appendix B, $F_{.05} \approx 2.60$. The rejection region is $F > 2.60$.

Since the observed value of the test statistic does not fall in the rejection region ($F = 2.26 \not> 2.60$), H_0 is not rejected. There is insufficient evidence to indicate a difference between the population variances.

b.   The p-value is $2P(F \geq 2.26)$. From Tables VIII and IX, with $v_1 = 26$ and $v_2 = 11$,

$$2(.05) < 2P(F \geq 2.26) < 2(.10) \Rightarrow .10 < 2P(F \geq 2.26) < .20$$

There is no evidence to reject H_0 for $\alpha \leq .10$.

7.47   a.   We must assume that the variances of the growth rates of the two sectors are the same.

b.   To determine if the variability of net income growth rates differ for the two industries, we test:

$$H_0: \sigma_1^2 = \sigma_2^2$$
$$H_a: \sigma_1^2 \neq \sigma_2^2$$

c.   The test statistic is $F = \dfrac{s_2^2}{s_1^2} = \dfrac{83.4^2}{15.7^2} = 28.22$

The rejection region requires $\alpha/2 = .05/2 = .025$ in the upper tail of the F-distribution with $v_1 = n_2 - 1 = 9 - 1 = 8$ and $v_2 = n_1 - 1 = 8 - 1 = 7$. From Table X, Appendix B, $F_{.025} = 4.90$. The rejection region is $F > 4.90$.

Since the observed value of the test statistic falls in the rejection region ($F = 28.22 > 4.90$), H_0 is rejected. There is sufficient evidence to indicate the variability of net income growth rates differ for the two industries at $\alpha = .05$.

d.   We must assume that:

1. Both samples populations are normally distributed
2 The samples are random and independent

Comparing Population Means

7.49   a.   Let σ_1^2 = variance in inspection errors for novice inspectors and σ_2^2 = variance in inspection errors for experienced inspectors. Since we wish to determine if the data support the belief that the variance is lower for experienced inspectors than for novice inspectors, we test:

$$H_0: \ \sigma_1^2 = \sigma_2^2$$
$$H_a: \ \sigma_1^2 > \sigma_2^2$$

The test statistic is $F = \dfrac{\text{Larger sample variance}}{\text{Smaller sample variance}} = \dfrac{s_1^2}{s_2^2} = \dfrac{8.643^2}{5.744^2} = 2.26$

The rejection region requires $\alpha = .05$ in the upper tail of the F-distribution with $\nu_1 = n_1 - 1 = 12 - 1 = 11$ and $\nu_2 = n_2 - 1 = 12 - 1 = 11$. From Table IX, Appendix B, $F_{.05} \approx 2.82$ (using interpolation). The rejection region is $F > 2.82$.

Since the observed value of the test statistic does not fall in the rejection region ($F = 2.26 \ngtr 2.82$), H_0 is not rejected. The sample data do not support her belief at $\alpha = .05$.

     b.   The p-value = $P(F \geq 2.26)$ with $\nu_1 = 11$ and $\nu_2 = 11$. Checking Tables VIII, IX, X, and XI in Appendix B, we find $F_{.10} = 2.23$ and $F_{.05} = 2.82$. Since the observed value of F exceeds $F_{.10}$ but is less than $F_{.05}$, the observed significance level for the test is less than .10. So $.05 < p\text{-value} < .10$.

7.51   a.   Let σ_1^2 = variance of the order-to-delivery times for the Persian Gulf War and σ_2^2 = variance of the order-to-delivery times for Bosnia.

To determine if the variances of the order-to-delivery times for the Persian Gulf and Bosnia shipments are equal, we test:

$$H_0: \ \frac{\sigma_1^2}{\sigma_2^2} = 1$$

$$H_a: \ \frac{\sigma_1^2}{\sigma_2^2} \neq 1$$

The test statistic is $F = 8.29$ (from printout).

The p-value is $p = 0.007$ (from printout). Since the p-value is less than α ($p = .007 < .05$), H_0 is rejected. There is sufficient evidence to indicate the variances of the order-to-delivery times for the Persian Gulf and Bosnia shipments differ at $\alpha = .05$.

     b.   No. One assumption necessary for the small sample confidence interval for $(\mu_1 - \mu_2)$ is that $\sigma_1^2 = \sigma_2^2$. For this problem, there is evidence to indicate that $\sigma_1^2 \neq \sigma_2^2$.

7.53   a.   The test statistic is T_2, the rank sum of population 2 (because $n_2 < n_1$).

The rejection region is $T_2 \leq 35$ or $T_2 \geq 67$, from Table XII, Appendix B, with $n_1 = 10$, $n_2 = 6$, and $\alpha = .10$.

b. The test statistic is T_1, the rank sum of population 1 (because $n_1 < n_2$).

The rejection region is $T_1 \geq 43$, from Table XII, Appendix B, with $n_1 = 5$, $n_2 = 7$, and $\alpha = .05$.

c. The test statistic is T_2, the rank sum of population 2 (because $n_2 < n_1$).

The rejection region is $T_2 \geq 93$, from Table XII, Appendix B, with $n_1 = 9$, $n_2 = 8$, and $\alpha = .025$.

d. Since $n_1 = n_2 = 15$, the test statistic is:

$$z = \frac{T_1 - \dfrac{n_1(n_1 + n_2 + 1)}{2}}{\sqrt{\dfrac{n_1 n_2(n_1 + n_2 + 1)}{12}}}$$

The rejection region is $z < -z_{\alpha/2}$ or $z > z_{\alpha/2}$. For $\alpha = .05$ and $\alpha/2 = .05/2 = .025$, $z_{.025} = 1.96$ from Table IV, Appendix B. The rejection region is $z < -1.96$ or $z > 1.96$.

7.55 The Wilcoxon rank sum test is a test of the location (center) of a distribution. The one-tailed test deals specifically with the center of one distribution being shifted in one direction (right or left) from the other distribution. The two-tailed test does not specify a particular direction of shift; we consider the possibility of a shift in either direction.

7.57 a. Some preliminary calculations:

Private Sector	Rank	Public Sector	Rank
2.58	10	5.40	15
5.05	13	2.55	9
0.05	1	9.00	16
2.10	5	10.55	17
4.30	12	1.02	2
2.25	6	5.11	14
2.50	8	12.42	18
1.94	4	1.67	3
2.33	7	3.33	11
	$T_1 = 66$		$T_2 = 105$

To determine if the distribution for public sector organizations is located to the right of the distribution for private sector firms, we test:

H_0: The two sampled populations have identical probability distributions
H_a: The probability distribution of the public sector is located to the right of that for the private sector

The test statistic is $T_2 = 105$.

The null hypothesis will be rejected if $T_2 \geq T_U$ where T_U corresponds to $\alpha = .025$ (one-tailed), and $n_1 = n_2 = 9$. From Table XII, Appendix B, $T_U = 108$. (There is no table for $\alpha = .01$. However, if we do not reject H_0 for $\alpha = .025$, we will not reject H_0 for $\alpha = .01$.)

Reject H_0 if $T_2 \geq 108$.

Since $T_2 = 105 \ngeq 108$, H_0 is not rejected. There is insufficient evidence to indicate that the distribution for public sector organizations is located to the right of the distribution for private sector firms at $\alpha = .01$.

b. The null hypothesis will be rejected if $T_2 \geq T_U$ where T_U corresponds to $\alpha = .05$ (one-tailed), and $n_1 = n_2 = 9$. From Table XII, Appendix B, $T_U = 105$. Since $T_1 = 105$, we would reject H_0. Thus, the p-value is less than or equal to $\alpha = .05$.

c. The assumptions necessary for the test are:

1. The two samples are random and independent.
2. The two probability distributions from which the samples were drawn are continuous.

7.59 a.

American Purchasing Managers		Mexican Purchasing Managers	
Sample 1	Rank	Sample 2	Rank
50	20.5	10	4.5
10	4.5	90	29
35	15.5	65	24
30	13.5	50	20.5
20	10.5	20	10.5
15	7.5	15	7.5
8	3	60	23
40	17.5	80	26.5
80	26.5	85	28
75	25	35	15.5
19	9	5	1.5
11	6	55	22
5	1.5	40	17.5
25	12	45	19
30	13.5	95	30
$T_1 = 186$		$T_2 = 279$	

These rank sums are the same as those found on the printout.

b. To determine whether American and Mexican purchasing managers perceive the given ethical situation differently, we test:

H_0: The two sampled populations have identical probability distributions

H_a: The probability distribution of the American managers is shifted to the right or left of the probability distribution of the Mexican managers.

The test statistic is $z = 1.908$ (from the printout)

The p-value is $p = .0564$. Since the p-value is greater than $\alpha = .05$, H_0 is not rejected. There is insufficient evidence to indicate American and Mexican purchasing managers perceive the given ethical situation differently at $\alpha = .05$.

c.	In order to use the *t*-test, we need to assume that the two populations being sampled from are normal and that the variances of the two populations are equal. To check these assumptions, we will use stem-and-leaf plots and dot plots.

The stem-and-leaf plots are:

```
Stem-and-leaf of Ethics    Managers = 1    N = 15
Leaf Unit = 1.0

     2      0 58
     6      1 0159
    (2)     2 05
     7      3 005
     4      4 0
     3      5 0
     2      6
     2      7 5
     1      8 0


Stem-and-leaf of Ethics    Managers = 2    N = 15
Leaf Unit = 1.0

     1      0 5
     3      1 05
     4      2 0
     5      3 5
     7      4 05
    (2)     5 05
     6      6 05
     4      7
     4      8 05
     2      9 05
```

Neither of these two stem-and-leaf plots look mound-shaped. The assumption that the populations are normal may not be valid.

The dot plots are:

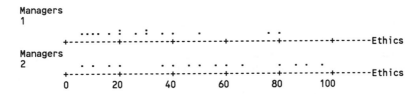

The spread of the two data sets look approximately equal. The assumption that the variances of the two populations are the same appears to be valid.

7.61

U.S. Plants		Japanese Plants	
Observation	Rank	Observation	Rank
7.11	9	3.52	4
6.06	7	2.02	2
8.00	10	4.91	6
6.87	8	3.22	3
4.77	5	1.92	1
	$T_1 = 39$		$T_2 = 16$

To determine if the distribution of American plants is shifted to the right of that for Japanese plants, we test:

H_0: The two sampled population have identical probability distributions
H_a: The probability distribution for American plants is shifted to the right of that for Japanese plants

The test statistic is $T_1 = 39$.
The rejection region is $T_1 \geq 36$ from Table XII, Appendix B, with $n_1 = n_2 = 5$, and $\alpha = .05$.

Since the observed value of the test statistic falls in the rejection region ($T_1 = 39 \geq 36$), H_0 is rejected. There is sufficient evidence to indicate the probability distribution for U.S. plants is shifted to the right of that for Japanese plants at $\alpha = .05$.

This result agrees with that from Exercise 7.18.

7.63 a. We first rank all the data:

Firms with Successful MIS (1)				Firms with Unsuccessful MIS (2)			
Score	Rank	Score	Rank	Score	Rank	Score	Rank
52	5	90	25.5	60	10.5	65	12.5
70	15	75	17	50	4	55	7
40	1.5	80	19	55	7	70	15
80	19	95	29.5	70	15	90	25.5
82	21	90	25.5	41	3	85	22
65	12.5	86	23	40	1.5	80	19
59	9	95	29.5	55	7	90	25.5
60	10.5	93	28				
	$T_1 = 290.5$				$T_2 = 174.5$		

To determine whether the distribution of quality scores for successfully implemented systems lies above that for unsuccessfully implemented systems, we test:

H_0: The two sampled populations have identical probability distributions
H_a: The probability distribution for successful MIS is shifted to the right of that for the unsuccessful MIS

The test statistic is $z = -1.75103$ (from printout).

The rejection region requires $\alpha = .05$ in the upper tail of the z-distribution. From Table IV, Appendix B, $z_{.05} = 1.645$. The rejection region is $z > -1.645$.

Since the observed value of the test statistic falls in the rejection region ($z = -1.75103 < -1.645$), H_0 is rejected. There is sufficient evidence to indicate the distribution of quality scores for successfully or good implemented systems lies above that for the unsuccessfully or poor implemented systems at $\alpha = .05$.

b. We could use the two-sample t-test if:

1. Both populations are normal.
2. The variances of the two populations are the same.

7.65   a.   The hypotheses are:

H_0: The two sampled populations have identical probability distributions
H_a: The probability distributions for population A is shifted to the right of that for population B

b.   Some preliminary calculations are:

| Treatment | | Difference | Rank of Absolute |
A	B	A − B	Difference
54	45	9	5
60	45	15	10
98	87	11	7
43	31	12	9
82	71	11	7
77	75	2	2.5
74	63	11	7
29	30	−1	1
63	59	4	4
80	82	−2	2.5
			$T_- = 3.5$

The test statistic is $T_- = 3.5$

The rejection region is $T_- \leq 8$, from Table XIII, Appendix B, with $n = 10$ and $\alpha = .025$.

Since the observed value of the test statistic falls in the rejection region ($T_- = 3.5 \leq 8$), H_0 is rejected. There is sufficient evidence to indicate the responses for A tend to be larger than those for B at $\alpha = .025$.

7.67   We assume that the probability distribution of differences is continuous so that the absolute differences will have unique ranks. Although tied (absolute) differences can be assigned average ranks, the number of ties should be small relative to the number of observations to assure validity.

7.69   a. H_0: The two sampled populations have identical probability distributions
H_a: The probability distribution for population 1 is located to the right of that for population 2

b.   The test statistic is:

$$z = \frac{T_+ - \dfrac{n(n+1)}{4}}{\sqrt{\dfrac{n(n+1)(2n+1)}{24}}} = \frac{354 - \dfrac{30(30+1)}{4}}{\sqrt{\dfrac{30(30+1)(60+1)}{24}}} = \frac{121.5}{48.6184} = 2.499$$

The rejection region requires $\alpha = .05$ in the upper tail of the z-distribution. From Table IV, Appendix B, $z = 1.645$. The rejection region is $z > 1.645$.

Since the observed value of the test statistic falls in the rejection region ($z = 2.499 > 1.645$), H_0 is rejected. There is sufficient evidence to indicate population 1 is located to the right of that for population 2 at $\alpha = .05$.

c.   The p-value $= P(z \geq 2.499) = .5 - .4938 = .0062$ (using Table IV, Appendix B).

d. The necessary assumptions are:

1. The sample of differences is randomly selected from the population of differences.
2. The probability distribution from which the sample of paired differences is drawn is continuous.

7.71

Operator	1999 Complaints	2000 Complaints	Difference	Rank of Absolute Difference
1	10	5	5	5.5
2	3	0	3	4
3	16	7	9	8
4	11	4	7	7
5	8	6	2	2.5
6	2	4	-2	2.5
7	1	2	-1	1
8	14	3	11	9
9	5	5	0	(eliminated)
10	6	1	5	5.5

Negative rank sum $T_- = 3.5$
Positive rank sum $T_+ = 41.5$

To determine if the distributions of the number of complaints differs for the two time periods, we test:

H_0: The distributions of the number of complaints for the two years are the same
H_a: The distribution of the number of complaints for 2000 is shifted to the right or left of the distribution for 1999

The test statistic is $T_- = 3.5$.

Since no α is given we will use $\alpha = .05$. The null hypothesis will be rejected if $T_- \leq T_0$ where T_0 corresponds to $\alpha = .05$ (two-tailed) and $n = 9$. From Table XIII, Appendix B, $T_0 = 6$.

Reject H_0 if $T_- \leq 6$.

Since the observed value of the test statistic falls in the rejection region ($T_- = 3.5 \leq 6$), H_0 is rejected. There is sufficient evidence to indicate the distributions of the complaints are different for the two years at $\alpha = .05$.

7.73   Some preliminary calculations are:

Employee	Before Flextime	After Flextime	Difference (B − A)	Difference
1	54	68	−14	7
2	25	42	−17	9
3	80	80	0	(Eliminated)
4	76	91	−15	8
5	63	70	−7	5
6	82	88	−6	3.5
7	94	90	4	2
8	72	81	−9	6
9	33	39	−6	3.5
10	90	93	−3	1
				$T_+ = 2$

To determine if the pilot flextime program is a success, we test:

H_0: The two probability distributions are identical
H_a: The probability distribution before is shifted to the left of that after

The test statistic is $T_+ = 2$.

The rejection region is $T_+ \le 8$, from Table XIII, Appendix B, with $n = 9$ and $\alpha = .05$.

Since the observed value of the test statistic falls in the rejection region ($T_+ = 2 \le 8$), H_0 is rejected. There is sufficient evidence to indicate the pilot flextime program has been a success at $\alpha = .05$.

7.75   To determine if one of the measuring facilities tends to read higher or lower than the other, we test:

H_0: The exhalation rate measurements for the two facilities are the same
H_a: The exhalation rate measurements by PCHD are shifted to the right or left of those by EERF

The test statistic is $z = -1.3631$ (from printout).

The p-value is $p = .1728$. Since the p-value is not less than α, ($p = .1728 \not< \alpha = .05$), H_0 is not rejected.

There is insufficient evidence to indicate a difference in the exhalation rate measurements for the two facilities at $\alpha = .05$.

7.77   In the second dot diagram **b**, the difference between the sample means is small relative to the variability within the sample observations. In the first dot diagram **a**, the values in each of the samples are grouped together with a range of 4, while in the second diagram **b**, the range of values is 8.

7.79 For each dot diagram, we want to test:

$$H_0: \mu_1 = \mu_2$$
$$H_a: \mu_1 \neq \mu_2$$

From Exercise 7.78,

Diagram a	Diagram b
$\bar{x}_1 = 9$	$\bar{x}_1 = 9$
$\bar{x}_2 = 14$	$\bar{x}_2 = 14$
$s_1^2 = 2$	$s_1^2 = 14.4$
$s_2^2 = 2$	$s_2^2 = 14.4$

a.

Diagram a	Diagram b
$s_p^2 = \dfrac{s_1^2 + s_2^2}{2}$	$s_p^2 = \dfrac{s_1^2 + s_2^2}{2}$
$= \dfrac{2 + 2}{2} = 2 \quad (n_1 = n_2)$	$= \dfrac{14.4 + 14.4}{2} = 14.4 \quad (n_1 = n_2)$
In Exercise 7.78, MSE $= 2$	In Exercise 7.78, MSE $= 14.4$

The pooled variance for the two-sample t-test is the same as the MSE for the F-test.

b.

Diagram a	Diagram b
$t = \dfrac{\bar{x}_1 - \bar{x}_2}{\sqrt{s_p^2\left[\dfrac{1}{n_1} + \dfrac{1}{n_2}\right]}} = \dfrac{9 - 14}{\sqrt{2\left[\dfrac{1}{6} + \dfrac{1}{6}\right]}}$	$t = \dfrac{\bar{x}_1 - \bar{x}_2}{\sqrt{s_p^2\left[\dfrac{1}{n_1} + \dfrac{1}{n_2}\right]}} = \dfrac{9 - 14}{\sqrt{14.4\left[\dfrac{1}{6} + \dfrac{1}{6}\right]}}$
$= -6.12$	$= -2.28$
In Exercise 7.78, $F = 37.5$	In Exercise 7.78, $F = 5.21$

The test statistic for the F-test is the square of the test statistic for the t-test.

c.

Diagram a	Diagram b
For the t-test, the rejection region requires $\alpha/2 = .05/2 = .025$ in each tail of the t-distribution with df $= n_1 + n_2 - 2 = 6 + 6 - 2 = 10$. From Table VI, Appendix B, $t_{.025} = 2.228$.	For the t-test, the rejection region is the same as Diagram a since we are using the same α, n_1, and n_2 for both tests.

The rejection region is $t < -2.228$ or $t > 2.228$.

In Exercise 7.78, the rejection region for both diagrams using the F-test is $F > 4.96$.

The tabled F value equals the square of the tabled t value.

d.

Diagram a	Diagram b
For the t-test, since the test statistic falls in the rejection region ($t = -6.12 < -2.228$), we would reject H_0. In Exercise 7.78, using the F-test, we rejected H_0.	For the t-test, since the test statistic falls in the rejection region ($t = -2.28 < -2.228$), we would reject H_0. In Exercise 7.78, using the F-test, we rejected H_0.

e. Assumptions for the *t*-test:

1. Both populations have relative frequency distributions that are approximately normal.
2. The two population variances are equal.
3. Samples are selected randomly and independently from the populations.

Assumptions for the *F*-test:

1. Both population probability distributions are normal.
2. The two population variances are equal.
3. Samples are selected randomly and independently from the respective populations.

The assumptions are the same for both tests.

7.81   a. The number of treatments is $3 + 1 = 4$. The total sample size is $37 + 1 = 38$.

b. To determine if the treatment means differ, we test:

H_0: $\mu_1 = \mu_2 = \mu_3 = \mu_4$
H_a: At least two treatment means differ

The test statistic is $F = 14.80$.

The rejection region requires $\alpha = .10$ in the upper tail of the *F*-distribution with $v_1 = p - 1 = 4 - 1 = 3$ and $v_2 = n - p = 38 - 4 = 34$. From Table VIII, Appendix B, $F_{.10} \approx 4.51$. The rejection region is $F > 4.51$.

Since the observed value of the test statistic falls in the rejection region ($F = 14.80 > 4.51$), H_0 is rejected. There is sufficient evidence to indicate differences among the treatment means at $\alpha = .10$.

c. We need the sample means to compare specific pairs of treatment means.

7.83   To determine if the mean ages of the powerful women differ among the three groups, we test:

H_0: $\mu_1 = \mu_2 = \mu_3$
H_a: At least two means differ

The test statistic is $F = 1.62$ (from the printout)

The *p*-value is $p = .209$. Since the *p*-value is not less than any reasonable value of α, H_0 is not rejected. There is insufficient evidence to indicate there is a difference in the mean ages of powerful women among the three groups.

The assumptions necessary include:

1. The samples are selected randomly and independently.

    The first assumption may not be met. The 50 most powerful women in America were selected. These 50 were not randomly selected.

2. The probability distributions of ages for each group are normal.

   Looking at the stem-and-leaf displays that accompany the ANOVA table, each of the three groups of ages look fairly mound shaped. Thus, the assumption of normality is probably valid.

3. The variances of the age probability distributions for each group are equal.

   Looking at the dot plots that accompany the ANOVA table, it appears that the spread of ages for group 3 is much smaller than the spread of ages for the other 2 groups. This indicates that the variances of the three populations may not be the same.

7.85   a.   To determine whether the mean scores of the four groups differ, we test:

H_0: $\mu_1 = \mu_2 = \mu_3 = \mu_4$
H_a: At least two treatment means differ

where μ_i represents the mean of the ith group.

For the variable "Infrequency":

The test statistic is $F = 155.8$.

The rejection region requires $\alpha = .05$ in the upper tail of the F-distribution with $v_1 = p - 1 = 4 - 1 = 3$ and $v_2 = n - p = 278 - 4 = 274$. Using Table IX, Appendix B, $F_{.05} \approx 2.60$. The rejection region is $F > 2.60$.

Since the observed value of the test statistic falls in the rejection region ($F = 155.5 > 2.60$), H_0 is rejected. There is sufficient evidence to indicate the mean scores on the "Infrequency" variable differ among the four groups at $\alpha = .05$.

For the variable "Obvious":

The test statistic is $F = 49.7$.

The rejection region is $F > 2.60$. (See above.)

Since the observed value of the test statistic falls in the rejection region ($F = 49.7 > 2.60$), H_0 is rejected. There is sufficient evidence to indicate the mean scores on the "Obvious" variable differ among the four groups at $\alpha = .05$.

For the variable "Subtle":

The test statistic is $F = 10.3$.

The rejection region is $F > 2.60$. (See above.)

Since the observed value of the test statistic falls in the rejection region ($F = 10.3 > 2.60$), H_0 is rejected. There is sufficient evidence to indicate the mean scores on the "Subtle" variable differ among the four groups at $\alpha = .05$.

For the variable "Obvious-Subtle":

The test statistic is $F = 45.4$.

The rejection region is $F > 2.60$. (See above.)

Since the observed value of the test statistic falls in the rejection region ($F = 45.4 > 2.60$), H_0 is rejected. There is sufficient evidence to indicate the mean scores on the "Obvious-Subtle" variable differ among the four groups at $\alpha = .05$.

For the variable "Dissimulation":

The test statistic is $F = 39.1$.

The rejection region is $F > 2.60$. (See above.)

Since the observed value of the test statistic falls in the rejection region ($F = 39.1 > 2.60$), H_0 is rejected. There is sufficient evidence to indicate the mean scores on the "Dissimulation" variable differ among the four groups at $\alpha = .05$.

b.   No. No information is provided on the sample means. The test of hypotheses performed in part **a** just indicate differences exist, but do not indicate where. Further analysis would be required.

7.87 a.   To determine if the mean level of trust differs among the six treatments, we test:

H_0: $\mu_1 = \mu_2 = \mu_3 = \mu_4 = \mu_5 = \mu_6$
H_a: At least one μ_i differs

b.   The test statistic is $F = 2.21$.

The rejection region requires α in the upper tail of the F-distribution with $\nu_1 = p - 1 = 6 - 1 = 5$ and $\nu_2 = n - p = 237 - 6 = 231$. From Table IX, Appendix B, $F_{.05} \approx 2.21$. The rejection region is $F > 2.21$.

Since the observed value of the test statistic does not fall in the rejection region ($F = 2.21 \not> 2.21$), H_0 is not rejected. There is insufficient evidence to indicate that at least two mean trusts differ at $\alpha = .05$.

c.   We must assume that all six samples are drawn from normal populations, the six population variances are the same, and that the samples are independent.

d.   I would classify this experiment as designed. Each subject was randomly assigned to receive one of the six scenarios.

7.89 a. $s_p^2 = \dfrac{(n_1 - 1)s_1^2 + (n_1 - 1)s_2^2}{n_1 + n_2 - 2} = \dfrac{11(74.2) + 13(60.5)}{12 + 14 - 2} = 66.7792$

$H_0: \mu_1 - \mu_2 = 0$
$H_a: \mu_1 - \mu_2 > 0$

The test statistic is $t = \dfrac{(\bar{x}_1 - \bar{x}_2) - 0}{\sqrt{s_p^2\left[\dfrac{1}{n_1} + \dfrac{1}{n_2}\right]}} = \dfrac{(17.8 - 15.3) - 0}{\sqrt{66.7792\left[\dfrac{1}{12} + \dfrac{1}{14}\right]}} = .78$

The rejection region requires $\alpha = .05$ in the upper tail of the t-distribution with df $= n_1 + n_2 - 2 = 12 + 14 - 2 = 24$. From Table VI, Appendix B, for df $= 24$, $t_{.05} = 1.711$. The rejection region is $t > 1.711$.

Since the observed value of the test statistic does not fall in the rejection region $(0.78 \not> 1.711)$, H_0 is not rejected. There is insufficient evidence to indicate that $\mu_1 > \mu_2$ at $\alpha = .05$.

b. For confidence coefficient .99, $\alpha = .01$ and $\alpha/2 = .01/2 = .005$. From Table VI, Appendix B, with df $= n_1 + n_2 - 2 = 12 + 14 - 2 = 24$, $t_{.005} = 2.797$. The confidence interval is:

$$(\bar{x}_1 - \bar{x}_2) \pm t_{.005}\sqrt{s_p^2\left[\dfrac{1}{n_1} + \dfrac{1}{n_2}\right]} \Rightarrow (17.8 - 15.3) \pm 2.797\sqrt{66.7792\left[\dfrac{1}{12} + \dfrac{1}{14}\right]}$$

$$\Rightarrow 2.50 \pm 8.99 \Rightarrow (-6.49, 11.49)$$

c. For confidence coefficient .99, $\alpha = .01$ and $\alpha/2 = .01/2 = .005$. From Table IV, Appendix B, $z_{.005} = 2.58$.

$$n_1 = n_2 = \dfrac{(z_{\alpha/2})^2(\sigma_1^2 + \sigma_2^2)}{B^2} = \dfrac{(2.58)^2(74.2 + 60.5)}{2^2} = 224.15 \approx 225$$

7.91 a. For confidence coefficient .90, $\alpha = .10$ and $\alpha/2 = .05$. From Table IV, Appendix B, $z_{.05} = 1.645$. The confidence interval is:

$$(\bar{x}_1 - \bar{x}_2) \pm z_{.05}\sqrt{\dfrac{s_1^2}{n_1} + \dfrac{s_2^2}{n_2}} \Rightarrow (12.2 - 8.3) \pm 1.645\sqrt{\dfrac{2.1}{135} + \dfrac{3.0}{148}}$$

$$\Rightarrow 3.90 \pm .31 \Rightarrow (3.59, 4.21)$$

b. $H_0: \mu_1 - \mu_2 = 0$
$H_a: \mu_1 - \mu_2 \neq 0$

The test statistic is $z = \dfrac{(\bar{x}_1 - \bar{x}_2 0) - 0}{\sqrt{\dfrac{s_1^2}{n_1} + \dfrac{s_2^2}{n_2}}} = \dfrac{(12.2 - 8.3) - 0}{\sqrt{\dfrac{2.1}{135} + \dfrac{3.0}{148}}} = 20.60$

The rejection region requires $\alpha/2 = .01/2 = .005$ in each tail of the z-distribution. From Table IV, Appendix B, $z_{.005} = 2.58$. The rejection region is $z < -2.58$ or $z > 2.58$.

Since the observed value of the test statistic falls in the rejection region $(20.60 > 2.58)$, H_0 is rejected. There is sufficient evidence to indicate that $\mu_1 \neq \mu_2$ at $\alpha = .01$.

c. For confidence coefficient .90, $\alpha = .10$ and $\alpha/2 = .05$. From Table IV, Appendix B, $z_{.05} = 1.645$.

$$n_1 = n_2 = \frac{(z_{\alpha/2})^2 (\sigma_1^2 + \sigma_2^2)}{B^2} = \frac{(1.645)^2 (2.1 + 3.0)}{.2^2} = 345.02 \approx 346$$

7.93 a. This is a paired difference experiment.

Pair	Difference (Pop. 1 − Pop. 2)
1	6
2	4
3	4
4	3
5	2

$$\bar{x}_D = \frac{\sum x_D}{n_D} = \frac{19}{5} = 3.8 \qquad s_D^2 = \frac{\sum x_D^2 - \frac{(\sum x_D)^2}{n_D}}{n_D - 1} = \frac{81 - \frac{19^2}{5}}{5 - 1} = 2.2$$

$$s_D = \sqrt{2.2} = 1.4832$$

$$H_0: \mu_D = 0$$
$$H_a: \mu_D \neq 0$$

The test statistic is $t = \dfrac{\bar{x}_D - 0}{s_D/\sqrt{n_D}} = \dfrac{3.8 - 0}{1.4832/\sqrt{5}} = 5.73$

The rejection region requires $\alpha/2 = .05/2 = .025$ in each tail of the t-distribution with df = $n - 1 = 5 - 1 = 4$. From Table VI, Appendix B, $t_{.025} = 2.776$. The rejection region is $t < -2.776$ or $t > 2.776$.

Since the observed value of the test statistic falls in the rejection region ($5.73 > 2.776$), H_0 is rejected. There is sufficient evidence to indicate that the population means are different at $\alpha = .05$.

b. For confidence coefficient .95, $\alpha = .05$ and $\alpha/2 = .025$. Therefore, we would use the same t value as above, $t_{.025} = 2.776$. The confidence interval is:

$$\bar{x}_D \pm t_{\alpha/2} \frac{s_D}{\sqrt{n_D}} \Rightarrow 3.8 \pm 3.8 \pm 2.776 \frac{1.4832}{\sqrt{5}} \Rightarrow 3.8 \pm 1.84 \Rightarrow (1.96, 5.64)$$

c. The sample of differences must be randomly selected from a population of differences which has a normal distribution.

7.95   Use the Wilcoxon signed rank test. Some preliminary calculations are:

Pair	X	Y	Difference	Rank of Absolute Difference
1	19	12	7	3
2	27	19	8	4.5
3	15	7	8	4.5
4	35	25	10	6
5	13	11	2	1.5
6	29	10	19	8
7	16	16	0	(eliminated)
8	22	10	12	7
9	16	18	−2	1.5
				$T_- = 1.5$

To determine if the probability distribution of x is shifted to the right of that for y, we test:

H_0:   The probability distributions are identical for the two variables
H_a: The probability distribution of x is shifted to the right of the probability distribution of y

The test statistic is $T = T_- = 1.5$

Reject H_0 if $T \le T_0$ where T_0 is based on $\alpha = .05$ and $n = 8$ (one-tailed):

Reject H_0 if $T \le 6$ (from Table XIII, Appendix B).

Since the observed value of the test statistic falls in the rejection region ($T = 1.5 \le 6$), reject H_0 at $\alpha = .05$. There is sufficient evidence to conclude that the probability distribution of x is shifted to the right of that for y.

7.97   a.   The data are collected as an independent samples design because five boxes of each size were randomly selected and tested.

b.   Yes. The confidence intervals surrounding each of the means do not overlap. This would indicate that there is a difference in the means for the two sizes.

c.   No. Several of the confidence intervals overlap. This would indicate that the mean compression strengths of the sizes that have intervals that overlap are not significantly different.

7.99   a.   To determine if there is a difference in the mean strength of the two types of shocks, we test:

H_0: $\mu_D = 0$
H_a $\mu_D \ne 0$

The test statistic is $t = 7.679$ (from the printout)

The p-value for the test is $p = .000597$. Since the p-value is less than $\alpha = .05$, H_0 is rejected. There is sufficient evidence to indicate a difference in the mean strength of the two types of shocks at $\alpha = .05$.

b.   The p-value for the test is $p = .000597$. Since the p-value is less than $\alpha = .05$, H_0 is rejected. There is sufficient evidence to indicate a difference in the mean strength of the two types of shocks at $\alpha = .05$.

c. The necessary assumptions are:

1. The population of difference is normal.
2. The differences are randomly and independently selected.

d. For confidence coefficient .95, $\alpha = 1 - .95 = .05$ and $\alpha/2 = .05/2 = .025$. From Table VI, Appendix B, with df $= 5$, $t_{.025} = 2.571$. The confidence interval is:

$$\bar{x}_D \pm t_{.025} \frac{s_D}{\sqrt{n_D}} \Rightarrow .4167 \pm 2.571 \left[\frac{.1329}{\sqrt{6}} \right] \Rightarrow .4167 \pm .1395 \Rightarrow (.2772, .5562)$$

We are 95% confident the difference in mean strength between the manufacturer's shock and that of the competitor's shock is between .2772 and .5562.

e. Some preliminary calculations are:

Car Number	Manufacturer's Shock	Competitor's Shock	Difference A $-$ B	Rank of Absolute Difference
1	8.8	8.4	.4	5
2	10.5	10.1	.4	3
3	12.5	12.0	.5	5
4	9.7	9.3	.4	3
5	9.6	9.0	.6	6
6	13.2	13.0	.2	1
				$T_- = 0$

To determine whether the distribution of the manufacturer's shock is different from the distribution of the competitor's shock, we test:

H_0: The two sampled populations have identical probability distributions
H_a: The probability distribution of the manufacturer's shocks is shifted to the right or left of that of the competitor's shock.

The test statistic is $T_- = 0$

The rejection region is $T_- \leq 1$, from Table XIII, Appendix B, with $n = 6$ and $\alpha = .05$.

Since the observed value of the test statistic falls in the rejection region ($T_- = 0 \leq 1$), H_0 is rejected. There is sufficient evidence to indicate a difference in location between the strengths of the two shocks at $\alpha = .05$.

7.101 a. Let $\mu_1 =$ mean GPA for traditional students and $\mu_2 =$ mean GPA for nontraditional students. To determine whether the mean GPAs of traditional and nontraditional students differ, we test:

H_0: $\mu_1 - \mu_2 = 0$
H_a: $\mu_1 - \mu_2 \neq 0$

b.   The test statistic is $z = \dfrac{(\bar{x}_1 - \bar{x}_2) - D_0}{\sqrt{\dfrac{s_1^2}{n_1} + \dfrac{s_2^2}{n_2}}} = \dfrac{(2.9 - 3.5) - 0}{\sqrt{\dfrac{.5^2}{94} + \dfrac{.5^2}{73}}} = -7.69$

The rejection region requires $\alpha/2 = .01/2 = .005$ in each tail of the z-distribution. From Table IV, Appendix B, $z_{.005} = 2.58$. The rejection region is $z < -2.58$ or $z > 2.58$.

Since the observed value of the test statistic falls in the rejection region ($z = -7.69 < -2.58$), H_0 is rejected. There is sufficient evidence to indicate that the mean GPAs of traditional and nontraditional students differ for $\alpha = .01$.

c.   We must assume that the two samples are randomly and independently selected from the populations of GPAs.

7.103   Let μ_1 = mean initial performance of stayers and μ_2 = mean initial performance of leavers.

To determine if the mean initial performance differs for stayers and leavers, we test:

H_0: $\mu_1 - \mu_2 = 0$
H_a: $\mu_1 - \mu_2 \neq 0$

The test statistic is $z = \dfrac{(\bar{x}_1 - \bar{x}_2) - 0}{\sqrt{\dfrac{s_1^2}{n_1} + \dfrac{s_2^2}{n_2}}} = \dfrac{(3.51 - 3.24) - 0}{\sqrt{\dfrac{.51^2}{174} + \dfrac{.52^2}{355}}} = 5.68$

Since no α is given, we will use $\alpha = .05$. The rejection region requires $\alpha/2 = .05/2 = .025$ in each tail of the z-distribution. For Table IV, Appendix B, $z_{.025} = 1.96$. The rejection region is $z < -1.96$ or $z > 1.96$.

Since the observed value of the test statistic falls in the rejection region ($z = 5.68 > 1.96$), H_0 is rejected. There is sufficient evidence to indicate the mean initial performance differs for stayers and leavers at $\alpha = .05$.

Let μ_1 = mean rate of career advancement of stayers and μ_2 = mean rate of career advancement of leavers.

To determine if the mean rate of career advancement differs for stayers and leavers, we test:

H_0: $\mu_1 - \mu_2 = 0$
H_a: $\mu_1 - \mu_2 \neq 0$

The test statistic is $z = \dfrac{(\bar{x}_1 - \bar{x}_2) - 0}{\sqrt{\dfrac{s_1^2}{n_1} + \dfrac{s_2^2}{n_2}}} = \dfrac{(0.43 - 0.31) - 0}{\sqrt{\dfrac{.20^2}{174} + \dfrac{.31^2}{355}}} = 5.36$

Since no α is given, we will use $\alpha = .05$. The rejection region is $z < -1.96$ or $z > 1.96$ (from above).

Since the observed value of the test statistic falls in the rejection region ($z = 5.36 > 1.96$), H_0 is rejected. There is sufficient evidence to indicate the mean rate of career advancement differs for stayers and leavers at $\alpha = .05$.

Let μ_1 = mean final performance appraisal of stayers and μ_2 = mean final performance appraisal of leavers.

To determine if the mean final performance appraisal differs for stayers and leavers, we test:

$$H_0: \mu_1 - \mu_2 = 0$$
$$H_a: \mu_1 - \mu_2 \neq 0$$

The test statistic is $z = \dfrac{(\bar{x}_1 - \bar{x}_2) - 0}{\sqrt{\dfrac{s_1^2}{n_1} + \dfrac{s_2^2}{n_2}}} = \dfrac{(3.78 - 3.15) - 0}{\sqrt{\dfrac{.62^2}{174} + \dfrac{.68^2}{355}}} = 10.63$

Since no α is given, we will use $\alpha = .05$. The rejection region is $z < -1.96$ or $z > 1.96$ (from above).

Since the observed value of the test statistic falls in the rejection region ($z = 10.63 > 1.96$), H_0 is rejected. There is sufficient evidence to indicate the mean final performance appraisal differs for stayers and leavers at $\alpha = .05$.

7.105 Some preliminary calculations are:

Supervisor	Difference (Pre-test − Post-test)
1	−15
2	1
3	−7
4	−8
5	−4
6	−13
7	−8
8	2
9	−10
10	−7

$$\bar{x}_D = \frac{\sum x_D}{n_D} = \frac{-69}{10} = -6.9$$

$$s_D^2 = \frac{\sum x_D^2 - \dfrac{\left(\sum x_D\right)^2}{n_D}}{n_D - 1} = \frac{741 - \dfrac{(-69)^2}{10}}{10 - 1} = \frac{264.9}{9} = 29.4333$$

$$s_D = \sqrt{29.4333} = 5.4252$$

a. To determine if the training program is effective in increasing supervisory skills, we test:

H_0: $\mu_D = 0$
H_a: $\mu_D < 0$

The test statistic is $t = \dfrac{\bar{x}_D - 0}{\dfrac{s_D}{\sqrt{n_D}}} = \dfrac{-6.9 - 0}{\dfrac{5.4252}{\sqrt{10}}} = -4.02$

The rejection region requires $\alpha = .10$ in the lower tail of the t-distribution with df $= n_D - 1 = 10 - 1 = 9$. From Table VI, Appendix B, $t_{.10} = 1.383$. The rejection region is $t < -1.383$.

Since the observed value of the test statistic falls in the rejection region ($t = -4.02 < -1.383$), H_0 is rejected. There is sufficient evidence to indicate the training program is effective in increasing supervisory skills at $\alpha = .10$.

b. From the printout, the p-value is $p = .0030$. The probability of observing a test statistic of -4.02 or anything lower is $.0030$ when H_0 is true. This is very unusual if H_0 is true. There is evidence to reject H_0 for $\alpha > .003$.

7.107 Some preliminary calculations are:

Before		After	
Observation	Rank	Observation	Rank
10	19	4	5.5
5	8.5	3	3.5
3	3.5	8	16.5
6	12	5	8.5
7	14.5	6	12
11	20	4	5.5
8	16.5	2	2
9	18	5	8.5
6	12	7	14.5
5	8.5	1	1
$T_{Before} = 132.5$		$T_{After} = 77.5$	

To determine if the situation has improved under the new policy, we test:

H_0: The two sampled population probability distributions are identical
H_a: The probability distribution associated with after the policy was instituted is shifted to the left of that before

The test statistic is $T_{Before} = 132.5$.

The rejection region is $T_{Before} \geq 127$ from Table XII, Appendix B, with $n_A = n_B = 10$ and $\alpha = .05$.

Since the observed value of the test statistic falls in the rejection region ($T_{Before} = 132.5 \geq 127$), H_0 is rejected. There is sufficient evidence to indicate the situation has improved under the new policy at $\alpha = .05$.

7.109   a.   For each of the three measures, let μ_1 = mean score for males seeing the first advertisement and μ_2 = mean score for males seeing the second advertisement. Also, let $\mu_D = \mu_1 - \mu_2$. To determine whether the first ad will be more effective when shown to males, we test:

$$H_0:\ \mu_D = 0$$
$$H_a:\ \mu_D > 0$$

b.   This experiment was a paired difference experiment. Each male was shown both advertisements.

c.   Attitude towards the Advertisement:

The p-value = .091. There is no evidence to reject H_0 for α = .05. There is no evidence to indicate the first ad will be more effective when shown to males for α = .05. There is evidence to reject H_0 for α = .10. There is evidence to indicate the first ad will be more effective when shown to males for α = .10.

Attitude toward Brand of Soft Drink:

The p-value = .032. There is evidence to reject H_0 for α > .032. There is evidence to indicate the first ad will be more effective when shown to males for α > .032.

Intention to Purchase the Soft Drink:

The p-value = .050. There is no evidence to reject H_0 for α = .05. There is no evidence to indicate the first ad will be more effective when shown to males for α = .05. There is evidence to reject H_0 for α > .050. There is evidence to indicate the first ad will be more effective when shown to males for α > .050.

d.   We must assume that the sample of differences is randomly selected.

8.1   Remember that $\hat{p}_1$ and $\hat{p}_2$ can be viewed as means of the number of successes per n trials in the respective samples. Therefore, when n_1 and n_2 are large, $\hat{p}_1 - \hat{p}_2$ is approximately normal by the Central Limit Theorem.

8.3   a.   The rejection region requires $\alpha = .01$ in the lower tail of the z-distribution. From Table IV, Appendix B, $z_{.01} = 2.33$. The rejection region is $z < -2.33$.

       b.   The rejection region requires $\alpha = .025$ in the lower tail of the z-distribution. From Table IV, Appendix B, $z_{.025} = 1.96$. The rejection region is $z < -1.96$.

       c.   The rejection region requires $\alpha = .05$ in the lower tail of the z-distribution. From Table IV, Appendix B, $z_{.05} = 1.645$. The rejection region is $z < -1.645$.

       d.   The rejection region requires $\alpha = .10$ in the lower tail of the z-distribution. From Table IV, Appendix B, $z_{.10} = 1.28$. The rejection region is $z < -1.28$.

8.5   For confidence coefficient .95, $\alpha = 1 - .95 = .05$ and $\alpha/2 = .05/2 = .025$. From Table IV, Appendix B, $z_{.025} = 1.96$. The 95% confidence interval for $p_1 - p_2$ is approximately:

   a. $(\hat{p}_1 - \hat{p}_2) \pm z_{\alpha/2}\sqrt{\dfrac{\hat{p}_1\hat{q}_1}{n_1} + \dfrac{\hat{p}_2\hat{q}_2}{n_2}} \Rightarrow (.65 - .58) \pm 1.96\sqrt{\dfrac{.65(1 - .65)}{400} + \dfrac{.58(1 - .58)}{400}}$

   $\Rightarrow .07 \pm .067 \Rightarrow (.003, .137)$

   b. $(\hat{p}_1 - \hat{p}_2) \pm z_{\alpha/2}\sqrt{\dfrac{\hat{p}_1\hat{q}_1}{n_1} + \dfrac{\hat{p}_2\hat{q}_2}{n_2}} \Rightarrow (.31 - .25) \pm 1.96\sqrt{\dfrac{.31(1 - .31)}{180} + \dfrac{.25(1 - .25)}{250}}$

   $\Rightarrow .06 \pm .086 \Rightarrow (-.026, .146)$

   c. $(\hat{p}_1 - \hat{p}_2) \pm z_{\alpha/2}\sqrt{\dfrac{\hat{p}_1\hat{q}_1}{n_1} + \dfrac{\hat{p}_2\hat{q}_2}{n_2}} \Rightarrow (.46 - .61) \pm 1.96\sqrt{\dfrac{.46(1 - .46)}{100} + \dfrac{.61(1 - .61)}{120}}$

   $\Rightarrow -.15 \pm .131 \Rightarrow (-.281, -.019)$

8.7 $\hat{p} = \dfrac{n_1\hat{p}_1 + n_2\hat{p}_2}{n_1 + n_2} = \dfrac{55(.7) + 65(.6)}{55 + 65} = \dfrac{78}{120} = .65$ $\qquad$ $\hat{q} = 1 - \hat{p} = 1 - .65 = .35$

   H_0: $p_1 - p_2 = 0$
   H_a: $p_1 - p_2 > 0$

   The test statistic is $z = \dfrac{(\hat{p}_1 - \hat{p}_2) - 0}{\sqrt{\hat{p}\hat{q}\left[\dfrac{1}{n_1} + \dfrac{1}{n_2}\right]}} = \dfrac{(.7 - .6) - 0}{\sqrt{.65(.35)\left[\dfrac{1}{55} + \dfrac{1}{65}\right]}} = \dfrac{.1}{.08739} = 1.14$

The rejection region requires $\alpha = .05$ in the upper tail of the z-distribution. From Table IV, Appendix B, $z_{.05} = 1.645$. The rejection region is $z > 1.645$.

Since the observed value of the test statistic does not fall in the rejection region ($z = 1.14 \ngtr 1.645$), H_0 is not rejected. There is insufficient evidence to indicate the proportion from population 1 is greater than that for population 2 at $\alpha = .05$.

8.9   a.   Let p_1 = death rate of Operation Crossroads sailors and p_2 = death rate of a comparable group of sailors. The parameter of interest for this problem is $p_1 - p_2$, or the difference in the death rates for the two groups.

        b.   "The increase was not statistically significant" means that even though the sample death rate of Operation Crossroads sailors is 4.6% higher than the sample death rate of a comparable group of sailors, we could not reject the null hypothesis that there is no difference in the death rates of the two groups of soldiers. For the given samples sizes, the test statistic did not fall in the rejection region.

8.11   a.   Let p_{1999} = proportion of adult Americans who would vote for a woman president in 1999 and p_{1975} = proportion of adult Americans who would vote for a woman president in 1975.

        b.   To see if the samples are sufficiently large:

$$\hat{p}_{1999} \pm 3\sigma_{\hat{p}_{1999}} \Rightarrow \hat{p}_{1999} \pm 3\sqrt{\frac{p_{1999}q_{1999}}{n_{1999}}} \Rightarrow \hat{p}_{1999} \pm 3\sqrt{\frac{\hat{p}_{1999}\hat{q}_{1999}}{n_{1999}}} \Rightarrow .92 \pm 3\sqrt{\frac{.92(.08)}{2000}}$$

$$\Rightarrow .92 \pm .02 \Rightarrow (.90, .94)$$

$$\hat{p}_{1975} \pm 3\sigma_{\hat{p}_{1975}} \Rightarrow \hat{p}_{1975} \pm 3\sqrt{\frac{p_{1975}q_{1975}}{n_{1975}}} \Rightarrow \hat{p}_{1975} \pm 3\sqrt{\frac{\hat{p}_{1975}\hat{q}_{1975}}{n_{1975}}} \Rightarrow .73 \pm 3\sqrt{\frac{.73(.27)}{1500}}$$

$$\Rightarrow .73 \pm .03 \Rightarrow (.70, .76)$$

Since both intervals are contained within the interval $(0, 1)$, the normal approximation will be adequate.

        c.   For confidence coefficient .90, $\alpha = .10$ and $\alpha/2 = .10/2 = .05$. From Table IV, Appendix B, $z_{.05} = 1.645$. The 90% confidence interval is:

$$(\hat{p}_1 - \hat{p}_2) \pm z_{.05}\sqrt{\frac{\hat{p}_1\hat{q}_1}{n_1} + \frac{\hat{p}_2\hat{q}_2}{n_2}} \Rightarrow (.92 - .73) \pm 1.645\sqrt{\frac{.92(.08)}{2000} + \frac{.73(.27)}{1500}}$$

$$\Rightarrow .19 \pm .02 \Rightarrow (.17, .21)$$

We are 90% confident that the difference in the proportions of adult Americans who would vote for a woman president between 1999 and 1975 is between .17 and .21.

d.   To see if the samples are sufficiently large:

$$\hat{p}_{1999} \pm 3\sigma_{\hat{p}_{1999}} \Rightarrow \hat{p}_{1999} \pm 3\sqrt{\frac{p_{1999}q_{1999}}{n_{1999}}} \Rightarrow \hat{p}_{1999} \pm 3\sqrt{\frac{\hat{p}_{1999}\hat{q}_{1999}}{n_{1999}}} \Rightarrow .92 \pm 3\sqrt{\frac{.92(.08)}{20}}$$
$$\Rightarrow .92 \pm .18 \Rightarrow (.74, 1.10)$$

$$\hat{p}_{1975} \pm 3\sigma_{\hat{p}_{1975}} \Rightarrow \hat{p}_{1975} \pm 3\sqrt{\frac{p_{1975}q_{1975}}{n_{1975}}} \Rightarrow \hat{p}_{1975} \pm 3\sqrt{\frac{\hat{p}_{1975}\hat{q}_{1975}}{n_{1975}}} \Rightarrow .73 \pm 3\sqrt{\frac{.73(.27)}{50}}$$
$$\Rightarrow .73 \pm .19 \Rightarrow (.54, .92)$$

Since the first interval is not contained within the interval $(0, 1)$, the normal approximation will not be adequate.

8.13   a.   Let p_1 = error rate for supermarkets and p_2 = error rate for department stores. To see if the samples are sufficiently large:

$$\hat{p}_1 \pm 3\sigma_{\hat{p}_1} \Rightarrow \hat{p}_1 \pm 3\sqrt{\frac{p_1 q_1}{n_1}} \Rightarrow \hat{p}_1 \pm 3\sqrt{\frac{\hat{p}_1 \hat{q}_1}{n_1}} \Rightarrow .0347 \pm 3\sqrt{\frac{.0347(.9653)}{800}}$$
$$\Rightarrow .0347 \pm .0194 \Rightarrow (.0153, .0541)$$

$$\hat{p}_2 \pm 3\sigma_{\hat{p}_2} \Rightarrow \hat{p}_2 \pm 3\sqrt{\frac{p_2 q_2}{n_2}} \Rightarrow \hat{p}_2 \pm 3\sqrt{\frac{\hat{p}_2 \hat{q}_2}{n_2}} \Rightarrow .0915 \pm 3\sqrt{\frac{.0915(.9085)}{900}}$$
$$\Rightarrow .0915 \pm .0288 \Rightarrow (.0627, .1203)$$

Since both intervals lie within the interval $(0, 1)$, the normal approximation will be adequate.

b.   For confidence coefficient .98, $\alpha = .02$ and $\alpha/2 = .02/2 = .01$. From Table IV, Appendix B, $z_{.01} = 2.33$. The 95% confidence interval is:

$$(\hat{p}_1 - \hat{p}_2) \pm z_{.01}\sqrt{\frac{\hat{p}_1 \hat{q}_1}{n_1} + \frac{\hat{p}_2 \hat{q}_2}{n_2}}$$
$$\Rightarrow (.0347 - .0915) \pm 2.33\sqrt{\frac{.0347(.9653)}{800} + \frac{.0915(.9085)}{900}} \Rightarrow -.0568 \pm .0270$$
$$\Rightarrow (-.0838, -.0298)$$

We are 98% confident that the difference in the error rates between supermarkets and department stores is between $-.0838$ and $-.0298$.

c.   We must assume that the sample sizes are sufficiently large and that the two samples were independently and randomly selected.

8.15   To determine if there is a difference in the proportions of consumer/commercial and industrial product managers who are at least 40 years old, we could use either a test of hypothesis or a confidence interval. Since we are asked only to determine if there is a difference in the proportions, we will use a test of hypothesis.

Let p_1 = proportion of consumer/commercial product managers at least 40 years old and p_2 = proportion of industrial product managers at least 40 years old.

$$\hat{p}_1 = .40 \qquad \hat{q}_1 = 1 - \hat{p}_1 = 1 - .40 = .60$$

$$\hat{p}_2 = .54 \qquad \hat{q}_2 = 1 - \hat{p}_2 = 1 - .54 = .46$$

$$\hat{p} = \frac{n_1\hat{p}_1 + n_2\hat{p}_2}{n_1 + n_2} = \frac{93(.40) + 212(.54)}{93 + 212} = .497 \qquad \hat{q} = 1 - \hat{p} = 1 - .497 = .503$$

To see if the samples are sufficiently large:

$$\hat{p}_1 \pm 3\sigma_{\hat{p}_1} \Rightarrow \hat{p}_1 \pm 3\sqrt{\frac{p_1 q_1}{n_1}} \Rightarrow \hat{p}_1 \pm 3\sqrt{\frac{\hat{p}_1 \hat{q}_1}{n_1}} \Rightarrow .40 \pm 3\sqrt{\frac{.40(.60)}{93}}$$

$$\Rightarrow .40 \pm .152 \Rightarrow (.248, .552)$$

$$\hat{p}_2 \pm 3\sigma_{\hat{p}_2} \Rightarrow \hat{p}_2 \pm 3\sqrt{\frac{p_2 q_2}{n_2}} \Rightarrow \hat{p}_2 \pm 3\sqrt{\frac{\hat{p}_2 \hat{q}_2}{n_2}} \Rightarrow .54 \pm 3\sqrt{\frac{.54(.46)}{212}}$$

$$\Rightarrow .54 \pm .103 \Rightarrow (.437, .643)$$

Since both intervals lie within the interval (0, 1), the normal approximation will be adequate.

To determine if there is a difference in the proportions of consumer/commercial and industrial product managers who are at least 40 years old, we test:

$$H_0: \ p_1 - p_2 = 0$$
$$H_a: \ p_1 - p_2 \neq 0$$

The test statistic is $z = \dfrac{(\hat{p}_1 - \hat{p}_2) - 0}{\sqrt{\hat{p}\hat{q}\left[\dfrac{1}{n_1} + \dfrac{1}{n_2}\right]}} = \dfrac{(.40 - .54) - 0}{\sqrt{.497(.503)\left[\dfrac{1}{93} + \dfrac{1}{212}\right]}} = -2.25$

We will use $\alpha = .05$. The rejection region requires $\alpha/2 = .05/2 = .025$ in each tail of the z-distribution. From Table IV, Appendix B, $z_{.025} = 1.96$. The rejection region is $z < -1.96$ or $z > 1.96$.

Since the observed value of the test statistic falls in the rejection region ($z = -2.25 < -1.96$), H_0 is rejected. There is sufficient evidence to indicate that there is a difference in the proportions of consumer/commercial and industrial product managers who are at least 40 years old at $\alpha = .05$.

Since the test statistic is negative, there is evidence to indicate that the industrial product managers tend to be older than the consumer/commercial product managers.

8.17   a.   For confidence coefficient .99, $\alpha = 1 - .99 = .01$ and $\alpha/2 = .01/2 = .005$. From Table IV, Appendix B, $z_{.005} = 2.58$.

$$n_1 = n_2 = \frac{(z_{\alpha/2})^2(p_1 q_1 + p_2 q_2)}{B^2} = \frac{2.58^2(.4(1 - .4) + .7(1 - .7))}{.01^2} = \frac{2.99538}{.0001}$$

$$= 29,953.8 \approx 29,954$$

b. For confidence coefficient .90, $\alpha = 1 - .90 = .10$ and $\alpha/2 = .10/2 = .05$. From Table IV, Appendix B, $z_{.05} = 1.645$. Since we have no prior information about the proportions, we use $p_1 = p_2 = .5$ to get a conservative estimate. For a width of .05, the bound is .025.

$$n_1 = n_2 = \frac{(z_{\alpha/2})^2 (p_1 q_1 + p_2 q_2)}{B^2} = \frac{(1.645)^2 (.5(1 - .5) + .5(1 - .5))}{.025^2} = 2164.82 \approx 2165$$

c. From part b, $z_{.05} = 1.645$.

$$n_1 = n_2 = \frac{(z_{\alpha/2})^2 (p_1 q_1 + p_2 q_2)}{B^2} = \frac{(1.645)^2 (.2(1 - .2) + .3(1 - .3))}{.03^2} = \frac{1.00123}{.0009}$$
$$= 1112.48 \approx 1113$$

8.19 a. For confidence coefficient .80, $\alpha = 1 - .80 = .20$ and $\alpha/2 = .20/2 = .10$. From Table IV, Appendix B, $z_{.10} = 1.28$. Since we have no prior information about the proportions, we use $p_1 = p_2 = .5$ to get a conservative estimate. For a width of .06, the bound is .03.

$$n_1 = n_2 = \frac{(z_{\alpha/2})^2 (p_1 q_1 + p_2 q_2)}{B^2} = \frac{(1.28)^2 (.5(1 - .5) + .5(1 - .5))}{.03^2} = 910.22 \approx 911$$

b. For confidence coefficient .90, $\alpha = 1 - .90 = .10$ and $\alpha/2 = .10/2 = .05$. From Table IV, Appendix B, $z_{.05} = 1.645$. Using the formula for the sample size needed to estimate a proportion,

$$n = \frac{(z_{\alpha/2})^2 pq}{B^2} = \frac{1.645^2 (.5(1 - .5))}{.02^2} = \frac{.6765}{.0004} = 1691.27 \approx 1692$$

No, the sample size from part a is not large enough.

8.21 For confidence coefficient .90, $\alpha = 1 - .90 = .10$ and $\alpha/2 = .10/2 = .05$. From Table IV, Appendix B, $z_{.05} = 1.645$. Since we want $n_1 = 2n_2$,

$$z_{\alpha/2} \sqrt{\frac{p_1 q_1}{n_1} + \frac{p_2 q_2}{n_2}} = B$$
$$\Rightarrow \frac{p_1 q_1}{2n_2} + \frac{p_2 q_2}{n_2} = \frac{B^2}{z_{\alpha/2}^2}$$
$$\Rightarrow \frac{p_1 q_1 + 2p_2 q_2}{2n_2} = \frac{B^2}{z_{\alpha/2}^2}$$
$$\Rightarrow n_2 = \frac{z_{\alpha/2}^2 (p_1 q_1 + 2p_2 q_2)}{B^2(2)} = \frac{1.645^2 (.2(.8) + 2(.2)(.8))}{.05^2(2)} = 259.78 \approx 260$$

Thus, $n_1 = 2n_2 = 2(260) = 520$ and $n_2 = 260$.

8.23 a. With df $= 10$, $\chi_{.05}^2 = 18.3070$

b. With df $= 50$, $\chi_{.990}^2 = 29.7067$

c. With df $= 16$, $\chi^2_{.10} = 23.5418$

d. With df $= 50$, $\chi^2_{.005} = 79.4900$

8.25 a. The characteristics of the multinomial experiment are:

  1. The experiment consists of n identical trials.
  2. There are k possible outcomes to each trial.
  3. The probabilities of the k outcomes, denoted $p_1, p_2, \ldots, p_k$, remain the same from trial to trial, where $p_1 + p_2 + \cdots + p_k = 1$.
  4. The trials are independent.
  5. The random variables of interest are the counts $n_1, n_2, \ldots, n_k$ in each of the k cells.

  The characteristics of the binomial are the same as those for the multinomial with $k = 2$.

  b. The sample size n will be large enough so that, for every cell, the expected cell count, $E(n_i)$, will be equal to 5 or more.

8.27 Some preliminary calculations are:

  If the probabilities are the same, $p_{1,0} = p_{2,0} = p_{3,0} = p_{4,0} = .25$

  $E(n_1) = np_{1,0} = 205(.25) = 51.25$
  $E(n_2) = E(n_3) = E(n_4) = 205(.25) = 51.25$

  a. To determine if the multinomial probabilities differ, we test:

  H_0: $p_1 = p_2 = p_3 = p_4 = .25$
  H_a: At least one of the probabilities differs from .25

  The test statistic is $\chi^2 = \sum \dfrac{[n_i - E(n_i)]^2}{E(n_i)}$

  $$= \frac{(43 - 51.25)^2}{51.25} + \frac{(56 - 51.25)^2}{51.25} + \frac{(59 - 51.25)^2}{51.25} + \frac{(47 - 51.25)^2}{51.25} = 3.293$$

  The rejection region requires $\alpha = .05$ in the upper tail of the χ^2 distribution with df $= k - 1$ $= 4 - 1 = 3$. From Table VII, Appendix B, $\chi^2_{.05} = 7.81473$. The rejection region is $\chi^2 > 7.81473$.

  Since the observed value of the test statistic does not fall in the rejection region ($\chi^2 = 3.293 \not> 7.81473$), H_0 is not rejected. There is insufficient evidence to indicate the multinomial probabilities differ at $\alpha = .05$.

  b. The Type I error is concluding the multinomial probabilities differ when, in fact, they do not.

  The Type II error is concluding the multinomial probabilities are equal, when, in fact, they are not.

8.29   a. $E(n_1) = np_{1,0} = 370(.30) = 111$

$E(n_2) = np_{2,0} = 370(.20) = 74$

$E(n_3) = np_{3,0} = 370(.20) = 74$

$E(n_4) = np_{4,0} = 370(.10) = 37$

$E(n_5) = np_{5,0} = 370(.10) = 37$

$E(n_6) = np_{6,0} = 370(.10) = 37$

b.   The test statistic is $\chi^2 = \sum \dfrac{[n_i - E(n_i)]^2}{E(n_i)}$

$$= \frac{(84 - 111)^2}{111} + \frac{(79 - 74)^2}{74} + \frac{(75 - 74)^2}{74} + \frac{(49 - 37)^2}{37}$$

$$+ \frac{(36 - 37)^2}{37} + \frac{(47 - 37)^2}{37} = 13.541$$

c.   To determine if the true percentages of the colors produced differ from the manufacturer's stated percentages, we test:

H_0: $p_1 = .30$, $p_2 = .20$, $p_3 = .20$, $p_4 = .10$, $p_5 = .10$, $p_6 = .10$
H_a: At least one p_i does not equal its hypothesized value.

The test statistic is $\chi^2 = 13.541$.

The rejection region requires $\alpha = .05$ in the upper tail of the χ^2 distribution with df $= k - 1$ $= 6 - 1 = 5$. From Table VII, Appendix B, $\chi^2_{.05} = 11.0705$. The rejection region is $\chi^2 > 11.0705$.

Since the observed value of the test statistic falls in the rejection region ($\chi^2 = 13.541 > 11.0705$), H_0 is rejected. There is sufficient evidence to indicate the true percentages of the colors produced differ from the manufacturer's stated percentages at $\alpha = .05$.

8.31   a.   To determine if the opinions are not evenly divided on the issue of national health insurance, we test:

H_0: $p_1 = p_2 = p_3 = 1/3$
H_a: At least one p_i differs from its hypothesized value.

The test statistic is $\chi^2 = 87.74$ (from the printout)

The observed p-value is $p = .0000$. Since the observed p-value is less than α ($p = .0000 < \alpha = .01$), H_0 is rejected. There is sufficient evidence to indicate the opinions are not evenly divided on the issue of national health insurance at $\alpha = .01$.

b. Let p_1 = proportion of heads of household in the U.S. population that favor national health insurance. Some preliminary calculations are:

$$\hat{p}_1 = \frac{n_1}{n} = \frac{234}{434} = .539$$

For confidence coefficient .95, $\alpha = .05$ and $\alpha/2 = .05/2 = .025$. From Table IV, Appendix B, $z_{.025} = 1.96$. The 95% confidence interval is:

$$\hat{p}_1 \pm z_{.025}\sqrt{\frac{\hat{p}_1(1 - \hat{p}_1)}{n}} \Rightarrow .539 \pm 1.96\sqrt{\frac{.539(1 - .539)}{434}}$$
$$\Rightarrow .539 \pm .047 \Rightarrow (.492, .586)$$

We are 95% confident that the true proportion of heads of household in the U.S. population that favor national health insurance is between .492 and .586.

8.33 a. To determine if the opinions of Internet users are evenly divided among the four categories, we test:

H_0: $p_1 = p_2 = p_3 = p_4 = .25$
H_a: At least one $p_i \neq .25$, for $i = 1, 2, 3, 4$

b. Some preliminary calculations are:

$E(n_1) = np_{1,0} = 328(.25) = 82$
$E(n_2) = E(n_3) = E(n_4) = 328(.25) = 82$

The test statistic is $\chi^2 = \sum \frac{[n_i - E(n_i)]^2}{E(n_i)}$

$$= \frac{(59 - 82)^2}{82} + \frac{(108 - 82)^2}{82} + \frac{(82 - 82)^2}{82} + \frac{(79 - 82)^2}{82} = 14.805$$

The rejection region requires $\alpha = .05$ in the upper tail of the χ^2 distribution with df $= k - 1$ $= 4 - 1 = 3$. From Table VII, Appendix B, $\chi^2_{.05} = 7.81473$. The rejection region is $\chi^2 > 7.81473$.

Since the observed value of the test statistic falls in the rejection region ($\chi^2 = 14.805 > 7.81473$), H_0 is rejected. There is sufficient evidence to indicate that the opinions of Internet users are not evenly divided among the four categories at $\alpha = .05$.

c. A Type I error would be to conclude that the opinions of Internet users are not evenly divided among the four categories when, in fact, they are evenly divided.

A Type II error would be to conclude that the opinions of Internet users are evenly divided among the four categories when, in fact, they are not evenly divided.

d. We must assume that:

1. A multinomial experiment was conducted. This is generally satisfied by taking a random sample from the population of interest.
2. The sample size n will be large enough so that, for every cell, the expected cell count, $E(n_i)$, will be equal to 5 or more.

8.35   To determine if the true percentages of ADEs in the five "cause" categories are different, we test:

H_0: $p_1 = p_2 = p_3 = p_4 = p_5 = .2$
H_a: At least one p_i differs from .2, $i = 1, 2, 3, 4, 5$

The test statistic is $\chi^2 = 16$ (from printout).

The p-value of the test is $p = .003019$.

Since the p-value is less than α ($p = .003019 < .10$), H_0 is rejected. There is sufficient evidence to indicate that at least one percentage of ADEs in the five "cause" categories is different at $\alpha = .10$.

8.37   a.   To determine if the number of overweight trucks per week is distributed over the 7 days of the week in direct proportion to the volume of truck traffic, we test:

H_0: $p_1 = .191, p_2 = .198, p_3 = .187, p_4 = .180, p_5 = .155, p_6 = .043, p_7 = .046$
H_a: At least one of the probabilities differs from the hypothesized value

$E(n_1) = np_{1,0} = 414(.191) = 79.074$
$E(n_2) = np_{2,0} = 414(.198) = 81.972$
$E(n_3) = np_{3,0} = 414(.187) = 77.418$
$E(n_4) = np_{4,0} = 414(.180) = 74.520$
$E(n_5) = np_{5,0} = 414(.155) = 64.170$
$E(n_6) = np_{6,0} = 414(.043) = 17.802$
$E(n_7) = np_{7,0} = 414(.046) = 19.044$

The test statistic is $\chi^2 = \sum \dfrac{[n_i - E(n_i)]^2}{E(n_i)} = \dfrac{(90 - 79.074)^2}{79.074} + \dfrac{(82 - 81.972)^2}{81.972}$

$+ \dfrac{(72 - 77.418)^2}{77.418} + \dfrac{(70 - 74.520)^2}{74.520} + \dfrac{(51 - 64.170)^2}{64.170} + \dfrac{(18 - 17.802)^2}{17.802}$

$+ \dfrac{(31 - 19.044)^2}{19.044} = 12.374$

The rejection region requires $\alpha = .05$ in the upper tail of the χ^2 distribution with df $= k - 1 = 7 - 1 = 6$. From Table VII, Appendix B, $\chi^2_{.05} = 12.5916$. The rejection region is $\chi^2 > 12.5916$.

Since the observed value of the test statistic does not fall in the rejection region ($\chi^2 = 12.374$ ≯ 12.5916), H_0 is not rejected. There is insufficient evidence to indicate the number of overweight trucks per week is distributed over the 7 days of the week is not in direct proportion to the volume of truck traffic at $\alpha = .05$.

b.   The p-value is $P(\chi^2 \geq 12.374)$. From Table VII, Appendix B, with df $= k - 1 = 7 - 1 = 6$, $.05 < P(\chi^2 \geq 12.374) < .10$.

8.39   a. H_0:   The row and column classifications are independent
H_a: The row and column classifications are dependent

b. The test statistic is $\chi^2 = \sum \sum \dfrac{\left[n_{ij} - \hat{E}(n_{ij})\right]^2}{\hat{E}(n_{ij})}$

The rejection region requires $\alpha = .01$ in the upper tail of the χ^2 distribution with df = $(r - 1)(c - 1) = (2 - 1)(3 - 1) = 2$. From Table VII, Appendix B, $\chi^2_{.01} = 9.21034$. The rejection region is $\chi^2 > 9.21034$.

c. The expected cell counts are:

$$\hat{E}(n_{11}) = \frac{r_1 c_1}{n} = \frac{96(25)}{167} = 14.37 \qquad E(n_{21}) = \frac{r_2 c_1}{n} = \frac{71(25)}{167} = 10.63$$

$$\hat{E}(n_{12}) = \frac{r_1 c_2}{n} = \frac{96(64)}{167} = 36.79 \qquad \hat{E}(n_{22}) = \frac{r_2 c_2}{n} = \frac{71(64)}{167} = 27.21$$

$$\hat{E}(n_{13}) = \frac{r_1 c_3}{n} = \frac{96(78)}{167} = 44.84 \qquad \hat{E}(n_{23}) = \frac{r_2 c_3}{n} = \frac{71(78)}{167} = 33.16$$

d. The test statistic is $\chi^2 = \sum \sum \dfrac{\left[n_{ij} - \hat{E}(n_{ij})\right]^2}{\hat{E}(n_{ij})}$

$$= \frac{(9 - 14.37)^2}{14.37} + \frac{(34 - 36.79)^2}{36.79} + \frac{(53 - 44.84)^2}{44.84} + \frac{(16 - 10.63)^2}{10.63}$$

$$+ \frac{(30 - 27.21)^2}{27.21} + \frac{(25 - 33.16)^2}{33.16} = 8.71$$

Since the observed value of the test statistic does not fall in the rejection region ($\chi^2 = 8.71$ $\nrightarrow$ 9.21034), H_0 is not rejected. There is insufficient evidence to indicate the row and column classifications are dependent at $\alpha = .01$.

e. To convert the frequencies to percentages, divide the numbers in each column by the column total and multiply by 100. Also, divide the row totals by the overall total and multiply by 100. The column totals are 25, 64, and 78, while the row totals are 96 and 71. The overall sample size is 165. The table of percentages are:

		Column		
	1	**2**	**3**	
Row 1	$\frac{9}{25} \cdot 100 = 36\%$	$\frac{34}{64} \cdot 100 = 53.1\%$	$\frac{53}{78} \cdot 100 = 67.9\%$	$\frac{96}{167} \cdot 100 = 57.5\%$
2	$\frac{16}{25} \cdot 100 = 64\%$	$\frac{30}{64} \cdot 100 = 46.9\%$	$\frac{25}{78} \cdot 100 = 32.1\%$	$\frac{71}{167} \cdot 100 = 42.5\%$

f.

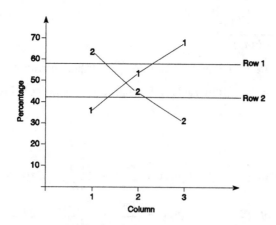

g. If the rows and columns are independent, the row percentages in each column would be close to the row total percentages. This pattern is not evident in the plot, implying the rows and columns are not independent.

8.41   Some preliminary calculations:

$$\hat{E}(n_{11}) = \frac{r_1 c_1}{n} = \frac{2,359(1,712)}{5,026} = 803.543 \qquad \hat{E}(n_{12}) = \frac{r_1 c_2}{n} = \frac{2,359(3,314)}{5,026} = 1,555.457$$

$$\hat{E}(n_{21}) = \frac{r_2 c_1}{n} = \frac{2,667(1,712)}{5,026} = 908.457 \qquad \hat{E}(n_{22}) = \frac{r_2 c_2}{n} = \frac{2,667(3,314)}{5,026} = 1,758.543$$

To determine if travelers who use the Internet to search for travel information are likely to be people who are college educated, we test:

H_0: Education and use of Internet for travel information are independent
H_a: Education and use of Internet for travel information are dependent

The test statistic is $\chi^2 = \sum \frac{\left[n_i - \hat{E}(n_i)\right]^2}{\hat{E}(n_i)}$

$$= \frac{(1,072 - 803.543)^2}{803.543} + \frac{(1,287 - 1,555.457)^2}{1,555.457} + \frac{(640 - 908.547)^2}{908.457} + \frac{(2,027 - 1,758.543)^2}{1,758.543}$$
$$= 256.336$$

The rejection region requires $\alpha = .05$ in the upper tail of the χ^2 distribution with df $= (r - 1)(c - 1) = (2 - 1)(2 - 1) = 1$. From Table VII, Appendix B, $\chi^2_{.05} = 3.84146$. The rejection region is $\chi^2 > 3.84146$.

Since the observed value of the test statistic falls in the rejection region ($\chi^2 = 256.336 > 3.814146$), H_0 is rejected. There is sufficient evidence to indicate that travelers who use the Internet to search for travel information and level of education are dependent at $\alpha = .05$. Since the proportion of college educated who use the Internet to search for travel information (1072/2359 = .45) is greater than the proportion of less than college educated (640/2667 = .24), the conclusion supports the researchers claim that travelers who use the Internet to search for travel information are likely to be people who are college educated.

The necessary assumptions are
    1.   The k observed counts are a random sample from the populations of interest.
    2.   The sample size, n, will be large enough so that, for every cell, the expected count, $E(n_{ij})$, will be equal to 5 or more.

8.43   a.   The sample proportion of injured Hispanic children who were not wearing seatbelts during the accident is:

$$\hat{p} = 283/314 = .901$$

b.   The sample proportion of injured non-Hispanic white children who were not wearing seatbelts during the accident is:

$$\hat{p} = 330/478 = .690$$

c.   Since the proportion of injured Hispanic children who were not wearing seatbelts during the accident (.901) is .211 higher than the proportion of injured non-Hispanic white children who were not wearing seatbelts during the accident (.690), the proportions probably differ.

d.   Some preliminary calculations are:

$$\hat{E}(n_{11}) = \frac{r_1 c_1}{n} = \frac{179(314)}{792} = 70.97 \qquad \hat{E}(n_{12}) = \frac{r_1 c_2}{n} = \frac{179(478)}{792} = 108.03$$

$$\hat{E}(n_{21}) = \frac{r_2 c_1}{n} = \frac{613(314)}{792} = 243.03 \qquad \hat{E}(n_{22}) = \frac{r_2 c_2}{n} = \frac{613(478)}{792} = 369.97$$

To determine whether seatbelt usage in motor vehicle accidents depends on ethnic status in the San Diego County Regionalized Trauma System, we test:

H_0:   Seatbelt usage in motor vehicle accidents and ethnic status in the San Diego County Regionalized Trauma System are independent

H_a: Seatbelt usage in motor vehicle accidents and ethnic status in the San Diego County Regionalized Trauma System are dependent

The test statistic is $\chi^2 = \sum\sum \dfrac{\left[n_{ij} - \hat{E}(n_{ij})\right]^2}{\hat{E}(n_{ij})^2}$

$$= \frac{(31 - 70.97)^2}{70.97} + \frac{(148 - 108.03)^2}{108.03} + \frac{(283 - 243.03)^2}{243.03} + \frac{(330 - 369.97)^2}{369.97} = 48.191$$

The rejection region requires $\alpha = .01$ in the upper tail of the χ^2 distribution with df = $(r - 1)(c - 1) = (2 - 1)(2 - 1) = 1$. From Table VII, Appendix B, $\chi^2_{.01} = 6.63490$. The rejection region is $\chi^2 > 6.63490$.

Since the observed value of the test statistic falls in the rejection region ($\chi^2 = 48.191 > 6.63490$), H_0 is rejected. There is sufficient evidence to indicate seatbelt usage in motor vehicle accidents depends on ethnic status in the San Diego County Regionalized Trauma System at $\alpha = .01$.

e.   For confidence coefficient .99, $\alpha = .01$ and $\alpha/2 = .01/2 = .005$. From Table IV, Appendix B, $z_{.005} = 2.58$. The confidence interval is:

$$(\hat{p}_1 - \hat{p}_2) \pm z_{.005}\sqrt{\frac{\hat{p}_1 \hat{q}_1}{n_1} + \frac{\hat{p}_2 \hat{q}_2}{n_2}} \Rightarrow (.901 - .690) \pm 2.58\sqrt{\frac{.901(.099)}{314} + \frac{.690(.310)}{478}}$$

$$\Rightarrow .211 \pm .070 \Rightarrow (.141, .281)$$

We are 99% confident that the difference in the proportion of injured Hispanic children who were not wearing seatbelts and the proportion of injured non-Hispanic white children who were not wearing seatbelts is between .141 and .281.

8.45   a.   Some preliminary calculations:

$$\hat{E}(n_{11}) = \frac{r_1 c_1}{n} = \frac{45(42)}{160} = 11.8 \quad \hat{E}(n_{12}) = \frac{r_1 c_2}{n} = \frac{45(30)}{160} = 8.4$$

$$\hat{E}(n_{13}) = \frac{r_1 c_3}{n} = \frac{45(88)}{160} = 24.8 \quad \hat{E}(n_{21}) = \frac{r_2 c_1}{n} = \frac{32(42)}{160} = 8.4$$

$$\hat{E}(n_{22}) = \frac{r_2 c_2}{n} = \frac{32(30)}{160} = 6.0 \quad \hat{E}(n_{23}) = \frac{r_2 c_3}{n} = \frac{32(88)}{160} = 17.6$$

$$\hat{E}(n_{31}) = \frac{r_3 c_1}{n} = \frac{83(42)}{160} = 21.8 \quad \hat{E}(n_{32}) = \frac{r_3 c_2}{n} = \frac{83(30)}{160} = 15.6$$

$$\hat{E}(n_{33}) = \frac{r_3 c_3}{n} = \frac{83(88)}{160} = 45.7$$

b.   To determine whether the movie reviews of the two critics are independent, we test:

H_0: The reviews of the two critics are independent
H_a: The reviews of the two critics are dependent

The test statistic is $\chi^2 = 45.357$ (from the printout)

The p-value is $p = .000$. Since the p-value is less than $\alpha = .01$, H_0 is rejected. There is sufficient evidence to indicate the movie reviews of the two critics are dependent at $\alpha = .01$.

8.47   a.   Some preliminary calculations are:

$$\hat{E}(n_{11}) = \frac{r_1 c_1}{n} = \frac{136(150)}{300} = 68 \qquad \hat{E}(n_{21}) = \frac{r_2 c_1}{n} = \frac{164(150)}{300} = 82$$

$$\hat{E}(n_{12}) = \frac{r_1 c_2}{n} = \frac{136(150)}{300} = 68 \qquad \hat{E}(n_{22}) = \frac{r_2 c_2}{n} = \frac{164(150)}{300} = 82$$

To determine whether audience gender and product identification are dependent factors for male spokespersons, we test:

H_0: Audience gender and product identification are independent factors
H_a: Audience gender and product identification are dependent factors

The test statistic is $\chi^2 = \sum\sum \dfrac{\left[n_{ij} - \hat{E}(n_{ij})\right]^2}{\hat{E}(n_{ij})^2}$

$$= \frac{(95 - 68)^3}{68} + \frac{(41 - 68)^2}{68} + \frac{(55 - 82)^2}{82} + \frac{(109 - 82)^2}{82} = 39.22$$

The rejection region requires $\alpha = .05$ in the upper tail of the χ^2 distribution with df $= (r - 1)(c - 1) = (2 - 1)(2 - 1) = 1$. From Table VII, Appendix B, $\chi^2_{.05} = 3.84146$. The rejection region is $\chi^2 > 3.84146$.

Since the observed value of the test statistic falls in the rejection region ($\chi^2 = 39.22 > 3.84146$), H_0 is rejected. There is sufficient evidence to indicate audience gender and product identification are dependent factors for $\alpha = .05$.

b. Some preliminary calculations are:

$$\hat{E}(n_{11}) = \frac{r_1 c_1}{n} = \frac{108(150)}{300} = 54 \qquad \hat{E}(n_{21}) = \frac{r_2 c_1}{n} = \frac{192(150)}{300} = 96$$

$$\hat{E}(n_{12}) = \frac{r_1 c_2}{n} = \frac{108(150)}{300} = 54 \qquad \hat{E}(n_{22}) = \frac{r_2 c_2}{n} = \frac{192(150)}{300} = 96$$

To determine whether audience gender and product identification are dependent factors for female spokespersons, we test:

H_0: Audience gender and product identification are independent factors

H_a: Audience gender and product identification are dependent factors

The test statistic is $\chi^2 = \sum\sum \dfrac{\left[n_{ij} - \hat{E}(n_{ij})\right]^2}{\hat{E}(n_{ij})}$

$$= \frac{(47 - 54)^2}{54} + \frac{(61 - 54)^2}{54} + \frac{(103 - 96)^2}{96} + \frac{(89 - 96)^2}{96} = 2.84$$

The rejection region requires $\alpha = .05$ in the upper tail of the χ^2 distribution with df $= (r - 1)(c - 1) = (2 - 1)(2 - 1) = 1$. From Table VII, Appendix B, $\chi^2_{.05} = 3.84146$. The rejection region is $\chi^2 > 3.84146$.

Since the observed value of the test statistic does not fall in the rejection region ($\chi^2 = 2.84 \ngtr 3.84146$), H_0 is not rejected. There is insufficient evidence to indicate audience gender and product identification are dependent factors for $\alpha = .05$.

c. When a male spokesperson is used in an advertisement, audience gender and product identification are dependent. Males tended to identify the product more frequently than females.

When a female spokesperson is used in an advertisement, there is no evidence that audience gender and product identification are dependent. Males and females tend to identify the product at the same rate.

8.49 First, we must set up the contingency table. The proportions given are the proportions of the whole group who show signs of stress and fall into a particular fitness level. Thus, the number of people showing signs of stress and falling in the poor fitness level is $np_1 = 549(.155) = 85$. The number of people showing signs of stress and falling in the average fitness level is $np_2 = 549(.133) = 73$. The number of people showing signs of stress and falling in the good fitness level is $np_3 = 549(.108) = 59$.

The contingency table is:

| Fitness level | Stress | | Total |
	No stress	Signs of stress	
Poor	157	85	242
Average	139	73	212
Good	36	59	95
Total	332	217	549

Some preliminary calculations are:

$$\hat{E}(n_{11}) = \frac{r_1 c_1}{n} = \frac{242(332)}{549} = 146.346 \qquad \hat{E}(n_{12}) = \frac{r_1 c_2}{n} = \frac{242(217)}{549} = 95.654$$

$$\hat{E}(n_{21}) = \frac{r_2 c_1}{n} = \frac{212(332)}{549} = 128.204 \qquad \hat{E}(n_{22}) = \frac{r_2 c_2}{n} = \frac{212(217)}{549} = 83.796$$

$$\hat{E}(n_{31}) = \frac{r_3 c_1}{n} = \frac{95(332)}{549} = 57.45 \qquad \hat{E}(n_{32}) = \frac{r_3 c_2}{n} = \frac{95(217)}{549} = 37.55$$

To determine whether the likelihood for stress is dependent on an employee's fitness level, we test:

H_0: Likelihood for stress is independent of an employee's fitness level
H_a: Likelihood for stress is dependent on an employee's fitness level

The test statistic is $\chi^2 = \sum \sum \dfrac{\left[n_{ij} - \hat{E}(n_{ij})\right]^2}{\hat{E}(n_{ij})}$

$$= \frac{(157 - 146.346)^2}{146.346} + \frac{(85 - 95.654)^2}{95.654} + \frac{(139 - 128.204)^2}{128.204}$$

$$+ \frac{(73 - 83.796)^2}{83.796} + \frac{(36 - 57.45)^2}{57.45} + \frac{(59 - 37.55)^2}{37.55} = 24.524$$

The rejection region requires $\alpha = .05$ in the upper tail of the χ^2 distribution with df = $(r - 1)(c - 1) = (3 - 1)(2 - 1) = 2$. From Table VII, Appendix B, $\chi^2_{.05} = 5.99147$. The rejection region is $\chi^2 > 5.99147$.

Since the observed value of the test statistic falls in the rejection region ($\chi^2 = 24.524 > 5.99147$), H_0 is rejected. There is sufficient evidence to indicate the likelihood for stress is dependent on an employee's fitness level for $\alpha = .05$.

8.51   a.   Some preliminary calculations are:

If all the categories are equally likely,

$$P_{1,0} = P_{2,0} = P_{3,0} = P_{4,0} = P_{5,0} = .2$$

$$E(n_1) = E(n_2) = E(n_3) = E(n_4) = E(n_5) = np_{1,0} = 150(.2) = 30$$

To determine if the categories are not equally likely, we test:

H_0: $p_1 = p_2 = p_3 = p_4 = p_5 = .2$
H_a: At least one probability is different from .2

The test statistic is $\chi^2 = \sum \dfrac{\left[n_i - E(n_i)\right]^2}{E(n_i)}$

$$= \frac{(28 - 30)^2}{30} + \frac{(35 - 30)^2}{30} + \frac{(33 - 30)^2}{30} + \frac{(25 - 30)^2}{30} + \frac{(29 - 30)^2}{30} = 2.133$$

The rejection region requires $\alpha = .10$ in the upper tail of the χ^2 distribution with df = $k - 1 = 5 - 1 = 4$. From Table VII, Appendix B, $\chi^2_{.10} = 7.77944$. The rejection region is $\chi^2 > 7.77944$.

Since the observed value of the test statistic does not fall in the rejection region ($\chi^2 = 2.133 \not> 7.77944$), H_0 is not rejected. There is insufficient evidence to indicate the categories are not equally likely at $\alpha = .10$.

b. $\hat{p}_2 = \dfrac{35}{150} = .233$

For confidence coefficient .90, $\alpha = .10$ and $\alpha/2 = .05$. From Table IV, Appendix B, $z_{.05} = 1.645$. The confidence interval is:

$$\hat{p}_2 \pm z_{.05}\sqrt{\dfrac{\hat{p}_2\hat{q}_2}{n_2}} \Rightarrow .233 \pm 1.645\sqrt{\dfrac{.233(.767)}{150}} \Rightarrow .233 \pm .057 \Rightarrow (.176, .290)$$

8.53   For confidence coefficient .90, $\alpha = 1 - .90 = .10$ and $\alpha = .10/2 = .05$. From Table IV, Appendix B, $z_{.05} = 1.645$. Since prior information is given about the values of p_1 and p_2, we will use these values as estimators. Thus, $p_1 = p_2 = .5$. A width of .10 means the bound is $.10/2 = .05$.

$$n_1 = n_2 = \dfrac{(z_{\alpha/2})^2(p_1q_1 + p_2q_2)}{B^2} = \dfrac{(1.645)^2(.5(.5) + .5(.5))}{.05^2} = 541.2 \approx 542$$

8.55   Some preliminary calculations are:

$$\hat{E}(n_{11}) = \dfrac{r_1c_1}{n} = \dfrac{419(380)}{703} = 226.49 \qquad \hat{E}(n_{21}) = \dfrac{r_2c_1}{n} = \dfrac{149(380)}{703} = 80.54$$

$$\hat{E}(n_{12}) = \dfrac{r_1c_2}{n} = \dfrac{419(323)}{703} = 192.51 \qquad \hat{E}(n_{22}) = \dfrac{r_2c_2}{n} = \dfrac{149(323)}{703} = 68.46$$

$$\hat{E}(n_{31}) = \dfrac{r_3c_1}{n} = \dfrac{87(380)}{703} = 47.03 \qquad \hat{E}(n_{41}) = \dfrac{r_4c_1}{n} = \dfrac{48(380)}{703} = 25.95$$

$$\hat{E}(n_{32}) = \dfrac{r_3c_2}{n} = \dfrac{87(323)}{703} = 39.97 \qquad \hat{E}(n_{42}) = \dfrac{r_4c_2}{n} = \dfrac{48(323)}{703} = 22.05$$

To determine whether retirement status of a traveler and the duration of a typical trip are dependent, we test:

H_0: Retirement status of a traveler and the duration of a typical trip are independent
H_a: Retirement status of a traveler and the duration of a typical trip are dependent

The test statistic is $\chi^2 = \displaystyle\sum\sum \dfrac{\left[n_{ij} - \hat{E}(n_{ij})\right]^2}{\hat{E}(n_{ij})}$

$$= \dfrac{(247 - 226.49)^2}{226.49} + \dfrac{(172 - 192.51)^2}{192.51} + \dfrac{(82 - 80.54)^2}{80.54} + \dfrac{(67 - 68.46)^2}{68.46}$$

$$+ \dfrac{(35 - 47.03)^2}{47.03} + \dfrac{(52 - 39.97)^2}{39.97} + \dfrac{(16 - 25.95)^2}{25.95} + \dfrac{(32 - 22.05)^2}{22.05} = 19.10$$

The rejection region requires $\alpha = .05$ in the upper tail of the χ^2 distribution with df $= (r - 1)(c - 1) = (4 - 1)(2 - 1) = 3$. From Table VII, Appendix B, $\chi^2_{.05} = 7.81473$. The rejection region is $\chi^2 > 7.81473$.

Since the observed value of the test statistic falls in the rejection region ($\chi^2 = 19.10 > 7.81473$), H_0 is rejected. There is sufficient evidence to indicate retirement status of a traveler and the duration of a typical trip are dependent for $\alpha = .05$.

8.57   For union members:

H_0: Level of confidence and job satisfaction are independent
H_a: Level of confidence and job satisfaction are dependent

The test statistic is $\chi^2 = 13.36744$ (from printout).

The rejection region requires $\alpha = .05$ in the upper tail of the χ^2 distribution with df = $(r - 1)(c - 1) = (3 - 1)(4 - 1) = 6$. From Table VII, Appendix B, $\chi^2_{.05} = 12.5916$. The rejection region is $\chi^2 > 12.5916$.

Since the observed value of the test statistic falls in the rejection region ($\chi^2 = 13.36 > 12.5916$), H_0 is rejected. There is sufficient evidence to indicate the level of confidence and job satisfaction are related at $\alpha = .05$ for union members.

Note: This test should be viewed cautiously since three cells have expected values less than 5.

For nonunion members:

H_0: Level of confidence and job satisfaction are independent
H_a: Level of confidence and job satisfaction are dependent

The test statistic is $\chi^2 = 9.63514$ (from printout).

The rejection region is $\chi^2 > 12.5916$. Since the observed value of the test statistic does not fall in the rejection region ($\chi^2 = 9.64 \not> 12.5916$), H_0 is not rejected. There is insufficient evidence to indicate the level of confidence and job satisfaction are related for nonunion workers at $\alpha = .05$.

8.59   a.   No. If January change is down, half the next 11-month changes are up and half are down.

b.   The percentages of years for which the 11-month movement is up based on January change are found by dividing the numbers in the first column by the corresponding row total and multiplying by 100. We also divide the first column total by the overall total and multiply by 100.

January Change:

Up $\dfrac{25}{35} \cdot 100 = 71.4\%$

Down $\dfrac{9}{18} \cdot 100 = 50\%$

Total $\dfrac{34}{53} \cdot 100 = 64.2\%$

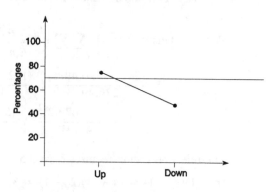

c. H_0: The January change and the next 11-month change are independent
H_a: The January change and the next 11-month change are dependent

d. Some preliminary calculations are:

$$\hat{E}(n_{11}) = \frac{r_1 c_1}{n} = \frac{35(34)}{53} = 22.453 \qquad \hat{E}(n_{12}) = \frac{35(19)}{53} = 12.547$$

$$\hat{E}(n_{21}) = \frac{18(34)}{53} = 11.547 \qquad \hat{E}(n_{22}) = \frac{18(19)}{53} = 6.453$$

The test statistic is $\chi^2 = \sum\sum \frac{\left[n_{ij} - \hat{E}(n_{ij})\right]^2}{\hat{E}(n_{ij})} = \frac{(25 - 22.453)^2}{22.453} + \frac{(9 - 11.547)^2}{11.547}$

$$+ \frac{(10 - 12.547)^2}{12.547} + \frac{(9 - 6.453)^2}{6.453} = 2.373$$

The rejection region requires $\alpha = .05$ in the upper tail of the χ^2 distribution with df = $(r - 1)(c - 1) = (2 - 1)(2 - 1) = 1$. From Table VII, Appendix B, $\chi^2_{.05} = 3.84146$. The rejection region is $\chi^2 > 3.84146$.

Since the observed value of the test statistic does not fall in the rejection region ($\chi^2 = 2.373 \not> 3.84146$), H_0 is not rejected. There is insufficient evidence to indicate the January change and the next 11-month change are dependent at $\alpha = .05$.

e. Yes. For $\alpha = .10$, the rejection region is $\chi^2 > \chi^2_{.10} = 2.70554$, from Table XIII, Appendix B, with df = 1. Since the observed value of the test statistic does not fall in the rejection region ($\chi^2 = 2.373 \not> 2.70554$), H_0 is not rejected. The conclusion is the same.

8.61 a. The contingency table is:

		Committee		
		Acceptable	Rejected	Totals
Inspector	Acceptable	101	23	124
	Rejected	10	19	29
	Totals	111	42	153

b. Yes. To plot the percentages, first convert frequencies to percentages by dividing the numbers in each column by the column total and multiplying by 100. Also, divide the row totals by the overall total and multiply by 100.

		Acceptable	Rejected	Totals
Inspector	Acceptable	$\frac{101}{111} \cdot 100 = 90.99\%$	$\frac{23}{42} \cdot 100 = 54.76\%$	$\frac{124}{153} \cdot 100 = 81.05\%$
	Rejected	$\frac{10}{111} \cdot 100 = 9.01\%$	$\frac{19}{42} \cdot 100 = 45.23\%$	$\frac{29}{153} \cdot 100 = 18.95\%$

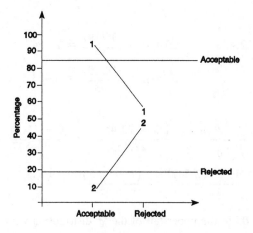

From the plot, it appears there is a relationship.

c.   Some preliminary calculations are:

$$\hat{E}(n_{11}) = \frac{r_1 c_1}{n} = \frac{124(111)}{153} = 89.691 \qquad \hat{E}(n_{12}) = \frac{r_1 c_2}{n} = \frac{124(42)}{153} = 34.039$$

$$\hat{E}(n_{21}) = \frac{r_2 c_1}{n} = \frac{29(111)}{153} = 21.039 \qquad \hat{E}(n_{22}) = \frac{r_2 c_2}{n} = \frac{29(42)}{153} = 7.961$$

To determine if the inspector's classifications and the committee's classifications are related, we test:

H_0: The inspector's and committee's classification are independent
H_a: The inspector's and committee's classifications are dependent

The test statistic is $\chi^2 = \sum \sum \dfrac{[n_{ij} - \hat{E}(n_{ij})]^2}{\hat{E}(n_{ij})}$

$$= \frac{(101 - 89.961)^2}{89.961} + \frac{(23 - 34.039)^2}{34.039} + \frac{(10 - 21.039)^2}{21.039} + \frac{(19 - 7.961)^2}{7.961}$$
$$= 26.034$$

The rejection region requires $\alpha = .05$ in the upper tail of the χ^2 distribution with df $= (r-1)(c-1) = (2-1)(2-1) = 1$. From Table VII, Appendix B, $\chi^2_{.05} = 3.84146$. The rejection region is $\chi^2 > 3.84146$.

Since the observed value of the test statistic falls in the rejection region ($\chi^2 = 26.034 > 3.84146$), H_0 is rejected. There is sufficient evidence to indicate the inspector's and committee's classifications are related at $\alpha = .05$. This indicates that the inspector and committee tend to make the same decisions.

8.65   a.   Some preliminary calculations are:

The contingency table is:

		Defectives	**Non-Defectives**	
	1	25	175	200
Shift	2	35	165	200
	3	80	120	200
		140	460	600

$$\hat{E}(n_{11}) = \frac{r_1 c_1}{n} = \frac{200(140)}{600} = 46.667$$

$$\hat{E}(n_{21}) = \hat{E}(n_{31}) = \frac{200(140)}{600} = 46.667$$

$$\hat{E}(n_{12}) = \hat{E}(n_{22}) = \hat{E}(n_{32}) = \frac{200(460)}{600} = 153.333$$

To determine if quality of the filters are related to shift, we test:

H_0:   Quality of filters and shift are independent
H_a: Quality of filters and shift are dependent

The test statistic is $\chi^2 = \sum\sum \frac{[n_{ij} - \hat{E}(n_{ij})]^2}{\hat{E}(n_{ij})} = \frac{(25 - 46.667)^2}{46.667} + \frac{(35 - 46.667)^2}{46.667}$

$+ \frac{(80 - 46.667)^2}{46.667} + \frac{(175 - 153.333)^2}{153.333} + \frac{(165 - 153.333)^2}{153.333} + \frac{(120 - 153.333)^2}{153.333}$

$= 47.98$

The rejection region requires $\alpha = .05$ in the upper tail of the χ^2 distribution with df $=$ $(r - 1)(c - 1) = (3 - 1)(2 - 1) = 2$. From Table VII, Appendix B, $\chi^2_{.05} = 5.99147$. The rejection region is $\chi^2 > 5.99147$.

Since the observed value of the test statistic falls in the rejection region ($\chi^2 = 47.98 > 5.99147$), H_0 is rejected. There is sufficient evidence to indicate quality of filters and shift are related at $\alpha = .05$.

b.   The form of the confidence interval for p is:

$$\hat{p}_1 \pm z_{\alpha/2}\sqrt{\frac{\hat{p}_1\hat{q}_1}{n}} \text{ where } \hat{p}_1 = \frac{25}{200} = .125$$

For confidence coefficient .95, $\alpha = 1 - .95 = .05$ and $\alpha/2 = .05/2 = .025$. From Table IV, Appendix B, $z_{.025} = 1.96$. The 95% confidence interval is:

$$.125 \pm 1.96\sqrt{\frac{.125(.875)}{200}} \Rightarrow .125 \pm .046 \Rightarrow (.079, .171)$$

9.1   a.

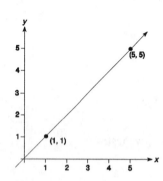

b.

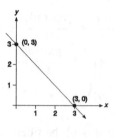

c.

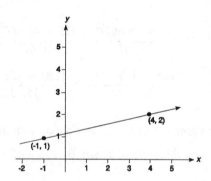

d.

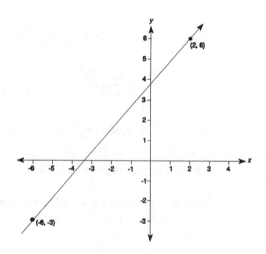

9.3   The two equations are:

$$4 = \beta_0 + \beta_1(-2) \text{ and } 6 = \beta_0 + \beta_1(4)$$

Subtracting the first equation from the second, we get

$$\begin{aligned} 6 &= \beta_0 + 4\beta_1 \\ -(4 &= \beta_0 - 2\beta_1) \\ \hline 2 &= 6\beta_1 \end{aligned} \Rightarrow \beta_1 = \frac{2}{6} = \frac{1}{3}$$

Substituting $\beta_1 = \frac{1}{3}$ into the first equation, we get:

$$4 = \beta_0 + \frac{1}{3}(-2) \Rightarrow \beta_0 = 4 + \frac{2}{3} = \frac{14}{3}$$

The equation for the line is $y = \frac{14}{3} + \frac{1}{3}x$.

9.5 To graph a line, we need two points. Pick two values for x, and find the corresponding y values by substituting the values of x into the equation.

a. Let $x = 0 \Rightarrow y = 4 + (0) = 4$
and $x = 2 \Rightarrow y = 4 + (2) = 6$

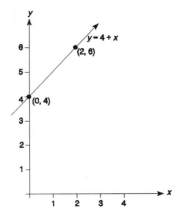

b. Let $x = 0 \Rightarrow y = 5 - 2(0) = 5$
and $x = 2 \Rightarrow y = 5 - 2(2) = 1$

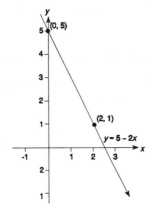

c. Let $x = 0 \Rightarrow y = -4 + 3(0) = -4$
and $x = 2 \Rightarrow y = -4 + 3(2) = 2$

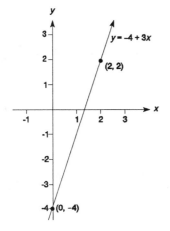

d. Let $x = 0 \Rightarrow y = -2(0) = 0$
and $x = 2 \Rightarrow y = -2(2) = -4$

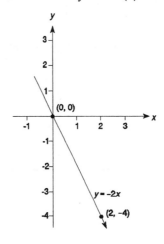

e. Let $x = 0 \Rightarrow y = 0$
and $x = 2 \Rightarrow y = 2$

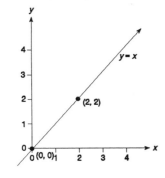

f. Let $x = 0 \Rightarrow y = .5 + 1.5(0) = .5$
and $x = 2 \Rightarrow y = .5 + 1.5(2) = 3.5$

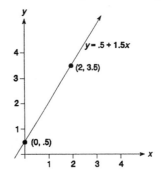

Simple Linear Regression

177

9.7   A deterministic model does not allow for random error or variation, whereas a probabilistic model does. An example where a deterministic model would be appropriate is:

Let y = cost of a 2 × 4 piece of lumber and
x = length (in feet)

The model would be $y = \beta_1 x$. There should be no variation in price for the same length of wood.

An example where a probabilistic model would be appropriate is:

Let y = sales per month of a commodity and
x = amount of money spent advertising

The model would be $y = \beta_0 + \beta_1 x + \epsilon$. The sales per month will probably vary even if the amount of money spent on advertising remains the same.

9.9   No. The random error component, ϵ, allows the values of the variable to fall above or below the line.

9.11   From Exercise 9.10, $\hat{\beta}_0 = 7.10$ and $\hat{\beta}_1 = -.78$.

The fitted line is $\hat{y} = 7.10 - .78x$. To obtain values for $\hat{y}$, we substitute values of x into the equation and solve for $\hat{y}$.

a.

x	y	$\hat{y} = 7.10 - .78x$	$(y - \hat{y})$	$(y - \hat{y})^2$
7	2	1.64	.36	.1296
4	4	3.98	.02	.0004
6	2	2.42	−.42	.1764
2	5	5.54	−.54	.2916
1	7	6.32	.68	.4624
1	6	6.32	−.32	.1024
3	5	4.76	.24	.0576

$\sum(y - \hat{y}) = 0.02$ SSE $= \sum(y - \hat{y})^2 = 1.2204$

b.

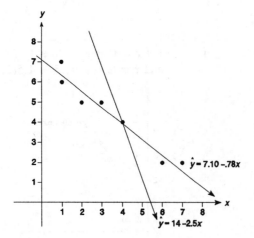

c.

x	y	$\hat{y} = 14 - 2.5x$	$(y - \hat{y})$	$(y - \hat{y})^2$
7	2	-3.5	5.5	30.25
4	4	4	0	0
6	2	-1	3	9
2	5	9	-4	16
1	7	11.5	-4.5	20.25
1	6	11.5	-5.5	30.25
3	5	6.5	-1.5	2.25
			$\Sigma(y - \hat{y}) = -7$	SSE $= 108.00$

9.13 a.

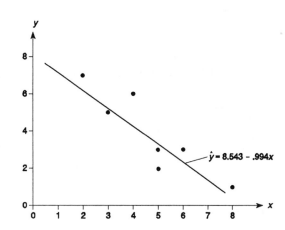

b. Looking at the scattergram, x and y appear to have a negative linear relationship.

c. From the printout, $\hat{\beta}_1 = -.9939$ and $\hat{\beta}_0 = 8.543$

d. The least squares line is $\hat{y} = 8.543 - .994x$. The line is plotted in part **a**. It appears to fit the data well.

e. $\hat{\beta}_0 = 8.543$ Since $x = 0$ is not in the observed range, $\hat{\beta}_0$ has no meaning other than the y-intercept.

  $\hat{\beta}_1 = -.994$ The estimated change in the mean value of y for each unit change in x is $-.994$. These interpretations are valid only for values of x in the range from 2 to 8.

9.15   a.   The slope should be positive. As batting averages increase, one would expect the number of games won to increase.

b.

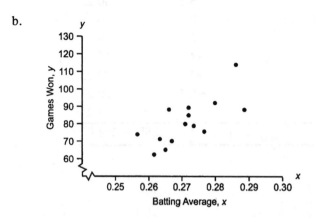

Yes; There appears to be a positive, linear relationship between y and x. As x increases, y tends to increase.

c. $\hat{\beta}_0 = -205.777174$, $\hat{\beta}_1 = 1057.367150$
The least squares line is $\hat{y} = -205.777174 + 1057.367150x$

d.

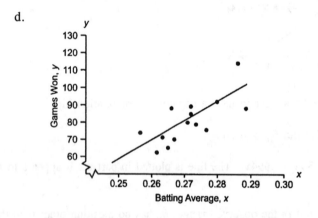

Yes; The least squares line seems to fit the points on the scattergram.

e.   Yes; The points on the scattergram are clustered fairly closely around the least squares line.

f. $\hat{\beta}_0 = -205.777174$. Since $x = 0$ is not in the observed range, $\hat{\beta}_0$ has no interpretation other than being the y-intercept.

$\hat{\beta}_1 = 1057.367150$. For each additional increase of 1 in batting average, the mean number of games won increases by an estimated 1057.367150 games. Since no one has a batting average of 1, a better interpretation would be as follows: For each additional increase of .01 in batting average, the mean number of games won is estimated to increase by 10.57367150 (or approximately 10 games).

9.17   a.   Using MINITAB, the scattergram is:

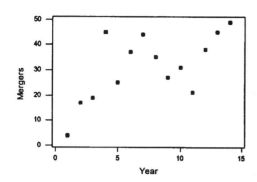

From the plot, it appears that there is a linear relationship between year and number of mergers. As the year increases, the number of mergers tends to increase.

b. $\sum x = 105$ $\sum y = 437$ $\sum xy = 3{,}721$ $\sum x^2 = 1{,}015$

$$\bar{x} = \frac{\sum x}{n} = \frac{105}{14} = 7.5 \qquad \bar{y} = \frac{\sum y}{n} = \frac{437}{14} = 31.21428571$$

$$SS_{xy} = \sum xy - \frac{(\sum x)(\sum y)}{n} = 3{,}721 - \frac{105(437)}{14} = 3{,}721 - 3{,}277.5 = 443.5$$

$$SS_{xx} = \sum x^2 - \frac{(\sum x)^2}{n} = 1{,}015 - \frac{105^2}{14} = 1{,}015 - 787.5 = 227.5$$

$$\hat{\beta}_1 = \frac{SS_{xy}}{SS_{xx}} = \frac{443.5}{227.5} = 1.949450549 \approx 1.949$$

$$\hat{\beta}_0 = \bar{y} - \hat{\beta}_1\bar{x} = 31.21428571 - 1.949450549(7.5) = 31.21428571 - 14.62087912$$
$$= 16.59340659 \approx 16.593$$

The fitted regression line is $\hat{y} = 16.593 + 1.949x$

c.   Using MINITAB, the least squares line is:

Regression Plot

Y = 16.5934 + 1.94945X
R-Sq = 39.2 %

d. For $x = 15$, $\hat{y} = 16.593 + 1.949(15) = 45.828$. This compares very favorably to the actual number of mergers in 1994 of 42.

9.19 a. It appears as salary increases, the retaliation index decreases.

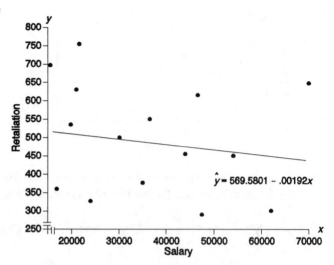

b. $\sum x = 544{,}100$ $\sum y = 7{,}497$ $\sum xy = 263{,}977{,}000$
$\sum x^2 = 23{,}876{,}290{,}000$

$$\bar{x} = \frac{\sum x}{n} = \frac{544{,}100}{15} = 36{,}273.333 \qquad \bar{y} = \frac{\sum y}{n} = \frac{7{,}497}{15} = 499.8$$

$$SS_{xy} = \sum xy - \frac{(\sum x)(\sum y)}{n} = 263{,}977{,}000 - \frac{(544{,}100)(7{,}497)}{15}$$
$$= 263{,}977{,}000 - 271{,}941{,}180 = -7{,}964{,}180$$

$$SS_{xx} = \sum x^2 - \frac{(\sum x)^2}{n} = 23{,}876{,}290{,}000 - \frac{(544{,}100)^2}{15}$$
$$= 23{,}876{,}290{,}000 - 19{,}736{,}320{,}670 = 4{,}139{,}969{,}330$$

$$\hat{\beta}_1 = \frac{SS_{xy}}{SS_{xx}} = \frac{-7{,}964{,}180}{4{,}139{,}969{,}330} = -.001923729 \approx -.00192$$

$$\hat{\beta}_0 = \bar{y} - \hat{\beta}_1\bar{x} = 499.8 - (-.001923729)(36{,}273.333)$$
$$= 499.8 + 69.78007144 = 569.5800714 \approx 569.5801$$

$$\hat{y} = 569.5801 - .00192x$$

c. The least squares line supports the answer because the line has a negative slope.

d. $\hat{\beta}_0 = 569.5801$ This has no meaning because $x = 0$ is not in the observed range.

e. $\hat{\beta}_1 = -.00192$ When the salary increases by \$1, the mean retaliation index is estimated to decrease by .00192. This is meaningful for the range of x from \$16,900 to \$70,000.

9.21 a. The plot of the data is:

It appears that as the age of the firm increases, the number of employees at fast-growing firms increases linearly. However, it does not appear to be a strong linear relationship. The points are not bunched very close to the line.

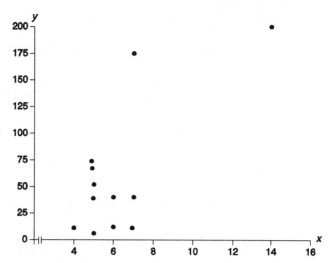

b. From the printout, $\hat{\beta}_0 = -51.361607$ and $\hat{\beta}_1 = 17.754464$.

$\hat{\beta}_0 = -51.361607$. Since $x = 0$ is not in the observed range, $\hat{\beta}_0$ is just an estimate of the y-intercept.

$\hat{\beta}_1 = 17.754464$. For each additional year of age, the mean number of employees is estimated to increase by 17.754464.

9.23 a. $s^2 = \dfrac{\text{SSE}}{n - 2} = \dfrac{8.34}{26 - 2} = .3475$

b. We would expect most of the observations to be within $2s$ of the least squares line. This is:
$$2s = 2\sqrt{.3475} \approx 1.179$$

9.25 $\text{SSE} = \text{SS}_{yy} - \hat{\beta}_1 \text{SS}_{xy}$

where $\text{SS}_{yy} = \sum y^2 - \dfrac{\left(\sum y\right)^2}{n}$

For Exercise 9.10,

$\sum y^2 = 159 \qquad \sum y = 31$

$\text{SS}_{yy} = 159 - \dfrac{31^2}{7} = 159 - 137.2857143 = 21.7142857$

$\text{SS}_{xy} = -26.2857143 \qquad \hat{\beta}_1 = -.779661017$

Therefore, $\text{SSE} = 21.7142857 - (-.779661017)(-26.2857143) = 1.22033896 \approx 1.2203$

$s^2 = \dfrac{\text{SSE}}{n - 2} = \dfrac{1.22033896}{7 - 2} = .244067792, \quad s = \sqrt{.244067792} = .4960$

We would expect most of the observations to fall within $2s$ or $2(.4940)$ or .988 units of the least squares prediction line.

For Exercise 9.13,

$$\sum x = 33 \qquad \sum y = 27 \qquad \sum xy = 104 \qquad \sum x^2 = 179 \qquad \sum y^2 = 133$$

$$SS_{xy} = \sum xy - \frac{\left(\sum x \sum y\right)}{n} = 104 - \frac{(23)(27)}{7} = 104 - 127.2857143 = -23.2857143$$

$$SS_{xx} = \sum x^2 - \frac{\left(\sum x\right)^2}{n} = 179 - \frac{(33)^2}{7} = 179 - 155.5714286 = 23.4285714$$

$$SS_{yy} = \sum y^2 - \frac{\left(\sum y\right)^2}{n} = 133 - \frac{(27)^2}{7} = 133 - 104.1428571 = 28.8571429$$

$$\hat{\beta}_1 = \frac{SS_{xy}}{S_{xx}} = \frac{-23.2857143}{23.4285714} = -.99390244$$

$$SSE = SS_{yy} - \hat{\beta}_1 SS_{xy} = 28.8571429 - (.99390244)(-23.2857143)$$
$$= 28.8571429 - 23.14372824 = 5.71341466$$

$$s^2 = \frac{SSE}{n-2} = \frac{5.71341466}{7-2} = 1.142682932 \qquad s = \sqrt{1.142682932} = 1.0690$$

We would expect most of the observations to fall within $2s$ or $2(1.0690)$ or 2.1380 units of the least squares prediction line.

9.27 a. $\hat{\beta}_1 = \dfrac{SS_{xy}}{SS_{xx}} = \dfrac{1,419,492.796}{3,809,368.452} = .372632055 \approx .373$

   $\hat{\beta}_0 = \bar{y} - \hat{\beta}_1 \bar{x} = 302.52 - .372632055(792.04) = 7.3805068 \approx 7.381$

   The least squares line is $\hat{y} = 7.381 + .373x$

   The graph of the data is:

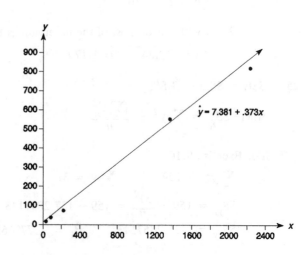

   b. For $x = \$1,600$ billion, $\hat{y} = 7.381 + .373(1,600) = \$604.181$ billion.

   c. $SSE = SS_{yy} - \hat{\beta}_1 SS_{xy} = 531,174.148 - .372632055(1,419,492.796) = 2225.6298$

   $s^2 = \dfrac{SSE}{n-2} = \dfrac{2225.6298}{5-2} = 741.8766 \qquad s = \sqrt{741.8766} = 27.237$

   d. We would expect almost all of the observed values of y to fall within $2s$ or $2(27.237)$ or 54.474 dollars of their least squares predicted values.

9.29 a. From the printout, SSE = 20,554.41518, s^2 = MSE = 2,055.44152, and s = ROOT MSE = 45.33698.

b. We would expect that most of the observations will fall within $2s$ or $2(45.33698) = 90.67396$ employees of their predicted values.

9.31 a. For confidence coefficient .95, $\alpha = 1 - .95 = .05$ and $\alpha/2 = .05/2 = .025$. From Table VI, Appendix B, with df $= n - 2 = 12 - 2 = 10$, $t_{.025} = 2.228$.

The 95% confidence interval for β_1 is:

$$\hat{\beta}_1 \pm t_{.025}s_{\hat{\beta}_1} \text{ where } s_{\hat{\beta}_1} = \frac{s}{\sqrt{SS_{xx}}} = \frac{3}{\sqrt{35}} = .5071$$

$$\Rightarrow 31 \pm 2.228(.5071) \Rightarrow 31 \pm 1.13 \Rightarrow (29.87, 32.13)$$

For confidence coefficient .90, $\alpha = 1 - .90 = .10$ and $\alpha/2 = .10/2 = .05$. From Table VI, Appendix B, with df $= 10$, $t_{.05} = 1.812$.

The 90% confidence interval for β_1 is:

$$\hat{\beta}_1 \pm t_{.05}s_{\hat{\beta}_1} \Rightarrow 31 \pm 1.812(.5071) \Rightarrow 31 \pm .92 \Rightarrow (30.08, 31.92)$$

b. $s^2 = \dfrac{SSE}{n-2} = \dfrac{1960}{18-2} = 122.5$, $s = \sqrt{s^2} = 11.0680$

For confidence coefficient, .95, $\alpha = 1 - .95 = .05$ and $\alpha/2 = .05/2 = .025$. From Table VI, Appendix B, with df $= n - 2 = 18 - 2 = 16$, $t_{.025} = 2.120$. The 95% confidence interval for β_1 is:

$$\hat{\beta}_1 \pm t_{.025}s_{\hat{\beta}_1} \text{ where } s_{\hat{\beta}_1} = \frac{s}{\sqrt{SS_{xx}}} = \frac{11.0680}{\sqrt{30}} = 2.0207$$

$$\Rightarrow 64 \pm 2.120(2.0207) \Rightarrow 64 \pm 4.28 \Rightarrow (59.72, 68.28)$$

For confidence coefficient .90, $\alpha = 1 - .90 = .10$ and $\alpha/2 = .10/2 = .05$. From Table VI, Appendix B, with df $= 16$, $t_{.05} = 1.746$.

The 90% confidence interval for β_1 is:

$$\hat{\beta}_1 \pm t_{.05}s_{\hat{\beta}_1} \Rightarrow 64 \pm 1.746(2.0207) \Rightarrow 64 \pm 3.53 \Rightarrow (60.47, 67.53)$$

c. $s^2 = \dfrac{SSE}{n-2} = \dfrac{146}{24-2} = 6.6364$, $s = \sqrt{s^2} = 2.5761$

For confidence coefficient .95, $\alpha = 1 - .95 = .05$ and $\alpha/2 = .05/2 = .025$. From Table VI, Appendix B, with df $= n - 2 = 24 - 2 = 22$, $t_{.025} = 2.074$. The 95% confidence interval for β_1 is:

$$\hat{\beta}_1 \pm t_{.025}s_{\hat{\beta}_1} \text{ where } s_{\hat{\beta}_1} = \frac{s}{\sqrt{SS_{xx}}} = \frac{2.5761}{\sqrt{64}} = .3220$$

$$\Rightarrow -8.4 \pm 2.074(.322) \Rightarrow -8.4 \pm .67 \Rightarrow (-9.07, -7.73)$$

For confidence coefficient .90, $\alpha = 1 - .90 = .10$ and $\alpha/2 = .10/2 = .05$. From Table VI, Appendix B, with df = 22, $t_{.05} = 1.717$.

The 90% confidence interval for β_1 is:

$$\hat{\beta}_1 \pm t_{.05}s_{\hat{\beta}_1} \Rightarrow -8.4 \pm 1.717(.322) \Rightarrow -8.4 \pm .55 \Rightarrow (-8.95, -7.85)$$

9.33   From Exercise 9.32, $\hat{\beta}_1 = .8214$, $s = 1.1922$, $SS_{xx} = 28$, and $n = 7$.

For confidence coefficient .80, $\alpha = 1 - .80 = .20$ and $\alpha/2 = .20/2 = .10$. From Table VI, Appendix B, with df = $n - 2 = 7 - 2 = 5$, $t_{.10} = 1.476$. The 80% confidence interval for β_1 is:

$$\hat{\beta}_1 \pm t_{.025}s_{\hat{\beta}_1} \text{ where } s_{\hat{\beta}_1} = \frac{s}{\sqrt{SS_{xx}}} = \frac{1.1922}{\sqrt{28}} = .2253$$

$$\Rightarrow .8214 \pm 1.476(.2253) \Rightarrow .8214 \pm .3325 \Rightarrow (.4889, 1.1539)$$

For confidence coefficient .98, $\alpha = 1 - .98 = .02$ and $\alpha/2 = .02/2 = .01$. From Table VI, Appendix B, with df = 5, $t_{.01} = 3.365$.

The 98% confidence interval for β_1 is:

$$\hat{\beta}_1 \pm t_{.01}s_{\hat{\beta}_1} \Rightarrow .8214 \pm 3.365(.2253) \Rightarrow .8214 \pm .7581 \Rightarrow (.0633, 1.5795)$$

9.35   a.   Using MINITAB, the scattergram is:

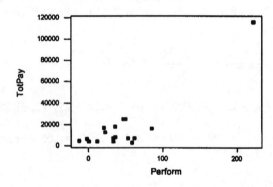

It appears that as performance increases, CEO pay tends to increase.

b. $\sum x = 755.9 \qquad \sum y = 285,412 \qquad \sum xy = 32,460,260 \qquad \sum x^2 = 77,402.39$

$\sum y^2 = 16,051,516,284$

$$\bar{x} = \frac{\sum x}{n} = \frac{755.9}{17} = 44.46470588 \qquad \bar{y} = \frac{\sum y}{n} = \frac{285,412}{17} = 16,788.94118$$

$$SS_{xy} = \sum xy - \frac{(\sum x)(\sum y)}{n} = 32,460,260 - \frac{755.9(285,412)}{17}$$

$$= 32,460,260 - 12,690,760.64 = 19,769,499.3$$

$$SS_{xx} = \sum x^2 - \frac{(\sum x)^2}{n} = 77{,}402.39 - \frac{755.9^2}{17}$$
$$= 77{,}402.39 - 33{,}619.87118 = 43{,}791.51882$$

$$\hat{\beta}_1 = \frac{SS_{xy}}{SS_{xx}} = \frac{19{,}769{,}499.3}{43{,}791.51882} = 451.4458483 \approx 451.446$$

$$\hat{\beta}_0 = \bar{y} - \hat{\beta}_1\bar{x} = 16{,}788.94118 - 451.4458483(44.46470588)$$
$$= 16{,}788.94118 - 20{,}073.40687 = -3{,}284.46569 \approx -3{,}284.466$$

The fitted regression line is $\hat{y} = -3{,}284.466 + 451.446x$

c. $$SS_{yy} = \sum y^2 - \frac{(\sum y)^2}{n} = 16{,}051{,}516{,}284 - \frac{285{,}412^2}{17}$$
$$= 16{,}051{,}516{,}284 - 4{,}791{,}765{,}279 = 11{,}259{,}751{,}005$$

$$SSE = SS_{yy} - \hat{\beta}_1 SS_{xy} = 11{,}259{,}751{,}005 - 451.4458483(19{,}769{,}499.3)$$
$$= 11{,}259{,}751{,}005 - 8{,}924{,}858{,}409 = 2{,}334{,}892{,}596$$

$$s^2 = \frac{SSE}{n-2} = \frac{2{,}334{,}892{,}596}{17-2} = 155{,}659{,}506.4$$

$$s = \sqrt{s^2} = \sqrt{155{,}659{,}506.4} = 12{,}476.3579$$

To determine if CEO compensation is related to company performance, we test:

$$H_0: \beta_1 = 0$$
$$H_a: \beta_1 \neq 0$$

The test statistic is $t = \dfrac{\hat{\beta}_1 - 0}{s_{\hat{\beta}_1}} = \dfrac{451.446 - 0}{\dfrac{12{,}476.3579}{\sqrt{43{,}791.51882}}} = 7.572$

The rejection region requires $\alpha/2 = .05/2 = .025$ in each tail of the t-distribution with df $= n - 2 = 17 - 2 = 15$. From Table VI, Appendix B, $t_{.025} = 2.131$. The rejection region is $t < -2.131$ or $t > 2.131$.

Since the observed value of the test statistic falls in the rejection region ($t = 7.572 > 2.131$), H_0 is rejected. There is sufficient evidence to indicate that CEO compensation is related to company performance at $\alpha = .05$.

d. $\hat{\beta}_1 = 451.446$. For each percent increase in performance, the mean total pay is estimated to increase by 451.446 thousand dollars.

e. For confidence coefficient .90, $\alpha = .10$ and $\alpha/2 = .10/2 = .05$. From Table VI, Appendix B, with df $= n - 2 = 17 - 2 = 15$, $t_{.05} = 1.753$. The 90% confidence interval is:

$$\hat{\beta}_1 \pm t_{.05}s_{\hat{\beta}_1} \Rightarrow 451.446 \pm 1.753\frac{12{,}476.3579}{\sqrt{43{,}791.51882}}$$
$$\Rightarrow 451.446 \pm 104.514 \Rightarrow (346.932, 555.960)$$

We are 90% confident that the change in the mean CEO compensation for each percent increase in performance is between 346.932 and 555.960 thousand dollars.

f. We would expect the variability among the CEO compensations in the same industry to be smaller than that from many industries. Thus, we would expect the width of the confidence interval in part **e** to be smaller.

9.37 a. Using MINITAB, the scattergram of the data is:

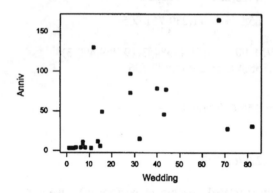

If the players' rankings remained the same, then the scattergram would be a straight line with a slope of 1. If the claim is true, then the scattergram would reveal points that would lie above this imaginary line. From the plot, there appears to be more points above this line than below it, which would support the claim.

b. $\sum x = 541$ $\qquad \sum y = 851$ $\qquad \sum xy = 32{,}145$ $\qquad \sum x^2 = 25{,}401$

$$\bar{x} = \frac{\sum x}{n} = \frac{541}{22} = 24.59090909 \qquad \bar{y} = \frac{\sum y}{n} = \frac{851}{22} = 38.68181818$$

$$SS_{xy} = \sum xy - \frac{(\sum x)(\sum y)}{n} = 32{,}145 - \frac{541(851)}{22}$$
$$= 32{,}145 - 20{,}926.86364 = 11{,}218.13636$$

$$SS_{xx} = \sum x^2 - \frac{(\sum x)^2}{n} = 25{,}401 - \frac{541^2}{22}$$
$$= 25{,}401 - 13{,}303.68182 = 12{,}097.31818$$

$$SS_{yy} = \sum y^2 - \frac{(\sum y)^2}{n} = 77{,}931 - \frac{851^2}{22}$$
$$= 77{,}931 - 32{,}918.22727 = 45{,}012.77273$$

$$\hat{\beta}_1 = \frac{SS_{xy}}{SS_{xx}} = \frac{11,218.13636}{12,097.31818} = .927324237 \approx .927$$

$$\hat{\beta}_0 = \bar{y} - \hat{\beta}_1\bar{x} = 38.68181818 - .927324237(24.59090909) = 15.87807217 \approx 15.878$$

The fitted model is: $\hat{y} = 15.878 + .927x$

c. $SSE = SS_{yy} - \hat{\beta}_1 SS_{xy} = 45,012.77273 - .927324237(11,218.13636)$

$\qquad = 45,012.77273 - 10,402.84974 = 34,609.92299$

$$s^2 = \frac{SSE}{n-2} = \frac{34,609.92299}{22-2} = 1730.49615 \qquad s = \sqrt{s^2} = \sqrt{1,730.49615} = 41.5992$$

To determine if the model contributes information for predicting players' rankings on their first anniversary, we test:

$H_0: \beta_1 = 0$
$H_a: \beta_1 \neq 0$

The test statistic is $t = \dfrac{\hat{\beta}_1 - 0}{s_{\hat{\beta}_1}} = \dfrac{.927 - 0}{\dfrac{41.5992}{\sqrt{12,097.31818}}} = 2.451$

The rejection region requires $\alpha/2 = .05/2 = .025$ in each tail of the t-distribution with df $= n - 2 = 22 - 2 = 20$. From Table VI, Appendix B, $t_{.025} = 2.086$. The rejection region is $t < -2.086$ or $t > 2.086$.

Since the observed value of the test statistic falls in the rejection region ($t = 2.451 > 2.086$), H_0 is rejected. There is sufficient evidence to indicate the model contributes information for predicting players' rankings on their first anniversary at $\alpha = .05$.

d. If there were no changes whatsoever in the rankings of the sample players after getting married, the true value of β_0 would be 0 and the true value of β_1 would be 1.

9.39 a. To determine if x and y are linearly related, we test:

$H_0: \beta_1 = 0$
$H_a: \beta_1 \neq 0$

The test statistic is $t = 4.98$.

The p-value is .001. Since the p-value is less than $\alpha = .01$, H_0 is rejected at $\alpha = .01$. There is sufficient evidence to indicate that x and y are linearly related.

b. Since the model is adequate, it is reasonable to use it to predict values of y.

For $x = 3$, $\hat{y} = .202 + .135x = .202 + .135(3) = .607$. This value is meaningful only if $x = 3$ is within the observed range.

9.41 a. To determine whether the number of employees is positively linearly related to age of a fast-growing firm, we test:

$$H_0: \beta_1 = 0$$
$$H_a: \beta_1 > 0$$

The test statistic is $t = \dfrac{\hat{\beta}_1 - 0}{s_{\hat{\beta}_1}} = 3.384$ (from printout).

The p-value is $.0070/2 = .0035$. Since the p-value is less than $\alpha = .01$, H_0 is rejected. There is sufficient evidence to indicate that the number of employees is positively linearly related to age of a fast-growing firm at $\alpha > .0035$.

b. For confidence coefficient .99, $\alpha = 1 - .99 = .01$ and $\alpha/2 = .005$. From Table VI, Appendix B, with df $= n - 2 = 12 - 2 = 10$, $t_{.005} = 3.169$. The confidence interval is:

$$\hat{\beta}_1 \pm t_{.005}\, s_{\hat{\beta}_1} \Rightarrow 17.754 \pm 3.169\,(5.2467) \Rightarrow 17.754 \pm 16.627 \Rightarrow (1.127, 34.381)$$

We are 99% confident that for each additional year of age, the mean number of employees will increase by anywhere from 1.127 to 34.381.

9.43 From Exercise 9.19,

$$SS_{xx} = 4,362,209,330 \qquad\qquad \hat{\beta}_1 = -.002186456$$
$$SS_{xy} = -9,537,780$$

$$\sum y = 7497 \qquad\qquad \sum y^2 = 4,061,063$$

$$SS_{yy} = \sum y^2 - \frac{(\sum y)^2}{n} = 4,061,063 - \frac{7497^2}{15} = 314062.4$$

$$SSE = SS_{yy} - \hat{\beta}_1 SS_{xy} = 314062.4 - (-.002186456)(-9,537,780) = 293208.4637$$

$$s^2 = \frac{SSE}{n-2} = \frac{293208.4637}{15-2} = 22554.49721 \quad s = \sqrt{22554.49721} = 150.1815$$

To determine if extent of retaliation is related to whistle blower's power, we test:

$$H_0: \beta_1 = 0$$
$$H_a: \beta_1 \neq 0$$

The test statistic is $t = \dfrac{\hat{\beta}_1 - 0}{s_{\hat{\beta}_1}} = \dfrac{-.0022}{\dfrac{150.1815}{\sqrt{4362209330}}} = -.96$

The rejection region requires $\alpha/2 = .05/2 = .025$ in each tail of the t-distribution with df $= n - 2 = 15 - 2 = 13$. From Table VI, Appendix B, $t_{.025} = 2.160$. The rejection region is $t > 2.160$ or $t < -2.160$.

Since the observed value of the test statistic does not fall in the rejection region ($t = -.96$ $\not< -2.160$), H_0 is not rejected. There is insufficient evidence to indicate the extent of retaliation is related to the whistle blower's power at $\alpha = .05$. This agrees with Near and Miceli.

9.45 a. If $r = .7$, there is a positive relationship between x and y. As x increases, y tends to increase. The slope is positive.

b. If $r = -.7$, there is a negative relationship between x and y. As x increases, y tends to decrease. The slope is negative.

c. If $r = 0$, there is a 0 slope. There is no relationship between x and y.

d. If $r^2 = .64$, then r is either .8 or $-.8$. The relationship between x and y could be either positive or negative.

9.47 a. From Exercises 9.10 and 9.25,

$$r^2 = 1 - \frac{SSE}{SS_{yy}} = 1 - \frac{1.22033896}{21.7142857} = 1 - .0562 = .9438$$

94.38% of the total sample variability around $\bar{y}$ is explained by the linear relationship between y and x.

b. From Exercises 9.13 and 9.25,

$$r^2 = 1 - \frac{SSE}{SS_{yy}} = 1 - \frac{5.71341466}{28.8571429} = .8020$$

80.20% of the total sample variability around $\bar{y}$ is explained by the linear relationship between y and x.

9.49 a. $r = .14$. Because this value is close to 0, there is a very weak positive linear relationship between math confidence and computer interest for boys.

b. $r = .33$. Because this value is fairly close to 0, there is a weak positive linear relationship between math confidence and computer interest for girls.

9.51 a. Using MINITAB, the scattergram is:

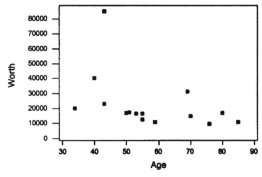

It appears from this scattergram that the relationship between net worth and age is a negative relationship. This relationship could be either linear or possibly quadratic. If any relationship exists, it is very weak.

b. $\sum x = 863$ $\sum y = 343,300$ $\sum xy = 18,039,900$ $\sum x^2 = 52,957$

$\sum y^2 = 12,876,790,000$

$$SS_{xy} = \sum xy - \frac{(\sum x)(\sum y)}{n} = 18,039,900 - \frac{863(343,300)}{15}$$
$$= 18,039,900 - 19,751,193.33 = -1,711,293.33$$

$$SS_{xx} = \sum x^2 - \frac{(\sum x)^2}{n} = 52{,}957 - \frac{863^2}{15}$$
$$= 52{,}957 - 49{,}651.26667 = 3{,}305.73333$$

$$SS_{yy} = \sum y^2 - \frac{(\sum y)^2}{n} = 12{,}876{,}790{,}000 - \frac{343{,}300^2}{15}$$
$$= 12{,}876{,}790{,}000 - 7{,}856{,}992{,}667 = 5{,}019{,}797{,}333$$

$$r = \frac{SS_{xy}}{\sqrt{SS_{xx}SS_{yy}}} = \frac{-1{,}711{,}293.33}{\sqrt{3{,}305.73333(5{,}019{,}797{,}333)}} = -.420$$

This correlation coefficient indicates that there is a moderately weak negative linear relationship between age and net worth.

c. The only change would be that the relationship would be a moderately weak positive linear relationship.

d. The coefficient of determination is: $r^2 = (-.420)^2 = .1764$. About 17.6% of the sample variation in net worth can be explained by the linear relationship between age and net worth.

9.53 From the printout, $r^2 =$ R-SQUARED $= 0.2935$.

29.4% of the sample variability around the sample mean S&P 500 stock composite average is explained by the linear relationship between the interest rate and the S&P 500 stock composite average.

From the printout, $r = -.5418$

The relationship between interest rate and S&P stock composite average is negative since $r < 0$. The relationship is not particularly strong because $-.5418$ is not that close to -1.

9.55 a. Using MINITAB, the scattergram is:

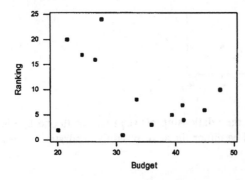

From the scattergram, there appears to be a relatively weak negative linear relationship between rankings and athletic department budget.

b. $\sum x = 435$ $\sum y = 123$ $\sum xy = 3{,}773.2$ $\sum x^2 = 15{,}547.3$

$\sum y^2 = 1{,}825$

$$SS_{xy} = \sum xy - \frac{(\sum x)(\sum y)}{n} = 3{,}773.2 - \frac{435(123)}{13}$$
$$= 3{,}773.2 - 4{,}115.769231 = -342.569231$$

$$SS_{xx} = \sum x^2 - \frac{(\sum x)^2}{n} = 15{,}547.3 - \frac{435^2}{13}$$
$$= 15{,}547.3 - 14{,}555.76923 = 991.53077$$

$$SS_{yy} = \sum y^2 - \frac{(\sum y)^2}{n} = 1{,}825 - \frac{123^2}{13}$$
$$= 1{,}825 - 1{,}163.769231 = 661.230769$$

$$r = \frac{SS_{xy}}{\sqrt{SS_{xx}SS_{yy}}} = \frac{-342.569231}{\sqrt{991.53077(661.230769)}} = -.423$$

This correlation coefficient indicates that there is a moderately weak negative linear relationship between AP rankings and athletic department budget.

9.57   a.

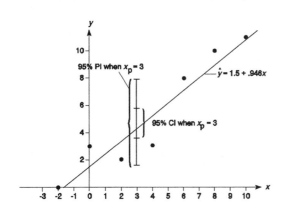

b.   Some preliminary calculations are:

$$\sum x = 28 \qquad \sum x^2 = 224 \quad \sum xy = 254 \qquad \sum y = 37 \quad \sum y^2 = 307$$

$$SS_{xy} = \sum xy - \frac{\sum x \sum y}{n} = 254 - \frac{28(37)}{7} = 106$$

$$SS_{xx} = \sum x^2 - \frac{(\sum x)^2}{n} = 224 - \frac{28^2}{7} = 112$$

$$SS_{yy} = \sum y^2 - \frac{(\sum y)^2}{n} = 307 - \frac{37^2}{7} = 111.4285714$$

$$\hat{\beta}_1 = \frac{SS_{xy}}{SS_{xx}} = \frac{106}{112} = .946428571$$

$$\hat{\beta}_0 = \bar{y} - \hat{\beta}_1 \bar{x} = \frac{37}{7} - .946428571\left[\frac{28}{7}\right] = 1.5$$

The least squares line is $\hat{y} = 1.5 + .946x$.

c. $SSE = SS_{yy} - \hat{\beta}_1 SS_{xy} = 111.4285714 - (.946428571)(106) = 11.1071429$

$s^2 = \dfrac{SSE}{n-2} = \dfrac{11.1071429}{7-2} = 2.22143$

d. The form of the confidence interval is:

$$\hat{y} \pm t_{\alpha/2} s \sqrt{\dfrac{1}{n} + \dfrac{(x_p - \bar{x})^2}{SS_{xx}}} \quad \text{where } s = \sqrt{s^2} = \sqrt{2.22143} = 1.4904$$

For $x_p = 3$, $\hat{y} = 1.5 + .946(3) = 4.338$ and $\bar{x} = \dfrac{28}{7} = 4$

For confidence coefficient .90, $\alpha = 1 - .90 = .10$ and $\alpha/2 = .10/2 = .05$. From Table VI, Appendix B, $t_{.05} = 2.015$ with df $= n - 2 = 7 - 2 = 5$.

The 90% confidence interval is:

$$4.338 \pm 2.015(1.4904) \sqrt{\dfrac{1}{7} + \dfrac{(3-4)^2}{112}} \Rightarrow 4.338 \pm 1.170 \Rightarrow (3.168, 5.508)$$

e. The form of the prediction interval is:

$$\hat{y} \pm t_{\alpha/2} s \sqrt{1 + \dfrac{1}{n} + \dfrac{(x_p - \bar{x})^2}{SS_{xx}}}$$

The 90% prediction interval is:

$$4.338 \pm 2.015(1.4904) \sqrt{1 + \dfrac{1}{7} + \dfrac{(3-4)^2}{112}} \Rightarrow 4.338 \pm 3.223 \Rightarrow (1.115, 7.561)$$

f. The 95% prediction interval for y is wider than the 95% confidence interval for the mean value of y when $x_p = 3$.

The error of predicting a particular value of y will be larger than the error of estimating the mean value of y for a particular x value. This is true since the error in estimating the mean value of y for a given x value is the distance between the least squares line and the true line of means, while the error in predicting some future value of y is the sum of two errors—the error of estimating the mean of y plus the random error that is a component of the value of y to be predicted.

9.59 a. The form of the confidence interval is:

$$\overline{y} \pm t_{\alpha/2}\frac{s}{\sqrt{n}} \text{ where } \overline{y} = \frac{\sum y}{n} = \frac{22}{10} = 2.2$$

$$s^2 = \frac{\sum y^2 - \frac{(\sum y)^2}{n}}{n-1} = \frac{82 - \frac{(22)^2}{10}}{10-1} = 3.7333 \text{ and } s = 1.9322$$

For confidence coefficient .95, $\alpha = 1 - .95 = .05$ and $\alpha/2 = .05/2 = .025$. From Table VI, Appendix B, $t_{.025} = 2.262$ with df $= n - 1 = 10 - 1 = 9$. The 95% confidence interval is:

$$2.2 \pm 2.262\frac{1.9322}{\sqrt{10}} \Rightarrow 2.2 \pm 1.382 \Rightarrow (.818, 3.582)$$

b.

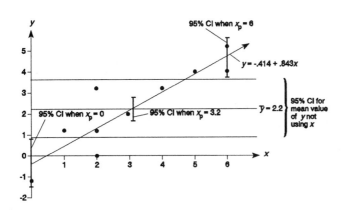

c. The confidence intervals computed in Exercise 9.58 are much narrower than that found in part **a**. Thus, x appears to contribute information about the mean value of y.

d. From Exercise 9.58, $\hat{\beta}_1 = .843$, $s = .8619$, $SS_{xx} = 38.9$, and $n = 10$.

$H_0: \beta_1 = 0$
$H_a: \beta_1 \neq 0$

The test statistic is $t = \dfrac{\hat{\beta}_1 - 0}{s_{\hat{\beta}_1}} = \dfrac{\hat{\beta}_1 - 0}{\dfrac{s}{\sqrt{SS_{xx}}}} = \dfrac{.843 - 0}{\dfrac{.8619}{\sqrt{38.9}}} = 6.10$

The rejection region requires $\alpha/2 = .05/2 = .025$ in each tail of the t-distribution with df $= n - 2 = 10 - 2 = 8$. From Table VI, Appendix B, $t_{.025} = 2.306$. The rejection region is $t > 2.306$ or $t < -2.306$.

Since the observed value of the test statistic falls in the rejection region ($t = 6.10 > 2.306$), H_0 is rejected. There is sufficient evidence to indicate the straight-line model contributes information for the prediction of y at $\alpha = .05$.

9.61   a.   Using MINITAB, the scattergram is:

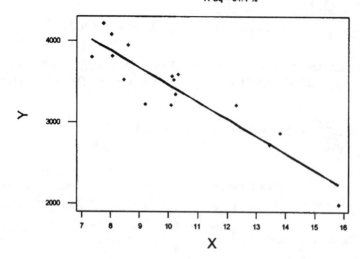

Regression Plot

Y = 5566.13 - 210.346X
R-Sq = 84.4 %

b.   From the printout, the least squares line is:

$\hat{y} = 5566.13 - 210.346x$

See the plot in part a.

c.   To determine if mortgage interest rates contribute information for the prediction of annual sales of existing single-family homes, we test:

$H_0: \beta_1 = 0$

$H_a: \beta_1 \neq 0$

The test statistic is $t = -8.69$ (from printout)

The p-value is $p = 0.0000$. Since the p-value is less than α ($p = 0.0000 < .05$), H_0 is rejected. There is sufficient evidence to indicate the mortgage interest rates contribute information for the prediction of annual sales of existing single-family homes at $\alpha = .05$.

d.   From the printout, $r^2 = .8437$.

84.37% of the sample variability of the annual sales of existing single-family homes about their means is explained by the linear relationship between interest rates and annual sales of existing single-family homes.

e.   From the printout, the confidence interval is: (3714.7, 4052.0)

We are 95% confident that the mean number of existing single-family homes sold when the average annual mortgage rate is 8% is between 3714.7 and 4052.0.

f.   From the printout, the prediction interval is: (3364.1, 4402.7)

We are 95% confident that the actual number of existing single-family homes sold when the average annual mortgage rate is 8% is between 3364.1 and 4402.7.

g.   The width of the prediction interval for an actual value of y is always larger than the width of the confidence interval for the mean value of y. The prediction interval takes into account two variances - the variance for locating the mean and the variance of y once the mean has been located. The confidence interval takes into account only one variance – the variance in locating the mean.

9.63   a.   For $x = 10$, the 95% prediction interval is (12.6384, 239.7). We are 95% confident that the actual number of employees for fast-growing firms is between 12.6384 and 239.7 when the age is 10 years.

b.   The width of the 95% confidence interval for the mean number of employees for fast-growing firms when the age is 10 years would be smaller than the 95% prediction interval.

c.   We would not recommend that this model be used to predict the number of employees for fast-growing firms when the age is 2 years, because 2 is outside the observed ages (ages ranged from 4 to 14). We have no idea if the relationship between the number of employees and age is the same outside the observed range.

9.65   a.   To determine if the average hourly wage rate contributes information to predict quit rates, we test:

H_0: $\beta_1 = 0$
H_a: $\beta_1 \neq 0$

The test statistic is $t = \dfrac{\hat{\beta}_1 - 0}{s_{\hat{\beta}_1}} = -5.91$ (from printout).

The rejection region requires $\alpha/2 = .05/2 = .025$ in each tail of the t-distribution with df $= n - 2 = 15 - 2 = 13$. From Table VI, Appendix B, $t_{.025} = 2.160$. The rejection region is $t < -2.160$ or $t > 2.160$.

Since the observed value of the test statistic falls in the rejection region ($t = -5.91 < -2.160$), H_0 is rejected. There is sufficient evidence to indicate that the average hourly wage rate contributes information to predict quit ratio at $\alpha = .05$.

Since the slope is negative ($\hat{\beta}_1 = -.3466$), the model suggests that x and y have a negative relationship. As the average hourly wage rate increases, the quit rate tends to decrease.

b.   From the printout, the 95% prediction interval is (.656, 2.829). We are 95% confident that the quit rate in an industry with an average hourly wage of $9.00 is between .656 and 2.829.

c.   From the printout, the 95% confidence interval is (1.467, 2.018). We are 95% confident that the mean quit rate of all industries with an average hourly wage of $9.00 is between 1.467 and 2.018.

9.67   a.   For $n = 22$, $P(r_s > .508) = .01$

b.   For $n = 28$, $P(r_s > .448) = .01$

c.   For $n = 10$, $P(r_s \leq .648) = 1 - .025 = .975$

d.   For $n = 8$, $P(r_s < -.738$ or $r_s > .738) = 2(.025) = .05$

9.69   Since there are no ties, we will use the shortcut formula.

a.   Some preliminary calculations are:

x Rank (u_i)	y Rank (v_i)	$d_i = u_i - v_i$	d_i^2
3	2	1	1
5	4	1	1
2	5	-3	9
1	1	0	0
4	3	1	1
			Total = 12

$$r_s = 1 - \frac{6 \sum d_i^2}{n(n^2 - 1)} = 1 - \frac{6(12)}{5(5^2 - 1)} = 1 - .6 = .4$$

b.

x Rank (u_i)	y Rank (v_i)	$d_i = u_i - v_i$	d_i^2
2	3	-1	1
3	4	-1	1
4	2	2	4
5	1	4	16
1	5	-4	16
			Total = 38

$$r_s = 1 - \frac{6 \sum d_i^2}{n(n^2 - 1)} = 1 - \frac{6(38)}{5(5^2 - 1)} = 1 - 1.9 = -.9$$

c.

x Rank (u_i)	y Rank (v_i)	$d_i = u_i - v_i$	d_i^2
1	2	-1	1
4	1	3	9
2	3	-1	1
3	4	-1	1
			Total = 12

$$r_s = 1 - \frac{6 \sum d_i^2}{n(n^2 - 1)} = 1 - \frac{6(12)}{4(4^2 - 1)} = 1 - 1.2 = -.2$$

d.

x Rank (u_i)	y Rank (v_i)	$d_i = u_i - v_i$	d_i^2
2	1	1	1
5	3	2	4
4	5	-1	1
3	2	1	1
1	4	-3	9
		Total $= 16$	

$$r_s = 1 - \frac{6\sum d_i^2}{n(n^2 - 1)} = 1 - \frac{6(16)}{5(5^2 - 1)} = 1 - .8 = .2$$

9.71   a.   Some preliminary calculations are:

x	u	y	v	u-sq	v-sq	uv
5.2	1	220	4.5	1	20.25	4.5
5.5	7	227	7.5	49	56.25	52.5
6.0	23.5	259	15.5	552.25	240.25	364.25
5.9	20.5	210	1	420.25	1	20.5
5.8	16	224	6	256	36	96
6.0	23.5	215	3	552.25	9	70.5
5.8	16	231	9	256	81	144
5.6	10	268	19	100	361	190
5.6	10	239	11	100	121	110
5.9	20.5	212	2	420.25	4	41
5.4	5	410	24	25	576	120
5.6	10	256	14	100	196	140
5.8	16	306	22	256	484	352
5.5	7	259	15.5	49	240.25	108.5
5.3	3	284	21	9	441	63
5.3	3	383	23	9	529	69
5.7	12.5	271	20	156.25	400	250
5.5	7	264	18	49	324	126
5.7	12.5	227	7.5	156.25	56.25	93.75
5.3	3	263	17	9	289	51
5.9	20.5	232	10	420.25	100	205
5.8	16	220	4.5	256	20.25	72
5.8	16	246	13	256	169	208
5.9	20.5	241	12	420.25	144	246
$\sum u = 300$		$\sum v = 300$		$\sum u^2 = 4878$	$\sum v^2 = 4898.5$	$\sum uv = 3197.5$

$$SS_{uv} = \sum uv - \frac{(\sum u)(\sum v)}{n} = 3197.5 - \frac{300(300)}{24} = -552.5$$

$$SS_{uu} = \sum u^2 - \frac{(\sum u)^2}{n} = 4878 - \frac{300^2}{24} = 1128$$

$$SS_{vv} = \sum v^2 - \frac{(\sum v)^2}{n} = 4898.5 - \frac{300^2}{24} = 1148.5$$

$$r_s = \frac{SS_{uv}}{\sqrt{SS_{uu}SS_{vv}}} = \frac{-552.5}{\sqrt{1128(1148.5)}} = -.4854$$

Since the magnitude of the correlation coefficient is not particularly large, there is a fairly weak negative relationship between sweetness index and pectin.

b. To determine if there is a negative association between the sweetness index and the amount of pectin, we test:

H_0: $\rho_s = 0$
H_a: $\rho_s < 0$

The test statistic is $r_s = -.4854$

Reject H_0 if $r_s < -r_{s,\alpha}$ where $\alpha = .01$ and $n = 24$.

Reject H_0 if $r_s < -.485$ (from Table XIV, Appendix B)

Since the observed value of the test statistic falls in the rejection region ($r_s = -.4854 < -.485$), H_0 is rejected. There is sufficient evidence to indicate there is a negative association between the sweetness index and the amount of pectin at $\alpha = .01$.

9.73   Some preliminary calculations:

Year	u_i	v_i	$d_i = u_i - v_i$	d_i^2
1980	1	4	−3	9
1985	2	7	−5	25
1990	3	6	−3	9
1994	4	3	1	1
1995	7	1	6	36
1996	5	2	3	9
1997	6	5	1	1
			$\sum d_i^2 = 90$	

$$r_s = 1 - \frac{6 \sum d_i^2}{n(n^2 - 1)} = 1 - \frac{6(90)}{7(7^2 - 1)} = 1 - 1.607 = -.607$$

b. To determine if there is an association between amount spent and number employed, we test:

H_0: $\rho_s = 0$
H_a: $\rho_s \neq 0$

The test statistic is $r_s = -.607$

Reject H_0 if $r_s < -r_{s,\alpha/2}$ or $r_s > r_{s,\alpha/2}$ where $\alpha/2 = .10/2 = .05$ and $n = 7$.

Reject H_0 if $r_s < -.714$ or $r_s > .714$ (from Table XIV, Appendix B)

Since the observed value of the test statistic does not fall in the rejection region ($r_s = -.607 \not< -.714$), H_0 is not rejected. There is insufficient evidence to indicate there is an association between amount spent and number of employees at $\alpha = .10$.

9.75   b.   Some preliminary calculations:

Involvement	u_i	v_i	Differences $d_i = u_i - v_i$	d_i^2
1	8	9	-1	1
2	6	7	-1	1
3	10	10	0	0
4	2	1	1	1
5	5	5	0	0
6	9	8	1	1
7	1	2	-1	1
8	4	4	0	0
9	7	6	1	1
10	11	11	0	0
11	3	3	0	1
				$\sum d_i^2 = 6$

$$r_s = 1 - \frac{6\sum d_i^2}{n(n^2 - 1)} = 1 - \frac{6(6)}{11(11^2 - 1)} = .972$$

To determine if a positive relationship exists between participation rates and cost savings rates, we test:

H_0: $\rho_s = 0$
H_a: $\rho_s > 0$

The test statistic is $r_s = .972$.

From Table XIV, Appendix B, $r_{s,.01} = .736$, with $n = 11$. The rejection region is $r_s > .736$.

Since the observed value of the test statistic does falls in the rejection region ($r_s = .972 > .736$), H_0 is rejected. There is sufficient evidence to indicate that a positive relationship exists between participation rates and cost savings rates at $\alpha = .01$.

c.   In order for the above test to be valid, we must assume:

1.   The sample is randomly selected.
2.   The probability distributions of both of the variables are continuous.

In order to use the Pearson correlation coefficient, we must assume that both populations are normally distributed. It is very unlikely that the data are normally distributed.

9.77 a. $\hat{\beta}_1 = \dfrac{SS_{xy}}{SS_{xx}} = \dfrac{-88}{55} = -1.6, \hat{\beta}_0 = \bar{y} - \hat{\beta}_1 \bar{x} = 35 - (-1.6)(1.3) = 37.08$

The least squares line is $\hat{y} = 37.08 - 1.6x$.

b.

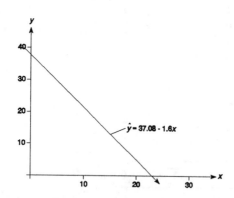

c. $SSE = SS_{yy} - \hat{\beta}_1 SS_{xy} = 198 - (-1.6)(-88) = 57.2$

d. $s^2 = \dfrac{SSE}{n-2} = \dfrac{57.2}{15-2} = 4.4$

e. For confidence coefficient .90, $\alpha = 1 - .90 = .10$ and $\alpha/2 = .10/2 = .05$. From Table VI, Appendix B, with df $= n - 2 = 15 - 2 = 13$, $t_{.05} = 1.771$. The 90% confidence interval for β_1 is:

$$\hat{\beta}_1 \pm t_{\alpha/2} \dfrac{s}{\sqrt{SS_{xx}}} \Rightarrow -1.6 \pm 1.771 \dfrac{\sqrt{4.4}}{\sqrt{55}} \Rightarrow -1.6 \pm .501 \Rightarrow (-2.101, -1.099)$$

We are 90% confident the change in the mean value of y for each unit change in x is between -2.101 and -1.099.

f. For $x_p = 15$, $\hat{y} = 37.08 - 1.6(15) = 13.08$

The 90% confidence interval is:

$$\hat{y} \pm t_{\alpha/2} s \sqrt{\dfrac{1}{n} + \dfrac{(x_p - \bar{x})^2}{SS_{xx}}} \Rightarrow 13.08 \pm 1.771(\sqrt{4.4}) \sqrt{\dfrac{1}{15} + \dfrac{(15 - 1.3)^2}{55}}$$

$$\Rightarrow 13.08 \pm 6.929 \Rightarrow (6.151, 20.009)$$

g. The 90% prediction interval is:

$$\hat{y} \pm t_{\alpha/2} s \sqrt{1 + \dfrac{1}{n} + \dfrac{(x_p - \bar{x})^2}{SS_{xx}}} \Rightarrow 13.08 \pm 1.771(\sqrt{4.4}) \sqrt{1 + \dfrac{1}{15} + \dfrac{(15 - 1.3)^2}{55}}$$

$$\Rightarrow 13.08 \pm 7.862 \Rightarrow (5.218, 20.942)$$

9.79   a.

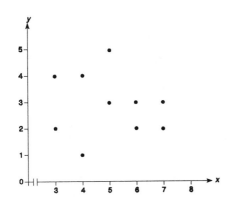

b.   Some preliminary calculations are:

$$\sum x = 50 \qquad \sum x^2 = 270 \qquad \sum xy = 143$$

$$\sum y = 29 \qquad \sum y^2 = 97$$

$$SS_{xy} = \sum xy - \frac{\sum x \sum y}{n} = 143 - \frac{50(29)}{10} = -2$$

$$SS_{xx} = \sum x^2 - \frac{(\sum x)^2}{n} = 270 - \frac{50^2}{10} = 20$$

$$SS_{yy} = \sum y^2 - \frac{(\sum y)^2}{n} = 97 - \frac{29^2}{10} = 12.9$$

$$r = \frac{SS_{xy}}{\sqrt{SS_{xx}SS_{yy}}} = \frac{-2}{\sqrt{20(12.9)}} -.1245$$

$$r^2 = 2(-.1245)^2 = .0155$$

c.   Some preliminary calculations are:

$$\hat{\beta}_1 = \frac{SS_{xy}}{SS_{xx}} = \frac{-2}{20} = -.1$$

$$SSE = SS_{yy} = \hat{\beta}_1 SS_{xy} = 12.9 - (-.1)(-2) = 12.7$$

$$s^2 = \frac{SSE}{n-2} = \frac{12.7}{10-2} = 1.5875 \quad s = \sqrt{1.5875} = 1.25996$$

To determine if x and y are linearly correlated, we test:

H_0: $\beta_1 = 0$
H_a: $\beta_1 \neq 0$

The test statistic is $t = \dfrac{\hat{\beta}_1 - 0}{\dfrac{s}{\sqrt{SS_{xx}}}} = \dfrac{-.1 - 0}{\dfrac{1.25996}{\sqrt{20}}} = -.35$

The rejection requires $\alpha/2 = .10/2 = .05$ in the each tail of the t-distribution with df $= n - 2$ $= 10 - 2 = 8$. From Table VI, Appendix B, $t_{.05} = 1.86$. The rejection region is $t > 1.86$ or $t < -1.86$.

Since the observed value of the test statistic does not fall in the rejection region ($t = -.35 \not<$ -1.86), H_0 is not rejected. There is insufficient evidence to indicate that x and y are linearly correlated at $\alpha = .10$.

d. Some preliminary calculations are:

x	u	y	v	u^2	v^2	uv
3	1.5	4	8.5	2.25	72.25	12.75
5	5.5	3	6	30.25	36	33
6	7.5	2	3	56.25	9	22.5
4	3.5	1	1	12.25	1	3.5
3	1.5	2	3	2.25	9	4.5
7	9.5	3	6	90.25	36	57.45
6	7.5	3	6	56.25	100	55
5	5.5	5	10	30.25	72.25	29.75
4	3.5	4	8.5	12.25	9	28.5
7	9.5	2	3	90.25		
$\sum u = 55$		$\sum v = 55$	$\sum u^2 = 382.5$		$\sum v^2 = 380.5$	$\sum uv = 291.5$

$$SS_{uv} = \sum uv - \frac{(\sum u)(\sum v)}{n} = 291.5 - \frac{55(55)}{10} = -11$$

$$SS_{uu} = \sum u^2 - \frac{(\sum u)^2}{n} = 382.5 - \frac{55^2}{10} = 80$$

$$SS_{vv} = \sum v^2 - \frac{(\sum v)^2}{n} = 380.5 - \frac{55^2}{10} = 78$$

$$r_s = \frac{SS_{uv}}{\sqrt{SS_{uu}SS_{vv}}} = \frac{-11}{\sqrt{80(78)}} = -.139$$

9.81 a. The plot of the data is:

It appears that there is a linear relationship between order size and time. As order size increases, the time tends to increase.

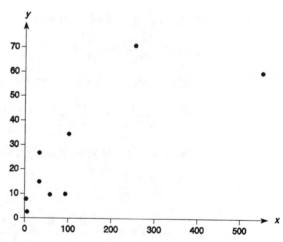

b. Some preliminary calculations are:

$$\sum x = 1149 \qquad \sum x^2 = 398,979 \qquad \sum xy = 58,102$$

$$\sum y = 239 \qquad \sum y^2 = 11,093$$

$$SS_{xy} = \sum xy - \frac{\sum x \sum y}{n} = 58,102 - \frac{1149(239)}{9} = 27,589.66667$$

$$SS_{xx} = \sum x^2 - \frac{(\sum x)^2}{n} = 398,979 - \frac{1149^2}{9} = 252,290$$

$$SS_{yy} = \sum y^2 - \frac{(\sum y)^2}{n} = 11,093 - \frac{239^2}{9} = 4746.222222$$

$$\hat{\beta}_1 = \frac{SS_{xy}}{SS_{xx}} = \frac{27,589.66667}{252,290} = .109356957 \approx .10936$$

$$\hat{\beta}_0 = \bar{y} - \hat{\beta}_1 \bar{x} = \frac{239}{9} - (.109356957)\frac{1149}{9} = 12.59431738 \approx 12.594$$

$$SSE = SS_{yy} - \hat{\beta}_1 SS_{xy} = 4746.222222 - (.109356957)(27,589.66667)$$
$$= 1729.10023$$

$$s^2 = \frac{SSE}{n-2} = \frac{1729.10023}{9-2} = 247.0143186 \qquad s = \sqrt{s^2} = 15.7167$$

The least squares line is $\hat{y} = 12.594 + .10936x$.

c. To determine if the mean time to fill an order increases with the size of the order, we test:

$H_0: \ \beta_1 = 0$
$H_a: \ \beta_1 > 0$

The test statistic is $t = \dfrac{\hat{\beta}_1 - 0}{s_{\hat{\beta}_1}} = \dfrac{.1094 - 0}{\dfrac{15.7167}{\sqrt{252,290}}} = 3.50$

The rejection region requires $\alpha = .05$ in the upper tail of the t-distribution. From Table VI, Appendix B, $t_{.05} = 1.895$, with df $= n - 2 = 9 - 2 = 7$. The rejection region is $t > 1.895$.

Since the observed value of the test statistic falls in the rejection region ($t = 3.50 > 1.895$), H_0 is rejected. There is sufficient evidence to indicate the mean time to fill an order increases with the size of the order for $\alpha = .05$.

d. For confidence coefficient .95, $\alpha = 1 - .95 = .05$ and $\alpha/2 = .05/2 = .025$. From Table VI, Appendix B, $t_{.025} = 2.365$ with df $= n - 2 = 9 - 2 = 7$.

The confidence interval is:

$$\hat{y} \pm t_{\alpha/2} s \sqrt{\frac{1}{n} + \frac{(x_p - \bar{x})^2}{SS_{xx}}}$$

For $x_p = 150$, $\hat{y} = 12.594 + .10936(150) = 28.998$, and $\bar{x} = \dfrac{1149}{9} = 127.6667$

$$28.988 \pm 2.365(15.7167)\sqrt{\frac{1}{9} + \frac{(150 - 127.6667)^2}{252,290}} \Rightarrow 28.988 \pm 12.500$$
$$\Rightarrow (16.498, 41.498)$$

9.83 Answers may vary. One possible answer may include:

The least squares line is $\hat{y} = -92.457684 + 8.346821x$. To determine if age can be used to predict market value, we test:

H_0: $\beta_1 = 0$
H_a: $\beta_1 \neq 0$

The test statistic is $t = 3.248$ with p-value $= .0021$. Reject the null hypothesis for levels of significance $\alpha > .0021$. There is sufficient evidence to indicate that age contributes information for the prediction of market value (y) at $\alpha > .0021$.

$r = .42$; Since this value is near .5, there is a moderate positive linear relationship between the value and age of the Beanie Baby.

9.85 a. $\sum x = 55$ $\sum x^2 = 899$ $\sum xy = 154.89$
$\sum y = 14.49$ $\sum y^2 = 42.0817$

$SS_{xy} = \sum xy - \dfrac{\sum x \sum y}{n} = 154.89 - \dfrac{55(14.49)}{5} = -4.55$

$SS_{xx} = \sum x^2 - \dfrac{(\sum x)^2}{n} = 899 - \dfrac{55^2}{5} = 294$

$SS_{yy} = \sum y^2 - \dfrac{(\sum y)^2}{n} = 42.0817 - \dfrac{14.49^2}{5} = .08968$

$\hat{\beta}_1 = \dfrac{SS_{xy}}{SS_{xx}} = \dfrac{-4.55}{294} = -.01547619 \approx -.015$

$\hat{\beta}_0 = \bar{y} - \hat{\beta}_1\bar{x} = 2.898 - (-.01547619)(11) = 3.068238095 \approx 3.068$

The least squares line is $\hat{y} = 3.068 - .015x$.

b.   The graph of the data is:

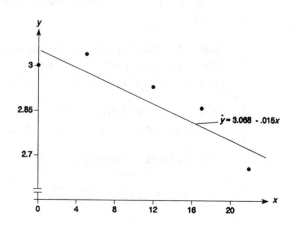

c. $SSE = SS_{yy} - \hat{\beta}_1 SS_{xy} = .08968 - (-.01547619)(-4.55) = .019263335$

$s^2 = \dfrac{SSE}{n-2} = \dfrac{.019263335}{5-2} = .0064211$ $s = \sqrt{.0064211} = .0801$

To determine whether the model is useful for predicting grade point average, we test:

H_0: $\beta_1 = 0$
H_a: $\beta_1 \neq 0$

The test statistic is $t = \dfrac{\hat{\beta}_1 - 0}{s_{\hat{\beta}_1}} = \dfrac{-.0155 - 0}{\dfrac{.0801}{\sqrt{294}}} = -3.32$

The rejection region requires $\alpha/2 = .10/2 = .05$ in each tail of the t-distribution. From Table VI, Appendix B, with df $= n - 2 = 5 - 2 = 3$, $t_{.05} = 2.353$. The rejection region is $t < -2.353$ or $t > 2.353$.

Since the observed value of the test statistic falls in the rejection region ($t = -3.32 < -2.353$), H_0 is rejected. There is sufficient evidence to indicate the model is useful for predicting grade point average at $\alpha = .10$.

d.   For $x = 10$, $\hat{y} = 3.068 - .0155(10) = 2.913$.

For confidence coefficient .90, $\alpha = .10$ and $\alpha/2 = .10/2 = .05$. From Table VI, Appendix B, with df $= n - 2 = 5 - 2 = 3$, $t_{.05} = 2.353$. The 90% prediction interval is:

$$\hat{y} \pm t_{\alpha/2}s\sqrt{1 + \frac{1}{n} + \frac{(x_p - \bar{x})^2}{SS_{xx}}} \Rightarrow 2.913 \pm 2.353(.0801)\sqrt{1 + \frac{1}{5} + \frac{(10 - 11)^2}{294}}$$

$\Rightarrow 2.913 \pm .207 \Rightarrow (2.706,\ 3.120)$

We are 90% confident that the actual grade point average of a high school student who works 10 hours per week is between 2.706 and 3.120.

9.87   a.   For confidence coefficient .90, $\alpha = 1 - .90 = .10$ and $\alpha/2 = .10/2 = .05$. From Table VI, Appendix B, $t_{.05} = 1.740$ with df $= n - 2 = 19 - 2 = 17$.

The prediction interval is:

$$\hat{y} \pm t_{\alpha/2}s\sqrt{1 + \frac{1}{n} + \frac{(x_p - \bar{x})^2}{SS_{xx}}} \quad \text{where } \hat{y} = 44.13 + .2366(55) = 57.143$$

$$\Rightarrow 57.143 \pm 1.74(19.40)\sqrt{1 + \frac{1}{19} + \frac{(55 - 44.1579)^2}{10,824.5263}} \Rightarrow 57.143 \pm 34.818$$

$$\Rightarrow (22.325,\ 91.961)$$

b.   The number of interactions with outsiders in the study went from 10 to 82. The value 110 is not within this interval. We do not know if the relationship between x and y is the same outside the observed range. Also, the farther x_p lies from $\bar{x}$ the larger will be the error of prediction. The prediction interval for a particular value of y will be very wide when $x_p = 110$.

c.   The prediction interval for a manager's success index will be narrowest when the number of contacts with people outside her work unit is $\bar{x} = 44.1579$ (44).

9.89

$$SS_{uv} = \sum uv - \frac{\sum uv}{n} = 2774.75 - \frac{210(210)}{20} = 569.75$$

$$SS_{uu} = \sum u^2 - \frac{\left(\sum u\right)^2}{n} = 2869.5 - \frac{210^2}{20} = 664.5$$

$$SS_{vv} = \sum v^2 - \frac{\left(\sum v\right)^2}{n} = 2869.5 - \frac{210^2}{20} = 664.5$$

$$r_s = \frac{SS_{uv}}{\sqrt{SS_{uu}SS_{vv}}} = \frac{569.75}{\sqrt{664.5(664.5)}} = .8574$$

Since $r_s = .8574$ is greater than 0, the relationship between current importance and ideal importance is positive. The relationship is fairly strong since r_s is close to 1. This implies the views on current importance and ideal importance are very similar.

9.91 Using MINITAB, the two regression analyses are:

Regression Analysis

The regression equation is
Ind.Costs = 301 + 10.3 Mach-Hours

Predictor	Coef	StDev	T	P
Constant	301.0	229.8	1.31	0.219
Mach-Hou	10.312	3.124	3.30	0.008

S = 170.5   R-Sq = 52.1%   R-Sq(adj) = 47.4%

Analysis of Variance

Source	DF	SS	MS	F	P
Regression	1	316874	316874	10.90	0.008
Residual Error	10	290824	29082		
Total	11	607698			

Regression Analysis
The regression equation is
Ind.Costs = 745 + 7.72 Direct-Hours

Predictor	Coef	StDev	T	P
Constant	744.7	217.6	3.42	0.007
Direct-H	7.716	5.396	1.43	0.183

S = 224.6   R-Sq = 17.0%   R-Sq(adj) = 8.7%

Analysis of Variance

Source	DF	SS	MS	F	P
Regression	1	103187	103187	2.05	0.183
Residual Error	10	504511	50451		
Total	11	607698			

Unusual Observations

Obs	Direct-H	Ind.Cost	Fit	StDev Fit	Residual	St Resid
9	70.0	1316.0	1284.8	181.9	31.2	0.24 X

X denotes an observation whose X value gives it large influence.

From these two cost functions, the model containing Machine-Hours should be used to predict Indirect Manufacturing Labor Costs. There is a significant linear relationship between Indirect Manufacturing Labor Costs and Machine-Hours $(t = 3.30, p = 0.008)$. There is not a significant linear relationship between Indirect Manufacturing Labor Costs and Direct Manufacturing Labor-Hours $(t = 1.43, p = 0.183)$. The r^2 for the first model is .521 while the r^2 for the second model is .170. In addition, the standard deviation for the first model is 170.5 while the standard deviation for the second model is 224.6. All of these lead to the better model as the model containing Machine-Hours as the independent variable.

Introduction to Multiple Regression

10.1   a. $E(y) = \beta_0 + \beta_1 x_1 + \beta_2 x_2$

      b. $E(y) = \beta_0 + \beta_1 x_1 + \beta_2 x_2 + \beta_3 x_3 + \beta_4 x_4$

      c. $E(y) = \beta_0 + \beta_1 x_1 + \beta_2 x_2 + \beta_3 x_3 + \beta_4 x_4 + \beta_5 x_5$

10.3   a.   We are given $\hat{\beta}_2 = 2.7$, $s_{\hat{\beta}_2} = 1.86$, and $n = 30$.

$$H_0: \beta_2 = 0$$
$$H_a: \beta_2 \neq 0$$

The test statistic is $t = \dfrac{\hat{\beta}_2 - 0}{s_{\hat{\beta}_2}} = \dfrac{2.7}{1.86} = 1.45$

The rejection region requires $\alpha/2 = .05/2 = .025$ in each tail of the t distribution with df $= n - (k + 1) = 30 - (3 + 1) = 26$. From Table VI, Appendix B, $t_{.025} = 2.056$. The rejection region is $t < -2.056$ or $t > 2.056$.

Since the observed value of the test statistic does not fall in the rejection region ($t = 1.45 \not> 2.056$), H_0 is not rejected. There is insufficient evidence to indicate $\beta_2 \neq 0$ at $\alpha = .05$.

      b.   We are given $\beta_3 = .93$, $s_{\hat{\beta}_3} = .29$, and $n = 30$.

Test $H_0: \beta_3 = 0$
             $H_a: \beta_3 \neq 0$

The test statistic is $t = \dfrac{\hat{\beta}_3 - 0}{s_{\hat{\beta}_3}} = \dfrac{.93}{.29} = 3.21$

The rejection region is the same as part **a**, $t < -2.056$ or $t > 2.056$.

Since the observed value of the test statistic falls in the rejection region ($t = 3.21 > 2.056$), H_0 is rejected. There is sufficient evidence to indicate $\beta_3 \neq 0$ at $\alpha = .05$.

      c. $\hat{\beta}_3$ has a smaller estimated standard error than $\hat{\beta}_2$. Therefore, the test statistic is larger for $\hat{\beta}_3$ even though $\hat{\beta}_3$ is smaller than $\hat{\beta}_2$.

10.5   The number of degrees of freedom available for estimating σ^2 is $n - (k + 1)$ where k is the number of independent variables in the regression model. Each additional independent variable placed in the model causes a corresponding decrease in the degrees of freedom.

10.7   a.   The first-order model is: $E(y) = \beta_0 + \beta_1 x_1 + \beta_2 x_2$

From the printout, the least squares prediction equation is:

$\hat{y} = -20.352 + 13.3504\, x_1 + 243.714\, x_2$

$\hat{\beta}_0 = -20.352$. This has no meaning since $x_1 = 0$ and $x_2 = 0$ are not in the observed range.

$\hat{\beta}_1 = 13.3504$. For each additional year of age, the mean annual earnings is predicted to increase by \$13.3504, holding hours worked per day constant.

$\hat{\beta}_2 = 243.714$. For each additional hour worked per day, the mean annual earnings is predicted to increase by \$243.714, holding age constant.

To determine if age is a useful predictor of annual earnings, we test:

$H_0: \beta_1 = 0$
$H_a: \beta_1 \neq 0$

The test statistic is $t = 1.74$.

The p-value is $p = .1074$. Since the p-value is greater than $\alpha = .01$ ($p = .1074 > \alpha = .01$), H_0 is not rejected. There is insufficient evidence to indicate that age is a useful predictor of annual earnings, adjusted for hours worked per day, at $\alpha = .01$.

e.   For confidence coefficient .99, $\alpha = .01$ and $\alpha/2 = .01/2 = .005$. From Table VI, Appendix B, with df $= n - 3 = 15 - 3 = 12$, $t = 3.055$. The 99% confidence interval is:

$\hat{\beta}_2 \pm t_{.005} s_{\hat{\beta}_2} \Rightarrow 243.714 \pm 3.055(63.5117) \Rightarrow 243.714 \pm 194.028 \Rightarrow (49.686, 437.742)$

We are 99% confident that the change in the mean annual earnings for each additional hour worked per day will be somewhere between \$49.686 and \$437.742, holding age constant.

10.9   a.   Using MINITAB, the output is:

Regression Analysis

```
The regression equation is
y = 20.9 + 0.261 x1 - 7.8 x2 + 0.0042 x3

Predictor      Coef      StDev         T        P
Constant      20.88      24.16      0.86    0.395
x1           0.2614     0.2394      1.09    0.284
x2            -7.85      13.28     -0.59    0.560
x3          0.00415    0.01042      0.40    0.693

S = 14.01      R-Sq = 9.7%      R-Sq(adj) = 0.0%

Analysis of Variance

Source           DF        SS       MS       F       P
Regression        3     566.5    188.8    0.96   0.425
Residual Error   27    5302.3    196.4
Total            30    5868.8

Source       DF     Seq SS
x1            1      372.3
x2            1      162.9
x3            1       31.2
```

The least squares prediction equation is:

$\hat{y} = 20.9 + 0.261 x_1 - 7.8 x_2 + .0042 x_3$.

      Chapter 10

b.　From the printout, the standard deviation is $s = 14.01$. Most of the observed values of the price of Ford stock will fall within $\pm 2s$ or $\pm 2(14.01)$ or ± 28.02 units of their predicted values.

c.　To determine if the price of Ford stock decreases as the yen rate increases, we test:

$$H_0: \beta_1 = 0$$
$$H_a: \beta_1 < 0$$

The test statistic is $t = 1.09$.

The p-value for the test is $p = 1 - .284/2 = 1 - .142 = .858$. Since the p-value is greater than $\alpha = .05$ ($p = .858 > \alpha = .05$), H_0 is not rejected. There is insufficient evidence to indicate that the price of Ford stock decreases as the yen rate increases holding the Deutsche Mark exchange rate and the S & P 500 Index constant at $\alpha = .05$.

d.　$\hat{\beta}_2 = -7.8$.　　The mean price of Ford stock is estimated to decrease by 7.8 for each unit increase in Deutsche mark exchange rate, holding the yen exchange rate and the S & P 500 Index constant.

10.11　a.　$\hat{y} = 12.2 - .0265x_1 - .458x_2$

b.　$\hat{\beta}_0 = 12.2 =$ the estimate of the y-intercept

$\hat{\beta}_1 = -.0265$. We estimate that the mean weight change will decrease by .0265% for each additional increase of 1% in digestion efficiency, with acid-detergent fibre held constant.

$\hat{\beta}_2 = -.458$. We estimate that the mean weight change will decrease by .458% for each additional increase of 1% in acid-detergent fibre, with digestion efficiency held constant.

c.　To determine if digestion efficiency is a useful predictor of weight change, we test:

$$H_0: \beta_1 = 0$$
$$H_a: \beta_1 \neq 0$$

The test statistic is $t = -.50$. The p-value is $p = .623$. Since the p-value is greater than α ($p = .623 > .01$), H_0 is not rejected. There is insufficient evidence to indicate that digestion efficiency is a useful linear predictor of weight change at $\alpha = .01$.

d.　For confidence coefficient .99, $\alpha = 1 - .99 = .01$ and $\alpha/2 = .01/2 = .005$. From Table VI, Appendix B, with df $= n - (k + 1) = 42 - (2 + 1) = 39$, $t_{.005} \approx 2.704$. The 99% confidence interval is:

$$\hat{\beta}_2 \pm t_{.005} s_{\hat{\beta}_2} \Rightarrow -.4578 \pm 2.704 \,(.1283) \Rightarrow -.4578 \pm .3469 \Rightarrow (-.8047, -.1109)$$

We are 99% confident that the change in mean weight change for each unit change in acid-detergent fiber, holding digestion efficiency constant is between $-.8047\%$ and $-.1109\%$.

e.　In order to include the qualitative variable Diet in the model, we must use dummy variables. Since there are only 2 levels of diet, we need 1 dummy variable.

Let $x_3 = \begin{cases} 1 \text{ if plants} \\ 0 \text{ otherwise} \end{cases}$

10.13 $\hat{\beta}_0 = 39.05 =$ the estimate of the y-intercept

$\hat{\beta}_1 = -5.41.$ We estimate that the mean operating margin will decrease by 5.41% for each additional increase of 1 unit of x_1, the state population divided by the total number of inns in the state (with all other variables held constant).

$\hat{\beta}_2 = 5.86.$ We estimate that the mean operating margin will increase by 5.86% for each additional increase of 1 unit of x_2, the room rate (with all other variables held constant).

$\hat{\beta}_3 = -3.09.$ We estimate that the mean operating margin will decrease by 3.09% for each additional increase of 1 unit of x_3, the square root of the median income of the area (with all other variables held constant).

$\hat{\beta}_4 = 1.75.$ We estimate that the mean operating margin will increase by 1.75% for each additional increase of 1 unit of x_4, the number of college students within four miles of the inn (with all other variables held constant).

10.15 a.   The SAS printout for the model is:

DEPENDENT VARIABLE: Y

SOURCE	DF	SUM OF SQUARES	MEAN SQUARE	F VALUE
MODEL	5	1052894700508.240	210578940101.648	190.75
ERROR	19	20975246806.001	1103960358.211	PR > F
CORRECTED TOTAL	24	1073869947314.240		0.0001

R-SQUARE	C.V.	ROOT MSE	Y MEAN
0.980468	11.4346	33225.899	290573.52000000

PARAMETER	ESTIMATE	T FOR HO: PARAMETER=0	PR > |T|	STD ERROR OF ESTIMATE
INTERCEPT	93073.85223495	3.24	0.0043	28720.89686205
X1	4152.20700875	2.78	0.0118	1491.62587008
X2	-854.94161450	-2.86	0.0099	298.44765134
X3	0.92424393	0.32	0.7515	2.87673442
X4	2692.46175182	1.71	0.1041	1577.28622584
X5	15.54276851	10.62	0.0001	1.46287006

The least squares prediction equation is:

$$\hat{y} = 93{,}074 + 4152x_1 - 855x_2 + .924x_3 + 2692x_4 + 15.5x_5$$

b. $s = $ ROOT MSE $= 33{,}225.9.$ We would expect about 95% of the observations to fall within $\pm 2s$ or $\pm 2(33{,}225.9)$ or $\pm 66{,}452$ units of the regression line.

c.   To determine if the value increases with the number of units, we test:

$H_0:\ \beta_1 = 0$
$H_a:\ \beta_1 > 0$

The test statistic is $t = \dfrac{\hat{\beta}_1 - 0}{s_{\hat{\beta}_1}} = \dfrac{4152 - 0}{1491.626} = 2.78$

The observed significance level or p-value is .0118/2 = .0059. Since this value is less than $\alpha = .05$, H_0 is rejected. There is sufficient evidence to indicate that the value increases as the number of units increases at $\alpha = .05$.

d. $\hat{\beta}_1$: We estimate the mean value will increase by \$4,152 for each additional apartment unit, all other variables held constant.

e. Using SAS, the plot is:

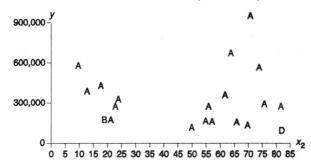

PLOT IF Y*X2   LEGEND: A = 1 OBS, B = 2 OBS, ETC.

It appears from the graph that there is not much of a linear relationship between value (y) and age (x_2).

f. H_0: $\beta_2 = 0$
H_a: $\beta_2 < 0$

The test statistic is $t = \dfrac{\hat{\beta}_2 - 0}{s_{\hat{\beta}_2}} = \dfrac{-855 - 0}{298.447} = -2.86$

The observed significance level or p-value is $.0099/2 = .00495$. Since this value is less than $\alpha = .01$, H_0 is rejected. There is sufficient evidence to indicate that the value and age are negatively related, all other variables in the model held constant, at $\alpha = .01$.

A one-tailed test is reasonable because the older the building, the lower the sales price (market value), at least for certain values of age.

g. The p-value is $.0099/2 = .00495$ (because we had a one-tailed test).

10.17 a. $R^2 = .8911$

89.11% of the total sample variability of y is explained by the linear relationship between y and the two independent variables x_1 and x_2.

b. $R_a^2 = .8775$

87.75% of the total sample variability of y is explained by the linear relationship between y and the two independent variables x_1 and x_2, adjusting for the sample size and the number of independent variables.

c. H_0: $\beta_1 = \beta_2 = 0$
H_a: At least one $\beta_i \neq 0$, for $i = 1, 2$

The test statistic is $F = \dfrac{R^2/k}{(1 - R^2)/[n - (k + 1)]} = \dfrac{.8911/2}{(1 - .8911)/[19 - (2 + 1)]} = 65.462$

The rejection region requires $\alpha = .05$ in the upper tail of the F distribution with df $= \nu_1 = k = 2$ and $\nu_2 = n - (k + 1) = 19 - (2 + 1) = 16$. From Table IX, Appendix B, $F_{.05} = 3.63$. The rejection region is $F > 3.63$.

Since the observed value of the test statistic falls in the rejection region ($F = 65.462 > 3.63$), H_0 is rejected. There is sufficient evidence to indicate the model is useful in predicting y at $\alpha = .05$.

The test statistic can also be calculated by $\dfrac{\text{MS(Model)}}{\text{MS(Error)}} = \dfrac{12.11167}{0.18497} = 65.479$

From the printout, $F = 65.478$.

d.   Prob $> F = p$-value $\le .0001$

The probability of observing a test statistic of 65.478 or anything higher is less than .0001. This is very unusual if H_0 is true. This is very significant. There is strong evidence to reject H_0 for $\alpha > .0001$.

10.19   a.   Some preliminary calculations are:

$$\text{SSE} = \sum (y_i - \hat{y}_i)^2 = 12.35, \; df = n - (k + 1) = 20 - (2 + 1) = 17$$

$$\text{SS(Total)} = \sum (y - \bar{y})^2 = 24.44, \; df = n - 1 = 20 - 1 = 19$$

$$\text{SS(Model)} = \text{SS(Total)} - \text{SSE} = 24.44 - 12.35 = 12.09, \; df = k = 2$$

$$\text{MS(Model)} = \frac{\text{SS(Model)}}{k} = \frac{12.09}{2} = 6.045$$

$$\text{MS(Error)} = \frac{\text{SSE}}{n - (k + 1)} = \frac{12.35}{17} = .72647$$

$$F = \frac{\text{MS(Model)}}{\text{MS(Error)}} = \frac{6.045}{.72647} = 8.321$$

$$R^2 = 1 - \frac{\text{SSE}}{\text{SS(Total)}} = 1 - \frac{12.35}{24.44} = .4947$$

$$R_a^2 = 1 - \left[\frac{n - 1}{n - (k + 1)} \right] (1 - R^2)$$

$$= 1 - \left[\frac{20 - 1}{20 - (2 + 1)} \right] (1 - .4947)$$

$$= .4352$$

The test statistic could also be calculated by:

$$F = \frac{R^2/k}{(1 - R^2)/[n - (k + 1)]} = \frac{.4947/2}{(1 - .4947)/17} = 8.32$$

The analysis of variance table is:

Source	df	SS	MS	F
Model	2	12.09	6.045	8.321
Error	17	12.35	.72647	
Total	19	24.44		

b. H_0: $\beta_1 = \beta_2 = 0$
H_a: At least one $\beta_i \neq 0$, $i = 1, 2$

The test statistic is $F = \dfrac{\text{MS(Model)}}{\text{MS(Error)}} = \dfrac{6.045}{.72647} = 8.321$

The rejection region requires $\alpha = .05$ in the upper tail of the F distribution with df $= \nu_1 = k$ $= 2$ and $\nu_2 = n - (k + 1) = 17$. From Table IX, Appendix B, $F_{.05} = 3.59$. The rejection region is $F > 3.59$.

Since the observed value of the test statistic falls in the rejection region ($F = 8.321 > 3.59$), H_0 is rejected. There is sufficient evidence to indicate the model is useful in predicting y at $\alpha = .05$.

10.21   a.   From the printout, R^2 = R Square = .8168492. This means that 81.68% of the sample variation of the total pay is explained by the linear relationship between total pay and the independent variables company performance and company sales.

b.   To determine if at least one of the variables in the model is useful in predicting total pay, we test:

H_0: $\beta_1 = \beta_2 = 0$
H_a: At least one of the coefficients is nonzero

c.   The test statistic is $F = \dfrac{\text{MS(Model)}}{\text{MSE}} = 31.22$

The p-value is $p = 6.91292E\text{-}06 = .0000069$.

d.   Since the p-value is less than $\alpha = .05$ ($p = .0000069 < .05$), H_0 is rejected. There is sufficient evidence to indicate the model is useful in predicting total pay at $\alpha = .05$.

10.23   a.   The least squares prediction equation is:

$$\hat{y} = -4.30 - .002x_1 + .336x_2 + .384x_3 + .067x_4 - .143x_5 + .081x_6 + .134x_7$$

b.   To determine if the model is adequate, we test:

H_0: $\beta_1 = \beta_2 = \beta_3 = \beta_4 = \beta_5 = \beta_6 = \beta_7 = 0$
H_a: At least one $\beta_i \neq 0$, $i = 1, 2, 3, ..., 7$

The test statistic is $F = 111.1$ (from table).

Since no α was given, we will use $\alpha = .05$. The rejection region requires $\alpha = .05$ in the upper tail of the F-distribution with $\nu_1 = k = 7$ and $\nu_2 = n - (k + 1) = 268 - (7 + 1) = 260$. From Table IX, Appendix B, $F_{.05} \approx 2.01$. The rejection region is $F > 2.01$.

Since the observed value of the test statistic falls in the rejection region ($F = 111.1 > 2.01$), H_0 is rejected. There is sufficient evidence to indicate that the model is adequate for predicting the logarithm of the audit fees at $\alpha = .05$.

c. $\hat{\beta}_3 = .384$.   For each additional subsidiary of the auditee, the mean of the logarithm of audit fee is estimated to increase by .384 units.

d.   To determine if the $\beta_4 > 0$, we test:

$$H_0: \beta_4 = 0$$
$$H_a: \beta_4 > 0$$

The test statistic is $t = 1.76$ (from table).

The p-value for the test is .079. Since the p-value is not less than α ($p = .079 \not< \alpha = .05$), H_0 is not rejected. There is insufficient evidence to indicate that $\beta_4 > 0$, holding all the other variables constant, at $\alpha = .05$.

e.   To determine if the $\beta_1 < 0$, we test:

$$H_0: \beta_1 = 0$$
$$H_a: \beta_1 < 0$$

The test statistic is $t = -0.049$ (from table).

The p-value for the test is .961. Since the p-value is not less than α ($p = .961 \not< \alpha = .05$), H_0 is not rejected. There is insufficient evidence to indicate that $\beta_1 < 0$, holding all the other variables constant, at $\alpha = .05$. There is insufficient evidence to indicate that the new auditors charge less than incumbent auditors.

10.25 a. $R^2 = .51$. 51% of the variability in the operating margins can be explained by the model containing these four independent variables.

b.   To determine if the model is adequate, we test:

$$H_0: \beta_1 = \beta_2 = \beta_3 = \beta_4 = 0$$
$$H_a: \text{At least one } \beta_i \neq 0, i = 1, 2, 3, 4$$

The test statistic is

$$F = \frac{R^2/k}{(1 - R^2)/[n - (k + 1)]} = \frac{.51/4}{(1 - .51)/[57 - (4 + 1)]} = 13.53$$

The rejection region requires $\alpha = .05$ in the upper tail of the F distribution with $v_1 = k = 4$ and $v_2 = n - (k + 1) = 57 - (4 + 1) = 52$. From Table IX, Appendix B, $F_{.05} \approx 2.61$. The rejection region is $F > 2.61$.

Since the observed value of the test statistic falls in the rejection region ($F = 13.53 > 2.61$), H_0 is rejected. There is sufficient evidence that the model is useful in predicting operating margins at $\alpha = .05$.

10.27 To determine if the model is useful, we test:

H_0: $\beta_1 = \beta_2 = \cdots = \beta_{18} = 0$
H_a: At least one $\beta_i \neq 0$, $i = 1, 2, \ldots, 18$

The test statistic is $F = \dfrac{R^2/k}{(1 - R^2)/[n - (k + 1)]} = \dfrac{.95/18}{(1 - .95)/[20 - (18 + 1)]} = 1.06$

The rejection region requires $\alpha = .05$ in the upper tail of the F distribution with $\nu_1 = k = 18$ and $\nu_2 = n - (k + 1) = 20 - (18 + 1) = 1$. From Table IX, Appendix B, $F_{.05} \approx 245.9$. The rejection region is $F > 245.9$.

Since the observed value of the test statistic does not fall in the rejection region ($F = 1.06 \not> 247$), H_0 is not rejected. There is insufficient evidence to indicate the model is adequate at $\alpha = .05$.

Note: Although R^2 is large, there are so many variables in the model that ν_2 is small.

10.29 a. $R^2 = .529$. 52.9% of the total variability of weight change is explained by the model containing the two independent variables.

$R_a^2 = .505$. This statistic has a similar interpretation to that of R^2, but is adjusted for both the sample size n and the number of β parameters in the model.

The R_a^2 statistic is the preferred measure of model fit because it takes into account the sample size and the number of β parameters.

b. H_0: $\beta_1 = \beta_2 = 0$
H_a: At least one $\beta_i \neq 0$, $i = 1, 2$

The test statistic is $F = 21.88$ with p-value $= .000$. Since the p-value is so small, H_0 is rejected for any $\alpha > .000$. There is sufficient evidence to indicate that the model is adequate.

10.31 a.   The 95% prediction interval is $(1,759.7, 4,275.4)$. We are 95% confident that the true actual annual earnings for a vendor who is 45 years old and who works 10 hours per day is between $1,759.7 and $4,275.4.

b.   The 95% confidence interval is $(2,620.3, 3,414.9)$. We are 95% confident that the true mean annual earnings for vendors who are 45 years old and who work 10 hours per day is between $2,620.3 and $3,414.9.

c.   Yes. The prediction interval for the ACTUAL value of y is always wider than the confidence interval for the MEAN value of y.

10.33 The first order model is:

$$E(y) = \beta_0 + \beta_1 x_1 + \beta_2 x_2 + \beta_3 x_5$$

We want to find a 95% prediction interval for the actual voltage when the volume fraction of the disperse phase is at the high level ($x_1 = 80$), the salinity is at the low level ($x_2 = 1$), and the amount of surfactant is at the low level ($x_5 = 2$).

Using MINITAB, the output is:

```
The regression equation is
y = 0.993 - 0.0243 x1 + 0.142 x2 + 0.385 x5

Predictor            Coef          StDev              T             P
Constant           0.9326         0.2482           3.76         0.002
x1             -0.024272        0.004900          -4.95         0.000
x2               0.14206         0.07573           1.88         0.080
x5               0.38457         0.09801           3.92         0.001

S = 0.4796        R-Sq = 66.6%       R-Sq(adj) = 59.9%

Analysis of Variance

Source               DF            SS             MS          F          P
Regression            3        6.8701         2.2900       9.95      0.001
Residual Error       15        3.4509         0.2301
Total                18       10.3210

Source        DF        Seq SS
x1             1        1.4016
x2             1        1.9263
x5             1        3.5422

Unusual Observations
Obs         x1          y          Fit     StDev Fit     Residual     St Resid
  3       40.0       3.200        2.068        0.239        1.132        2.72R

R denotes an observation with a large standardized residual

Predicted Values

    Fit     StDev Fit          95.0% CI              95.0% PI
 -0.098        0.232     ( -0.592,   0.396)     ( -1.233,   1.038)
```

The 95% prediction interval is $(-1.233, 1.038)$. We are 95% confident that the actual voltage is between -1.233 and 1.038 kw/cm when the volume fraction of the disperse phase is at the high level $(x_1 = 80)$, the salinity is at the low level $(x_2 = 1)$, and the amount of surfactant is at the low level $(x_5 = 2)$.

10.35   a.   To determine if the model is useful for predicting the number of man-hours needed, we test:

H_0: $\beta_1 = \beta_2 = \beta_3 = \beta_4 = 0$
H_a: At least one $\beta_i \neq 0$, $i = 1, 2, 3, 4$

The test statistic is $F = 72.11$ with p-value $= .000$. Since the p-value is less than $\alpha = .01$, we can reject H_0. There is sufficient evidence that the model is useful for predicting man-hours at $\alpha = .01$.

b.   The confidence interval is (1449, 2424).

With 95% confidence, we can conclude that the mean number of man-hours for all boilers with characteristics $x_1 = 150,000$, $x_2 = 500$, $x_3 = 1$, $x_4 = 0$ will fall between 1449 hours and 2424 hours.

The prediction interval is (47, 3825).

With 95% confidence, we can conclude that the number of man-hours for an individual boiler with characteristics $x_1 = 150,000$, $x_2 = 500$, $x_3 = 1$, $x_4 = 0$ will fall between 47 hours and 3825 hours.

10.37   Yes. x_2 and x_4 are highly correlated (.93), as well as x_4 and x_5 (.86). When highly correlated independent variables are present in a regression model, the results can be confusing. The researcher may want to include only one of the variables.

10.39   When independent variables that are highly correlated with each other are included in a regression model, the results may be confusing. Highly correlated independent variables contribute overlapping information in the prediction of the dependent variable. The overall global test can indicate that the model is useful in predicting the dependent variable, while the individual t-tests on the independent variables can indicate that none of the independent variables are significant. This happens because the individual t-tests tests for the significance of an independent variable after the other independent variables are taken into account. Usually, only one of the independent variables that are highly correlated with each other is included in the regression model.

10.41   a.   Using MINITAB, all pairwise correlations are found among the independent variables.

Correlations (Pearson)

	Estimate	LowBid	Status	No_Bid	Days	Length	Asphalt	Base
LowBid	-0.082							
	0.172							
Status	-0.191	0.559						
	0.001	0.000						
No_Bid	0.352	-0.394	-0.487					
	0.000	0.000	0.000					
Days	0.805	-0.042	-0.199	0.304				
	0.000	0.489	0.001	0.000				
Length	-0.071	-0.074	0.055	-0.002	-0.135			
	0.299	0.279	0.418	0.978	0.047			
Asphalt	-0.321	0.027	0.133	-0.291	-0.476	0.461		
	0.000	0.652	0.026	0.000	0.000	0.000		
Base	0.147	-0.230	-0.252	0.453	0.226	-0.075	-0.513	
	0.014	0.000	0.000	0.000	0.000	0.272	0.000	
Excav	0.422	-0.113	-0.157	0.338	0.427	-0.253	-0.669	0.397
	0.000	0.059	0.009	0.000	0.000	0.000	0.000	0.000
Mobil	-0.069	0.356	0.343	-0.320	-0.062	-0.141	-0.135	-0.191
	0.249	0.000	0.000	0.000	0.304	0.038	0.024	0.001
Struct	0.280	-0.049	-0.207	0.290	0.500	-0.335	-0.653	0.120
	0.000	0.419	0.001	0.000	0.000	0.000	0.000	0.045
Traffic	-0.262	0.353	0.476	-0.391	-0.241	-0.181	-0.102	-0.168
	0.000	0.000	0.000	0.000	0.000	0.008	0.090	0.005
Subcont	0.121	-0.049	-0.006	-0.037	0.125	-0.060	0.001	0.013
	0.044	0.418	0.923	0.539	0.037	0.381	0.983	0.825

	Excav	Mobil	Struct	Traffic
Mobil	-0.031			
	0.611			
Struct	0.274	0.007		
	0.000	0.911		
Traffic	-0.113	0.392	0.001	
	0.060	0.000	0.989	
Subcont	-0.081	-0.019	0.038	0.018
	0.175	0.752	0.532	0.760

Cell Contents: Correlation
               P-Value

Of all the pairwise correlations, only 6 have correlations greater than .5 in magnitude. These include the correlation between status of contract and ratio of low bid price to DOT engineer's estimate of fair price (.559), correlation between estimated number of days to complete work and the DOT engineer's estimate of fair contract price (.805), correlation between percentage of costs allocated to excavation and percentage of costs allocated to liquid asphalt (-.669), correlation between percentage of costs allocated to structures and percentage of costs allocated to liquid asphalt (-.653), correlation between percentage of costs allocated to base material and percentage of costs allocated to liquid asphalt (-.513), and correlation between percentage of costs allocated to structures and estimated days to complete work (.500).

The best model might include only one of the two variables in the any of the above pairs. One method that could be used to select which variables to include is a stepwise regression. To do the stepwise regression, we did not include the variable Ratio of low (winning) bid price to DOT engineer's estimate of fair price since the value of this variable would not be known without knowing the actual price. In addition, dummy variables need to be created for the qualitative variable District. Using MINITAB, the results of the stepwise regression are:

Stepwise Regression

```
F-to-Enter:      3.00    F-to-Remove:      3.00

Response is  Price   on 16 predictors, with N =  217
N(cases with missing observations) =  62 N(all cases) =  279

      Step         1        2        3
  Constant     33.25   -12.93   -59.80

  Estimate    0.9064   0.9123   0.8862
  T-Value      94.11    94.89    57.08

  Status                  132      139
  T-Value                3.19     3.38

  Days                            0.36
  T-Value                         2.14

  S              279      273      271
  R-Sq         97.63    97.74    97.79
```

From this analysis, only three variables were selected – DOT engineer's estimate of fair contract price, status of the contract and estimated number of days to complete work. From the correlations, the DOT engineer's estimate of fair contract price and the estimated number of days to complete work had a very high correlation (.805). However, the stepwise correlation still selected both variables for the model.

b.   Using MINITAB, the analysis of fitting a first order model is:

Regression Analysis

```
The regression equation is
Price = - 55.9 + 0.912 Estimate + 139 Status + 0.272 Days

Predictor         Coef       StDev          T        P
Constant        -55.88       30.90      -1.81    0.072
Estimate       0.91159     0.01468      62.10    0.000
Status          139.39       37.53       3.71    0.000
Days            0.2716      0.1580       1.72    0.087

S = 282.3      R-Sq = 97.6%     R-Sq(adj) = 97.6%

Analysis of Variance

Source            DF          SS          MS         F        P
Regression         3   902557034   300852345   3775.20    0.000
Residual Error   275    21915236       79692
Total            278   924472271

Source     DF       Seq SS
Estimate    1    901293659
Status      1      1027885
Days        1       235490
```

Even though the stepwise regression selected days for the model, we see from this analysis that days would not be included at the $\alpha = .05$ level.

Eliminating days as a predictor variable, the results are:

Regression Analysis

```
The regression equation is
Price = - 22.5 + 0.932 Estimate + 134 Status

Predictor         Coef       StDev          T        P
Constant        -22.47       24.11      -0.93    0.352
Estimate      0.931701    0.008899     104.70    0.000
Status          134.36       37.54       3.58    0.000

S = 283.3      R-Sq = 97.6%     R-Sq(adj) = 97.6%

Analysis of Variance

Source            DF          SS          MS         F        P
Regression         2   902321545   451160772   5621.50    0.000
Residual Error   276    22150726       80256
Total            278   924472271

Source     DF       Seq SS
Estimate    1    901293659
Status      1      1027885
```

c. Using MINITAB, the stem-and-leaf display of the residuals is:

```
Stem-and-leaf of RESI1     N  = 279
Leaf Unit = 100

     1   -2 2
     1   -1
     2   -1 1
     9   -0 9977765
   130   -0 433333221111111111111111111111111111110000000000000000000000000000+
  (139)   0 0000000000000000000000000000000000000000000000000000000000000000000+
    10    0 55666779
     2    1 0
     1    1 7
```

Since the display looks fairly mound-shaped, the assumption of normality appears to be valid.

A plot of the residuals versus ESTIMATE is:

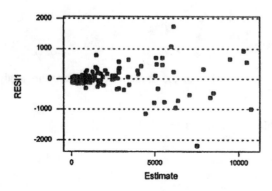

There are 7 observations that are more than 3 standard deviations from the mean. Thus, there are 7 possible outliers. There is no apparent curve to the plot, so adding Estimate squared to the model would not appear to improve it. The residuals have an apparent cone-shape. Thus it appears that the residuals increase as the estimate increases. The assumption of constant variance appears to be violated.

Since status has only two levels, the plot of the residuals versus STATUS will not provide any useful information.

10.43 a. No; Income and household size do not seem to be highly correlated. The correlation coefficient between income and household size is $-.137$.

b. Yes; The residuals versus income and residuals versus homesize exhibit a curved shape. Such a pattern could indicate that a second-order model may be more appropriate.

c. No; The residuals versus the predicted values reveals varying spreads for different values of $\hat{y}$. This implies that the variance of $\in$ is not constant for all settings of the x's.

d. Yes; The outlier shows up in several plots and is the 26th household (Food consumption = $7500, income = $7300 and household size = 5).

e. No; The frequency distribution of the residuals shows that the outlier skews the frequency distribution to the right.

10.45 In multiple regression, as in simple regression, the confidence interval for the mean value of y is narrower than the prediction interval of a particular value of y.

10.47   a.   The least squares equation is $\hat{y} = 90.1 - 1.836x_1 + .285x_2$

   b. $R^2 = .916$. About 91.6% of the sample variability in the y's is explained by the model $E(y) = \beta_0 + \beta_1 x_1 + \beta_2 x_2$

   c.   To determine if the model is useful for predicting y, we test:

$$H_0: \beta_1 = \beta_2 = 0$$
$$H_a: \text{At least one } \beta_i \neq 0, \ i = 1, 2$$

The test statistic is $F = \dfrac{MSR}{MSE} = \dfrac{7400}{114} = 64.91$

The rejection region requires $\alpha = .05$ in the upper tail of the F distribution with $\nu_1 = k = 2$ and $\nu_2 = n - (k + 1) = 15 - (2 + 1) = 12$. From Table IX, Appendix B, $F_{.05} = 3.89$. The rejection region is $F > 3.89$.

Since the observed value of the test statistic falls in the rejection region ($F = 64.91 > 3.89$), H_0 is rejected. There is sufficient evidence to indicate the model is useful for predicting y at $\alpha = .05$.

   d. $H_0: \beta_1 = 0$
$H_a: \beta_1 \neq 0$

The test statistic is $t = \dfrac{\hat{\beta}_1}{s_{\hat{\beta}_1}} = \dfrac{-1.836}{.367} = -5.01$

The rejection region requires $\alpha/2 = .05/2 = .025$ in each tail of the t distribution with df $= n - (k + 1) = 15 - (2 + 1) = 12$. From Table VI, Appendix B, $t_{.025} = 2.179$. The rejection region is $t < -2.179$ or $t > 2.179$.

Since the observed value of the test statistic falls in the rejection region ($t = -5.01 < -2.179$), H_0 is rejected. There is sufficient evidence to indicate β_1 is not 0 at $\alpha = .05$.

   e.   The standard deviation is $\sqrt{MSE} = \sqrt{114} = 10.677$. We would expect about 95% of the observations to fall within $2(10.677) = 21.354$ units of the fitted regression line.

10.49 Even though SSE $= 0$, we cannot estimate σ^2 because there are no degrees of freedom corresponding to error. With three data points, there are only two degrees of freedom available. The degrees of freedom corresponding to the model is $k = 2$ and the degrees of freedom corresponding to error is $n - (k + 1) = 3 - (2 + 1) = 0$. Without an estimate for σ^2, no inferences can be made.

10.51 a. The first order model for $E(y)$ as a function of the first five independent variables is:

$$E(y) = \beta_0 + \beta_1 x_1 + \beta_2 x_2 + \beta_3 x_3 + \beta_4 x_4 + \beta_5 x_5$$

b. To test the utility of the model, we test:

H_0: $\beta_1 = \beta_2 = \beta_3 = \beta_4 = \beta_5 = 0$
H_a: At least one $\beta_i \neq 0$, $i = 1, 2, 3, 4, 5$

The test statistic is $F = 34.47$.

The p-value is $p < .001$. Since the p-value is so small, there is sufficient evidence to indicate the model is useful for predicting GSI at $\alpha > .001$.

$R^2 = .469$.   46.9% of the variability in the GSI scores is explained by the model including the first five independent variables.

c. The first order model for $E(y)$ as a function of all seven independent variables is:

$$E(y) = \beta_0 + \beta_1 x_1 + \beta_2 x_2 + \beta_3 x_3 + \beta_4 x_4 + \beta_5 x_5 + \beta_6 x_6 + \beta_7 x_7$$

d. $R^2 = .603$ 60.3% of the variability in the GSI scores is explained by the model including the first seven independent variables.

e. Since the p-values associated with the variables DES and PDEQ-SR are both less than .001, there is evidence that both variables contribute to the prediction of GSI, adjusted for all the other variables already in the model for $\alpha > .001$.

10.53 The correlation coefficient between Importance and Replace is .2682. This correlation coefficient is fairly small and would not indicate a problem with multicollinearity between Importance and Replace. The correlation coefficient between Importance and Support is .6991. This correlation coefficient is fairly large and would indicate a potential problem with multicollinearity between Importance and Support. Probably only one of these variables should be included in the regression model. The correlation coefficient between Replace and Support is $-.0531$. This correlation coefficient is very small and would not indicate a problem with multicollinearity between Replace and Support. Thus, the model could probably include Replace and one of the variables Support or Importance.

10.55 a. The model is:
$$E(y) = \beta_0 + \beta_1 x$$

where y = market share and
$$x = \begin{cases} 1 \text{ if L} \\ 0 \text{ otherwise} \end{cases}$$

We assume that the error terms (ϵ_i) or y's are normally distributed at each exposure level, with a common variance. Also, we assume that the ϵ_i's have a mean of 0 and are independent.

b. Using MINITAB, the printout is:

Regression Analysis

```
The regression equation is
y = 11.5 - 0.917 x

Predictor        Coef       StDev          T          P
Constant       11.4917     0.1918      59.92      0.000
x              -0.9167     0.2712      -3.38      0.003

S = 0.6644      R-Sq = 34.2%      R-Sq(adj) = 31.2%

Analysis of Variance

Source           DF          SS          MS          F          P
Regression        1      5.0417      5.0417      11.42      0.003
Residual Error   22      9.7117      0.4414
Total            23     14.7533

Unusual Observations
Obs        x           y          Fit   StDev Fit    Residual      St      Resid
 18     0.00      12.900       11.492       0.192       1.408             2.21R

R denotes an observation with a large standardized residual
```

The fitted model is $\hat{y} = 11.5 - 0.917x$.

To determine if the expected market share differs for the two levels of advertising exposure, we test:

H_0: $\beta_1 = 0$
H_a: $\beta_1 \neq 0$

The test statistic is $t = -3.38$.

The p-value is $p = 0.003$. Since the p-value is so small, H_0 is rejected. There is sufficient evidence to indicate the expected market share differs for the two levels of advertising exposure at $\alpha = .05$.

10.57 a. $\hat{\beta}_1 = .02573$. The mean GPA is estimated to increase by .02573 for each 1 percentile point increase in verbal score, mathematics score held constant.

$\hat{\beta}_2 = .03361$. The mean GPA is estimated to increase by .03361 for each 1 percentile point increase in mathematics score, verbal score held constant.

b. The standard deviation is $\sqrt{\text{MSE}} = \sqrt{.16183} = .40228$. We would expect about 95% of the observations to fall within $2(.40228) = .80456$ units of their predicted values.

$R_a^2 = .66382$. About 66% of the sample variability in GPA's is explained by the model containing verbal and mathematics scores, adjusting for sample size and the number of parameters in the model.

c.   To determine if the model is useful for predicting GPA, we test:

H_0: $\beta_1 = \beta_2 = 0$
H_a: At least one $\beta_i \neq 0$, $i = 1, 2$

The test statistic is $F = \dfrac{MSR}{MSE} = \dfrac{6.39297}{.16183} = 39.505$

The p-value is .0000. Since the p-value is so small, H_0 is rejected. There is sufficient evidence to indicate the model is useful for predicting GPA.

d.   For $x_2 = 60$, $\hat{y} = -1.57 + .026x_1 + .034(60) = .47 + .026x_1$
     For $x_2 = 75$, $\hat{y} = -1.57 + .026x_1 + .034(75) = .98 + .026x_1$
     For $x_2 = 90$, $\hat{y} = -1.57 + .026x_1 + .034(90) = 1.49 + .026x_1$

The plot is:

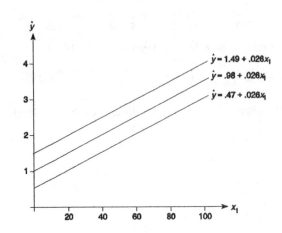

e.   From the plot of the residuals against x_1, there is a general mound shape to the residuals. This indicates that adding x_1^2 to the model might provide significant information to the model. From the plot of the residuals against x_2, there is a general mound or bowl shape. This indicates that adding x_2^2 to the model might add significantly to the model.

11.1   A control chart is a time series plot of individual measurements or means of a quality variable to which a centerline and two other horizontal lines called control limits have been added. The center line represents the mean of the process when the process is in a state of statistical control. The upper control limit and the lower control limit are positioned so that when the process is in control the probability of an individual measurement or mean falling outside the limits is very small. A control chart is used to determine if a process is in control (only common causes of variation present) or not (both common and special causes of variation present). This information helps us to determine when to take action to find and remove special causes of variation and when to leave the process alone.

11.3   When a control chart is first constructed, it is not known whether the process is in control or not. If the process is found not to be in control, then the centerline and control limits should not be used to monitor the process in the future.

11.5   Even if all the points of an $\bar{x}$-chart fall within the control limits, the process may be out of control. Nonrandom patterns may exist among the plotted points that are within the control limits, but are very unlikely if the process is in control. Examples include six points in a row steadily increasing or decreasing and 14 points in a row alternating up and down.

11.7   Rule 1:   One point beyond Zone A: No points are beyond Zone A.
        Rule 2:   Nine points in a row in Zone C or beyond: No sequence of nine points are in Zone C (on one side of the centerline) or beyond.
        Rule 3:   Six points in a row steadily increasing or decreasing: No sequence of six points steadily increase or decrease.
        Rule 4:   Fourteen points in a row alternating up and down: This pattern does not exist.
        Rule 5:   Two out of three points in Zone A or beyond: There are no groups of three consecutive points that have two or more in Zone A or beyond.
        Rule 6:   Four out of five points in a row in Zone B or beyond: Points 18 through 21 are all in Zone B or beyond. This indicates the process is out of control.

        Thus, rule 6 indicates this process is out of control.

11.9   Using Table XV, Appendix B:

        a.   With $n = 3$, $A_2 = 1.023$

        b.   With $n = 10$, $A_2 = 0.308$

        c.   With $n = 22$, $A_2 = 0.167$

11.11 a. For each sample, we compute $\bar{x} = \dfrac{\sum x}{n}$ and R = range = largest measurement − smallest measurement. The results are listed in the table:

Sample No.	$\bar{x}$	R	Sample No.	$\bar{x}$	R
1	20.225	1.8	11	21.225	3.2
2	19.750	2.8	12	20.475	0.9
3	20.425	3.8	13	19.650	2.6
4	19.725	2.5	14	19.075	4.0
5	20.550	3.7	15	19.400	2.2
6	19.900	5.0	16	20.700	4.3
7	21.325	5.5	17	19.850	3.6
8	19.625	3.5	18	20.200	2.5
9	19.350	2.5	19	20.425	2.2
10	20.550	4.1	20	19.900	5.5

b. $\bar{\bar{x}} = \dfrac{\bar{x}_1 + \bar{x}_2 + \cdots + \bar{x}_{20}}{k} = \dfrac{402.325}{20} = 20.11625$

$\bar{R} = \dfrac{R_1 + R_2 + \cdots + R_{20}}{k} = \dfrac{66.2}{20} = 3.31$

c. *Centerline* $= \bar{\bar{x}} = 20.116$

From Table XV, Appendix B, with $n = 4$, $A_2 = .729$.

Upper control limit $= \bar{\bar{x}} + A_2\bar{R} = 20.116 + .729(3.31) = 22.529$
Lower control limit $= \bar{\bar{x}} - A_2\bar{R} = 20.116 - .729(3.31) = 17.703$

d. *Upper* A−B *boundary* $= \bar{\bar{x}} + \dfrac{2}{3}(A_2\bar{R}) = 20.116 + \dfrac{2}{3}(.729)(3.31) = 21.725$

Lower A−B *boundary* $= \bar{\bar{x}} - \dfrac{2}{3}(A_2\bar{R}) = 20.116 - \dfrac{2}{3}(.729)(3.31) = 18.507$

Upper B−C *boundary* $= \bar{\bar{x}} + \dfrac{1}{3}(A_2\bar{R}) = 20.116 + \dfrac{1}{3}(.729)(3.31) = 20.920$

Lower B−C *boundary* $= \bar{\bar{x}} - \dfrac{1}{3}(A_2\bar{R}) = 20.116 - \dfrac{1}{3}(.729)(3.31) = 19.312$

e.   The $\bar{x}$-chart is:

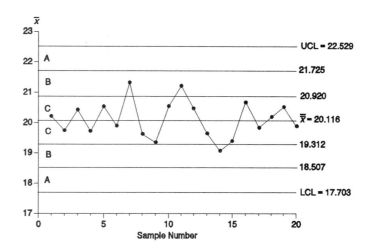

Rule 1:   One point beyond Zone A: No points are beyond Zone A.
Rule 2:   Nine points in a row in Zone C or beyond: No sequence of nine points are in Zone C (on one side of the centerline) or beyond.
Rule 3:   Six points in a row steadily increasing or decreasing: No sequence of six points steadily increase or decrease.
Rule 4:   Fourteen points in a row alternating up and down: This pattern does not exist.
Rule 5:   Two out of three points in Zone A or beyond: There are no groups of three consecutive points that have two or more in Zone A or beyond.
Rule 6:   Four out of five points in a row in Zone B or beyond: No sequence of five points has four or more in Zone B or beyond.

The process appears to be in control.

11.13   a. $\bar{\bar{x}} = \dfrac{\bar{x}_1 + \bar{x}_2 + \cdots + \bar{x}_{20}}{k} = \dfrac{479.942}{20} = 23.9971$

$\bar{R} = \dfrac{R_1 + R_2 + \cdots + R_{20}}{k} = \dfrac{3.63}{20} = .1815$

Centerline $= \bar{\bar{x}} = 23.9971$

From Table XV, Appendix B, with $n = 5$, $A_2 = .577$.

Upper control limit $= \bar{\bar{x}} + A_2\bar{R} = 23.9971 + .577(.1815) = 24.102$
Lower control limit $= \bar{\bar{x}} - A_2\bar{R} = 23.9971 - .577(.1815) = 23.892$

Upper A$-$B *boundary* $= \bar{\bar{x}} + \dfrac{2}{3}(A_2\bar{R}) = 23.9971 + \dfrac{2}{3}(.577)(.1815) = 24.067$

Lower A$-$B *boundary* $= \bar{\bar{x}} - \dfrac{2}{3}(A_2\bar{R}) = 23.9971 - \dfrac{2}{3}(.577)(.1815) = 23.927$

Upper B$-$C *boundary* $= \bar{\bar{x}} + \dfrac{1}{3}(A_2\bar{R}) = 23.9971 + \dfrac{1}{3}(.577)(.1815) = 24.032$

Lower B$-$C *boundary* $= \bar{\bar{x}} - \dfrac{1}{3}(A_2\bar{R}) = 23.9971 - \dfrac{1}{3}(.577)(.1815) = 23.962$

The $\bar{x}$-chart is:

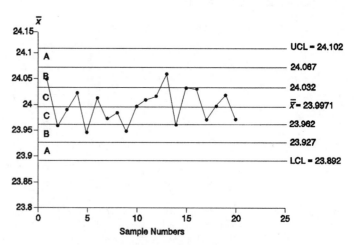

b.   To determine if the process is in or out of control, we check the six rules:

      Rule 1:   One point beyond Zone A: No points are beyond Zone A.

      Rule 2:   Nine points in a row in Zone C or beyond: No sequence of nine points are in Zone C (on one side of the centerline) or beyond.

      Rule 3:   Six points in a row steadily increasing or decreasing: No sequence of six points steadily increase or decrease.

      Rule 4:   Fourteen points in a row alternating up and down: This pattern does not exist.

      Rule 5:   Two out of three points in Zone A or beyond: There are no groups of three consecutive points that have two or more in Zone A or beyond.

      Rule 6:   Four out of five points in a row in Zone B or beyond: No sequence of five points has four or more in Zone B or beyond.

The process appears to be in control.

c.   Since the process is in control, these limits should be used to monitor future process output.

d.   The rational subgrouping strategy used by K-Company will facilitate the identification of process variation caused by differences in the two shifts. All observations within one sample are from the same shift. The shift change is at 3:00 P.M. The samples are selected at 8:00 A.M., 11:00 A.M., 2:00 P.M., 5:00 P.M., and 8:00 P.M. No samples will contain observations from both shifts.

11.15   a.   First, we must compute the range for each sample. The range = R = largest measurement − smallest measurement. The results are listed in the table:

Sample No.	R	Sample No.	R	Sample No.	R
1	2.0	25	4.6	49	4.0
2	2.1	26	3.0	50	4.9
3	1.8	27	3.4	51	3.8
4	1.6	28	2.3	52	4.6
5	3.1	29	2.2	53	7.1
6	3.1	30	3.3	54	4.6
7	4.2	31	3.6	55	2.2
8	3.6	32	4.2	56	3.6
9	4.6	33	2.4	57	2.6
10	2.6	34	4.5	58	2.0
11	3.5	35	5.6	59	1.5
12	5.3	36	4.9	60	6.0
13	5.5	37	10.2	61	5.7
14	5.6	38	5.5	62	5.6
15	4.6	39	4.7	63	2.3
16	3.0	40	4.7	64	2.3
17	4.6	41	3.6	65	2.6
18	4.5	42	3.0	66	3.8
19	4.8	43	2.2	67	2.8
20	5.4	44	3.3	68	2.2
21	5.5	45	3.2	69	4.2
22	3.8	46	0.8	70	2.6
23	3.6	47	4.2	71	1.0
24	2.5	48	5.6	72	1.9

$$\bar{\bar{x}} = \frac{\bar{x}_1 + \bar{x}_2 + \cdots + \bar{x}_{72}}{k} = \frac{3537.3}{72} = 49.129$$

$$\bar{R} = \frac{R_1 + R_2 + \cdots + R_{72}}{k} = \frac{268.8}{72} = 3.733$$

Centerline = $\bar{\bar{x}}$ = 49.13

From Table XV, Appendix B, with $n = 6$, $A_2 = .483$.

Upper control limit = $\bar{\bar{x}} + A_2\bar{R}$ = 49.129 + .483(3.733) = 50.932
Lower control limit = $\bar{\bar{x}} - A_2\bar{R}$ = 49.129 − .483(3.733) = 47.326

Upper A−B *boundary* = $\bar{\bar{x}} + \frac{2}{3}(A_2\bar{R})$ = 49.129 + $\frac{2}{3}$(.483)(3.733) = 50.331

Lower A−B *boundary* = $\bar{\bar{x}} - \frac{2}{3}(A_2\bar{R})$ = 49.129 − $\frac{2}{3}$(.483)(3.733) = 47.927

Upper B−C *boundary* = $\bar{\bar{x}} + \frac{1}{3}(A_2\bar{R})$ = 49.129 + $\frac{1}{3}$(.483)(3.733) = 49.730

Lower B−C *boundary* = $\bar{\bar{x}} - \frac{1}{3}(A_2\bar{R})$ = 49.129 − $\frac{1}{3}$(.483)(3.733) = 48.528

The $\bar{x}$-chart is:

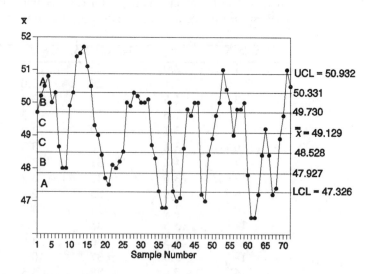

b. To determine if the process is in or out of control, we check the six rules:

Rule 1: One point beyond Zone A: There are a total of 17 points beyond Zone A.

Rule 2: Nine points in a row in Zone C or beyond: No sequence of nine points are in Zone C (on one side of the centerline) or beyond.

Rule 3: Six points in a row steadily increasing or decreasing: There is one sequence of seven points that are steadily increasing—Points 15 through 21.

Rule 4: Fourteen points in a row alternating up and down: This pattern does not exist.

Rule 5: Two out of three points in Zone A or beyond: There are four groups of at least three points in Zone A or beyond—Points 12–16, Points 35–37, Points 39–41, and Points 60–63.

Rule 6: Four out of five points in a row in Zone B or beyond: There are four groups of points that satisfy this rule—Points 10–16, Points 19–24, Points 26–32, and Points 60–64.

The process appears to be out of control. Rules 1, 3, 5, and 6 indicate that the process is out of control.

c. No. The problem does not give the times of the shifts. However, suppose we let the first shift be from 6:00 A.M. to 2:00 P.M., the second shift be from 2:00 P.M. to 10:00 P.M., and the third shift be from 10:00 P.M. to 6:00 A.M. If this is the case, the major problems are during the second shift.

11.17 a. $\bar{\bar{x}} = \dfrac{\bar{x}_1 + \bar{x}_2 + \cdots + \bar{x}_{20}}{k} = \dfrac{1{,}052.933333}{20} = 52.6467$

$\bar{R} = \dfrac{R_1 + R_2 + \cdots + R_{20}}{k} = \dfrac{15.1}{20} = .755$

Centerline $= \bar{\bar{x}} = 52.6467$

From Table XV, Appendix B, with $n = 3$, $A_2 = 1.023$

$Upper\ control\ limit = \overline{\overline{x}} + A_2\overline{R} = 52.6467 + 1.023(.755) = 53.419$

$Lower\ control\ limit = \overline{\overline{x}} - A_2\overline{R} = 52.6467 - 1.023(.755) = 51.874$

$Upper\ A - B\ boundary = \overline{\overline{x}} + \dfrac{2}{3}(A_2\overline{R}) = 52.6467 + \dfrac{2}{3}(1.023)(.755) = 53.162$

$Lower\ A - B\ boundary = \overline{\overline{x}} - \dfrac{2}{3}(A_2\overline{R}) = 52.6467 - \dfrac{2}{3}(1.023)(.755) = 52.132$

$Upper\ B - C\ boundary = \overline{\overline{x}} + \dfrac{1}{3}(A_2\overline{R}) = 52.6467 + \dfrac{1}{3}(1.023)(.755) = 52.904$

$Lower\ B - C\ boundary = \overline{\overline{x}} - \dfrac{1}{3}(A_2\overline{R}) = 52.6467 - \dfrac{1}{3}(1.023)(.755) = 52.389$

The $\overline{x}$-chart is:

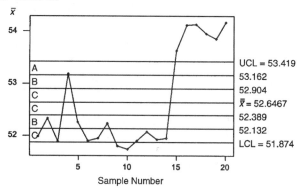

b. To determine if the process is in or out of control, we check the six rules:

Rule 1: One point beyond Zone A: Eight points are beyond Zone A.
Rule 2: Nine points in a row in Zone C or beyond: Data points 5 through 14 (10 points) are in Zone C (on one side of the centerline) or beyond.
Rule 3: Six points in a row steadily increasing or decreasing: No sequence of six points steadily increase or decrease.
Rule 4: Fourteen points in a row alternating up and down: This pattern does not exist.
Rule 5: Two out of three points in Zone A or beyond: There are several sets of three consecutive points that have two points in Zone A or beyond.
Rule 6: Four out of five points in a row in Zone B or beyond: There are several sets of five points where four or more are in Zone B or beyond.

Special causes of variation appear to be present. The process appears to be out of control. Rules 1, 2, 5, and 6 indicate the process is out of control.

c. Processes that are out of control exhibit variation that is the result of both common causes and special causes of variation. Common causes affect all output of the process. Special causes typically affect only local areas or operations within a process.

d. Since the process is out of control, the control limits and centerline should not be used to monitor future output.

11.19   The control limits of the $\bar{x}$-chart are a function of and reflect the variation in the process. If the variation were unstable (i.e., out of control), the control limits would not be constant. Under these circumstances, the fixed control limits of the $\bar{x}$-chart would have little meaning. We use the R-chart to determine whether the variation of the process is stable. If it is, the $\bar{x}$-chart is meaningful. Thus, we interpret the R-chart prior to the $\bar{x}$-chart.

11.21   a.   From Exercise 11.10, $\bar{R} = \dfrac{R_1 + R_2 + \cdots + R_{25}}{k} = \dfrac{198.7}{25} = 7.948$

Centerline $= \bar{R} = 7.948$

From Table XV, Appendix B, with $n = 5$, $D_4 = 2.114$ and $D_3 = 0$.

Upper control limit $= \bar{R}D_4 = 7.948(2.114) = 16.802$

Since $D_3 = 0$, the lower control limit is negative and is not included on the chart.

   b.   From Table XV, Appendix B, with $n = 5$, $d_2 = 2.326$, and $d_3 = .864$.

Upper A$-$B boundary $= \bar{R} + 2d_3\dfrac{\bar{R}}{d_2} = 7.948 + 2(.864)\dfrac{7.948}{2.326} = 13.853$

Lower A$-$B boundary $= \bar{R} - 2d_3\dfrac{\bar{R}}{d_2} = 7.948 - 2(.864)\dfrac{7.948}{2.326} = 2.043$

Upper B$-$C boundary $= \bar{R} + d_3\dfrac{\bar{R}}{d_2} = 7.948 + (.864)\dfrac{7.948}{2.326} = 10.900$

Lower B$-$C boundary $= \bar{R} - d_3\dfrac{\bar{R}}{d_2} = 7.948 - (.864)\dfrac{7.948}{2.326} = 4.996$

   c.   The R-chart is:

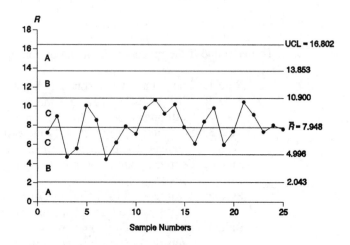

To determine if the process is in or out of control, we check the four rules:

Rule 1:   One point beyond Zone A: No points are beyond Zone A.
Rule 2:   Nine points in a row in Zone C or beyond: No sequence of nine points are in Zone C (on one side of the centerline) or beyond.
Rule 3:   Six points in a row steadily increasing or decreasing: No sequence of six points steadily increase or decrease.
Rule 4:   Fourteen points in a row alternating up and down: This pattern does not exist.

The process appears to be in control.

11.23   First, we construct an R-chart.

$$\bar{R} = \frac{R_1 + R_2 + \cdots + R_{20}}{k} = \frac{80.6}{20} = 4.03$$

Centerline $= \bar{R} = 4.03$

From Table XV, Appendix B, with $n = 7$, $D_4 = 1.924$ and $D_3 = .076$.

Upper control limit $= \bar{R}D_4 = 4.03(1.924) = 7.754$
Lower control limit $= \bar{R}D_3 = 4.03(0.076) = 0.306$

From Table XV, Appendix B, with $n = 7$, $d_2 = 2.704$ and $d_3 = .833$.

Upper A−B *boundary* $= \bar{R} + 2d_3\frac{\bar{R}}{d_2} = 4.03 + 2(.833)\frac{4.03}{2.704} = 6.513$

Lower A−B *boundary* $= \bar{R} - 2d_3\frac{\bar{R}}{d_2} = 4.03 - 2(.833)\frac{4.03}{2.704} = 1.547$

Upper B−C *boundary* $= \bar{R} + d_3\frac{\bar{R}}{d_2} = 4.03 + (.833)\frac{4.03}{2.704} = 5.271$

Lower B−C *boundary* $= \bar{R} - d_3\frac{\bar{R}}{d_2} = 4.03 - (.833)\frac{4.03}{2.704} = 2.789$

The R-chart is:

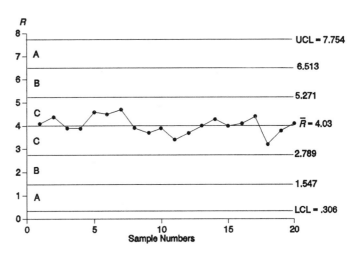

To determine if the process is in or out of control, we check the four rules:

Rule 1:   One point beyond Zone A: No points are beyond Zone A.
Rule 2:   Nine points in a row in Zone C or beyond: No sequence of nine points are in Zone C (on one side of the centerline) or beyond.
Rule 3:   Six points in a row steadily increasing or decreasing: No sequence of six points steadily increase or decrease.
Rule 4:   Fourteen points in a row alternating up and down: This pattern does not exist.

The process appears to be in control. Since the process variation is in control, it is appropriate to construct the $\bar{x}$-chart.

To construct an $\bar{x}$-chart, we first calculate the following:

$$\bar{\bar{x}} = \frac{\bar{x}_1 + \bar{x}_2 + \cdots + \bar{x}_{20}}{k} = \frac{434.56}{20} = 21.728$$

$$\bar{R} = \frac{R_1 + R_2 + \cdots + R_{20}}{k} = \frac{80.6}{20} = 4.03$$

Centerline $= \bar{\bar{x}} = 21.728$

From Table XV, Appendix B, with $n = 7$, $A_2 = .419$.

Upper control limit $= \bar{\bar{x}} + A_2\bar{R} = 21.728 + .419(4.03) = 23.417$
Lower control limit $= \bar{\bar{x}} - A_2\bar{R} = 21.728 - .419(4.03) = 20.039$

Upper A−B *boundary* $= \bar{\bar{x}} + \frac{2}{3}(A_2\bar{R}) = 21.728 + \frac{2}{3}(.419)(4.03) = 22.854$

Lower A−B *boundary* $= \bar{\bar{x}} - \frac{2}{3}(A_2\bar{R}) = 21.728 - \frac{2}{3}(.419)(4.03) = 20.602$

Upper B−C *boundary* $= \bar{\bar{x}} + \frac{1}{3}(A_2\bar{R}) = 21.728 + \frac{1}{3}(.419)(4.03) = 22.291$

Lower B−C *boundary* $= \bar{\bar{x}} - \frac{1}{3}(A_2\bar{R}) = 21.728 - \frac{1}{3}(.419)(4.03) = 21.165$

The $\bar{x}$-chart is:

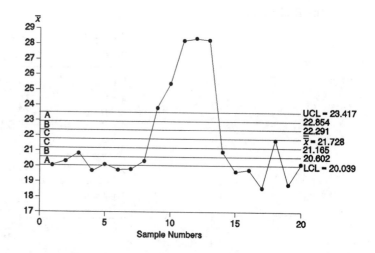

To determine if the process is in or out of control, we check the six rules:

Rule 1: One point beyond Zone A: There are 12 points beyond Zone A. This indicates the process is out of control.

Rule 2: Nine points in a row in Zone C or beyond: No sequence of nine points are in Zone C (on one side of the centerline) or beyond.

Rule 3: Six points in a row steadily increasing or decreasing: Points 6 through 12 steadily increase. This indicates the process is out of control.

Rule 4: Fourteen points in a row alternating up and down: This pattern does not exist.

Rule 5: Two out of three points in Zone A or beyond: There are several groups of three consecutive points that have two or more in Zone A or beyond. This indicates the process is out of control.

Rule 6: Four out of five points in a row in Zone B or beyond: Several sequences of five points have four or more in Zone B or beyond. This indicates the process is out of control.

Rules 1, 3, 5, and 6 indicate that the process is out of control.

11.25   a.   Yes. Because all five observations in each sample were selected from the same dispenser, the rational subgrouping will enable the company to detect variation in fill caused by differences in the carbon dioxide dispensers.

      b.   For each sample, we compute the range $= R =$ largest measurement $-$ smallest measurement. The results are listed in the table:

Sample No.	R	Sample No.	R
1	.05	13	.05
2	.06	14	.04
3	.06	15	.05
4	.05	16	.05
5	.07	17	.06
6	.07	18	.06
7	.09	19	.05
8	.08	20	.08
9	.08	21	.08
10	.11	22	.12
11	.14	23	.12
12	.14	24	.15

$$\bar{R} = \frac{R_1 + R_2 + \cdots + R_{24}}{k} = \frac{1.91}{24} = .0796$$

Centerline $= \bar{R} = .0796$

From Table XV, Appendix B, with $n = 5$, $D_4 = 2.114$, and $D_3 = 0$.

Upper control limit $= \bar{R}D_4 = .0796(2.114) = .168$

Since $D_3 = 0$, the lower control limit is negative and is not included on the chart.

From Table XV, Appendix B, with $n = 5$, $d_2 = 2.326$, and $d_3 = .864$.

Upper A$-$B boundary $= \bar{R} + 2d_3 \dfrac{\bar{R}}{d_2} = .0796 + 2(.864)\dfrac{.0796}{2.326} = .139$

Lower A$-$B boundary $= \bar{R} - 2d_3 \dfrac{\bar{R}}{d_2} = .0796 - 2(.864)\dfrac{.0796}{2.326} = .020$

Upper B$-$C boundary $= \bar{R} + d_3 \dfrac{\bar{R}}{d_2} = .0796 + (.864)\dfrac{.0796}{2.326} = .109$

Lower B$-$C boundary $= \bar{R} - d_3 \dfrac{\bar{R}}{d_2} = .0796 - (.864)\dfrac{.0796}{2.326} = .050$

The *R*-chart is:

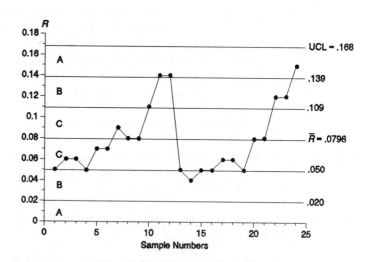

c. To determine if the process is in or out of control, we check the four rules:

Rule 1: One point beyond Zone A: No points are beyond Zone A.
Rule 2: Nine points in a row in Zone C or beyond: No sequence of nine points are in Zone C (on one side of the centerline) or beyond.
Rule 3: Six points in a row steadily increasing or decreasing: No sequence of six points steadily increase or decrease.
Rule 4: Fourteen points in a row alternating up and down: This pattern does not exist.

The process appears to be in control.

d. Since the process variation is in control, the *R*-chart should be used to monitor future process output.

e. The $\bar{x}$-chart should be constructed. The control limits of the $\bar{x}$-chart depend on the variation of the process. (In particular, they are constructed using $\bar{R}$.) If the variation of the process is in control, the control limits of the $\bar{x}$-chart are meaningful.

11.27 a. $$\bar{R} = \frac{R_1 + R_2 + \cdots + R_{25}}{k} = \frac{2.5 + 1.5 + \cdots + 2.0}{25} = \frac{52}{25} = 2.08$$

Centerline $= \bar{R} = 2.08$

From Table XV, Appendix B, with $n = 5$, $D_4 = 2.114$ and $D_3 = 0$.

Upper control limit $= \bar{R}D_4 = 2.08(2.114) = 4.397$

Since $D_3 = 0$, the lower control limit is negative and is not included on the chart.

From Table XV, Appendix B, with $n = 5$, $d_2 = 2.326$ and $d_3 = 0.864$.

Upper A – B *boundary* $= \bar{R} + 2d_3\dfrac{\bar{R}}{d_2} = 2.08 + 2(.864)\dfrac{2.08}{2.326} = 3.625$

$$Lower\ A-B\ boundary = \bar{R} - 2d_3\frac{\bar{R}}{d_2} = 2.08 - 2(.864)\frac{2.08}{2.326} = .535$$

$$Upper\ B-C\ boundary = \bar{R} + d_3\frac{\bar{R}}{d_2} = 2.08 + (.864)\frac{2.08}{2.326} = 2.853$$

$$Lower\ B-C\ boundary = \bar{R} - d_3\frac{\bar{R}}{d_2} = 2.08 - (.864)\frac{2.08}{2.326} = 1.307$$

The R-chart is:

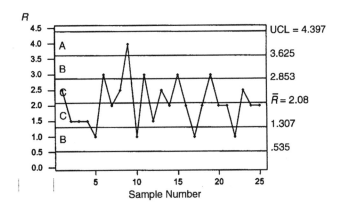

To determine if the process is in or out of control, we check the four rules:

Rule 1: One point beyond Zone A: No points are beyond Zone A.
Rule 2: Nine points in a row in Zone C or beyond: No sequence of nine points are in Zone C (on one side of the centerline) or beyond.
Rule 3: Six points in a row steadily increasing or decreasing: No sequence of six points steadily increase or decrease.
Rule 4: Fourteen points in a row alternating up and down: This pattern does not exist.

The process appears to be in control since none of the out-of-control signals are observed. No special causes of variation appear to be present.

Since the process appears to be under control, it is appropriate to construct an $\bar{x}$-chart for the data.

c. $$\bar{R} = \frac{R_1 + R_2 + \cdots + R_{25}}{k} = \frac{2.5 + 0.0 + \cdots + 2.5}{25} = \frac{42.5}{25} = 1.7$$

$$Centerline = \bar{R} = 1.7$$

From Table XV, Appendix B, with $n = 5$, $D_4 = 2.114$ and $D_3 = 0$.

$$Upper\ control\ limit = \bar{R}D_4 = 1.7(2.114) = 3.594$$

Since $D_3 = 0$, the lower control limit is negative and is not included on the chart.

From Table XV, Appendix B, with $n = 5$, $d_2 = 2.326$ and $d_3 = 0.864$.

$$\text{Upper A - B boundary} = \bar{R} + 2d_3\frac{\bar{R}}{d_2} = 1.7 + 2(.864)\frac{1.7}{2.326} = 2.963$$

$$\text{Lower A - B boundary} = \bar{R} - 2d_3\frac{\bar{R}}{d_2} = 1.7 - 2(.864)\frac{1.7}{2.326} = .437$$

$$\text{Upper B - C boundary} = \bar{R} + d_3\frac{\bar{R}}{d_2} = 1.7 + (.864)\frac{1.7}{2.326} = 2.331$$

$$\text{Lower B - C boundary} = \bar{R} - d_3\frac{\bar{R}}{d_2} = 1.7 - (.864)\frac{1.7}{2.326} = 1.069$$

The R chart is:

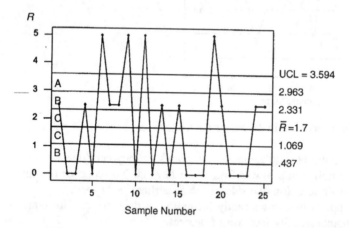

To determine if the process is in or out of control, we check the four rules:

Rule 1: One point beyond Zone A: Four points are beyond Zone A.
Rule 2: Nine points in a row in Zone C or beyond: No sequence of nine points are in Zone C (on one side of the centerline) or beyond.
Rule 3: Six points in a row steadily increasing or decreasing: No sequence of six points steadily increase or decrease.
Rule 4: Fourteen points in a row alternating up and down: This pattern does not exist.

The process appears to be out of control. Rule 1 indicates the process is out of control. Since this process is out of control, it is not appropriate to construct an $\bar{x}$-chart for the data.

d. We get two different answers as to whether this process is in control, depending on the accuracy of the data. When the data were measured to an accuracy of .5 gram, the process appears to be in control. However, when the data were measured to an accuracy of only 2.5 grams, the process appears to be out of control. The same data were used for each chart – just measured to different accuracies.

11.29 a. $\bar{R} = \dfrac{R_1 + R_2 + \cdots + R_{16}}{k} = \dfrac{.4 + 1.4 + \cdots + 2.6}{16} = \dfrac{44.1}{16} = 2.756$

$Centerline = \bar{R} = 2.756$

From Table XV, Appendix B, with $n = 5$, $D_4 = 2.114$ and $D_3 = 0$.

$Upper\ control\ limit = \bar{R}D_4 = 2.756(2.114) = 5.826$

Since $D_3 = 0$, the lower control limit is negative and is not included on the chart.

From Table XV, Appendix B, with $n = 5$, $d_2 = 2.326$ and $d_3 = 0.864$.

$Upper\ \text{A – B}\ boundary = \bar{R} + 2d_3\dfrac{\bar{R}}{d_2} = 2.756 + 2(.864)\dfrac{2.756}{2.326} = 4.803$

$Lower\ \text{A – B}\ boundary = \bar{R} - 2d_3\dfrac{\bar{R}}{d_2} = 2.756 - 2(.864)\dfrac{2.756}{2.326} = .709$

$Upper\ \text{B – C}\ boundary = \bar{R} + d_3\dfrac{\bar{R}}{d_2} = 2.756 + (.864)\dfrac{2.756}{2.326} = 3.780$

$Lower\ \text{B – C}\ boundary = \bar{R} - d_3\dfrac{\bar{R}}{d_2} = 2.756 - (.864)\dfrac{2.756}{2.326} = 1.732$

The R-chart is:

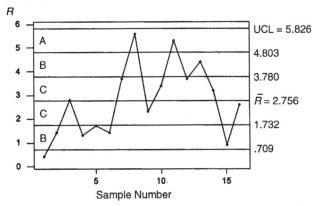

b. The R-chart is designed to monitor the process variation.

c. To determine if the process is in or out of control, we check the four rules:

Rule 1: One point beyond Zone A: No points are beyond Zone A.
Rule 2: Nine points in a row in Zone C or beyond: No sequence of nine points are in Zone C (on one side of the centerline) or beyond.
Rule 3: Six points in a row steadily increasing or decreasing: No sequence of six points steadily increases or decreases.
Rule 4: Fourteen points in a row alternating up and down: This pattern does not exist.

The process appears to be in control. None of the out-of-control signals are present. There is no indication that special causes of variation present.

11.31 The sample size is determined as follows:

$$n > \frac{9(1 - p_0)}{p_0} = \frac{9(1 - .08)}{.08} = 103.5 \approx 104$$

11.33 **a.** We must first calculate $\bar{p}$. To do this, it is necessary to find the total number of defectives in all the samples. To find the number of defectives per sample, we multiple the proportion by the sample size, 150. The number of defectives per sample are shown in the table:

Sample No.	p	No. Defectives	Sample No.	p	No. Defectives
1	.03	4.5	11	.07	10.5
2	.05	7.5	12	.04	6.0
3	.10	15.0	13	.06	9.0
4	.02	3.0	14	.05	7.5
5	.08	12.0	15	.07	10.5
6	.09	13.5	16	.06	9.0
7	.08	12.0	17	.07	10.5
8	.05	7.5	18	.02	3.0
9	.07	10.5	19	.05	7.5
10	.06	9.0	20	.03	4.5

Note: There cannot be a fraction of a defective. The proportions presented in the exercise have been rounded off. I have used the fractions to minimize the roundoff error.

To get the total number of defectives, sum the number of defectives for all 20 samples. The sum is 172.5. To get the total number of units sampled, multiply the sample size by the number of samples: $150(20) = 3000$.

$$\bar{p} = \frac{\text{Total defective in all samples}}{\text{Total units sampled}} = \frac{172.5}{3000} = .0575$$

$Centerline = \bar{p} = .0575$

$$Upper\ control\ limit = \bar{p} + 3\sqrt{\frac{\bar{p}(1 - \bar{p})}{n}} = .0575 + 3\sqrt{\frac{.0575(.9425)}{150}} = .1145$$

$$Lower\ control\ limit = \bar{p} - 3\sqrt{\frac{\bar{p}(1 - \bar{p})}{n}} = .0575 - 3\sqrt{\frac{.0575(.9425)}{150}} = .0005$$

b.

$$Upper\ A-B\ boundary = \bar{p} + 2\sqrt{\frac{\bar{p}(1 - \bar{p})}{n}} = .0575 + 2\sqrt{\frac{.0575(.9425)}{150}} = .0955$$

$$Lower\ A-B\ boundary = \bar{p} - 2\sqrt{\frac{\bar{p}(1 - \bar{p})}{n}} = .0575 - 2\sqrt{\frac{.0575(.9425)}{150}} = .0195$$

$$Upper\ B-C\ boundary = \bar{p} + \sqrt{\frac{\bar{p}(1 - \bar{p})}{n}} = .0575 + \sqrt{\frac{.0575(.9425)}{150}} = .0765$$

$$Lower\ B-C\ boundary = \bar{p} - \sqrt{\frac{\bar{p}(1 - \bar{p})}{n}} = .0575 - \sqrt{\frac{.0575(.9425)}{150}} = .0385$$

c.   The *p*-chart is:

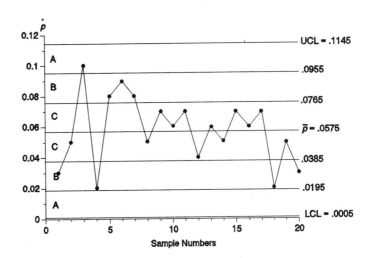

d.   To determine if the process is in or out of control, we check the four rules:

    Rule 1:   One point beyond Zone A: No points are beyond Zone A.

    Rule 2:   Nine points in a row in Zone C or beyond: No sequence of nine points are in Zone C (on one side of the centerline) or beyond.

    Rule 3:   Six points in a row steadily increasing or decreasing: No sequence of six points steadily increase or decrease.

    Rule 4:   Fourteen points in a row alternating up and down: Points 7 through 20 alternate up and down. This indicates the process is out of control.

Rule 4 indicates that the process is out of control.

e.   Since the process is out of control, the centerline and control limits should not be used to monitor future process output. The centerline and control limits are intended to represent the behavior of the process when it is under control.

11.35 a.   Yes. The minimum sample size necessary so the lower control limit is not negative is:

$$n > \frac{9(1 - p_0)}{p_0}$$

From the data, $p_0 \approx .01$

Thus, $n > \dfrac{9(1 - .01)}{.01} = 891$. Our sample size was 1000.

b.   *Upper control limit* $= \bar{p} + 3\sqrt{\dfrac{\bar{p}(1 - \bar{p})}{n}} = .01047 + 3\sqrt{\dfrac{.01047(.98953)}{1000}} = .02013$

    Lower control limit $= \bar{p} - 3\sqrt{\dfrac{\bar{p}(1 - \bar{p})}{n}} = .01047 - 3\sqrt{\dfrac{.01047(.98953)}{1000}} = .00081$

c. To determine if special causes are present, we must complete the *p*-chart.

$$\text{Upper A$-$B boundary} = \bar{p} + 2\sqrt{\frac{\bar{p}(1-\bar{p})}{n}} = .01047 + 2\sqrt{\frac{.01047(.98953)}{1000}} = .01691$$

$$\text{Lower A$-$B boundary} = \bar{p} - 2\sqrt{\frac{\bar{p}(1-\bar{p})}{n}} = .01047 - 2\sqrt{\frac{.01047(.98953)}{1000}} = .00403$$

$$\text{Upper B$-$C boundary} = \bar{p} + \sqrt{\frac{\bar{p}(1-\bar{p})}{n}} = .01047 + \sqrt{\frac{.01047(.98953)}{1000}} = .01369$$

$$\text{Lower B$-$C boundary} = \bar{p} - \sqrt{\frac{\bar{p}(1-\bar{p})}{n}} = .01047 - \sqrt{\frac{.01047(.98953)}{1000}} = .00725$$

To determine if the process is in control, we check the four rules.

Rule 1: One point beyond Zone A: No points are beyond Zone A.
Rule 2: Nine points in a row in Zone C or beyond: There are not nine points in a row in Zone C (on one side of the centerline) or beyond.
Rule 3: Six points in a row steadily increasing or decreasing: No sequence of six points steadily increase or decrease.
Rule 4: Fourteen points in a row alternating up and down: This pattern does not exist.

It appears that the process is in control.

d. The rational subgrouping strategy says that samples should be chosen so that it gives the maximum chance for the measurements in each sample to be similar and so that it gives the maximum chance for the samples to differ. By selecting 1000 consecutive chips each time, this gives the maximum chance for the measurements in the sample to be similar. By selecting the samples every other day, there is a relatively large chance that the samples differ.

11.37 a. To compute the proportion of defectives in each sample, divide the number of defectives by the number in the sample, 100:

$$\hat{p} = \frac{\text{No. of defectives}}{\text{No. in sample}}$$

The sample proportions are listed in the table:

Sample No.	$\hat{p}$	Sample No.	$\hat{p}$
1	.02	16	.02
2	.04	17	.03
3	.10	18	.07
4	.04	19	.03
5	.01	20	.02
6	.01	21	.03
7	.13	22	.07
8	.09	23	.04
9	.11	24	.03
10	.00	25	.02
11	.03	26	.02
12	.04	27	.00
13	.02	28	.01
14	.02	29	.03
15	.08	30	.04

To get the total number of defectives, sum the number of defectives for all 30 samples. The sum is 120. To get the total number of units sampled, multiply the sample size by the number of samples: $100(30) = 3000$.

$$\bar{p} = \frac{\text{Total defective in all samples}}{\text{Total units sampled}} = \frac{120}{3000} = .04$$

The centerline is $\bar{p} = .04$

$$\text{Upper control limit} = \bar{p} + 3\sqrt{\frac{\bar{p}(1 - \bar{p})}{n}} = .04 + 3\sqrt{\frac{.04(1 - .04)}{100}} = .099$$

$$\text{Lower control limit} = \bar{p} - 3\sqrt{\frac{\bar{p}(1 - \bar{p})}{n}} = .04 - 3\sqrt{\frac{.04(1 - .04)}{100}} = -.019$$

$$\text{Upper A−B boundary} = \bar{p} + 2\sqrt{\frac{\bar{p}(1 - \bar{p})}{n}} = .04 + 2\sqrt{\frac{.04(1 - .04)}{100}} = .079$$

$$\text{Lower A−B boundary} = \bar{p} - 2\sqrt{\frac{\bar{p}(1 - \bar{p})}{n}} = .04 - 2\sqrt{\frac{.04(1 - .04)}{100}} = .001$$

$$\text{Upper B−C boundary} = \bar{p} + \sqrt{\frac{\bar{p}(1 - \bar{p})}{n}} = .04 + \sqrt{\frac{.04(1 - .04)}{100}} = .060$$

$$\text{Lower B−C boundary} = \bar{p} - \sqrt{\frac{\bar{p}(1 - \bar{p})}{n}} = .04 - \sqrt{\frac{.04(1 - .04)}{100}} = .020$$

The *p*-chart is:

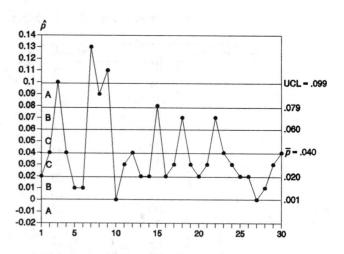

b. To determine if the process is in or out of control, we check the four rules for the *R*-chart.

Rule 1: One point beyond Zone A: There are 3 points beyond Zone A—points 2, 7, and 9.

Rule 2: Nine points in a row in Zone C or beyond: No sequence of nine points are in Zone C (on one side of the centerline) or beyond.

Rule 3: Six points in a row steadily increasing or decreasing: This pattern is not present.

Rule 4: Fourteen points in a row alternating up and down: This pattern does not exist.

The process does not appear to be in control. Rule 1 indicates that the process is out of control.

c. No. Since the process is not in control, then these control limits are meaningless.

11.39 The quality of a good or service is indicated by the extent to which it satisfies the needs and preferences of its users. Its eight dimensions are: performance, features, reliability, conformance, durability, serviceability, aesthetics, and other perceptions that influence judgments of quality.

11.41 A process is a series of actions or operations that transform inputs to outputs. A process produces output over time. Organizational process: Manufacturing a product. Personnel Process: Balancing a checkbook.

11.43 The six major sources of process variation are: people, machines, materials, methods, measurements, and environment.

11.45 Common causes of variation are the methods, materials, equipment, personnel, and environment that make up a process and the inputs required by the process. That is, common causes are attributable to the design of the process. Special causes of variation are events or actions that are not part of the process design. Typically, they are transient, fleeting events that affect only local areas or operations within the process for a brief period of time. Occasionally, however, such events may have a persistent or recurrent effect on the process.

11.47   Control limits are a function of the natural variability of the process. The position of the limits is a function of the size of the process standard deviation. Specification limits are boundary points that define the acceptable values for an output variable of a particular product or service. They are determined by customers, management, and/or product designers. Specification limits may be either two-sided, with upper and lower limits, or one-sided with either an upper or lower limit. Specification limits are not dependent on the process in any way. The process may not be able to meet the specification limits even when it is under statistical control.

11.49   a.   The centerline is:

$$\bar{x} = \frac{\sum x}{n} = \frac{96}{15} = 6.4$$

The time series plot is:

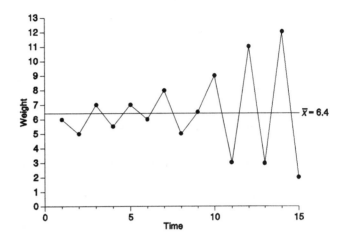

b.   The type of variation best described by the pattern in this plot is increasing variance. The spread of the measurements increases with the passing of time.

11.51   To determine if the process is in or out of control, we check the six rules:

Rule 1:   One point beyond Zone A: No points are beyond Zone A.

Rule 2:   Nine points in a row in Zone C or beyond: Points 8 through 16 are in Zone C (on one side of the centerline) or beyond. This indicates the process is out of control.

Rule 3:   Six points in a row steadily increasing or decreasing: No sequence of six points steadily increase or decrease.

Rule 4:   Fourteen points in a row alternating up and down: This pattern does not exist.

Rule 5:   Two out of three points in Zone A or beyond: No group of three consecutive points have two or more in Zone A or beyond.

Rule 6:   Four out of five points in a row in Zone B or beyond: No sequence of five points has four or more in Zone B or beyond.

Rule 2 indicates that the process is out of control. A special cause of variation appears to be present.

11.53 a. To compute the range, subtract the larger score minus the smaller score. The ranges for the samples are listed in the table:

Sample No.	R	$\bar{x}$	Sample No.	R	$\bar{x}$
1	4	343.0	11	5	357.5
2	3	329.5	12	10	330.0
3	12	349.0	13	2	349.0
4	1	351.5	14	1	336.5
5	12	354.0	15	16	337.0
6	6	339.0	16	7	354.5
7	3	329.5	17	1	352.5
8	0	344.0	18	6	337.0
9	25	346.5	19	6	338.0
10	15	353.5	20	13	351.5

The centerline is $\bar{R} = \dfrac{\sum R}{k} = \dfrac{148}{20} = 7.4$

From Table XV, Appendix B, with $n = 2$, $D_3 = 0$, and $D_4 = 3.267$.

Upper control limit $= \bar{R}D_4 = 7.4(3.267) = 24.1758$

Since $D_3 = 0$, the lower control limit is negative and is not included on the chart.

From Table XV, Appendix B, with $n = 2$, $d_2 = 1.128$, and $d_3 = .853$.

Upper A–B boundary $= \bar{R} + 2d_3 \dfrac{\bar{R}}{d_2} = 7.4 + 2(.853)\dfrac{(7.4)}{1.128} = 18.5918$

Lower A–B boundary $= \bar{R} - 2d_3 \dfrac{\bar{R}}{d_2} = 7.4 - 2(.853)\dfrac{(7.4)}{1.128} = -3.7918$

Upper B–C boundary $= \bar{R} + d_3 \dfrac{\bar{R}}{d_2} = 7.4 + (.853)\dfrac{(7.4)}{1.128} = 12.9959$

Lower B–C boundary $= \bar{R} - d_3 \dfrac{\bar{R}}{d_2} = 7.4 - (.853)\dfrac{(7.4)}{1.128} = 1.8041$

The R-chart is:

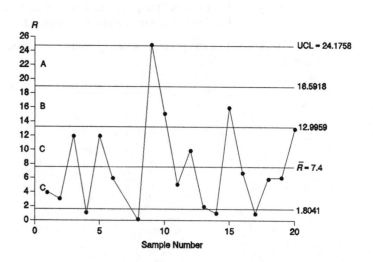

To determine if the process is in control, we check the four rules.

Rule 1: One point beyond Zone A: Point 9 is beyond Zone A. This indicates the process is out of control.

Rule 2: Nine points in a row in Zone C or beyond: There are not nine points in a row in Zone C (on one side of the centerline) or beyond.

Rule 3: Six points in a row steadily increasing or decreasing: No sequence of six points steadily increase or decrease.

Rule 4: Fourteen points in a row alternating up and down: This pattern does not exist.

Rule 1 indicates that the process is out of control. We should not use this to construct the $\bar{x}$-chart.

b. We will construct the $\bar{x}$-chart even though the R-chart indicates the variation is out of control. First, compute the mean for each sample by adding the 2 observations and dividing by 2. These values are in the table in part **a.**

The centerline is $\bar{\bar{x}} = \dfrac{\sum \bar{x}}{k} = \dfrac{6883}{20} = 344.15$

From Table XV, Appendix B, with $n = 2$, $A_2 = 1.880$.

$\bar{\bar{x}} = 344.15$ and $\bar{R} = 7.4$

Upper control limit $= \bar{\bar{x}} + A_2\bar{R} = 344.15 + 1.88(7.4) = 358.062$
Lower control limit $= \bar{\bar{x}} - A_2\bar{R} = 344.15 - 1.88(7.4) = 330.238$

Upper A–B *boundary* $= \bar{\bar{x}} + \dfrac{2}{3}\left(A_2\bar{R}\right) = 344.15 + \dfrac{2}{3}(1.88)(7.4) = 353.425$

Lower A–B *boundary* $= \bar{\bar{x}} - \dfrac{2}{3}\left(A_2\bar{R}\right) = 344.15 - \dfrac{2}{3}(1.88)(7.4) = 334.875$

Upper B–C *boundary* $= \bar{\bar{x}} + \dfrac{1}{3}\left(A_2\bar{R}\right) = 344.15 + \dfrac{1}{3}(1.88)(7.4) = 348.787$

Lower B–C *boundary* $= \bar{\bar{x}} - \dfrac{1}{3}\left(A_2\bar{R}\right) = 344.15 - \dfrac{1}{3}(1.88)(7.4) = 339.513$

The $\bar{x}$-chart is:

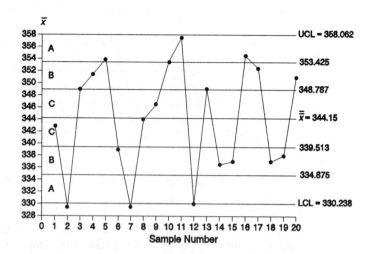

To determine if the process is in control, we check the six rules.

Rule 1:   One point beyond Zone A: Points 2 and 7 are beyond Zone A. This indicates the process is out of control.

Rule 2:   Nine points in a row in Zone C or beyond: There are nine points (Points 9 through 17) in a row in Zone C (on one side of the centerline) or beyond. This indicates that the process is out of control.

Rule 3:   Six points in a row steadily increasing or decreasing: No sequence of six points steadily increase or decrease.

Rule 4:   Fourteen points in a row alternating up and down: This pattern does not exist.

Rule 5:   Two out of three points in Zone A or beyond: Points 10 and 11 are in Zone 3 or beyond. This indicates that the process is out of control.

Rule 6:   Four out of five points in a row in Zone B or beyond: No sequence of five points has four or more in Zone B or beyond.

Rules 1 and 5 indicate the process is out of control. The $\bar{x}$-chart should not be used to monitor the process.

c.   These control limits should not be used to monitor future output because both processes are out of control. One or more special causes of variation are affecting the process variation and process mean. These should be identified and eliminated in order to bring the processes into control.

d.   Of the 40 patients sampled, 10 received care that did not conform to the hospital's requirement. The proportion is 10/40 = .25.

11.55 a. For each sample, we compute $\bar{x} = \dfrac{\sum x}{n}$ and R = range = largest measurement − smallest measurement. The results are listed in the table:

Sample No.	$\bar{x}$	R	Sample No.	$\bar{x}$	R
1	4.36	7.1	11	3.32	4.8
2	5.10	7.7	12	4.02	4.8
3	4.52	5.0	13	5.24	7.8
4	3.42	5.8	14	3.58	3.9
5	2.62	6.2	15	3.48	5.5
6	3.94	3.9	16	5.00	3.0
7	2.34	5.3	17	3.68	6.2
8	3.26	3.2	18	2.68	3.9
9	4.06	8.0	19	3.66	4.4
10	4.96	7.1	20	4.10	5.5

$$\bar{\bar{x}} = \frac{\bar{x}_1 + \bar{x}_2 + \cdots + \bar{x}_{20}}{k} = \frac{77.34}{20} = 3.867$$

$$\bar{R} = \frac{R_1 + R_2 + \cdots + R_{20}}{k} = \frac{109.1}{20} = 5.455$$

First, we construct an R-chart.

Centerline $= \bar{R} = 5.455$

From Table XV, Appendix B, with $n = 5$, $D_4 = 2.114$, and $D_3 = 0$.

Upper control limit $= \bar{R}D_4 = 5.455(2.114) = 11.532$

Since $D_3 = 0$, the lower control limit is negative and is not included on the chart.

Upper A–B *boundary* $= \bar{R} + 2d_3\dfrac{\bar{R}}{d_2} = 5.455 + 2(.864)\dfrac{(5.455)}{2.326} = 9.508$

Lower A–B *boundary* $= \bar{R} - 2d_3\dfrac{\bar{R}}{d_2} = 5.455 - 2(.864)\dfrac{(5.455)}{2.326} = 1.402$

Upper B–C *boundary* $= \bar{R} + d_3\dfrac{\bar{R}}{d_2} = 5.455 + (.864)\dfrac{(5.455)}{2.326} = 7.481$

Lower B–C *boundary* $= \bar{R} - d_3\dfrac{\bar{R}}{d_2} = 5.455 - (.864)\dfrac{(5.455)}{2.326} = 3.429$

Methods for Quality Improvement

The *R*-chart is:

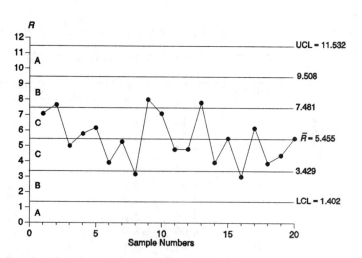

b. To determine if the process is in or out of control, we check the four rules:

Rule 1: One point beyond Zone A: No points are beyond Zone A.
Rule 2: Nine points in a row in Zone C or beyond: No sequence of nine points are in Zone C (on one side of the centerline) or beyond.
Rule 3: Six points in a row steadily increasing or decreasing: No sequence of six points steadily increase or decrease.
Rule 4: Fourteen points in a row alternating up and down: This pattern does not exist.

The process appears to be in control. Since the process variation is in control, it is appropriate to construct the $\bar{x}$-chart.

c. In order for the $\bar{x}$-chart to be valid, the process variation must be in control. The *R*-chart checks to see if the process variation is in control. For more details, see the answer to Exercise 11.19.

d. To construct an $\bar{x}$-chart, we first calculate the following:

$$\bar{\bar{x}} = \frac{\bar{x}_1 + \bar{x}_2 + \cdots + \bar{x}_{20}}{k} = \frac{77.34}{20} = 3.867$$

$$\bar{R} = \frac{R_1 + R_2 + \cdots + R_{20}}{k} = \frac{109.1}{20} = 5.455$$

Centerline $= \bar{\bar{x}} = 3.867$

From Table XV, Appendix B, with $n = 5$, $A_2 = .577$.

Upper control limit $= \bar{\bar{x}} + A_2\bar{R} = 3.867 + .577(5.455) = 7.015$
Lower control limit $= \bar{\bar{x}} - A_2\bar{R} = 3.867 - .577(5.455) = .719$

Upper A–B *boundary* $= \bar{\bar{x}} + \frac{2}{3}\left(A_2\bar{R}\right) = 3.867 + \frac{2}{3}(.577)(5.455) = 5.965$

Lower A–B *boundary* $= \bar{\bar{x}} - \frac{2}{3}\left(A_2\bar{R}\right) = 3.867 - \frac{2}{3}(.577)(5.455) = 1.769$

$$\text{Upper B–C boundary} = \overline{\overline{x}} + \frac{1}{3}\left(A_2\overline{R}\right) = 3.867 + \frac{1}{3}(.577)(5.455) = 4.916$$

$$\text{Lower B–C boundary} = \overline{\overline{x}} - \frac{1}{3}\left(A_2\overline{R}\right) = 3.867 - \frac{1}{3}(.577)(5.455) = 2.818$$

The $\overline{x}$-chart is:

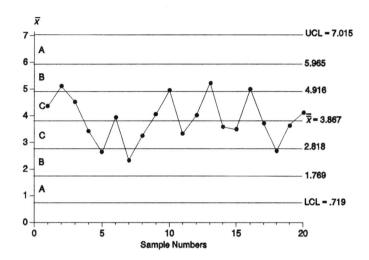

e. To determine if the process is in or out of control, we check the six rules:

    Rule 1: One point beyond Zone A: No points are beyond Zone A.

    Rule 2: Nine points in a row in Zone C or beyond: No sequence of nine points are in Zone C (on one side of the centerline) or beyond.

    Rule 3: Six points in a row steadily increasing or decreasing: No sequence of six points steadily increases or decreases.

    Rule 4: Fourteen points in a row alternating up and down: This pattern does not exist.

    Rule 5: Two out of three points in Zone A or beyond: There are no groups of three consecutive points that have two or more in Zone A or beyond.

    Rule 6: Four out of five points in a row in Zone B or beyond: No sequence of five points has four or more in Zone B or beyond.

The process appears to be in control.

f. Since both the R-chart and the $\overline{x}$-chart are in control, these control limits should be used to monitor future process output.

11.57   a. The sample size is determined by the following:

$$n > \frac{9(1 - p_0)}{p_0} = \frac{9(1 - .06)}{.06} = 141$$

The minimum sample size is 141. Since the sample size of 150 was used, it is large enough.

b. To compute the proportion of defectives in each sample, divide the number of defectives by the number in the sample, 150:

$$\hat{p} = \frac{\text{No. of defectives}}{\text{No. in sample}}$$

The sample proportions are listed in the table:

Sample No.	$\hat{p}$	Sample No.	$\hat{p}$
1	.060	11	.047
2	.073	12	.040
3	.080	13	.080
4	.053	14	.067
5	.067	15	.073
6	.040	16	.047
7	.087	17	.040
8	.060	18	.080
9	.073	19	.093
10	.033	20	.067

To get the total number of defectives, sum the number of defectives for all 20 samples. The sum is 189. To get the total number of units sampled, multiply the sample size by the number of samples: 150(20) = 3000.

$$\bar{p} = \frac{\text{Total defectives in all samples}}{\text{Total units sampled}} = \frac{189}{3000} = .063$$

Centerline $= \bar{p} = .063$

$$\textit{Upper control limit} = \bar{p} + 3\sqrt{\frac{\bar{p}(1 - \bar{p})}{n}} = .063 + 3\sqrt{\frac{.063(.937)}{150}} = .123$$

$$\textit{Lower control limit} = \bar{p} - 3\sqrt{\frac{\bar{p}(1 - \bar{p})}{n}} = .063 - 3\sqrt{\frac{.063(.937)}{150}} = .003$$

$$\textit{Upper A}-\textit{B boundary} = \bar{p} + 2\sqrt{\frac{\bar{p}(1 - \bar{p})}{n}} = .063 + 2\sqrt{\frac{.063(.937)}{150}} = .103$$

$$\textit{Lower A}-\textit{B boundary} = \bar{p} - 2\sqrt{\frac{\bar{p}(1 - \bar{p})}{n}} = .063 - 2\sqrt{\frac{.063(.937)}{150}} = .023$$

$$\textit{Upper B}-\textit{C boundary} = \bar{p} + \sqrt{\frac{\bar{p}(1 - \bar{p})}{n}} = .063 + \sqrt{\frac{.063(.937)}{150}} = .083$$

$$\textit{Lower B}-\textit{C boundary} = \bar{p} - \sqrt{\frac{\bar{p}(1 - \bar{p})}{n}} = .063 - \sqrt{\frac{.063(.937)}{150}} = .043$$

The *p*-chart is:

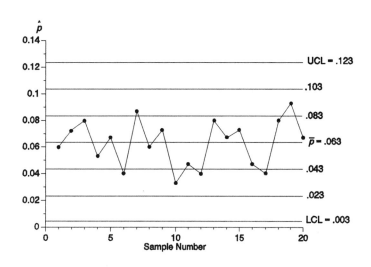

c. To determine if the process is in or out of control, we check the four rules.

Rule 1: One point beyond Zone A: No points are beyond Zone A.
Rule 2: Nine points in a row in Zone C or beyond: No sequence of nine points are in Zone C (on one side of the centerline) or beyond.
Rule 3: Six points in a row steadily increasing or decreasing: No sequence of six points steadily increase or decrease.
Rule 4: Fourteen points in a row alternating up and down: Points 2 through 16 alternate up and down. This indicates the process is out of control.

Rule 4 indicates the process is out of control. Special causes of variation appear to be present.

e. Since the process is out of control, the control limits should not be used to monitor future process output. It would not be appropriate to evaluate whether the process is in control using control limits determined during a period when the process was out of control.

ENGLISH SKILLS

FIFTH EDITION

JOHN LANGAN

Atlantic Community College

McGRAW-HILL, INC.

New York   St. Louis   San Francisco   Auckland
Bogotá   Caracas   Lisbon   London   Madrid
Mexico   Milan   Montreal   New Delhi   Paris
San Juan   Singapore   Sydney   Tokyo   Toronto

ENGLISH SKILLS

3 4 5 6 7 8 9 0 DOC DOC 9 0 9 8 7 6 5 4 3

ISBN 0-07-036393-5

This book was set in Times Roman by Monotype Composition Company.
The editors were Lesley Denton and Susan Gamer;
the designer was Rafael Hernandez;
the production supervisor was Friederich W. Schulte.
R. R. Donnelley & Sons Company was printer and binder.

Library of Congress Cataloging-in-Publication Data

Langan, John, (date).
    English skills / John Langan. — 5th ed.
      p.   cm.
    Includes index.
    ISBN 0-07-036393-5
      1. English language—Rhetoric.   I. Title.
PE1408.L318   1993
808′.042—dc20   92-26672

ABOUT
THE AUTHOR

John Langan has taught reading and writing at Atlantic Community College near Atlantic City, New Jersey, for over twenty years. The author of a popular series of college textbooks on both subjects, he enjoys the challenge of developing materials that teach skills in an especially clear and lively way. Before teaching, he earned advanced degrees in writing at Rutgers University and in reading at Glassboro State College. He also spent a year writing fiction that, he says, ''is now at the back of a drawer waiting to be discovered and acclaimed posthumously.'' While in school, he supported himself by working as a truck driver, machinist, battery assembler, hospital attendant, and apple packer. He presently lives with his wife, Judith Nadell, near Philadelphia. Among his everyday pleasures are running, working on his Macintosh computer, and watching Philadelphia sports teams or *L.A. Law* on TV. He also loves to read: newspapers at breakfast, magazines at lunch, and a chapter or two of a recent book (''preferably an autobiography'') at night.

TO MY WIFE,
JUDITH NADELL

CONTENTS

TO
THE
INSTRUCTOR

English Skills will help students learn and apply the basic principles of effective composition as well as master important grammar, punctuation, and usage skills. It is a nuts-and-bolts book based on a number of assumptions or beliefs about the writing process:

■ First of all, *English Skills* assumes that four principles in particular are keys to effective writing: unity, support, coherence, and sentence skills. These four principles are highlighted on the inside front cover and reinforced throughout the book. Part One focuses on the first three principles; Part Five fully treats sentence skills. The rest of the book shows how the four principles apply in different types of paragraph development (Part Two), in several-paragraph essays (Part Three), and in specialized types of writing (Part Four). The success of previous editions of *English Skills* supports the belief that these four principles are easily grasped, remembered, and followed by students.

■ The book also reflects a belief that, in addition to these four principles, there are other important factors in writing effectively. The second chapter discusses prewriting, rewriting, and editing. Besides encouraging students to see writing as a process, the chapter also asks students to examine their attitude toward writing, to write on what they know about or can learn about, to consider keeping a writing journal, and to make outlining a part of the writing process.

■ *English Skills* assumes that the best way to begin writing is with personal experience. After students have learned to support a point by providing material from their own experience, they are ready to develop an idea by drawing on their own reasoning abilities and on information in notes, articles, and books. In Parts Two and Three, students are asked to write on both experiential and objective topics. Part Four offers guidance in many practical writing situations: exam essays, summaries, reports, the job application letter, and the research paper.

■ The book also assumes that beginning writers are more likely to learn composition skills through lively, engaging, and realistic models than through materials remote from the common experiences that are part of everyday life. For example, when a writer argues that proms should be banned, or catalogs ways to harass an instructor, or talks about why some teenagers take drugs, students will be more apt to remember and follow the writing principles that are involved. A related assumption is that students are especially interested in and challenged by the writing of their peers. After reading vigorous papers composed by other students and understanding the power that good writing can have, students will be more encouraged to aim for a similar honesty, realism, and detail in their own work.

■ Another premise of *English Skills* is that mastery of the paragraph should precede work on the several-paragraph essay. Thus Part One illustrates the basic principles of composition using paragraph models, and the assignments in Part Two aim at developing the ability to support ideas within a variety of paragraph forms. The essential principles of paragraph writing are then applied to the several-paragraph essays in Part Three.

■ Other parts of the book reflect additional beliefs about the needs an English text should address. Part Four includes skills that will help in a variety of writing situations. Among the skills covered are writing reports and summaries, using the library, and writing and documenting a research paper. Also, the grammar, punctuation, and usage skills that make up Part Five are explained clearly and directly, without unnecessary technical terms. Here, as elsewhere, abundant exercise material is provided, especially for the mistakes that are most likely to interfere with clear communication.

■ A final assumption is that, since no two people will use an English text in exactly the same way, the material should be organized in a highly accessible manner. Because each of the five parts of the book deals with a distinct area of writing, instructors can turn quickly and easily to the skills they want to present. At the same time, ideas for sequencing material are provided by three boxes titled ''Some Suggestions on What to Do Next''; these boxes appear at the ends of the opening chapters. And a detailed syllabus is provided in the Instructor's Manual.

NOTES ON THE FIFTH EDITION

With pleasure and gratitude, I have watched the audience for *English Skills* expand each year. Instructors continue to say that the four bases really do help students learn to write effectively. And they continue to comment that students find the activities, assignments, and model passages especially interesting and worthwhile.

At the same time, more and more instructors have said that the book would benefit from an earlier emphasis on the writing process. Accordingly, in this edition I have expanded the treatment of prewriting and other important factors in writing and relocated this material: it now appears in the second chapter of the book. Instructors who are more comfortable with the previous format, however, can easily skip this chapter, moving directly to the third chapter and its treatment of the first two steps in effective writing. The material skipped could then be worked into the course later, a bit at a time.

Here is an overview of what is new in the fifth edition:

- The first chapter, "Getting Started," now introduces students to the basic principles of effective writing in more detail. Almost immediately, students read and discuss a model paragraph; they are then asked to write a paragraph of their own. This "baseline" paragraph provides the instructor and the student with a standard of comparison that can be used to measure progress in writing during the semester.

- The second chapter, "Important Factors in Writing," includes material that made up Part Two in earlier editions. Some of that material has been revised, and there are new sections on keeping a journal and prewriting in the form of diagramming or "mapping."

- "Introduction to Paragraph Development"—the chapter that begins Part Two—has been expanded to include writing for a specific purpose and audience, using peer review, and using a personal checklist.

- In Part Four, the chapter "Using the Library" has been substantially revised. It emphasizes the basic steps that students need to take in exploring possible topics for a research paper. It also describes the computerized search facilities that are often a part of today's libraries.

- Also in Part Four, the chapter "Writing a Research Paper" has been expanded to include a complete model research paper.

- Part Five, "Sentence Skills," has been enlarged to include two new chapters— "Pronoun Types" and "Adjectives and Adverbs"—as well as two new editing tests.

- Many smaller changes appear throughout the book. For example, there is a new introduction to transitions in Part One; the chapter on run-ons now includes subordination as a method of correction; two rhetorical chapters ("Explaining a Process" and "Examining Cause and Effect") and three sentence-skills chapters ("Misplaced Modifiers," "Dangling Modifiers," and "Faulty Parallelism") have been resequenced.

SUPPLEMENTS

A newly designed *Instructor's Manual and Test Bank* includes, whenever possible, separate answer sheets for each skill. Instructors can easily copy the appropriate sheets and distribute them to students for self-teaching. And an expanded set of *ditto masters*, free to instructors adopting the book, provides more tests and activities than were previously available.

Both of these supplements, along with a computer disk of mastery tests (now in both IBM and Macintosh formats), are available from the local McGraw-Hill representative or by writing to the College English Editor, 43d Floor, McGraw-Hill, Inc., 1221 Avenue of the Americas, New York, New York 10020.

ACKNOWLEDGMENTS

Reviewers who have provided assistance include Barbara Colavecchio, Community College of Rhode Island; Julie Draus, Charles County Community College; Karen Gleeman, Normandale Community College; Mary Joan Hoff, Valencia Community College; Gloria John, Catonsville Community College; Alice Lyon, Community College of Rhode Island; Zira Piltch, Iona College; Ann Pope Stone, Santa Monica College; Ed Sams, Gavilan College; and Paige Wilson, Pasadena City College.

I am also grateful for Janet M. Goldstein's help in preparing the Instructor's Manual. And I am endlessly thankful to my students. The vitality and the specialness of their lives are amply demonstrated on many pages of this text. My last acknowledgment goes to my parents—if for nothing in particular, then for much in general.

JOHN LANGAN

ENGLISH
SKILLS

PART ONE

BASIC PRINCIPLES OF EFFECTIVE WRITING

Part One begins by introducing you to the book and to paragraph form. As you work through the brief activities in the introduction, you will gain a quick understanding of the book's purpose, the way it is organized, and how it will help you develop your writing skills. After presenting a series of important general factors that will help you create good papers, Part One then describes four basic steps that can make you an effective writer. The four steps are:

1   Make a point.
2   Support the point with specific evidence.
3   Organize and connect the specific evidence.
4   Write clear, error-free sentences.

Explanations, examples, and activities are provided to help you master the first three steps. (You will be referred to Part Five of the book for a detailed treatment of the fourth step.) After seeing how these steps can help you write a competent paper, you will learn how they lead to four standards, or "bases," of effective writing: unity, support, coherence, and sentence skills. You will then practice evaluating a number of papers in terms of these four bases.

GETTING STARTED

This chapter will

- Introduce you to the basic principles of effective writing
- Ask you to write a simple paragraph
- Explain how the book is organized
- Suggest a sequence for using the book

English Skills grows out of experiences I had when learning how to write. My early memories of writing in school are not pleasant. In the middle grades I remember getting back paper after paper on which the only comment was "Handwriting very poor." In high school, the night before a book report was due, I recall working anxiously at a card table in my bedroom. I was nervous and sweaty because I felt so out of my element, like a person who knows only how to open a can of soup being asked to cook a five-course meal. The act of writing was hard enough, and my feeling that I wasn't any good at it made me hate the process all the more.

Luckily, in college I had an instructor who changed my negative attitude about writing. During my first semester in composition, I realized that my instructor repeatedly asked two questions of any paper I wrote: "What is your point?" and "What is your support for that point?" I learned that sound writing consists basically of making a point and then providing evidence to support or develop that point. As I understood, practiced, and mastered these and other principles, I began to write effective papers. By the end of the semester, much of my uneasiness and bad feelings about writing had disappeared. I knew that competent writing is a skill that I or anyone can learn with practice. It is a nuts-and-bolts process consisting of a number of principles and skills that can be studied and mastered. Further, I learned that while there is no alternative to the work required for competent writing, there is satisfaction to be gained through such work. I no longer feared or hated writing, for I knew I could work at it and be good at it.

English Skills explains in a clear and direct way the basic principles and skills you must learn to write effectively. And it provides a number of practice materials so that you can work on these skills enough to make them habits. This chapter will introduce the most basic principles of effective writing. The chapter will also show you how the rest of the book is organized and how it can help you become an effective writer.

AN INTRODUCTION TO WRITING

Point and Support:
An Important Difference between Writing and Talking

In everyday conversation, you make all kinds of points, or assertions. You say, for example, ''I hate my job''; ''Sue's a really generous person''; or ''That exam was unfair.'' The points that you make concern such personal matters as well as, at times, larger issues: ''A lot of doctors are arrogant''; ''The death penalty should exist for certain crimes''; ''Tobacco and marijuana are equally dangerous.''

The people you are talking with do not always challenge you to give reasons for your statements. They may know why you feel as you do, or they may already agree with you, or they simply may not want to put you on the spot; and so they do not always ask ''Why?'' But the people who *read* what you write may not know you, agree with you, or feel in any way obliged to you. If you want to communicate effectively with readers, you must provide solid evidence for any point you make. An important difference, then, between writing and talking is this: *In writing, any idea that you advance must be supported with specific reasons or details.*

Think of your readers as reasonable people. They will not take your views on faith, but they *are* willing to accept what you say as long as you support it. Therefore, remember to support any statement that you make with specific evidence.

Point and Support in a Paragraph

A *paragraph* is a short paper of 150 words or more. It usually consists of an opening point called a *topic sentence* followed by a series of specifics, in the form of sentences, that support the point. Most of the writing featured in this book will be paragraphs.

A Sample Paragraph: Below is a paragraph on why the writer plans not to go out with Tony anymore.

Good-Bye, Tony

I have decided not to go out with Tony anymore. First of all, he was late for our first date. He said that he would be at my house by 8:30, but he did not arrive until 9:20. Second, he was bossy. He told me that it would be too late to go to the new Steve Martin movie I wanted to see, and that we would go to a horror classic, The Night of the Living Dead, instead. I told him that I didn't like gruesome movies, but he said that I could shut my eyes during the gory parts. Only because it was a first date did I let him have his way. Finally, he was abrupt. After the movie, rather than suggesting a hamburger or a drink, he drove right out to a back road near Oakcrest High School and started necking with me. What he did a half hour later angered me most of all. He cut his finger on a pin I was wearing and immediately said we had to go right home. He was afraid the scratch would get infected if he didn't put Bactine and a Band-Aid on it. When he dropped me off, I said, "Good-bye, Tony," in a friendly enough way, but in my head I thought, "Good-bye forever, Tony."

Notice what the details in this paragraph have done. They have provided you, the reader, with a basis for understanding why the writer made the decision she did. Through specific evidence, the writer has explained and communicated her point successfully. The evidence that supports the point in a paragraph often consists of a series of reasons introduced by signal words (*First of all, Second,* and the like) and followed by examples and details that support the reasons. That is true of the sample paragraph above: three reasons are provided, followed by examples and details that back up those reasons.

Activity 1

Complete the following outline of the sample paragraph. Summarize in a few words the details that develop each reason, rather than writing the details out in full.

Point: _____

_____

Reason 1: _____

  Details that develop reason 1: _____

  _____

  _____

Reason 2: _____

    Details that develop reason 2: _____

    _____

    _____

Reason 3: _____

    Details that develop reason 3: _____

    _____

    _____

Activity 2

See if you can complete the statements below.

1. An important difference between writing and talking is that in writing we absolutely must _____ any statement we make.

2. A _____ is a collection of specifics that support a point.

Writing a Paragraph: An excellent way to get a feel for the paragraph is to write one. Your instructor may ask you to do that now. The only guidelines you need to follow are the ones described here. There is an advantage to doing a paragraph right away, at a point where you have had almost no instruction. This first paragraph will give a quick sense of your writing needs and will provide a baseline — a standard of comparison that you and your instructor can use to measure your writing progress during the semester.

Activity

Here, then, is your topic: write a paragraph on the best or worst job you have ever had. Provide three reasons why your job was the best or the worst, and give plenty of details to develop each of your three reasons. Note that the sample paragraph, "Good-Bye, Tony," has the same format your paragraph should have: the author (1) states a point in her first sentence, (2) gives three reasons to support the point, (3) clearly introduces each reason with signal words (*First of all, Second,* and *Finally*), and then (4) provides details that develop each of the three reasons. Write your paragraph on a separate sheet of paper.

AN INTRODUCTION TO THIS BOOK

How the Book is Organized

English Skills is divided into five parts. Each part will be discussed briefly below. Brief questions appear as well, not to test you but simply to introduce you to the central ideas in the text and the organization of the book. Your instructor may ask you to write in the answers or just to note the answers in your head.

Part One (Pages 1–118):   A good way to get a quick sense of any part of a book is to look at the table of contents. Turn back to the contents at the start of this book (pages vii – xi) and answer the following questions:

- What is the title of Part One? _____
- "Getting Started" is the opening chapter of Part One. What is the title of the *next* chapter in Part One, and what are the first seven subheads after the title?

   Title _____

   Subhead _____

   Subhead _____

   Subhead _____

   Subhead _____

   Subhead _____

   Subhead _____

   Subhead _____

These seven headings refer to important general factors that will help you become an effective writer.

- The title of the third chapter in Part One is "The First and Second Steps in Writing." According to the subheads, what are the first and second steps in writing?

   _____

   _____

- The title of the fourth chapter in Part One is "The Third and Fourth Steps in Writing." According to the subheads, what are the third and fourth steps in writing?

   _____

   _____

Part One describes three of the four steps in writing. The fourth step, which includes all the skills involved in writing clear, error-free sentences, has been placed in a later part of the book, where these sentence skills can be treated in detail and can be easily referred to as needed. Use the table of contents (pages vii – xi) to answer the following question:

■ In what part of the book are sentence skills treated?

_____

■ The title of the final chapter in Part One is "Four Bases for Evaluating Writing." Look at the contents (pages vii – xi) again and fill in the first four subheads following the title.

Subhead _____

Subhead _____

Subhead _____

Subhead _____

Inside Front Cover: Turn now to the inside front cover. You will see there a (*fill in the missing word*) _____ of the four bases of effective writing. These four standards can be used as a guide for every paper that you write. They are summarized on the inside front cover for easy reference. If you follow them, you are almost sure to write effective papers.

Part Two (Pages 119 – 217): The title of Part Two is _____.
Part Two, as the title explains, is concerned with different ways to develop paragraphs. Read the preview on page 120 and record here how many types of paragraph development are presented: _____.
Turn to the first method of paragraph development, "Providing Examples," on page 127. You will see that the chapter opens with a brief introduction followed by several paragraphs written by students. Then you will see a series of six (*fill in the missing word*) _____ to help you evaluate the example paragraphs in terms of unity, support, and coherence. Finally, some writing topics that can be developed by means of examples are presented. The same format is used for each of the other methods of paragraph development in Part Two.

Part Three (Pages 219 – 249): The title of Part Three is _____

_____

As the preview on page 220 notes, in Part Two you were asked to write single paragraphs; in Part Three, you are asked to write papers of more than one (*fill in the missing word*) _____.

Part Four (Pages 251–324):   The title of Part Four is _____

_____.

Part Four gives you advice on a number of important skills that are related to writing. You can refer to this part of the book whenever the need arises.

■   Which chapter will help you write effective answers to essay questions?

_____

■   Which chapters will help you prepare papers on books or articles you must read for other classes?

_____

_____

_____

Part Five (Pages 325–557):   The title of Part Five is _____

_____.

Part Five is the largest part of the book. It gives you practice in skills needed to write clear and effective sentences. You will note from the table of contents (pages vii–xi) that it contains a diagnostic test, the skills themselves, editing activities, and an achievement test. The skills are grouped into four sections:

"Grammar," "Mechanics," (*fill in the missing word*) "_____," and "Word Use."

Inside Back Cover: On the inside back cover is an alphabetical list of (*fill in the missing words*) _____.
Your instructor may use these symbols in marking your papers. In addition, you can use the page numbers in the list for quick reference to a specific sentence skill.

Charts in the Book: In addition to the guides on the inside front and back covers, several charts have been provided in the book to help you take responsibility for your own learning.

■   What are the names of the charts on pages 565–568?

_____

_____

_____

How to Use the Book

Here is a suggested sequence for using this book if you are working on your own.

1 After completing this introduction, read the next four chapters in Part One and work through as many of the activities as you need to master the ideas in these chapters. Your instructor may give you answer sheets so that you can check your answers. At that point, you will have covered all the basic theory needed to write effective papers.

2 Turn to Part Five and take the diagnostic test. The test will help you determine what sentence skills you need to review. Study those skills one or two at a time while you continue to work on other parts of the book.

3 What you do next depends on course requirements, individual needs, or both. You will want to practice at least several different kinds of paragraph development in Part Two. If your time is limited, be sure to include ''Providing Examples,'' ''Explaining a Process,'' ''Comparing or Contrasting,'' and ''Arguing a Position.'' After that, you could logically go on to write one or more of the several-paragraph essays described in Part Three. And the writing-related skills in Part Four can be referred to whenever they are needed.

AS YOU BEGIN . . .

English Skills will help you learn, practice, and apply the writing skills you need to communicate clearly and effectively. But the starting point must be your determination to do the work needed to become an independent writer. If you decide — *and only you can decide* — that you want to learn to write effectively, this book will help you reach that goal.

IMPORTANT FACTORS IN WRITING

This chapter will discuss the importance of

- Your attitude about writing
- Writing for a specific purpose and audience
- Knowing or discovering your subject
- Keeping a journal
- Prewriting
- Outlining
- Revising, editing, and proofreading

The preceding chapter introduced you to the paragraph form, and the chapters that follow in Part One will explain the basic steps in writing a paragraph and basic standards for evaluating a paragraph. The purpose of this chapter is to describe a number of important general factors that will help you create good papers. These factors are (1) having the right attitude about writing, (2) writing for a specific purpose and audience, (3) knowing or discovering your subject, (4) keeping a journal, (5) prewriting, (6), outlining, and (7) revising, editing, and proofreading.

Your Attitude about Writing

One way to wreck your chances of learning how to write competently is to believe that writing is a natural gift. People with this attitude think that they are the only ones for whom writing is unbearably difficult. They feel that everyone else finds writing easy or at least tolerable. Such people typically say, ''I'm not any good at writing'' or ''English was not one of my good subjects.'' They imply that they simply do not have a talent for writing, while others do. As a result of this attitude, they do not do their best when they write, or — even worse — they hardly try at all. Their self-defeating attitude becomes a reality; their writing fails chiefly because they have brainwashed themselves into thinking that they don't have the natural talent needed to write. Until their attitude changes, they probably will not learn how to write effectively.

A realistic attitude about writing, rather than the mistaken notion that writing is a ''natural gift,'' should build on two crucial ideas.

1   *Writing is hard work for almost everyone.* It is difficult to do the intense and active thinking that clear writing demands. (Perhaps television has made us all so passive that the active thinking necessary in both writing and reading now seems doubly hard.) It is frightening to sit down before a blank sheet of paper and know that an hour later, nothing on it may be worth keeping. It is frustrating to discover how much of a challenge it is to transfer thoughts and feelings from one's head onto a sheet of paper. It is upsetting to find that an apparently simple writing subject often turns out to be complicated. But writing is not an automatic process; we will not get something for nothing; and we cannot expect something for nothing. Competent writing results only from plain hard work — determination, sweat, and head-on battle.

2   *Writing is a skill.* Writing is a skill, like driving, typing, or cooking. Like any skill, it can be learned — if you decide that you are going to learn it, and if you then really work at it. This book will give you the extensive practice needed to develop your writing skills.

Activity

Answering these questions will help you evaluate your attitude about writing.

1.   How much practice were you given in writing compositions in high school?

_____ Much   _____ Some   _____ Little

2.   How much feedback on your compositions (positive or negative comments) did your teachers give you?

_____ Much   _____ Some   _____ Little

3. How did your teachers seem to regard your writing?

_____ Good   _____ Fair   _____ Poor

4. Do you feel that some people simply have a gift for writing and others do not?

_____ Yes   _____ Sometimes   _____ No

5. When do you start writing a paper?

_____ Several days before it is due

_____ About a day before it is due

_____ At the last possible minute

Many people who answer *Little* to questions 1 and 2 also answer *Poor* to question 3, *Yes* to question 4, and *At the last possible minute* to question 5. On the other hand, people who answer *Much* or *Some* to questions 1 and 2 also tend to give more favorable responses to the other questions. The point is that people with little practice in writing often have understandably negative feelings about their ability to write. They need not have such feelings, however, because writing is a skill that they can learn with practice.

Writing for a Specific Purpose and Audience

The three most common purposes of writing are to inform, to entertain, and to persuade. Most of the writing you will do in this book will involve some form of persuasion. You will advance a point or topic sentence and then support it in a variety of ways. To some extent, also, you will write papers to inform — to provide readers with information about a particular subject.

Your audience will be primarily your instructor, and sometimes other students as well. Your instructor is really a symbol of the larger audience you should see yourself as writing for — an educated, adult audience that expects you to present your ideas in a clear, direct, organized way. If you can learn to write to persuade or inform such a general audience, you will have accomplished a great deal.

It will also be helpful for you to write some papers for a more specific audience. By so doing, you will develop an ability to choose words and adopt a tone of voice that is just right for a given purpose and a given group of people. For example, Part Two of this book includes assignments asking you to write with a very specific purpose in mind, and for a very specific audience.

Knowing or Discovering Your Subject

KNOWING YOUR SUBJECT

Whenever possible, try to write on a subject which interests you. You will then find it easier to put more time into your work. Even more important, try to write on a subject that you already know something about. If you do not have direct experience with the subject, you should at least have indirect experience — knowledge gained through thinking, prewriting (to be explained on pages 17–22), reading, or talking about the subject.

If you are asked to write on a topic about which you have no experience or knowledge, you should do whatever research is required to gain the information you need. The chapter ''Using the Library'' on pages 291–307 will show you how to use the library to look up relevant information. Without direct or indirect experience, or the information you gain through research, you will not be able to provide the specific evidence needed to develop whatever point you are trying to make. Your writing will be starved for specifics.

DISCOVERING YOUR SUBJECT

At times you will not know your subject when you begin to write. Instead, you will discover it in the actual process of writing. For example, when a student named Gene sat down to write a paper about a memorable job (see page 33), he thought for a while that his topic was going to be an especially depressing moment on that job. As he began to accumulate details, however, he realized that his topic was really the job itself and all the drawbacks it entailed. Gene only *thought* he knew the focus of his paper when he began to write. In fact, he *discovered his subject in the course of writing.*

Another student, Rhonda, talking afterwards about a paper she wrote, explained that at first her topic was how she relaxed with her children. But as she accumulated details, she realized after a page of writing that the words *relax* and *children* simply did not go together. Her details were really examples of how she *enjoyed* her children, not how she *relaxed* with them. She sensed that the real focus of her writing should be what she did by herself to relax, and then she thought suddenly that the best time of her week was Thursdays after school. ''A light clicked on in my head,'' she explained. ''I knew I had my paper.'' Then it was a matter of detailing exactly what she did to relax on Thursday evenings. Her paper, ''How I Relax,'' is on page 73.

The moral of these examples is that sometimes you must write a bit in order to find out just what you want to write. Writing can help you think about and explore your topic and decide just what direction your paper will finally take. The techniques presented in ''Prewriting''—the section starting on page 17—will suggest specific ways to discover and develop a subject.

One related feature of the writing process bears mention. Do not feel that you must proceed in a linear fashion when you write. That is, do not assume that the writing process is a railroad track going straight from your central point to supporting detail 1 to supporting detail 2 to supporting detail 3 to your concluding paragraph. Instead, as you draft the paper, proceed in whatever way seems most comfortable. You may want to start by writing the closing section or by developing your third supporting detail.

Do whatever is easiest; as you get material down on the page, it will make what you have left to do a bit easier. And sometimes, of course, as you work on one section, it may happen that a new focal point for your paper will emerge. That's fine: if your writing tells you that it wants to be something else, then revise or start over as needed to take advantage of that discovery. Your goal is to wind up with a paper that solidly makes and supports a point. Be ready and open to change direction and to make whatever adjustments are needed to reach your goal.

Activity 1

Answer the following questions.

1. What are three ways of gaining the knowledge you need to write about a subject?   a. _____   b. _____   c. _____

2. A student begins to write a paper about her favorite vacation. After writing for a half hour, she realizes that the most vivid details coming to her are of her worst vacation. What has happened in the process of writing?

   _____

   _____

   _____

3. Suppose you want to write a paper on different kinds of drivers. You think you can discuss slowpoke drivers, high-speed drivers, and sensible-speed drivers. You feel you have the most details about high-speed drivers. Should you start with that type of driver, or should you start with one of the other two types? _____

   _____

   _____

Activity 2

Write for five minutes about the house, dormitory, or apartment where you live. Simply write down whatever details come to you. Don't worry about being neat; just pile up as many details as you can.

Afterward, go through the material. Try to find a potential focus within all those details. Do the details suggest a simple point that you could make about the place where you live? If so, you've seen a small example of how writing about a topic can be an excellent way of discovering a point about that topic.

Keeping a Journal

Because writing is a skill, it makes sense that the more you practice writing, the better you will write. One excellent way to get practice in writing is to keep a daily or almost daily journal.

At some point during the day — perhaps in a study period after your last class, perhaps before dinner, or perhaps before going to bed — spend fifteen minutes or so writing in your journal. Keep in mind that you do not have to prepare what to write or be in the mood or worry about making mistakes; just write down whatever words come out. As a minimum, you should complete at least one page in each writing session.

You may want to use a notebook that you can easily carry with you for on-the-spot writing. Or you may decide to write on looseleaf paper that can be transferred later to a journal folder or binder on your desk. No matter how you proceed, be sure to date all your entries.

The content of your journal should be some of the specific happenings, thoughts, and feelings of the day. Your starting point may be a comment by an instructor, a classmate, or a family member; a gesture or action that has amused, angered, confused, or depressed you; something you have read or seen on television – anything, really, that has caught your attention and that you decided to explore a bit in writing. Some journal entries may focus on a single subject; others may wander from one topic to another.

Your instructor may ask you to make journal entries a set number of times a week, for a set number of weeks. He or she may ask you to turn in your journal every so often for review and feedback. If you are keeping the journal on your own, try to make entries three to five times a week, every week of the semester.

Your journal can serve as a sourcebook of ideas for possible papers. More important, keeping a journal will help you develop the habit of thinking on paper, and it can help you make writing a familiar part of your life.

Following is an excerpt from one student's journal. (Sentence-skills mistakes have been corrected to improve readability.) As you read, look for a general point and supporting material that could be the basis for an interesting paper.

October 6

Today a woman came into our department at the store and wanted to know if we had any scrap lumber ten feet long. Ten feet! "Lady," I said, "anything we have that's ten feet long sure as heck isn't scrap." When the boss heard me say that, he almost canned me. My boss is a company man, down to his toe tips. He wants to make a big impression on his bosses, and he'll run us around like mad all night to make himself look good. He's the most ambitious man I've ever met. If I don't transfer out of Hardware soon, I'm going to go crazy on this job. I'm not ready to quit, though. The time is not right. I want to be here for a year and have another job lined up and have other things right before I quit. It's good the boss wasn't around tonight when another customer wanted me to carry a bookcase he had bought out to his car. He didn't ask me to help him — he <u>expected</u> me to help him. I hate that kind of "You're my servant" attitude, and I told him that carrying stuff out to cars wasn't my job. Ordinarily I go out of my way to give people a hand, but not guys like him. . . .

- If the writer of this journal was looking for an idea for an essay, he could probably find several in this single entry. For example, he might write a narrative supporting the point that "In my sales job I have to deal with some irritating customers." See if you can find another idea in this entry that might be the basis for an interesting paragraph. Write your point in the space below.

_____

- Take fifteen minutes to prepare a journal entry right now on this day in your life. On a separate sheet of paper, just start writing about anything that you have seen, said, heard, thought, or felt, and let your thoughts take you where they may.

Prewriting

If you are like many people, you may sometimes have trouble getting started writing. A mental block may develop when you sit down before a blank sheet of paper. You may not be able to think of a topic or an interesting slant on a topic. Or you may have trouble coming up with interesting and relevant details to support your topic. Even after starting a paper, you may hit snags — moments when you wonder "Where do I go next?"

The following pages describe five techniques that will help you think about and develop a topic and get words down on paper: (1) brainstorming, (2) freewriting, (3) making a list, (4) diagramming, and (5) making a scratch outline. These techniques, which are often called *prewriting techniques,* are a central part of the writing process.

TECHNIQUE 1: BRAINSTORMING

In *brainstorming*, you generate ideas and details by asking as many questions as you can think of about your subject. Such questions include *What? When? Why? How? Where?* and *Who?*

Following is an example of how one student, Sal, used brainstorming to generate material for a paper. Sal felt that he could write about a painful moment he had experienced, but he was having trouble getting started. So he asked himself a series of brainstorming questions about the experience. As a result, he accumulated a series of details that provided the basis for the paper he finally wrote.

Here are the questions Sal asked and the answers he wrote:

Questions	Answers
Where did the experience happen?	In my girlfriend's dorm room at Penn State.
When did it happen?	A week before Thanksgiving.
Who was involved?	My girlfriend, her roommate (briefly), and I.
What happened?	I discovered my girlfriend was dating someone else.
Why was the experience so painful?	Bonnie and I were engaged. She had never mentioned Blake. My surprise visit turned into a terrible surprise for me.
How did Bonnie react?	She was nervous and tried to avoid answering my questions.
How did I react?	I felt sick and angry. I wanted to do something violent. I wanted to tear up the poster with Blake's name on it. I wanted to slam the door, but I walked out quietly. My knees were shaking.

After brainstorming, Sal's next step was to prepare a scratch outline. He then prepared several drafts of the paper. The effective paragraph that eventually resulted from Sal's prewriting techniques appears on page 192.

Activity

To get a sense of brainstorming, use a sheet of paper to ask yourself a series of questions about a pleasant diner you have visited. See how many details you can accumulate about that diner in ten minutes.

TECHNIQUE 2: FREEWRITING

When you do not know what to write about a subject or when you are blocked in writing, freewriting sometimes helps. In *freewriting,* you write on your topic for ten minutes. You do not worry about spelling, punctuation, erasing mistakes, or finding exact words. You just write without stopping. If you get stuck for words, you write ''I am looking for something to say'' or repeat words until something comes. There is no need to feel inhibited, since mistakes do not count and you do not have to hand in your paper.

Freewriting will limber up your writing muscles and make you familiar with the act of writing. It is a way to break through mental blocks about writing and the fear of making errors. As you do not have to worry about making mistakes, you can concentrate on discovering what you want to say about a subject. Your initial ideas and impressions will often become clearer after you have gotten them down on paper. Through continued practice in freewriting, you will develop the habit of thinking as you write. And you will learn a technique that is a helpful way to get started on almost any paper that you write.

Here is the freewriting that one student did to accumulate details for a paper on why he stopped smoking:

> I was way overdue to stop smoking cigarettes and I finally did. I had a friend who went to the hospital with lung cancer. No one can say that he's going to recover. He's in Eagleville Hospital. When I heard about him, it was the last straw for me. Smoking is a life-and-death matter. My friend is the one who brought this message home to me. Smoking is a life-and-death matter just like the ads say. When I think about it, I hated the fact that I was helping corporations make a lot of money all the while that I smoked. The corporations produced all this slick advertising and I felt I was one of the puppets who listened to it. I marched to their tune. I didn't want to make wealthy corporations even richer, and I hated it every time I gave over hard-earned dollars for a carton of cigarettes. Cigarettes were a very expensive habit. I can hardly say how much a year I had to put out for them. You could see I smoked as you walked through my house. There were ashtrays in the living room, dining room, bathroom, and kitchen. My wife said there was a smell of smoke in the house. I couldn't tell. I had a nose so clogged that I couldn't smell much at all. Cigarettes were a bum trip that I am not going to take any longer.

The writer's next step was to use the freewriting as the basis for a scratch outline. The effective paper that eventually resulted from the author's freewriting, a scratch outline, and a good deal of rewriting appears on page 144.

Activity

To get a sense of freewriting, use a sheet of paper to freewrite about your everyday worries. See how many ideas and details you can accumulate in ten minutes.

TECHNIQUE 3: MAKING A LIST

Another way to get started is to make a list of as many different items as you can think of concerning your topic. Do not worry about repeating yourself, about sorting out major details from minor ones, or about spelling or punctuating correctly. Simply make a list of everything about your subject that occurs to you. Your aim is to generate details and to accumulate as much raw material for writing as possible.

Following is a list prepared by one student, Linda, who was gathering details for a paper on abuse of public parks. Her first stage in doing the paper was simply to make a list of thoughts and details that occurred to her about the topic. Here is her list:

Messy picnickers (most common)
Noisy radios
Graffiti on buildings and fences
Frisbee games that disturb others
Dumping car ashtrays
Stealing park property
Nude sunbathing
Destroying flowers
Damaging fountains and statues
Litter
Muggings

Notice that Linda puts in parentheses a note to herself that messy picnickers are the most common type of park abusers. Very often as you make a list, ideas about how to develop a paper will occur to you. Jot them down.

Making a list is an excellent way to get started. Often you can then go on to make a scratch outline and write the first draft of your paper. A scratch outline for Linda's list is shown on page 22.

Activity

To get a sense of making a list, use a sheet of paper to list specific problems you will face this semester. See how many ideas and details you can accumulate in ten minutes.

TECHNIQUE 4: DIAGRAMMING

Diagramming, also known as *mapping* or *clustering,* is another prewriting activity that can help you generate ideas and details about a topic. In diagramming, you use lines, boxes, arrows, and circles to show relationships among the ideas and details that come to you.

Diagramming is helpful to people who like to do their thinking in a visual way. Whether you use a diagram, and just how you proceed with it, is up to you.

Here is the diagram that one student, Mel, prepared for a paper on differences between his job as he imagined it and as it turned out to be. The diagram, with its clear picture of relationships, was especially helpful for the comparison-contrast paper that Mel was doing. His final essay appears on page 158.

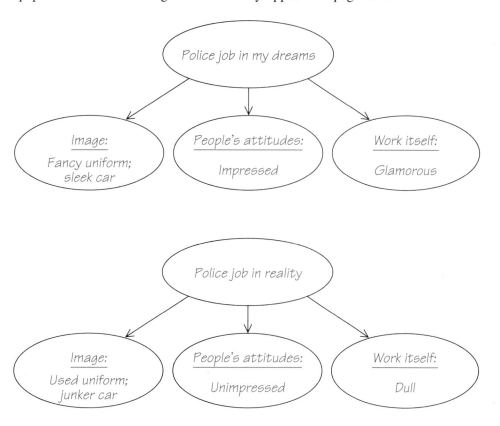

Activity

To get a sense of diagramming, use a sheet of paper to make a diagram of differences between two instructors or two jobs. See how many ideas and details you can accumulate in ten minutes.

TECHNIQUE 5: PREPARING A SCRATCH OUTLINE

A scratch outline can often be the *single most helpful technique* for writing a good paper. It is an excellent complement to the prewriting techniques already mentioned: brainstorming, freewriting, making a list, and diagramming. In a scratch outline, you think carefully about the exact point you are making, about the exact items you will use to support that point, and about the exact order in which you will arrange those items. The scratch outline is a plan or blueprint that will help you achieve a unified, supported, and organized composition.

Here is the scratch outline that Linda prepared from her general list about abuse of parks:

Some people abuse public parks.
1. Cleaning out cars
   a. Ashtrays
   b. Litter bags
2. Defacing park property
3. Stealing park property
   a. Flowers, trees, shrubs
   b. Park sod
4. Not cleaning up after picnics
   a. Paper trash
   b. Bottles and cans

This scratch outline enabled Linda to think about her paper — to decide exactly which items to include and in what order. Without writing more than one sentence, she has taken a giant step toward a paper that is unified (she has left out items that are not related); supported (she has added items that develop her point); and organized (she has arranged the items in a logical way — here, in emphatic order). The effective paragraph that eventually resulted from Linda's list and scratch outline is on page 51 (paragraph A).

Activity

To get a sense of preparing a scratch outline, make an outline of reasons why you did well or did not do well in high school. See how many ideas and details you can accumulate in ten minutes.

USING ALL FIVE TECHNIQUES

Very often a scratch outline follows brainstorming, freewriting, diagramming, and making a list. At other times, the scratch outline may be substituted for the other four techniques. Also, you may use several techniques almost simultaneously when writing a paper. You may, for example, ask questions while making a list; you may diagram and outline the list as you write it; you may ask yourself questions and then freewrite answers to them. The five techniques are all ways to help you go about the process of writing a paper.

Activity 1

Answer the following questions.

1. Which of the prewriting techniques do you already practice?

   _____ Brainstorming

   _____ Making a list

   _____ Freewriting

   _____ Making a scratch outline

   _____ Diagramming

2. Which prewriting technique involves asking questions about your topic?

   _____

3. Which prewriting technique shows in a visual way the relationship between ideas and details?

   _____

4. Which prewriting technique involves writing quickly about your topic without being concerned about grammar or spelling?

   _____

5. Which prewriting technique is almost always part of doing an essay?

   _____

6. Which techniques do you think will work best for you?

   _____

Activity 2

Following are examples of how the five prewriting techniques could be used to develop the topic "Inconsiderate Drivers." Identify each technique by writing B (for *brainstorming*), F (for *freewriting*), D (for the *diagram*), L (for the *list*), or SO (for the *scratch outline*) in the answer space.

_____   High beams on
Weave in and out at high speeds
Treat street like a trash can
Open car door onto street without looking
Stop on street looking for an address
Don't use turn signals
High speeds in low-speed zones
Don't take turns merging
Use horn when they don't need to
Don't give walkers the right of way

_____   What is one example of an inconsiderate driver?   A person who suddenly turns without using a signal to let the drivers behind know in advance.

When does this happen?   At city intersections or on smaller country roads.

Why is this dangerous?   You have to be alert to slow down yourself to avoid rear-ending the car in front.

What is another example of inconsideration on the road?   Drivers who come toward you at night with their high beams on.

_____   Some people are inconsiderate drivers.
  1. In city:
      a. Stop in middle of street
      b. Turn without signaling
  2. On highway:
      a. Leave high beams on
      b. Stay in passing lane
      c. Cheat during a merge
  3. Both in city and on highway: Throw trash out the windows

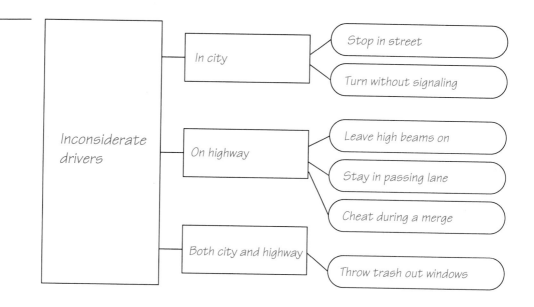

I was driving home last night after class and had three people try to blind me by coming at me with their high beams on. I had to zap them all with my high beams. Rude drivers make me crazy. The worst are the ones that use the road as a trash can. People who throw butts and cups and hamburger wrappings and other stuff out the car windows should be tossed into a trash dumpster. If word got around that this was the punishment maybe they would wise up. Other people do dumb things as well. I hate the person who will just stop in the middle of the street and try to figure out directions or look for a house address. Why don't they pull over to the side of the street? That hardly seems like too much to ask. Instead, they stop all traffic while doing their own thing. Then there are the people who keep what they want to do a secret. They're not going to tell you they plan to make a right- or left-hand turn. Instead, you've got to figure it out yourself when they suddenly slow down in front of you.

Outlining

As already mentioned, often the best way to write an effective paragraph is to outline it first. (At times, you may have to do a fair amount of writing first to discover your topic, but a stage will come when outlining, or reoutlining, will help.) Outlining is an organizational skill that will develop your ability to think in a clear and logical manner. An outline lets you work on, and see, the bare bones of a paper, without the distraction of a clutter of words and sentences. Outlining provides a quick check on whether your paper is *unified*. It suggests right at the start whether your paper will be adequately *supported*. And it shows you how to plan a paper that is *well organized*.

The following series of exercises will help develop the outlining skills that are so important to writing an effective paper.

Activity 1

One key to effective outlining is the ability to distinguish between major ideas and the details that fit under those ideas. This exercise will develop your ability to generalize from a list of details and to determine a major thought. In each case on the opposite page, write in the heading that accurately describes the list provided. Note the two examples below.

Examples   *Signs of a cold*

Headache

Runny nose

Fever

Chills

Fuels

Wood

Oil

Gas

Kerosene

1. _____
   Vanilla fudge
   Strawberry
   Chocolate
   Butter almond

2. _____
   Lincoln
   Reagan
   Jefferson
   Roosevelt

3. _____
   Birthday
   Get well
   Anniversary
   Graduation

4. _____
   Anacin
   Bufferin
   Alka-Seltzer
   Tylenol

5. _____
   Robbery
   Murder
   Assault
   Kidnapping

6. _____
   Loafers
   Moccasins
   Sneakers
   Sandals

7. _____
   Russian
   Oil and vinegar
   Blue cheese
   French

8. _____
   Washing dishes
   Taking out trash
   Preparing meals
   Dusting

9. _____
   Cash
   Check
   Money order
   Credit card

10. _____
    Writing
    Speaking
    Listening
    Reading

Activity 2

Major and minor ideas are mixed together in the two paragraphs outlined below. Put the ideas in logical order by filling in the outlines.

1. **Topic sentence:** People can be classified by how they treat their cars.

Seldom wax or vacuum car

Keep every mechanical item in top shape

Protective owners

Deliberately ignore needed maintenance

Indifferent owners

Wash and polish car every week

Accelerate too quickly and brake too hard

Abusive owners

Inspect and service car only when required by state law

a. _____

   (1) _____

   (2) _____

b. _____

   (1) _____

   (2) _____

c. _____

   (1) _____

   (2) _____

2. **Topic sentence:** Living with an elderly parent has many benefits.

Advantages for elderly person

Live-in baby-sitter

Learn about the past

Advantages for adult children

Serve useful role in family

Help with household tasks

Advantages for grandchildren

Stay active and interested in young people

More attention from adults

a. _____

   (1) _____

   (2) _____

b. _____

   (1) _____

   (2) _____

c. _____

   (1) _____

   (2) _____

Activity 3

Again, major and minor ideas are mixed together. In addition, in each outline one of the three major ideas is missing and must be added. Put the ideas in logical order by filling in the outlines that follow and adding a third major idea.

1. **Topic sentence:** Extending the school day would have several advantages.

Help children academically

Parents know children are safe at the school

More time to spend on basics

Less pressure to cover subjects quickly

More time for extras like art, music, and sports

Help working parents

More convenient to pick up children at 4 or 5 P.M.

Teachers' salaries would be raised

a. _____

  (1) _____

  (2) _____

b. _____

  (1) _____

  (2) _____

c. _____

  (1) _____

  (2) _____

2. **Topic sentence:** Living in a mobile home has many disadvantages.

Cost of site rental can go up every year

Tiny baths and bedrooms

Few closets or storage areas

Crowded conditions in park

Lack of space

Noise from neighbors

Resale value low compared with houses

No privacy outside in tiny yards

a. _____

  (1) _____

  (2) _____

b. _____

  (1) _____

  (2) _____

c. _____

  (1) _____

  (2) _____

Activity 4

Read the following two paragraphs. Then outline each one in the space provided. Write out the topic sentence in each case and summarize in a few words the primary and secondary supporting material that fits under the topic sentence.

1.   Why I'm a Stay-at-Home Baseball Fan

      I'd much rather stay at home and watch ball games on television than go to the ball park. First of all, it's cheaper to watch a game at home. I don't have to spend $8 for a ticket and another $5 for a parking space. If I want some refreshments, I can have what's already in the refrigerator instead of shelling out another $3.50 for a limp, lukewarm hot dog and a watery Coke. Also, it's more comfortable at home. I avoid a bumper-to-bumper drive to the ball park and pushy crowds who want to go through the same gate I do. I can lie quietly on my living-room sofa instead of sitting on a hard stadium seat with noisy people all around me. Most of all, watching a game on television is more informative. Not only do I see all the plays which I might not from my $8 seat, but I see some of them two and three times on instant replay. In addition, I get each play explained to me in glorious detail. If I were at the ball park, I wouldn't know that the pitch our third baseman hit was a high and inside slider or that his grand-slam home run was a record-setting seventh in his career. So I'll let the other fans spend their money, put up with traffic, crowds, and hard seats, and guess at the plays. I'll take my baseball lying down--at home.

Topic sentence: _____

_____

a. _____

    (1) _____

    (2) _____

b. _____

    (1) _____

    (2) _____

c. _____

    (1) _____

    (2) _____

2.

Why Teenagers Take Drugs

There are several reasons why teenagers take drugs. First of all, it is easy for young people to get drugs. Drugs are available almost anywhere, from a school cafeteria to a movie line to a football game. Teens don't have to risk traveling to the slums or dealing with shady types on street corners. It is also easy to get drugs because today's teens have spending money, which comes from allowances or earnings from part-time jobs. Teens can use their money to buy the luxuries they want--records, makeup, clothes, or drugs. Second, teens take drugs because the adolescent years are filled with psychological problems. For a teenager, one of these problems is facing the pressure of making important life decisions, such as choosing a career path. Another problem is establishing a sense of self. The teen years are the time when young people must become more independent from their parents and form their own values. The enormous mental pressures of these years can make some people turn to drugs. A final, and perhaps most important, reason why teenagers take drugs is peer pressure to conform. Teens often become very close to special friends, for one thing, and they will share a friend's interests, even if one of them is drugs. Teenagers also attend parties and other social events where it's all-important to be one of the crowd, to be "cool." Even the most mature teenager might be tempted to use drugs rather than risk being an outcast. For all these reasons, drugs are a major problem facing teenagers.

Topic sentence: _____

_____

a. _____

    (1) _____

    (2) _____

b. _____

    (1) _____

    (2) _____

c. _____

    (1) _____

    (2) _____

Revising, Editing, and Proofreading

Writing an effective paper is almost never done all at once. Rather, it is a step-by-step process in which you take your paper through a series of stages—from prewriting to final draft.

In the first stage, you get your initial ideas and impressions about the subject down on paper; *you accumulate raw material.* You do this through brainstorming, freewriting, and making lists and scratch outlines.

In the second stage, *you shape, add to, and perhaps subtract from your raw material* as you take your paper through a series of two or three or four rough drafts. You work to make clear the single point of your paper, to develop fully the specific evidence needed to support that point, and to organize and connect the specific evidence. For example, perhaps in the second draft you will concentrate on adding details that will further support the main idea of your paper. At the same time you also may eliminate details that, you realize, do not truly back up your main point. And perhaps in the next draft, you will work on reorganizing the details and adding connections between sentences so that your material will hold together.

In the last stage, you *edit* and *proofread.* You *edit* the next-to-final draft — that is, you check it carefully for the correctness of sentence skills — grammar, mechanics, punctuation, and usage. Then you *proofread* the final copy of the paper for any typing or handwriting mistakes. Editing and proofreading are important steps that some people avoid, often because they have worked so hard (or so little) on the previous stages.

Ideally, you should have enough time to set your paper aside for a while, so that you can check it later from a fresh point of view. Remember that locating and correcting sentence-skills mistakes can turn an average paper into a better one and a good paper into an excellent one. A later section of this book will give you practice in editing and proofreading in the form of a series of editing tests (pages 539–551).

Practice in Seeing the Entire Writing Process

This section will show you the stages that can be involved in writing an effective paper. You will see what one student, Gene, does in preparing a paper on his worst job.

There is no single sequence that all people follow in writing a composition. However, the different stages in composing that Gene goes through in writing his paper should give you some idea of what to expect. As you'll see, Gene does not just sit down and proceed neatly from start to middle to finish. Writing seldom works like that.

STAGE 1:
THINKING AND PREWRITING ABOUT YOUR TOPIC

In retrospect, here is what Gene says about his initial writing topic and his reaction to it:

> "The assignment was to write about a memorable job. I rejected several ideas; I wanted something I had strong feelings about. Then I thought of a job that I had really hated, working in an apple plant. I remembered a moment when I thought I was at the end of the world. It was a cold winter morning about 5 A.M. I had just loaded apple juice all night and was now cleaning one of the apple vats. The vat was an old gasoline truck body. I was inside it, slipping on its rounded stainless steel floor. It was dark in there, the only light coming from a porthole entrance that I had used to crawl in. Apple juice residue was dripping onto my head, and I was using Ajax and a scrub brush to clean off the residue. I felt incredibly depressed. I didn't have a girlfriend then, and my parents were always fighting, and I was incredibly lonely. I felt I was at the bottom of the barrel in my life, and I was never going to get out.
>
> "All this is what I wanted to write about, I thought, and I scribbled down a lot of the details I just mentioned. After I had over two pages of material, I began to think, 'This is too large a topic. It involves a whole terrible phase in my life. I need to narrow my topic down.' Then I decided to focus on the job itself and what I didn't like about it. I felt I was on my way."

STAGE 2: MAKING A LIST

At this point Gene makes up an initial list of details about the job. The list is shown below:

Apple factory job—worst one I ever had

Boss was a madman
Working conditions were poor
Went to work at 5 P.M., got back at 7 A.M.
Lifted cartons of apple juice for ten hours
Slept, ate, and worked--no social life
Gas money to and from work
Loaded onto wooden skids in a truck
Short breaks but breakneck pace
No real companions at work
Cold outside
Floors of trucks cold metal
Had to clean apple vats

■ Comments and Activity

Fill in the missing words: Gene is fortunate enough to know almost from the start what the _____ in his paper will be. Most of his work can thus go into developing details to support the point. Details seldom come automatically; they must be dug for, and Gene's sketchy list of unpleasant items about the job is an early stage in the development of his subject. Making a _____ is an excellent way to get started.

Note that, in his list, Gene is not concerned about ordering the details in any way, or about deciding whether any detail really fits, or even about repeating himself. He is just doing first things first: getting raw material down on paper. In the second stage, Gene will also concentrate on accumulating raw material and will start to give attention to shaping that material.

STAGE 3: ADDITIONAL PREWRITING

After making a list, Gene continues on to a partial draft of the paper.

Note: To keep Gene's drafts as readable as possible, his spelling and sentence-skills mistakes have been corrected. Ordinarily, a number of such mistakes might be present, and editing a paper for them would be a part of the writing process.

> I hated my job in the apple factory. I hated it because the work was hard. I loaded cartons of apple juice onto wooden skids in tractor trailers. Two parts to job: ten hours on line; two hours cleaning. I hated the job because the working conditions were bad and the pay was poor.

Why were working conditions bad?	Why was pay poor?
Outside weather cold	Two dollars an hour (minimum wage at the time)
Usually zero degrees	
Floor of tractor trailer was cold steel	Quarter more for working the second shift
Breaks were limited-- 10 minutes every $2\frac{1}{2}$ hours	Only money was in overtime--when you got time-and-a-half
$\frac{1}{2}$ hour for lunch	No double time
	I would work twenty hours Friday through Saturday to get as much overtime as possible

1   Work hard
3   Working conditions were poor
   (temperature outside, cleaning of the vats)
2   Money poor

■ **Comments and Activity**

Fill in the missing words: The second stage of Gene's paper is a mix of freewriting, brainstorming, and a scratch _____. Gene uses all these techniques as he continues to draw out and accumulate _____.
At the same time, he also has realized how to organize his details. He decides to focus *not* on his unpleasant boss but on the job itself. In a rough scratch outline, he lists his three reasons (hard work, bad working conditions, poor wages) for hating the job. He then tentatively decides on working conditions as the worst part of the job and numbers the reasons 1, 2, 3 in the order in which he might develop them. Keep in mind that as you accumulate and develop details, you should, like

Gene, be thinking of a way to _____ them.

STAGE 4: WRITING A DRAFT

Gene puts his work aside for the day and then continues writing the next morning. He now moves to a fuller draft:

> Working in an apple plant was the worst job I ever had. The work was physically hard. For ~~a long time~~ ten hours a night, I stacked cartons in a tractor trailer. The cartons rolled down a metal track. ~~Each carton was very heavy Each carton was heavy with with cans or bottles of apple juice~~ Each carton contained twelve thirty-two-ounce cans or bottles of apple juice, and they were heavy. ~~At the same time, I had to keep a mental count of all the cartons I had loaded.~~ The pay for the job was another bad feature. I was getting the minimum wage at that time plus a quarter extra for night shift. I worked ~~long hours~~ over sixty hours a week. I still did not take home much ~~money~~ more than $100. Working conditions were poor at the plant. During work we were limited to ~~short breaks~~ two ten-minute breaks and unpaid lunch. . . . The truck-loading dock had zero temperatures. . . . Lonely on the job . . . no interests with other loaders . . . worked by myself at the end of the shift . . . cleaned up the apple vats.

◼ **Comments and Activity**

Fill in the missing words: At this stage, Gene has enough details to write the initial draft of his paper. Notice that he continues to accumulate specific supporting details as he writes the draft. For example, he crosses out and replaces ''long hours'' with the more specific _____; he crosses out and replaces ''short breaks'' with the more specific _____. He also works to improve some of his sentences (for instance, he writes three different versions of the _____ sentence in the paragraph). In addition, he crosses out and eliminates a sentence about a _____ because he realizes it does not develop his first supporting point that the work was physically hard.

Toward the end of the paper, Gene either can't find the right words to say what he wants to say or he doesn't quite know yet what he wants to say. So he freewrites (shown by the ellipses . . .), putting down on paper all the impressions that come into his head. He knows that the technique of _____ may help him move closer to the right thought and the right words.

In a second and a third draft, Gene continues to work on and improve his paper. He then edits carefully his next-to-final draft, and the result is the final draft that follows.

STAGE 5: WRITING THE FINAL VERSION

Thanks to the work done on the earlier stages, Gene can now progress to a final draft of the paper:

My Job in an Apple Plant

Working in an apple plant was the worst job I ever had. First of all, the work was physically hard. For ten hours a night, I took cartons that rolled down a metal track and stacked them onto wooden skids in a tractor trailer. Each carton contained twelve thirty-two-ounce cans or bottles of apple juice, and they were heavy. The second bad feature of the job was the pay. I was getting the minimum wage at that time, two dollars an hour, plus the minimum of a quarter extra for working the night shift. Even after working over sixty hours a week, I still did not take home much more than $100. The worst feature of the apple plant job was the working conditions. During work we were limited to two ten-minute breaks and an unpaid half hour for lunch. Most of my time was spent outside on the truck-loading dock in near-zero-degree temperatures. And I was very lonely on the job, since I had no interests in common with the other truck loaders. I felt this isolation especially when the production line shut down for the night, and I worked by myself for two hours cleaning the apple vats. The vats were an ugly place to be on a cold morning, and the job was a bitter one to have.

■ **Comments and Activity**

Fill in the missing words: Notice the many improvements that Gene has made as a result of his second and third drafts. He has added transitional words that mark clearly the three supporting points of his paper. The transitional words are ''first

of all,'' ''_____'', and ''_____''. He has sharpened his details, improved the phrasing of his sentences, and found the words needed to complete the last section of his paper. He has also edited and proofread his paper carefully, checking the spelling of words he was unsure about and correcting several sentence-skills mistakes.

Almost every effective writer, like Gene, is engaged in a continuing process of moving toward a completely realized paper. The final version is seldom, almost never, attained all at once. Instead, it is the end result of a series of

_____. All too often, people stop writing when they are only part way through the writing process; they turn in a paper that is really only an early draft. They have the mistaken notion that a paper is something you should be able to do ''all at once.'' But for almost everyone, writing means hard work and lots

of _____. Be sure, then, to take your paper through the entire series of drafts that you probably will need to write an effective composition.

Activity 1

Answering the questions below will help you evaluate your attitude about revising, editing, and proofreading.

1. When do you typically start to work on a paper?

    _____ Several nights before it's due

    _____ Night before it's due

    _____ Day it's due

2. How many drafts do you typically write when doing a paper?

    _____ One _____ Two _____ Three _____ Four or more

3. How would you describe your editing (checking the next-to-final draft for sentence-skills mistakes)?

    _____ Do little or no editing

    _____ Look quickly for and correct obvious errors

    _____ Consult a sentence-skills handbook and a dictionary about all possible errors

4. How would you describe your proofreading (checking the final draft for typing or handwriting errors)?

    _____ Do not look at paper again after the last word is written

    _____ May glance quickly through the paper

    _____ Read paper over carefully to find mistakes

5. Do you ever get back papers marked for obvious errors?

    _____ Frequently _____ Sometimes _____ Almost never _____ Never

Activity 2

Listed in the box below are five different stages in the process of composing a paragraph titled ''Dangerous Places'':

> 1. Prewriting (list)
> 2. Prewriting (list, brainstorming, and outline)
> 3. First draft
> 4. Second draft
> 5. Final draft

The five stages appear in scrambled order below and on the next page. Write the number 1 in the blank space in front of the first stage of development and number the remaining stages in sequence.

_____
    There are some places where I never feel safe. For example, bus stations. The people there look ~~strange look~~ tired and scared. The security guards there make me feel ~~something bad is going to happen~~ a fight is going to start. I'm also afraid in parking lots. ~~Late at night, I don't like walking in the lot After class, I don't like the parking lot~~. When I leave my night class or the shopping mall late the walk to the car is scary. ~~Most parking lots have large lights which make me feel at least a little better~~. I feel least safe in our laundry room . . . It is a depressing place . . . bars on the windows . . . pipes making noises . . . cement steps the only way out. . . .

_____
Dangerous Places
Highways
Cars—especially parking lots
Feel frightened in our laundry room
Big crowds—concerts, movies
Closed-in places
Bus and train stations
Airplane
Elevators and escalators

_____
                    Dangerous Places
    There are some places where I never feel completely safe. For example, I never feel safe in bus stations. The people in bus stations often look tired and scared as they keep one hand protectively on their suitcases. The security guards roaming around bus stations add to my feeling that a fight is about to break out, that someone is going to be mugged, or that a man will force me into a dark corner. I also feel unsafe in large, dark parking lots. When I leave my night class a little late, or I am one of the few leaving the mall at 10 P.M., I dread the walk to my car. I am afraid that someone may be lurking behind another car, ready to mug me. And I fear that my car will not start, leaving me stuck in the dark parking lot. The place where I feel least safe is the basement laundry room in our apartment building. No matter what time I do my laundry, I seem to be the only person there. The windows are barred, and the only exit is a steep flight of cement steps. While I'm folding the clothes, I feel trapped. If anyone unfriendly came down those steps, I would have nowhere to go. The pipes in the room make sudden gurgles, clanks, and hisses, adding to my uneasiness. Places like bus stations, dark parking lots, and our laundry room give me the shivers.

_____   There are some places where I never feel completely safe. For example, I never feel safe in bus stations. The people in bus stations often look tired and scared as they hang on to their suitcases. The security guards add to my feeling that a fight is about to start. I also feel unsafe in large, dark parking lots. When I leave my night class a little late or I leave the mall at 10 P.M., the walk to the car is scary. I'm afraid that someone may be behind a car. Also that my car won't start. The place where I feel least safe is the basement laundry room in our apartment building. No matter when I do the laundry, I'm the only person there. The windows are barred and there are steep steps. I feel trapped when I fold the clothes. The pipes in the room make frightening noises such as hisses and clanks. Places like bus stations, parking lots, and our laundry room give me the shivers.

_____   Some places seem dangerous and unsafe to me. For example, last night I stayed till 10:15 after night class and walked out to parking lot alone. Very scary. Also, other places I go to every day, such as places in my apartment building. Also frightened by big crowds and lonely bus stations.

Why was the parking lot scary?	What places in my building scare me?
Dark	Laundry room (especially)
Only a few cars	Elevators
No one else in lot	Lobby at night sometimes
Could be someone behind a car	Outside walkway at night
Cold	

2   Parking lots
3   Laundry room
1   Bus stations

Activity 3

The author of ''Dangerous Places'' in Activity 2 made a number of editing changes between the second draft and the final draft. Compare the two drafts and identify five of the changes in the spaces provided below.

1. _____

2. _____

3. _____

4. _____

5. _____

A Review of the Chapter

Fill in the missing words in the following summary of this chapter.

Having the Right Attitude: Some people feel that in order to be a good writer, writing must come easily. This idea is false and can interfere with the ability to make progress in writing. A more realistic and productive attitude includes the understanding that, for most people, writing is _____. In addition, it helps to realize that, like driving or typing, writing is a _____ that can be learned with lots of _____ .

Writing for a Specific Purpose and Audience: The three most common purposes of writing are to _____ , to _____ , and to _____ . In this book, you will have a lot of practice in writing to persuade people and some practice in writing to provide information.

Your main audience in class is your _____ . But he or she is really a symbol for a general audience of well-educated adults. Writing for this audience should be clear and well organized.

This book will also give you practice in writing for a specific audience. That practice will help you develop the knack of choosing the words and tone of voice that suit specific purposes and people.

Knowing or Discovering Your Subject: It is best to write about a subject that _____ you and that you know something about, either directly or indirectly. When you must write on a topic about which you have little or no background, you should do _____ to gain the necessary knowledge. The library is one good place to do that.

There are times, however, when you won't know your exact subject until after you have written for a while. Writing will help you think about and explore your material. On occasion you will write for a page or two and discover that it will make sense to change the _____ of your paper.

As you work on a paper, remember that it is not necessary to write a paper straight through from _____ to end. You should proceed in whatever way seems easiest, including starting at the middle or even the end. Make whatever adjustments are needed to reach your goal of a paper that makes and _____ a point.

Prewriting: There are five prewriting techniques. One technique, called

_____ , is a process of generating ideas by asking questions about your subject. Such questions include *What? When?* and *Why?*

Freewriting is a second prewriting technique. It involves writing on your topic

for ten minutes without _____ or worrying about being correct. In this process, your thoughts about your paper often become clearer.

Making a _____ is a third excellent prewriting technique for getting started on a paper. The goal is to generate many possible details for your paper and maybe even ways of developing that paper.

Diagramming, also known as _____ or _____ , is a fourth prewriting activity. Here you use lines, boxes, arrows, and circles to show relationships among the ideas and details that come to you.

Fifth, perhaps the most helpful technique for writing a good paper is preparing a scratch outline. It is an excellent follow-up to the other prewriting techniques. Sometimes you may even skip the other techniques and concentrate on this one. In a scratch outline, you think about the specific point you will make in your

paper, the exact _____ that will support that point, and the exact

_____ in which you will arrange those items.

Outlining: Often the best way to write an effective paragraph is to

_____ it. Outlining develops your ability to think clearly and logically. It helps you to see and work with the fundamental ideas of a paper, and it helps you to focus on producing a paper that is unified, well supported, and well organized.

Revising, Editing, and Proofreading: Writing a paper is usually a step-by-step process. It begins with prewriting, during which you accumulate raw material. In the second stage, you shape your paper by writing and revising it several times. Finally, you edit and proofread. Editing involves checking your paper for mistakes

in sentence _____ . Proofreading involves checking the final copy

of your paper for typing or handwriting _____ .

THE FIRST
AND SECOND
STEPS
IN WRITING

This chapter will show you how to

- Begin a paper by making a point of some kind
- Provide specific evidence to support that point
- Write a simple paragraph

The four basic steps in writing an effective paragraph are as follows:

1 Make a point.
2 Support the point with specific evidence.
3 Organize and connect the specific evidence.
4 Write clear, error-free sentences.

This chapter will present the first two steps, and the chapter that follows (see page 72) will present the last two.

Step 1: Make a Point

Your first step in writing is to decide what point you want to make and to write that point in a single sentence. The point is commonly known as a *topic sentence*. As a guide to yourself and to the reader, put that point in the first sentence of your paragraph. Everything else in the paragraph should then develop and support in specific ways the single point given in the first sentence.

Activity

Read the two paragraphs below, written by students on the topic "Cheating in Everyday Life." Which paragraph starts with a clear, single point and goes on to support that point? Which paragraph fails to start with a clear point and rambles on in many directions, introducing a number of ideas but developing none of them?

Paragraph A

Cheating

Cheating has always been a part of life, and it will be so in the future. An obvious situation is that students have many ways of cheating in school. This habit can continue after school is over and become part of their daily lives. There are steps that can be taken to prevent cheating, but many teachers do not seem to care. Maybe they are so burned out by their jobs that they do not want to bother. The honest student is often the one hurt by the cheating of others. Cheating at work occurs also. This cheating may be more dangerous, because employers watch out for it more. Businesses have had to close down because cheating by employees took away a good deal of their profits. A news story recently concerned a waiter who was fired for taking a steak home from the restaurant where he worked, but his taking the steak may have been justified. Cheating in the sense of being unfaithful to a loved one is a different story because emotions are involved. People will probably never stop cheating unless there is a heavy penalty to be paid.

Paragraph B

Everyday Cheating

Cheating is common in everyday life. For one thing, cheating at school is common. Many students will borrow a friend's homework and copy it in their own handwriting. During a test, students will use a tiny sheet of answers stored in their pockets or sit near a friend to copy answers. People also cheat on the job. They use the postal meter at work for personal mail or take home office supplies such as tape, paper, or pens. Some people who are not closely supervised or who are out on the road may cheat an employer by taking dozens of breaks or using work time for personal chores. Finally, many people cheat when they deal with large businesses. For instance, few customers will report an incorrect bill in their favor. Visitors in a hotel may take home towels, and restaurant patrons may take home silverware. A customer in a store may change price tags because "This is how much the shirt cost last month." For many people, daily cheating is an acceptable way to behave.

Complete the following statement: Paragraph _____ is effective because it makes a clear, single point in the first sentence and goes on in the remaining sentences to support that single point.

Paragraph B starts with the single idea — that people cheat in everyday life — and then supports that idea with several different examples. But paragraph A does not begin by making a definite point. Instead, we get two broad, obvious statements — that cheating "has always been a part of life" and "will be so in the future." Because the author has not focused on a clear, single point, what happens in this paragraph is inevitable.

The line of thought in paragraph A swerves about like a car without a steering wheel. In the second sentence, we read that ". . . students have many ways of cheating in school," and we think for a moment that this will be the author's point: he or she will give us supporting details about different ways students cheat in school. But the next sentence makes another point: that after school is over, students may continue to cheat as "part of their daily lives." We therefore expect the author to give us details backing up the idea that students who cheat continue to cheat after they leave or finish school. However, the next sentence makes two additional points: "There are steps that can be taken to prevent cheating, but many teachers do not seem to care." These are two more ideas that could be — but are not — the focus of the paragraph. By now we are not really surprised at what happens in the following sentences. Several more points are made: "The honest student is often the one hurt by the cheating of others," cheating at work "may be more dangerous," an employee who stole a steak "may have been justified," and cheating by being unfaithful is different "because emotions are involved." No single idea is developed; the result is confusion.

Step 2: Support the Point with Specific Evidence

The first essential step in writing effectively is to start with a clearly stated point. The second basic step is to support that point with specific evidence. Following are the two examples of supported points that you've already read.

Point 1

I've decided not to go out with Torly anymore.

Support for Point 1

1. Late for our date
2. Bossy
3. Abrupt

Point 2

Cheating is common in everyday life.

Support for Point 2

1. At school
   a. Copying homework
   b. Cheating on tests
2. At work
   a. Using postage meter
   b. Stealing office supplies
   c. Taking breaks and doing errands on company time
3. With large businesses
   a. Not reporting error on bill
   b. Stealing towels and silverware
   c. Switching price tags

The supporting evidence is needed so that we can *see and understand for ourselves* that each writer's point is a sound one. By providing us with particulars about Tony's actions, the first writer shows why she has decided not to go out with him anymore. We can see that she has made a sound point. Likewise, the author of "Everyday Cheating" has supplied specific supporting examples of how cheating is common in everyday life. That paragraph, too, has provided the evidence that is needed for us to understand and agree with the writer's point.

Activity

Both of the paragraphs that follow resulted from an assignment to ''Write a paper that details your reasons for being in college.'' Both writers make the point that they have various reasons for attending college. Which paragraph then goes on to provide plenty of specific evidence to back up its point? Which paragraph is vague and repetitive and lacks the concrete details needed to show us exactly why the author decided to attend college?

Hint: Imagine that you were asked to make a short film based on each paragraph. Which one suggests specific pictures, locations, words, and scenes you could shoot?

Paragraph A

Reasons for Going to College

I decided to attend college for various reasons. One reason is self-respect. For a long time now, I have felt little self-respect. I spent a lot of time doing nothing, just hanging around or getting into trouble, and eventually I began to feel bad about it. Going to college is a way to start feeling better about myself. By accomplishing things, I will improve my self-image. Another reason for going to college is that things happened in my life that made me think about a change. For one thing, I lost the part-time job I had. When I lost the job, I realized I would have to do something in life, so I thought about school. I was in a rut and needed to get out of it but did not know how. But when something happens out of your control, then you have to make some kind of decision. The most important reason for college, though, is to fulfill my dream. I know I need an education, and I want to take the courses I need to reach the position that I think I can handle. Only by qualifying yourself can you get what you want. Going to college will help me fulfill this goal. These are the main reasons why I am attending college.

Paragraph B

Why I'm in School

There are several reasons I'm in school. First of all, my father's attitude made me want to succeed in school. One night last year, after I had come in at 3 A.M., my father said, ''Mickey, you're a bum. When I look at my son, all I see is a good-for-nothing bum.'' I was angry, but I knew my father was right in a way. I had spent the last two years working odd jobs at a pizza parlor and luncheonette, taking ''uppers'' and ''downers'' with my friends. That night, though, I decided I would prove my father wrong. I would go to college and be a success. Another reason I'm in college is my girlfriend's encouragement. Marie has already been in school for a year, and she is doing well in her computer courses. Marie helped me fill out my application and register for courses. She even lent me sixty-five dollars for textbooks. On her day off, she lets me use her car so I don't have to take the college bus. The main reason I am in college is to fulfill a personal goal: I want to finish something for the first time in my life. For example, I quit high school in the eleventh grade. Then I enrolled in a government job-training program, but I dropped out after six months. I tried to get a high school equivalency diploma, but I started missing classes and eventually gave up. Now I am in a special program where I will earn my high school degree by completing a series of five courses. I am determined to accomplish this goal and to then go on and work for a degree in hotel management.

Complete the following statement: Paragraph _____ provides clear, vividly detailed reasons why the writer decided to attend college.

Paragraph B is the one that solidly backs up its point. The writer gives us specific reasons he is in school. On the basis of such evidence, we can clearly understand his opening point. The writer of paragraph A offers only vague, general reasons for being in school. We do not get specific examples of how the writer was ''getting into trouble,'' what events occurred that forced the decision, or even what kind of job he or she wants to qualify for. We sense that the feeling expressed is sincere; but without the benefit of particular examples, we cannot really see why the writer decided to attend college.

THE IMPORTANCE OF SPECIFIC DETAILS

The point that opens a paper is a general statement. The evidence that supports a point is made up of specific details, reasons, examples, and facts.

Specific details have two key functions. First of all, details *excite the reader's interest*. They make writing a pleasure to read, for we all enjoy learning particulars about other people—what they do and think and feel. Second, details *support and explain a writer's point;* they give the evidence needed for us to see and understand a general idea. For example, the writer of ''Good-Bye, Tony'' provides details that make vividly clear her decision not to see Tony anymore. She specifies the exact time Tony was supposed to arrive (8:30) and when he actually arrived (9:20). She mentions the kind of film she wanted to see (a new Steve Martin movie) and the one that Tony took her to instead (*The Night of the Living Dead*). She tells us what she may have wanted to do after the movie (have a hamburger or a drink) and what Tony did instead (went necking); she even specifies the exact location of the place Tony took her (a back road near Oakcrest High School). She explains precisely what happened next (Tony ''cut his finger on a pin I was wearing'') and even mentions by name (Bactine and a Band-Aid) the treatments he planned to use.

The writer of "Why I'm in School" provides equally vivid details. He gives clear reasons for being in school (his father's attitude, his girlfriend's encouragement, and his wish to fulfill a personal goal) and backs up each reason with specific details. His details give us many sharp pictures. For instance, we hear the exact words his father spoke: "Mickey, you're a bum." He tells us exactly how he was spending his time ("working odd jobs at a pizza parlor and luncheonette, taking 'uppers' and 'downers' with my friends"). He describes how his girlfriend helped him (filling out the college application, lending money and her car). Finally, instead of stating generally that "you have to make some kind of decision," as the writer of "Reasons for Going to College" does, he specifies that he has a strong desire to finish college because he dropped out of many schools and programs in the past: high school, a job-training program, and a high school equivalency course.

In both "Good-Bye, Tony" and "Why I'm in School," then, the vivid, exact details capture our interest and enable us to share in the writer's experience. We see people's actions and hear their words; the details provide pictures that make each of us feel "I am there." The particulars also allow us to understand each writer's point clearly. We are shown exactly why the first writer has decided not to see Tony anymore and exactly why the second writer is attending college.

Activity

Each of the five points below is followed by two selections. Write S (for *specific*) in the space next to the selection that provides specific support for the point. Write X in the space next to the selection that lacks supporting details.

1. My two-year-old son was in a stubborn mood today.

    _____ a. When I asked him to do something, he gave me nothing but trouble. He seemed determined to make things difficult for me, for he had his mind made up.

    _____ b. When I asked him to stop playing in the yard and come indoors, he looked me square in the eye and shouted "No!" and then spelled it out, "N . . . O!"

2. The prices in the amusement park were outrageously high.

    _____ a. The food seemed to cost twice as much as it would in a supermarket and was sometimes of poor quality. The rides also cost a lot, and so I had to tell the children that they were limited to a certain number of them.

    _____ b. The cost of the log flume, a ride that lasts roughly 3 minutes, was $4.75 a person. Then I had to pay $1.50 for an 8-ounce cup of Coke and $3.25 for a hot dog.

3. My brother-in-law is accident-prone.

_____ a. Once he tried to open a tube of Krazy Glue with his teeth. When the cap came loose, glue squirted out and sealed his lips shut. They had to be pried open in a hospital emergency room.

_____ b. Even when he does seemingly simple jobs, he seems to get into trouble. This can lead to hilarious, but sometimes dangerous, results. Things never seem to go right for him, and he often needs the help of others to get out of one predicament or another.

4. The so-called "bargains" at the yard sale were junk.

_____ a. The tables at the yard sale were filled with useless stuff no one could possibly want. They were the kinds of things that should be thrown away, not sold.

_____ b. The "bargains" at the yard sale included two headless dolls, blankets filled with holes, scorched pot holders, and a plastic Christmas tree with several branches missing.

5. The key to success in college is organization.

_____ a. Knowing what you're doing, when you have to do it, and so on is a big help for a student. A system is crucial in achieving an ordered approach to study. Otherwise, things become very disorganized, and it is not long before grades will begin to drop.

_____ b. Organized students never forget paper or exam dates, which are marked on a calendar above their desks. And instead of having to cram for exams, they study their clear, neat classroom and textbook notes on a daily basis.

Comments: The specific support for the first point is answer *b*. The writer does not just tell us that the little boy was stubborn but provides an example that shows us. In particular, the detail of the son's spelling out "N . . . O!" makes his stubbornness vividly real for the reader. For the second point, answer *b* gives specific prices ($4.75 for a ride, $1.50 for a Coke, and $3.25 for a hot dog) to support the idea that the amusement park was expensive. For the third point, answer *a* vividly backs up the idea that the brother-in-law is accident-prone by detailing an accident with Krazy Glue. The fourth point is supported by answer *b*, which lists specific examples of useless items that were offered for sale—from headless dolls to a broken plastic Christmas tree. We cannot help agreeing with the writer's point that the items were not bargains but junk. The fifth point is backed up by answer *b*, which identifies two specific strategies of organized students: they mark important dates on calendars above their desks, and they take careful notes and study them on a daily basis.

In each of the five cases, then, specific evidence is presented to enable us to *see for ourselves* that the writer's point is valid.

THE IMPORTANCE OF ADEQUATE DETAILS

One of the most common and most serious problems in students' writing is inadequate development. You must provide *enough* specific details to support fully a point you are making. You could not, for example, submit a paragraph about how your brother-in-law is accident-prone and provide only a short example. You would have to add several other examples or provide an extended example of your brother-in-law's ill luck. Without such additional support, your paragraph would be underdeveloped.

At times, students try to disguise an undersupported point by using repetition and wordy generalities. You saw this, for example, in the paragraph titled "Reasons for Going to College" on page 47. Be prepared to do the plain hard work needed to ensure that each of your paragraphs has full and solid support.

Activity

The following paragraphs were written on the same topic, and each has a clear opening point. Which one is adequately developed? Which one has only several particulars and uses mostly vague, general, wordy sentences to conceal the fact that it is starved for specific details?

Paragraph A

Abuse of Public Parks

Some people abuse public parks. Instead of using the park for recreation, they go there, for instance, to clean their cars. Park caretakers regularly have to pick up the contents of dumped ashtrays and car litter bags. Certain juveniles visit parks with cans of spray paint to deface buildings, fences, fountains, and statues. Other offenders are those who dig up and cart away park flowers, shrubs, and trees. One couple were even arrested for stealing park sod, which they were using to fill in their back lawn. Perhaps the most widespread offenders are the people who use park tables and benches and fireplaces but do not clean up afterward. Picnic tables are littered with trash, including crumpled bags, paper plates smeared with catsup, and paper cups half-filled with stale soda. On the ground are empty beer bottles, dented soda cans, and sharp metal pop tops. Parks are made for people, and yet--ironically--their worst enemy is "people pollution."

Paragraph B

Mistreatment of Public Parks

Some people mistreat public parks. Their behavior is evident in many ways, and the catalog of abuses could go on almost without stopping. Different kinds of debris are left by people who have used the park as a place for attending to their automobiles. They are not the only individuals who mistreat public parks, which should be used with respect for the common good of all. Many young people come to the park and abuse it, and their offenses can occur in any season of the year. The reason for their inconsiderate behavior is best known only to themselves. Other visitors have a lack of personal cleanliness in their personal habits when they come to the park, and the park suffers because of it. Such people seem to have the attitude that someone else should clean up after them. It is an undeniable fact that people are the most dangerous thing that parks must contend with.

Complete the following statement: Paragraph _____ provides an adequate number of specific details to support its point.

Paragraph A offers a series of well-detailed examples of how people abuse parks. Paragraph B, on the other hand, is underdeveloped. Paragraph B speaks only of ''different kinds of debris,'' while paragraph A refers specifically to ''dumped ashtrays and car litter bags''; paragraph B talks in a general way of young people abusing the park, while paragraph A supplies such particulars as ''cans of spray paint'' and defacing ''buildings, fences, fountains, and statues.'' And there is no equivalent in paragraph B for the specifics in paragraph A about people who steal park property and litter park grounds. In summary, paragraph B lacks the full, detailed support needed to develop its opening point convincingly.

■ Review Activity

To check your understanding of the chapter so far, see if you can answer the following questions.

1. It has been observed: ''To write well, the first thing that you must do is decide what nail you want to drive home.'' What is meant by *nail*?

    _____

2. How do you drive home the nail in a paper?

    _____

3. What are the two reasons for using specific details in your writing?

    a. _____

    b. _____

Practice in Making and Supporting a Point

You now know the two most important steps in competent writing: (1) making a point and (2) supporting that point with specific evidence. The purpose of this section is to expand and strengthen your understanding of these two basic steps.

You will first work through a series of activities on *making* a point:

1 Identifying Common Errors in Topic Sentences

2 Understanding the Two Parts of a Topic Sentence

3 Writing a Topic Sentence: I

4 Writing a Topic Sentence: II

You will then sharpen your understanding of specific details by working through a series of activities on *supporting* a point:

5 Making Words and Phrases Specific

6 Making Sentences Specific

7 Providing Specific Evidence

8 Identifying Adequate Supporting Evidence

9 Adding Details to Complete a Paragraph

Finally, you will practice writing a paragraph of your own:

10 Writing a Simple Paragraph

1   IDENTIFYING COMMON ERRORS IN TOPIC SENTENCES

When writing a point, or topic sentence, people sometimes make one of several mistakes that undermine their chances of producing an effective paper. One mistake is to substitute an announcement of the topic for a true topic sentence. Other mistakes include writing statements that are too broad or too narrow. On the following page are examples of all three errors, along with contrasting examples of effective topic sentences.

Announcement

My Ford Escort is the concern of this paragraph.

The statement above is a simple announcement of a subject, rather than a topic sentence in which an idea is expressed about the subject.

Statement That Is Too Broad

Many people have problems with their cars.

The statement above is too broad to be supported adequately with specific details in a single paragraph.

Statement That Is Too Narrow

My car is a Ford Escort.

The statement above is too narrow to be expanded into a paragraph. Such a narrow statement is sometimes called a *dead-end statement* because there is no place to go with it. It is a simple fact that does not need or call for any support.

Effective Topic Sentence

I hate my Ford Escort.

The statement above expresses an opinion that could be supported in a paragraph. The writer could offer a series of specific supporting reasons, examples, and details to make it clear why he or she hates the car.

Here are additional examples:

Announcements

The subject of this paper will be my apartment.

I want to talk about increases in the divorce rate.

Statements That Are Too Broad

The places where people live have definite effects on their lives.

Many people have trouble getting along with others.

Statements That Are Too Narrow

I have no hot water in my apartment at night.

Almost one of every two marriages ends in divorce.

Effective Topic Sentences

My apartment is a terrible place to live.

The divorce rate is increasing for several reasons.

Activity 1

In each pair of sentences below, write A beside the sentence that *only announces* a topic. Write OK beside the sentence that *advances an idea* about the topic.

1. _____ a. This paper will deal with flunking math.

   _____ b. I flunked math last semester for several reasons.

2. _____ a. I am going to write about my job as a gas station attendant.

   _____ b. Working as a gas station attendant was the worst job I ever had.

3. _____ a. Obscene phone calls are the subject of this paragraph.

   _____ b. People should know what to do when they receive an obscene phone call.

4. _____ a. In several ways, my college library is inconvenient to use.

   _____ b. This paragraph will deal with the college library.

5. _____ a. My paper will discuss the topic of procrastinating.

   _____ b. The following steps will help you stop procrastinating.

Activity 2

In each pair of sentences below, write TN beside the statement that is *too narrow* to be developed into a paragraph. (Such a narrow statement is also known as a *dead-end sentence.*) Write OK beside the statement in each pair that calls for support or development of some kind.

1. _____ a. I do push-ups and sit-ups each morning.

   _____ b. Exercising every morning has had positive effects on my health.

2. _____ a. José works nine hours a day and then goes to school three hours a night.

   _____ b. José is an ambitious man.

3. _____ a. I started college after being away from school for seven years.

   _____ b. Several of my fears about returning to school have proved to be groundless.

4. _____ a. Parts of Walt Disney's *Bambi* make it a frightening movie for children.

   _____ b. Last summer I visited Disneyland in Anaheim, California.

5. _____ a. My brother was depressed yesterday for several reasons.

   _____ b. Yesterday my brother had to pay fifty-two dollars for a motor tune-up.

Activity 3

In each pair of sentences below, write TB beside the statement that is *too broad* to be supported adequately in a short paper. Write OK beside the statement that makes a limited point.

1. _____ a. Professional football is a dangerous sport.

   _____ b. Professional sports are violent.

2. _____ a. Married life is the best way of living.

   _____ b. Teenage marriages often end in divorce for several reasons.

3. _____ a. Aspirin can have several harmful side effects.

   _____ b. Drugs are dangerous.

4. _____ a. I've always done poorly in school.

   _____ b. I flunked math last semester for several reasons.

5. _____ a. Computers are changing our society.

   _____ b. Using computers to teach schoolchildren is a mistake.

2   UNDERSTANDING THE TWO PARTS OF A TOPIC SENTENCE

As stated earlier, the point that opens a paragraph is often called a *topic sentence*. When you look closely at a point, or topic sentence, you can see that it is made up of two parts:

1 The *limited topic*
2 The writer's *attitude* about the limited topic

The writer's attitude or point of view or idea is usually expressed in a *key word* or *words*. All the details in a paragraph should support the idea expressed in the key words. In each of the topic sentences below, a single line appears under the topic and a double line under the idea about the topic (expressed in a key word or words):

My girlfriend is very aggressive.
Highway accidents are often caused by absentmindedness.
The kitchen is the most widely used room in my house.
Voting should be required by law in America.
My pickup truck is the most reliable vehicle I have ever owned.

In the first sentence, the topic is *girlfriend,* and the key word that expresses the writer's idea about his topic is that his girlfriend is *aggressive.* In the second sentence, the topic is *highway accidents,* and the key word that determines the focus of the paragraph is that such accidents are often caused by *absentmindedness.* Notice each topic and key word or words in the other three sentences as well.

Activity

For each point below, draw a single line under the topic and a double line under the idea about the topic.

1. Billboards should be abolished.
2. My boss is an ambitious man.
3. The middle child is often a neglected member of the family.
4. The apartment needed repairs.
5. Television commercials are often insulting.
6. My parents have rigid racial attitudes.
7. The language in many movies today is offensive.
8. Homeowners today are more energy-conscious than ever before.
9. My friend Debbie, who is only nineteen, is extremely old-fashioned.
10. Looking for a job can be a degrading experience.
11. Certain regulations in the school cafeteria should be strictly enforced.
12. My car is a temperamental machine.
13. Living in a one-room apartment has its drawbacks.
14. The city's traffic-light system has both values and drawbacks.
15. Consumers' complaints can often have positive results.

3 WRITING A TOPIC SENTENCE: I

Activity

The activity on the following pages will give you practice in writing an accurate point, or topic sentence—one that is neither too broad nor too narrow for the supporting material in a paragraph. Sometimes you will construct your topic sentence after you have decided what details you want to discuss. An added value of this activity is that it shows you how to write a topic sentence that will exactly match the details you have developed.

1.   ***Topic sentence:*** _____

_____

   a. Some are caused by careless people tossing matches out of car windows.
   b. A few are started when lightning strikes a tree.
   c. Some result from campers who fail to douse cooking fires.
   d. The majority of forest fires are deliberately set by arsonists.

2.   ***Topic sentence:*** _____

_____

   a. We had to wait a half hour even though we had reserved a table.
   b. Our appetizers and main courses all arrived at the same time.
   c. The busboy ignored our requests for more water.
   d. The wrong desserts were delivered to us.

3.   ***Topic sentence:*** _____

_____

   a. My phone goes dead at certain times of the day.
   b. When I talk long distance, I hear conversations in the background.
   c. The line to the phone service center is busy for hours.
   d. My telephone bill includes three calls I never made.

4.   ***Topic sentence:*** _____

_____

   a. The crowd scenes were crudely spliced from another film.
   b. Mountains and other background scenery were just painted cardboard cutouts.
   c. The ''sync'' was off, so that you heard voices even when the actors' lips were not moving.
   d. The so-called monster was just a spider that had been filmed through a magnifying lens.

5. | ***Topic sentence:*** _____

   _____

   a. In early grades we had spelling bees, and I would be among the first ones sitting down.
   b. In sixth-grade English, my teacher kept me busy diagramming sentences on the board.
   c. In tenth grade we had to recite poems, and I always forgot my lines.
   d. In my senior year, my compositions had more red correction marks than anyone else's.

4   WRITING A TOPIC SENTENCE: II

Often you will start with a general topic or a general idea of what you want to write about. You may, for example, want to write a paragraph about some aspect of school life. To come up with a point about school life, begin by limiting your topic. One way to do this is to make a list of all the limited topics you can think of that fit under the general topic.

Activity

On the following pages are five general topics and a series of limited topics that fit under them. Make a point out of one of the limited topics in each group.

Hint: To create a topic sentence, ask yourself, "What point do I want to make about _____ (*my limited topic*)?"

Example   Recreation

- Movies
- Dancing
- TV shows
- Reading
- Sports parks

Your point: *Sports parks today have some truly exciting games.*

1. Your school

   ■ Instructor
   ■ Cafeteria
   ■ Specific class
   ■ Particular room or building
   ■ Particular policy (attendance, grading, etc.)
   ■ Classmate

   Your point: _____

2. Job

   ■ Pay
   ■ Boss
   ■ Working conditions
   ■ Duties
   ■ Coworkers
   ■ Customers or clients

   Your point: _____

3. Money

   ■ Budgets
   ■ Credit cards
   ■ Dealing with a bank
   ■ School expenses
   ■ Ways to get it
   ■ Ways to save it

   Your point: _____

4. Cars

   ■ First car
   ■ Driver's test
   ■ Road conditions
   ■ Accident
   ■ Mandatory speed limit
   ■ Safety problems

   Your point: _____

5. Sports
   - A team's chances
   - At your school
   - Women's teams
   - Recreational versus spectator
   - Favorite team
   - Outstanding athlete

   Your point: _____

5 MAKING WORDS AND PHRASES SPECIFIC

To be an effective writer, you must use specific, rather than general, words. Specific words create pictures in the reader's mind. They help capture interest and make your meaning clear.

Activity

This activity will give you practice at changing vague, indefinite words into sharp, specific words. Add three or more specific words to replace the general word or words underlined in each sentence. Make changes in the wording of a sentence as necessary.

Example   My bathroom cabinet contains <u>many drugs</u>.

My bathroom cabinet contains aspirin, antibiotics, tranquilizers,

and codeine cough medicine.

1. At the shopping center, we visited <u>several stores</u>.

   _____

   _____

2. Sunday is my day to take care of <u>chores</u>.

   _____

   _____

3. Lola enjoys <u>various activities</u> in her spare time.

   _____

   _____

4. I spent most of my afternoon doing <u>homework</u>.

   _____

   _____

5. We returned home from vacation to discover that <u>several pests</u> had invaded the house.

   _____

   _____

6 MAKING SENTENCES SPECIFIC

Again, you will practice changing vague, indefinite writing into lively, image-filled writing that captures your reader's interest and makes your meaning clear. Compare the following sentences:

General	*Specific*
The boy came down the street.	Jerry ran down Woodlawn Avenue.
A bird appeared on the grass.	A blue jay swooped down on the frost-covered lawn.
She stopped the car.	Wanda slammed on the brakes of her Escort.

The specific sentences create clear pictures in your reader's mind. The details *show* readers exactly what has happened.

Here are four ways to make your words and sentences specific:

1 Use exact names.

She loves her *motorbike*.
Lola loves her *Honda*.

2 Use lively verbs.

The garbage truck *went* down Front Street.
The garbage truck *rumbled* down Front Street.

3 Use descriptive words (modifiers) before nouns.

A girl peeked out the window.
A *chubby, six-year-old* girl peeked out the *dirty kitchen* window.

4 Use words that relate to the five senses: sight, hearing, taste, smell, and touch.

That woman is a karate expert.
That *tiny, silver-haired* woman is a karate expert. (*Sight*)

When the dryer stopped, a signal sounded.
When the *whooshing* dryer stopped, a *loud buzzer* sounded. (*Hearing*)

Lola offered me an orange slice.
Lola offered me a *sweet, juicy* orange slice. (*Taste*)

The real estate agent opened the door of the closet.
The real estate agent opened the door of the *cedar-scented* closet. (*Smell*)

I pulled the blanket around me to fight off the wind.
I pulled the *scratchy* blanket around me to fight off the *chilling* wind. (*Touch*)

Activity

With the help of the methods described above, add specific details to any eight of the ten sentences that follow. Use separate paper.

Examples The person got out of the car.

The elderly man painfully lifted himself out of the white Buick station wagon.

The fans enjoyed the victory.
Many of the fifty thousand fans stood, waved blankets, and cheered wildly when Barnes scored the winning touchdown.

1. The lunch was not very good.
2. The animal ran away.
3. An accident occurred.
4. The instructor came into the room.
5. The machine did not work.
6. The crowd grew restless.
7. I relaxed.
8. The room was cluttered.
9. The child threw the object.
10. The driver was angry.

7 PROVIDING SPECIFIC EVIDENCE

Activity

Provide three details that logically support each of the following points, or topic sentences. Your details can be drawn from your own experience, or they can be invented. In each case, the details should show in a specific way what the point expresses in only a general way. State your details briefly in several words rather than in complete sentences.

Example The student had several ways of passing time during the dull lecture.

1. *Shielded his eyes with his hand and dozed awhile.*
2. *Read the sports magazine he had brought to class.*
3. *Made an elaborate drawing on a page of his notebook.*

1. I could tell I was coming down with the flu.

   _____

   _____

   _____

2. The food at the cafeteria was terrible yesterday.

   _____

   _____

   _____

3. I had car problems recently.

   _____

   _____

   _____

4. When your money gets tight, there are several ways to economize.

   Don't go out much

   only get what you need

   _____

5. Some people have dangerous driving habits.

   _____

   _____

   _____

8 IDENTIFYING ADEQUATE SUPPORTING EVIDENCE

Activity

Two of the following paragraphs provide sufficient details to support their topic sentences convincingly. Write AD, for *adequate development*, beside those paragraphs. There are also three paragraphs that, for the most part, use vague, general, or wordy sentences as a substitute for concrete details. Write U, for *underdeveloped,* beside those paragraphs.

_____ 1.

My Husband's Stubbornness

My husband's worst problem is his stubbornness. He simply will not let any kind of weakness show. If he isn't feeling well, he refuses to admit it. He will keep on doing whatever he is doing and will wait until the symptoms get almost unbearable before he will even hint that anything is the matter with him. Then things are so far along that he has to spend more time recovering than he would if he had a different attitude. He also hates to be wrong. If he is wrong, he will be the last to admit it. This happened once when we went shopping, and he spent an endless amount of time going from one place to the next. He insisted that one of them had a fantastic sale on things he wanted. We never found a sale, but the fact that this situation happened will not change his attitude. Finally, he never listens to anyone else's suggestions on a car trip. He always knows he's on the right road, and the results have led to a lot of time wasted getting back in the right direction. Every time one of these incidents happens, it only means it is going to happen again in the future.

  2.

Dangerous Games

Because they feel compelled to show off in front of their friends, some teenagers play dangerous games. In one incident, police found a group of boys performing a dangerous stunt with their cars. The boys would perch on the hoods of cars going thirty-five or forty miles an hour. Then the driver would brake sharply, and the boy who flew the farthest off the car would win. Teenagers also drive their cars with the lights off and pass each other on hills or curves as ways of challenging each other. Water, as well as cars, seems to tempt young people to invent dangerous contests. Some students dared each other to swim through a narrow pipe under a four-lane highway. The pipe carried water from a stream to a pond, and the swimmer would have to hold his or her breath for several minutes before coming out on the other side. Another contest involved diving off the rocky sides of a quarry. Because large stones sat under the water in certain places, any dive could result in a broken neck. But the students would egg each other on to go ''rock diving.'' Playing deadly games like these is a horrifying phase of growing up for some teenagers.

3.

Attitudes about Food

Attitudes about food that we form as children are not easily changed. In some families, food is love. Not all families are like this, but some children grow up with this attitude. Some families think of food as something precious and not to be wasted. The attitudes children pick up about food are hard to change in adulthood. Some families celebrate with food. If a child learns an attitude, it is hard to break this later. Someone once said: "As the twig is bent, so grows the tree." Children are very impressionable, and they can't really think for themselves when they are small. Children learn from the parent figures in their lives, and later from their peers. Some families have healthy attitudes about food. It is important for adults to teach their children these healthy attitudes. Otherwise, the children may have weight problems when they are adults.

4.

Qualities in a Friend

There are several qualities I look for in a friend. A friend should give support and security. A friend should also be fun to be around. Friends can have faults, like anyone else, and sometimes it is hard to overlook them. But a friend can't be dropped because he or she has faults. A friend should stick by you, even in bad times. There is a saying that "a friend in need is a friend indeed." I believe this means that there are good friends and fair-weather friends. The second type is not a true friend. He or she is the kind of person who runs when there's trouble. Friends don't always last a lifetime. Someone you believed to be your best friend may lose contact with you if you move to a different area or go around with a different group of people. A friend should be generous and understanding. A friend does not have to be exactly like you. Sometimes friends are opposites, but they still like each other and get along. Since I am a very quiet person, I can't say that I have many friends. But these are the qualities I believe a friend should have.

5.

A Dangerous Place

We play touch football on a dangerous field. First of all, the grass on the field is seldom mowed. The result is that we have to run through tangled weeds that wrap around our ankles like trip wires. The tall grass also hides some gaping holes lurking beneath. The best players know the exact positions of all the holes and manage to detour around them like soldiers zigzagging across a minefield. Most of us, though, endure at least one sprained ankle per game. Another danger is the old baseball infield that we use as the last twenty yards of our gridiron. This area is covered with stones and broken glass. No matter how often we clean it up, we can never keep pace with the broken bottles hurled on the field by the teenagers we call the "night shift." These people apparently hold drinking parties every night in the abandoned dugout and enjoy throwing the empties out on the field. During every game, we try to avoid falling on especially big chunks of Budweiser bottles. Finally, encircling the entire field is an old, rusty chain-link fence full of tears and holes. Being slammed into the fence during the play can mean a painful stabbing by the jagged wires. All these dangers have made us less afraid of opposing teams than of the field where we play.

9   ADDING DETAILS TO COMPLETE A PARAGRAPH

Activity

Each of the following paragraphs needs specific details to back up its supporting points. In the spaces provided, add a sentence or two of realistic details for each supporting point. The more specific you are, the more convincing your details are likely to be.

1.

A Pushover Instructor

We knew after the first few classes that the instructor was a pushover. First of all, he didn't seem able to control the class.

_____

_____

In addition, he made some course requirements easier when a few

students complained. _____

_____

_____

_____

Finally, he gave the easiest quiz we had ever taken. _____

_____

_____

_____

2.

Helping a Parent in College

There are several ways a family can help a parent who is attending college. First, family members can take over some of the household chores that the parent usually does. _____

_____

_____

_____

Also, family members can make sure that the student has some quiet study time. _____

_____

_____

_____

Last, families can take an interest in the student's problems and accomplishments. _____

_____

_____

_____

10   WRITING A SIMPLE PARAGRAPH

You know now that an effective paragraph does two essential things: (1) it makes a point, and (2) it provides specific details to support that point. You have considered a number of paragraphs that have been effective because they followed these two basic steps or ineffective because they failed to follow them.

You are ready, then, to write a simple paragraph of your own. Choose one of the three assignments below, and follow carefully the guidelines provided.

■ Assignment 1

Turn back to the activity on page 64 and select the point for which you have the best supporting details. Develop the point into a paragraph by following these steps:

a If necessary, rewrite the point so that the first sentence is more specific or suits your purpose more exactly. For example, you might want to rewrite the second point so that it includes a specific time and place: "Dinner at the Union Building Cafeteria was terrible yesterday."

b Provide several sentences of information to develop each of your three supporting details fully. Make sure that all the information in your paragraph truly supports your point. As an aid, use the paragraph form on page 568.

c Use the words *First of all, Second,* and *Finally* to introduce each of your three supporting details.

d Conclude your paragraph with a sentence that refers to your opening point. This last sentence "rounds off" the paragraph and lets the reader know that your discussion is complete. For example, the second paragraph about cheating on page 44 begins with "Cheating is common in everyday life." It closes with a statement that refers to, and echoes, the opening point: "For many people, daily cheating is an acceptable way to behave."

e Supply a title based on the point. For instance, the fourth point might have the title "Ways to Economize."

Use the following list to check your paragraph for each of the above items:

Yes *No*

_____ _____ Do you begin with a point?

_____ _____ Do you provide relevant, specific details that support the point?

_____ _____ Do you use the words *First of all, Second,* and *Finally* to introduce each of your three supporting details?

_____ _____ Do you have a closing sentence?

_____ _____ Do you have a title based on the point?

_____ _____ Are your sentences clear and free from obvious errors?

■ **Assignment 2**

In this chapter you have read two paragraphs (page 47) on reasons for being in college. For this assignment, write a paragraph describing your own reasons for being in college. You might want to look first at the following list of common reasons students give for going to school. Use the ones that apply to you (making them as specific as possible) or supply your own. Select three of your most important reasons for being in school and generate specific supporting details for each reason.

Before starting, reread paragraph B on page 47. *You must provide comparable specific details of your own.* Make your paragraph truly personal; do not fall back on vague generalities like those in paragraph A on page 47. Use the checklist for Assignment 1 as a guideline as you work on the paragraph.

Apply in My Case	*Reasons Students Go to College*
_____	■ To have some fun before getting a job
_____	■ To prepare for a specific career
_____	■ To please their families
_____	■ To educate and enrich themselves
_____	■ To be with friends who are going to college
_____	■ To take advantage of an opportunity they didn't have before
_____	■ To find a husband or wife
_____	■ To see if college has anything to offer them
_____	■ To do more with their lives than they've done so far
_____	■ To take advantage of Veterans' Administration benefits or other special funding
_____	■ To earn the status that they feel comes with a college degree
_____	■ To get a new start in life

■ **Assignment 3**

Write a paragraph about stress in your life. Choose three of the following areas of stress and provide specific examples and details to develop each area.

Stress at school

Stress at work

Stress at home

Stress with a friend or friends

Use the checklist for Assignment 1 as a guideline while working on the paragraph.

Some Suggestions on What to Do Next

1   Work through the next chapter in Part One: ''The Third and Fourth Steps in Writing'' (page 72).

2   Read ''Providing Examples'' (page 127) in Part Two and do the first writing assignment.

3   Work through ''Using the Dictionary'' (page 481) and ''Improving Spelling'' (page 489) in Part Five.

4   Read the first two skills under ''Developing Key Study Skills'' (page 253) in Part Four.

5   Take the ''Sentence-Skills Diagnostic Test'' (page 327) in Part Five and begin working on the sentence skills you need to review.

THE THIRD
AND FOURTH
STEPS
IN WRITING

This chapter will show you how to

- Organize specific evidence in a paper by using a clear method of organization
- Connect the specific evidence by using transitions and other connecting words
- Write clear, error-free sentences by referring to the rules in Part Five of this book

The third and fourth steps in effective writing are

3   Organize and connect the specific evidence
4   Write clear, error-free sentences

You know from the previous chapter that the first two steps in writing an effective paragraph are stating a point and supporting it with specific evidence. The third step is organizing and connecting the specific evidence. Most of this chapter will deal with the chief ways to organize and connect this supporting information in a paper. The chapter will then look briefly at the sentence skills that make up the fourth step in writing a successful paper.

Step 3: Organize and Connect the Specific Evidence

At the same time that you are generating the specific details needed to support a point, you should be thinking about ways to organize and connect those details. All the details in your paper must cohere, or stick together; when they do, your reader is able to move smoothly and clearly from one bit of supporting information to the next. This chapter will discuss the following ways to organize and connect supporting details: (1) common methods of organization, (2) transitions, and (3) other connecting words.

COMMON METHODS OF ORGANIZATION: TIME ORDER AND EMPHATIC ORDER

Time order and emphatic order are common methods used to organize the supporting material in a paper. You will learn more specialized methods of development in Part Two of the book.

Time order simply means that details are listed as they occur in time. *First* this is done; *next* this; *then* this; *after* that, this; and so on. Here is a paragraph that organizes its details through time order:

How I Relax

The way I relax when I get home from school on Thursday night is, first of all, to put my three children to bed. Next, I run hot water in the tub and put in lots of perfumed bubble bath. As the bubbles rise, I undress and get into the tub. The water is relaxing to my tired muscles, and the bubbles are tingly on my skin. I lie back and put my feet on the water spigots, with everything but my hair under the water. I like to stick my big toe up the spigot and spray water over the tub. After about ten minutes of soaking, I wash myself with scented soap, get out and dry myself off, and put on my nightgown. Then I go downstairs and make myself two ham, lettuce, and tomato sandwiches on white bread and pour myself a tall glass of iced tea with plenty of sugar and ice cubes. I carry these into the living room and turn on the television. To get comfortable, I sit on the couch with a pillow behind me and my legs under me. I enjoy watching the Tonight Show or a late movie. The time is very peaceful after a long, hard day of housecleaning, cooking, washing, and attending night class.

Fill in the missing words: "How I Relax" uses the following words to help show time order: _____, _____, _____, _____, and _____.

Emphatic order is sometimes described as "save-the-best-till-last" order. It means that the most interesting or important detail is placed in the last part of a paper. (In cases where all the details seem equal in importance, the writer should impose a personal order that seems logical or appropriate to the details in question.) The last position in a paper is the most emphatic position because the reader is most likely to remember the last thing read. *Finally, last of all,* and *most important* are typical words showing emphasis. The following paragraph organizes its details through emphatic order.

The National Enquirer

There are several reasons why the National Enquirer is so popular. First of all, the paper is heavily advertised on television. In the ads, attractive-looking people say, with a smile, "I want to know!" as they scan the pages of the Enquirer. The ads reassure people that it's all right to want to read stories such as "Grace Kelly's Ghost Haunts Her Family" or "Burt's Secret Affair with Schoolgirl." In addition, the paper is easily available. In supermarkets, convenience stores, and drugstores, the Enquirer is always placed in racks close to the cash register. As customers wait in line, they can't help being attracted by the paper's glaring headlines. Then, on impulse, customers will add the paper to their other purchases. Most of all, people read the Enquirer because of a love of gossip. We find other people's lives fascinating, especially if those people are rich and famous. We want to see and read about their homes; their clothes; their friends, lovers, and families. We also take a kind of mean delight in their problems and mistakes, perhaps because we're jealous of them. It's hard to resist buying a paper that promises to show "Liz's Fabulous Jewels," "The Husband Barbra Dumped," or even--though we may feel ashamed of our interest-- "Michael's Secret Love." The Enquirer knows how to get us interested and make us buy.

Fill in the missing words: The paragraph lists a total of _____ different reasons people read the *National Enquirer*. The writer of the paragraph feels that the most important reason is _____.

He or she signals this reason by using the emphasis words _____.

Some paragraphs use a *combination of time order and emphatic order*. For example, "Good-Bye, Tony" on page 5 includes time order: it moves from the time Tony arrived to the end of the evening. In addition, the writer uses emphatic order, ending with her most important reason (signaled by the words "most of all") for not wanting to see Tony anymore.

TRANSITIONS

Transitions are signal words that help readers follow the direction of the writer's thought. They show the relationship between ideas, connecting one thought with the next. They can be compared to signs on the road that guide travelers.

To see the value of transitions, look at the following pairs of examples. Put a check beside the example in each pair that is easier and clearer to read and understand.

1. _____ a. Our landlord recently repainted our apartment. He replaced our faulty air conditioner.

   _____ b. Our landlord recently repainted our apartment. Also, he replaced our faulty air conditioner.

2. _____ a. I carefully inserted a disk into the computer. I turned on the power button.

   _____ b. I carefully inserted a disk into the computer. Then I turned on the power button.

3. _____ a. Moviegoers usually dislike film monsters. Filmgoers pitied King Kong and even shed tears at his death.

   _____ b. Moviegoers usually dislike film monsters. However, filmgoers pitied King Kong and even shed tears at his death.

You should have checked the second example in each pair. The transitional words in those sentences — *Also, Then,* and *However* — make the relationship between the sentences clear. Like all effective transitions, they help connect the writer's thoughts.

In the following box are common transitional words and phrases, grouped according to the kind of signal they give readers. Note that certain words provide more than one kind of signal. In the paragraphs you write, you will most often use addition signals: words like *first of all, also, another,* and *finally* will help you move from one supporting reason or detail to the next.

study

Transitions

Addition signals: first of all, for one thing, second, the third reason, also, next, another, and, in addition, moreover, furthermore, finally, last of all

Time signals: first, then, next, after, as, before, while, meanwhile, now, during, finally

Space signals: next to, across, on the opposite side, to the left, to the right, in front, in back, above, below, behind, nearby

Change-of-direction signals: but, however, yet, in contrast, other-wise, still, on the contrary, on the other hand

Illustration signals: for example, for instance, specifically, as an illustration, once, such as

Conclusion signals: therefore, consequently, thus, then, as a result, in summary, to conclude, last of all, finally

Activity

1. Underline the three *addition* signals in the following selection:

I am opposed to state-supported lotteries for a number of reasons. First of all, by supporting lotteries, states are supporting gambling. I don't see anything morally wrong with gambling, but it is a known cause of suffering for many people who do it to excess. The state should be concerned with relieving suffering, not causing it. Another objection I have to the state lotteries is the kind of advertising they do on television. The commercials promote the lotteries as an easy way to get rich. In fact, the odds against getting rich are astronomical. Last, the lotteries take advantage of the people who can least afford them. Studies have shown that people with lower incomes are more likely to play the lottery than people with higher incomes. This is the harshest reality of the lotteries: the state is encouraging people of limited means not to save their money but to throw it away on a state-supported pipe dream.

2. Underline the four *time* signals in the following selection:

    It is often easy to spot bad drivers on the road because they usually make more than one mistake: they make their mistakes in series. First, for example, you notice that a man is tailgating you. Then, almost as soon as you notice, he has passed you in a no-passing zone. That's two mistakes already in a matter of seconds. Next, almost invariably, you see him speed down the road and pass someone else. Finally, as you watch in disbelief, glad that he's out of your way, he speeds through a red light or cuts across oncoming traffic in a wild left turn.

3. Underline the three *space* signals in the following selection:

    Standing in the burned-out shell of my living room was a shocking experience. Above my head were charred beams, all that remained of our ceiling. In front of me, where our television and stereo had once stood, were twisted pieces of metal and chunks of blackened glass. Strangely, some items seemed little damaged by the fire. For example, I could see the TV tuner knob and a dusty record under the rubble. I walked through the gritty ashes until I came to what was left of our sofa. Behind the sofa had been a wall of family photographs. Now, the wall and the pictures were gone. I found only a water-logged scrap of my wedding picture.

4. Underline the four *change-of-direction* signals in the following selection:

    In some ways, train travel is superior to air travel. People always marvel at the speed with which airplanes can zip from one end of the country to another. Trains, on the other hand, definitely take longer. But sometimes longer can be better. Traveling across the country by train allows you to experience the trip more completely. You get to see the cities and towns, mountains and prairies that too often pass by unnoticed when you fly. Another advantage of train travel is comfort. Traveling by plane means wedging yourself into a narrow seat with your knees bumping the back of the seat in front of you and being handed a "snack" consisting of a bag of ten roasted peanuts. In contrast, the seats on most trains are spacious and comfortable, permitting even the most long-legged traveler to stretch out and watch the scenery just outside the window. And when train travelers grow hungry, they can get up and stroll to the dining car, where they can order anything from a simple snack to a gourmet meal. There's no question that train travel is definitely slow and old-fashioned compared with air travel. However, in many ways it is much more civilized.

5. Underline the three *illustration* signals in the following selection:

> Status symbols are all around us. The cars we drive, for instance, say
> something about who we are and how successful we have been. The auto
> makers depend on this perception of automobiles, designing their
> commercials to show older, well-established people driving Cadillacs and
> young, fun-loving people driving to the beach in sports cars. Television, too,
> has become something of a status symbol. Specifically, schoolchildren are
> often rated by their classmates according to whether or not their family has a
> cable television hookup. Another example of a status symbol is the video
> cassette recorder. This device, not so long ago considered a novelty, is now
> considered as common as the television set itself. Being without a VCR
> today is like having a car without whitewalls in the fifties.

6. Underline the *conclusion* signal in the following selection:

> A hundred years ago, miners used to bring caged canaries down into the
> mines with them to act as warning signals. If the bird died, the miner knew
> that the oxygen was running out. The smaller animal would be affected much
> more quickly than the miners. In the same way, animals are acting as
> warning signals to us today. Baby birds die before they can hatch because
> pesticides in the environment cause the adults to lay eggs with paper-thin
> shells. Fish die when lakes are contaminated with acid rain or poisonous
> mercury. The dangers in our environment will eventually affect all life on
> earth, including humans. Therefore, we must pay attention to these early
> warning signals. If we don't, we will be as foolish as a miner who ignored a
> dead canary--and we will die.

OTHER CONNECTING WORDS

In addition to transitions, there are three other kinds of connecting words that help
tie together the specific evidence in a paper: repeated words, pronouns, and
synonyms. Each will be discussed in turn.

Repeated Words

Many of us have been taught by English instructors — correctly so — not to repeat
ourselves in our writing. On the other hand, repeating key words can help tie a
flow of thought together. In the selection that follows, the word *retirement* is
repeated to remind readers of the key idea on which the discussion is centered.
Underline the word the five times it appears.

> Oddly enough, retirement can pose more problems for the spouse than
> for the retired person. For a person who has been accustomed to a
> demanding job, retirement can mean frustration and a feeling of

uselessness. This feeling will put pressure on the spouse to provide challenges at home equal to those of the workplace. Often, these tasks will disrupt the spouse's well-established routine. Another problem arising from retirement is filling up all those empty hours. The spouse may find himself or herself in the role of social director or tour guide, expected to come up with a new form of amusement every day. Without sufficient challenges or leisure activities, a person can become irritable and take out the resulting boredom and frustration of retirement on the marriage partner. It is no wonder that many of these partners wish their spouses would come out of retirement and do something--anything--just to get out of the house.

Pronouns

Pronouns (*he, she, it, you, they, this, that,* and others) are another way to connect ideas as you develop a paper. Using pronouns to take the place of other words or ideas can help you avoid needless repetition. (Be careful, though, to use pronouns with care in order to avoid the unclear or inconsistent pronoun references described on pages 402–405 of this book.) Underline the eight pronouns in the passage below, noting at the same time the words that the pronouns refer to.

A professor of nutrition at a major university recently advised his students that they could do better on their examinations by eating lots of sweets. He told them that the sugar in cakes and candy would stimulate their brains to work more efficiently, and that if the sugar was eaten for only a month or two, it would not do them any harm.

Synonyms

Using synonyms — words that are alike in meaning — can also help move the reader clearly from one thought to the next. In addition, the use of synonyms increases variety and interest by avoiding needless repetition of the same words. Underline the three words used as synonyms for *fallacies* in the following selection.

There are many fallacies about suicide. One false idea is that a person who talks about suicide never follows through. The truth is that about three out of every four people who commit suicide notify one or more other persons ahead of time. Another misconception is that a person who commits suicide is poor or downtrodden. Actually, poverty appears to be a deterrent to suicide rather than a predisposing factor. A third myth about suicide is that people bent on suicide will eventually take their lives one way or another, whether or not the most obvious means of suicide is removed from their reach. In fact, since an attempt at suicide is a kind of cry for help, removing a convenient means of taking one's life, such as a gun, shows people bent on suicide that someone cares enough about them to try to prevent it.

Activity

Read the selection below and then answer the questions about it that follow.

My Worst Experience of the Week

¹The registration process at State College was a nightmare. ²The night before registration officially began, I went to bed anxious about the whole matter, and nothing that happened the next day served to ease my tension. ³First, even though I had paid my registration fee early last spring, the people at the bursar's office had no record of my payment. ⁴And for some bizarre reason, they wouldn't accept the receipt I had. ⁵Consequently, I had to stand in line for two hours, waiting for someone to give me a slip of paper which stated that I had, in fact, paid my registration fee. ⁶The need for this new receipt seemed ludicrous to me since, all along, I had proof that I had paid. ⁷I was next told that I had to see my adviser in the Law and Justice Department and that the department was in Corridor C of the Triad Building. ⁸I had no idea what or where the Triad was. ⁹But, finally, I found my way to the ugly, gray-white building. ¹⁰Then I began looking for Corridor C. ¹¹When I found it, everyone there was a member of the Communications Department. ¹²No one seemed to know where Law and Justice had gone. ¹³Finally, one instructor said she thought Law and Justice was in Corridor A. ¹⁴"And where is Corridor A?" I asked. ¹⁵"I don't know," the teacher answered. ¹⁶"I'm new here." ¹⁷She saw the bewildered look on my face and said sympathetically, "You're not the only one who's confused." ¹⁸I nodded and walked numbly away. ¹⁹I felt as if I were fated to spend the rest of the semester trying to complete the registration process, and I wondered if I would ever become an official college student.

Questions

1. How many times is the key idea *registration* repeated? _____

2. Write here the pronoun that is used for *people at the bursar's office* (sentence 4): _____; *Corridor C* (sentence 11): _____; *instructor* (sentence 17): _____.

3. Write here the words that are used as a synonym for *receipt* (sentence 5):

   _____;

   the words that are used as a synonym for *Triad* (sentence 9):

   _____;

   the word that is used as a synonym for *instructor* (sentence 15):

   _____.

Step 4: Write Clear, Error-Free Sentences

The fourth step in writing an effective paper is to follow the agreed-upon rules, or conventions, of written English. These conventions — or, as they are called in this book, *sentence skills* — must be followed if your sentences are to be clear and error-free. Here are some of the most important of these skills.

1 Write complete sentences rather than fragments.
2 Do not write run-on sentences.
3 Use verb forms and tenses correctly and consistently.
4 Make sure that subjects and verbs agree.
5 Use pronoun forms and types correctly.
6 Use adjectives and adverbs correctly.
7 Eliminate faulty modifiers and faulty parallelism.
8 Use correct paper format.
9 Use capital letters where needed.
10 Use numbers and abbreviations correctly.
11 Use the following punctuation marks correctly: apostrophe, quotation marks, comma, colon, semicolon, dash, hyphen, parentheses.
12 Use the dictionary as necessary.
13 Eliminate spelling errors.
14 Use words accurately by developing your vocabulary and distinguishing between commonly confused words.
15 Choose words effectively to avoid slang, clichés, and wordiness.
16 Vary your sentences.
17 Edit and proofread to eliminate careless errors.

The sentence skills are explained in detail, and activities are provided, in Part Five, where they can be referred to easily as needed. A diagnostic test on page 327 will help you identify skills you may need to review. Your instructor will also identify such skills in marking your papers and may use the correction symbols shown on the inside back cover. Note that the correction symbols, and also the checklist of sentence skills on the inside front cover, include page references, so that you can turn quickly to those skills that give you problems.

■ **Review Activity**

Complete the following statements.

1. The four steps in writing a paper are:

    a. _____

    b. _____

    c. _____

    d. _____

2. *Time order* means _____

    _____

3. *Emphatic order* means _____

    _____

4. _____ are signal words that help readers follow the direction

    of a writer's thought.

5. In addition to transitions, three other kinds of connecting words that help

    link sentences and ideas are repeated words, _____, and

    _____.

Practice in Organizing and Connecting Specific Evidence

You now know the third step in effective writing: organizing the specific evidence used to support the main point of a paper. You also know that the fourth step — writing clear, error-free sentences—will be treated in detail in Part Five of the book. This section will expand and strengthen your understanding of the third step in writing.

You will work through the following series of activities:

1 Organizing through Time Order
2 Organizing through Emphatic Order
3 Organizing through a Combination of Time Order and Emphatic Order
4 Identifying Transitions
5 Providing Transitions
6 Identifying Transitions and Other Connecting Words

1 ORGANIZING THROUGH TIME ORDER

Activity

Use time order to organize the scrambled list of sentences below. Write the number 1 beside the point that all the other sentences support. Then number each supporting sentence as it occurs in time.

___10___ The table is right near the garbage pail.

___6___ So you reluctantly select a gluelike tuna-fish sandwich, a crushed-in apple pie, and watery hot coffee.

___12___ You sit at the edge of the table, away from the garbage pail, and gulp down your meal.

___1___ Trying to eat in the cafeteria is an unpleasant experience.

___9___ Suddenly you spot a free table in the corner.

___13___ With a last swallow of the lukewarm coffee, you get up and leave the cafeteria as rapidly as possible.

___11___ Flies are flitting in and out of the pail.

___4___ By the time it is your turn, the few things that are almost good are gone.

___8___ There does not seem to be a free table anywhere.

___3___ Unfortunately, there is a line in the cafeteria.

___5___ The hoagies, coconut-custard pie, and iced tea have all disappeared.

___7___ You hold your tray and look for a place to sit down.

___2___ You have a class in a few minutes, and so you run in to grab something to eat quickly.

2 ORGANIZING THROUGH EMPHATIC ORDER

Activity

Use emphatic order (order of importance) to arrange the scrambled list of sentences on the following page. Write the number 1 beside the point that all the other sentences support. Then number each supporting sentence, starting with what seems the least important detail and ending with the most important detail.

_4_ The people here are all around my age and seem to be genuinely friendly and interested in me.

_1_ The place where I live has several important advantages.

_10_ The schools in this neighborhood have a good reputation, so I feel that my daughter is getting a good education.

_9_ The best thing of all about this area, though, is the school system.

_~~8~~ 7_ Therefore, I don't have to put up with public transportation or worry about how much it's going to cost to park each day.

_11_ The school also has an extended day-care program, so I know my daughter is in good hands until I come home from work.

_2_ First of all, I like the people who live in the other apartments near mine.

_~~4~~ 5_ Another positive aspect of this area is that it's close to where I work.

_3_ That's more than I can say for the last place I lived, where people stayed behind locked doors.

_6_ The office where I'm a receptionist is only a six-block walk from my house.

_8_ In addition, I save a lot of wear and tear on my car.

3 ORGANIZING THROUGH A COMBINATION OF TIME ORDER AND EMPHATIC ORDER

Activity

Use a combination of time and emphatic order to arrange the scrambled list of sentences below. Write the number 1 beside the point that all the other sentences support. Then number each supporting sentence. Paying close attention to transitional words and phrases will help you organize and connect the supporting sentences.

3 I did not see the spider but visited my friend in the hospital, where he suffered through a week of nausea and dizziness because of the poison.

7 We were listening to the radio when we discovered that nature was calling.

_9_

_18_

_2_

_10_

_1_

_8_

_14_

_4_

_16 2_

_17 13_

_6_

_12_

_5_

_15_

_13 13_

_14_

As I got back into the car, I sensed, rather than felt or saw, a presence on my left hand.

After my two experiences, I suspect that my fear of spiders will be with me until I die.

The first experience was when my best friend received a bite from a black widow spider.

I looked down at my hand, but I could not see anything because it was so dark.

I had two experiences when I was sixteen that are the cause of my *arachniphobia,* or terrible and uncontrollable fear of spiders.

We stopped the car at the side of the road, walked into the woods a few feet, and watered the leaves.

My friend then entered the car, putting on the dashboard light, and I almost passed out with horror.

I saw the bandage on his hand and the puffy swelling when the bandage was removed.

Then it flew off my hand and into the dark bushes nearby.

I sat in the car for an hour afterward, shaking and sweating and constantly rubbing the fingers of my hand to reassure myself that the spider was no longer there.

But my more dramatic experience with spiders happened one evening when another friend and I were driving around in his car.

Almost completely covering my fingers was a monstrous brown spider, with white stripes running down each of a seemingly endless number of long, furry legs.

Most of all, I saw the ugly red scab on his hand and the yellow pus that continued oozing from under the scab for several weeks.

I imagined my entire hand soon disappearing as the behemoth relentlessly devoured it.

At the same time I cried out "Arghh!" and flicked my hand violently back and forth to shake off the spider.

For a long, horrible second it clung stickily, as if intertwined for good among the fingers of my hand.

4 IDENTIFYING TRANSITIONS

Activity

Locate the major transitions used in the following two selections. Then write the transitions in the spaces provided. Mostly you will find addition words such as *another* and *also*. You will also find several change-of-direction words such as *but* and *however*.

1.   Watching TV Football

Watching a football game on television may seem like the easiest thing in the world. However, like the game of football itself, watching a game correctly is far more complicated than it appears. First is the matter of the company. The ideal number of people depends on the size of your living room. Also, at least one of your companions should be rooting for the opposite team. There's nothing like a little rivalry to increase the enjoyment of a football game. Next, you must attend to the refreshments. Make sure to have on hand plenty of everyone's favorite drinks, along with the essential chips, dips, and pretzels. You may even want something more substantial on hand, like sandwiches or pizza. If you do, make everyone wait until the moment of kickoff before eating. Waiting will make everything taste much better. Finally, there is one last piece of equipment you should have on hand: a football. The purpose of this object is not to send lamps hurtling from tables or to smash the television screen, but to toss around--outside-- during halftime. If your team happens to be getting trounced, you may decide not to wait until halftime.

a. _First_

b. _Also_

c. _Next_

d. _However_

e. _Finally_

2.   Avoidance Tactics

Getting down to studying for an exam or writing a paper is hard, and so it is tempting for students to use one of the following five avoidance tactics in order to put the work aside. For one thing, students may say to themselves, "I can't do it." They adopt a defeatist attitude at the start and give up without a struggle. They could get help with their work by using such college services as tutoring programs and skills labs. However, they refuse even to try. A second avoidance technique is to say, "I'm too busy." Students may take on an extra job, become heavily involved in social activities, or allow family problems to become so time-consuming that they cannot concentrate on their studies. Yet if college really matters to a student, he or she will make sure that there is enough time to do the required work. Another avoidance technique is expressed by the phrase "I'm too tired." Typically, sleepiness occurs when it is time to study or go to class and then vanishes when the school pressure is off. This sleepiness is a sign of work avoidance. A fourth excuse is to say, "I'll do it later." Putting things off until the last minute is practically a guarantee of poor grades on tests and papers. When everything else seems more urgent than studying--watching TV, calling a friend, or even cleaning the oven--a student may simply be escaping academic work. Last, some students avoid work by saying to themselves, "I'm here and that's what counts." Such students live under the dangerous delusion that, since they possess a college ID, a parking sticker, and textbooks, the course work will somehow take care of itself. But once a student has a college ID in a pocket, he or she has only just begun. Doing the necessary studying, writing, and reading will bring real results: good grades, genuine learning, and a sense of accomplishment.

a. _____

b. _____

c. _____

d. _____

e. _____

f. _____

g. _____

h. _____

5 PROVIDING TRANSITIONS

Activity

In the spaces provided, add logical transitions to tie together the sentences and ideas in the following paragraphs. Use the words in the boxes that precede each paragraph.

1.

however	a second	last of all
for one thing	also	on the other hand

Why School May Frighten a Young Child

Schools may be frightening to young children for a number of reasons. _For one thing_, the regimented environment may be a new and disturbing experience. At home children may have been able to do what they wanted when they wanted to do it. In school, _however_, they are given a set time for talking, working, playing, eating, and even going to the toilet. _A second_ source of anxiety may be the public method of discipline that some teachers use. Whereas at home children are scolded in private, in school they may be held up to embarrassment and ridicule in front of their peers. "Bonnie," the teacher may say, "why are you the only one in the class who didn't do your homework?" Or, "David, why are you the only one who can't work quietly at your seat?" Children may _also_ be frightened by the loss of personal attention. Their little discomforts or mishaps, such as tripping on the stairs, may bring instant sympathy from a parent; in school, there is often no one to notice, or the teacher is frequently too busy to care and just says, "Go do your work. You'll be all right." _On the other hand_, a child may be scared by the competitive environment of the school. At home, one hopes, such competition for attention is minimal. In school, _last of all_, children may vie for the teacher's approving glance or tone of voice, or for stars on a paper, or for favored seats in the front row. For these and other reasons, it is not surprising that children may have difficulty adjusting to school.

2.

for example	finally	first of all
but	such as	as a result
	another	

Job Burnout

Job burnout has several causes. _First of All_, successful workers may be given more to do just because they do their jobs well. Soon they become overloaded and must work even harder just to keep up with the pace. The work load becomes impossible, and exhaustion sets in. _Another_ cause of burnout is conflicting demands. Many career women, _for example_ find themselves trapped between one set of expectations in the workplace and another at home. They are expected to perform competently for eight hours a day and then come home to cook a gourmet meal or help a child or spouse with a problem. ~~As A Result~~ _Finally_ certain occupations entail a high risk of burnout. People in the service professions, _Such as_ nurses, social workers, and teachers, begin their careers filled with idealism and commitment. _But_ the long hours, heavy case loads or enrollments, and miles of red tape become overwhelming, and the rewards--the few people they can help--are all too few. ~~Finally~~ _As A Result_, burnout for these people is almost inevitable.

6 IDENTIFYING TRANSITIONS AND OTHER CONNECTING WORDS

Activity

The selections on the following page use transitions, repeated words, synonyms, and pronouns to help tie ideas together. The connecting words you are to identify have been underlined. In the space provided, write T for *transition*, RW for *repeated word*, S for *synonym*, and P for *pronoun*.

_____ 1. I decided to pick up a drop-add form from the registrar's office. <u>However</u>, I changed my mind when I saw the long line of students waiting there.

_____ 2. We absorb radiation from many sources in our environment. Our color television sets and microwave ovens, among other things, give off low-level <u>radiation</u>.

_____ 3. I checked my car's tires, oil, water, and belts before the trip. But the ungrateful <u>machine</u> blew a gasket about fifty miles from home.

_____ 4. At the turn of the century, bananas were still an oddity in America. Some people even attempted to eat <u>them</u> with the skins on.

_____ 5. Many researchers believe that people have weight set-points their bodies try to maintain. This may explain why many dieters return to their original <u>weight</u>.

_____ 6. Women's clothes, in general, use less material than men's clothes. Yet women's <u>garments</u> are usually more expensive than men's.

_____ 7. In England, drivers use the left-hand side of the road. <u>Consequently</u>, steering wheels are on the right-hand side of their cars.

_____ 8. At the end of the rock concert, thousands of fans held up Bic lighters in the darkened arena. The sea of lights signaled that the <u>fans</u> wanted an encore.

_____ 9. The temperance movement in this country sought to ban alcohol. Drinking <u>liquor</u>, movement leaders said, led to violence, poverty, prostitution, and insanity.

_____ 10. Crawling babies will often investigate new objects by putting them in their mouths. <u>Therefore</u>, parents should be alert for any pins, tacks, or other dangerous items on floors and carpets.

_____ 11. One technique that advertisers use is to have a celebrity endorse a product. The consumer <u>then</u> associates the star qualities of the celebrity with the product.

_____ 12. Canning vegetables is easy and economical. <u>It</u> can also be very dangerous.

_____ 13. For me, apathy quickly sets in when the weather becomes hot and humid. This <u>listlessness</u> disappears when the humidity decreases.

_____ 14. Establishing credit is important for a woman. A good <u>credit</u> history is often necessary when applying for a loan or charge account.

_____ 15. The restaurant table must have had uneven legs. Every time we tried to eat, <u>it</u> wobbled like a seesaw.

Some Suggestions on What to Do Next

1   Work through the final chapter in Part One: ''Four Bases for Evaluating Writing.''

2   Read ''Explaining a Process'' (page 135) in Part Two and do the first writing assignment.

3   Read the rest of ''Developing Key Study Skills'' in Part Four and ''Vocabulary Development'' in Part Five.

4   Continue your review of sentence skills in Part Five. If you plan to make a general review of all the skills, here is an appropriate sequence to follow: (1) Paper Format, (2) Capital Letters, (3) Subjects and Verbs, (4) Sentence Fragments, (5) Run-Ons, (6) Standard English Verbs, (7) Irregular Verbs, (8) Subject-Verb Agreement, (9) Apostrophe, (10) Comma, (11) Quotation Marks, (12) Sentence Variety.

FOUR BASES
FOR EVALUATING
WRITING

This chapter will show you how to evaluate a paper for

- Unity
- Support
- Coherence
- Sentence skills

In the preceding two chapters, you learned four essential steps in writing an effective paper. The box below shows how these steps lead to four bases, or standards, you can use in evaluating a paper.

Four Steps ————————————→ *Four Bases*

1 If you make one point and stick to that point,   your writing will have *unity*.

2 If you back up the point with specific evidence,   your writing will have *support*.

3 If you organize and connect the specific evidence,   your writing will have *coherence*.

4 If you write clear, error-free sentences,   your writing will reflect effective *sentence skills*.

This chapter will discuss the four bases of unity, support, coherence, and sentence skills and will show how these four bases can be used to evaluate writing.

Base 1: Unity

Activity

The following two paragraphs were written by students on the topic "Why Students Drop Out of College." Read them and decide which one makes its point more clearly and effectively, and why.

Paragraph A

Why Students Drop Out

Students drop out of college for many reasons. First of all, some students are bored in school. These students may enter college expecting nonstop fun or a series of fascinating courses. When they find out that college is often routine, they quickly lose interest. They do not want to take dull required courses or spend their nights studying, and so they drop out. Students also drop out of college because the work is harder than they thought it would be. These students may have made decent grades in high school simply by showing up for class. In college, however, they may have to prepare for two-hour exams, write fifteen-page term papers, or make detailed presentations to a class. The hard work comes as a shock, and students give up. Perhaps the most common reason students drop out is because they are having personal or emotional problems. Younger students, especially, may be attending college at an age when they are also feeling confused, lonely, or depressed. These students may have problems with roommates, family, boyfriends, or girlfriends. They become too unhappy to deal with both hard academic work and emotional troubles. For many types of students, dropping out seems to be the only solution they can imagine.

Paragraph B

Student Dropouts

There are three main reasons students drop out of college. Some students, for one thing, are not really sure they want to be in school and lack the desire to do the work. When exams come up, or when a course requires a difficult project or term paper, these students will not do the required studying or research. Eventually, they may drop out because their grades are so poor they are about to flunk out anyway. Such students sometimes come back to school later with a completely different attitude about school. Other students drop out for financial reasons. The pressures of paying tuition, buying textbooks, and possibly having to support themselves can be overwhelming. These students can often be helped by the school because financial aid is available, and some schools offer work-study programs.

Finally, students drop out because they have personal problems. They cannot concentrate on their courses because they are unhappy at home, they are lonely, or they are having trouble with boyfriends or girlfriends. Instructors should suggest that such troubled students see counselors or join support groups. If instructors would take a more personal interest in their students, more students would make it through troubled times.

Fill in the blanks: Paragraph _____ makes its point more clearly and effectively because _____

_____

_____ .

UNDERSTANDING UNITY

Paragraph A is more effective because it is *unified*. All the details in this paragraph are *on target;* they support and develop the single point expressed in the first sentence — that there are many reasons students drop out of college. On the other hand, paragraph B contains some details irrelevant to the opening point — that there are three main reasons students drop out. These details should be omitted in the interest of paragraph unity. Go back to paragraph B and cross out the sections that are off target — the sections that do not support the opening idea.

You should have crossed out the following sections: ''Such students sometimes . . . attitude about school''; ''These students can often . . . work-study programs''; and ''Instructors should suggest . . . through troubled times.''

The difference between these two paragraphs leads us to the first base, or standard, of effective writing: *unity*. To achieve unity is to have all the details in your paper related to the single point expressed in the topic sentence, the first sentence. Each time you think of something to put in, ask yourself whether it relates to your main point. If it does not, leave it out. For example, if you were writing about a certain job as the worst job you ever had and then spent a couple of sentences talking about the interesting people that you met there, you would be missing the first and most essential base of good writing. The pages ahead will consider the other three bases that you must touch in order to ''score'' in your writing.

CHECKING FOR UNITY

To check a paper for unity, ask yourself these questions:

1 Is there a clear opening statement of the point of the paper?
2 Is all the material on target in support of the opening point?

Base 2: Support

Activity

The following student paragraphs were written on the topic "A Quality of Some Person You Know." Both are unified, but one communicates more clearly and effectively. Which one, and why?

Paragraph A

My Quick-Tempered Father

My father is easily angered by normal everyday mistakes. One day my father told me to wash the car and cut the grass. I did not hear exactly what he said, and so I asked him to repeat it. Then he went into a hysterical mood and shouted, "Can't you hear?" Another time he asked my mother to go to the store and buy groceries with a fifty-dollar bill, and he told her to spend no more than twenty dollars. She spent twenty-two dollars. As soon as he found out, he immediately took the change from her and told her not to go anywhere else for him; he did not speak to her the rest of the day. My father even gives my older brothers a hard time with his irritable moods. One day he told them to be home from their dates by midnight; they came home at 12:15. He informed them that they were grounded for three weeks. To my father, making a simple mistake is like committing a crime.

Paragraph B

My Generous Grandfather

My grandfather is the most generous person I know. He has given up a life of his own in order to give his grandchildren everything they want. Not only has he given up many years of his life to raise his children properly, but he is now sacrificing many more years to his grandchildren. His generosity is also evident in his relationship with his neighbors, friends, and the members of his church. He has been responsible for many good deeds and has always been there to help all the people around him in times of trouble. Everyone knows that he will gladly lend a helping hand. He is so generous that you almost have to feel sorry for him. If one day he suddenly became selfish, it would be earthshaking. That's my grandfather.

Fill in the blanks: Paragraph _____ makes its point more clearly and effectively

because _____

_____

_____.

UNDERSTANDING SUPPORT

Paragraph A is more effective, for it offers specific examples that show us the father in action. We see for ourselves why the writer describes the father as quick-tempered. The second writer, on the other hand, gives us no specific evidence. The writer tells us repeatedly that the grandfather is generous but never shows us examples of that generosity. Just how, for instance, did the grandfather sacrifice his life for his children and grandchildren? Did he hold two jobs so that his son could go to college, or so that his daughter could have her own car? Does he give up time with his wife and friends to travel every day to his daughter's house to baby-sit, go to the store, and help with the dishes? Does he wear threadbare suits and coats and eat Hamburger Helper and other inexpensive meals (with no desserts) so that he can give money to his children and toys to his grandchildren? We want to see and judge for ourselves whether the writer is making a valid point about the grandfather, but without specific details we cannot do so. In fact, we have almost no picture of him at all.

Consideration of these two paragraphs leads us to the second base of effective writing: *support*. After realizing the importance of specific supporting details, one student writer revised a paper she had done on a restaurant job as the worst job she ever had. In the revised paper, instead of talking about "unsanitary conditions in the kitchen," she referred to such specifics as "green mold on the bacon" and "ants in the potato salad." All your papers should include many vivid details!

CHECKING FOR SUPPORT

To check a paper for support, ask yourself these questions:

1   Is there *specific* evidence to support the opening point?
2   Is there *enough* specific evidence?

Base 3: Coherence

Activity

The following two paragraphs were written on the topic ''The Best or Worst Job You Ever Had.'' Both are unified and both are supported. However, one communicates more clearly and effectively. Which one, and why?

Paragraph A

Pantry Helper

My worst job was as a pantry helper in one of San Diego's well-known restaurants. I had an assistant from three to six in the afternoon who did little but stand around and eat the whole time she was there. She kept an ear open for the sound of the back door opening, which was a sure sign the boss was coming in. The boss would testily say to me, ''You've got a lot of things to do here, Alice. Try to get a move on.'' I would come in at two o'clock to relieve the woman on the morning shift. If her day was busy, that meant I would have to prepare salads, slice meat and cheese, and so on. Orders for sandwiches and cold platters would come in and have to be prepared. The worst thing about the job was that the heat in the kitchen, combined with my nerves, would give me an upset stomach by seven o'clock almost every night. I might be going to the storeroom to get some supplies, and one of the waitresses would tell me she wanted a bacon, lettuce, and tomato sandwich on white toast. I would put the toast in and head for the supply room, and a waitress would holler out that her customer was in a hurry. Green flies would come in through the torn screen in the kitchen window and sting me. I was getting paid only $3.60 an hour. At five o'clock when the dinner rush began, I would be dead tired. Roaches scurried in all directions whenever I moved a box or picked up a head of lettuce to cut.

Paragraph B

My Worst Job

The worst job I ever had was as a waiter at the Westside Inn. First of all, many of the people I waited on were rude. When a baked potato was hard inside or a salad was flat or their steak wasn't just the way they wanted it, they blamed me, rather than the kitchen. Or they would ask me to light their cigarettes, or chase flies from their tables, or even take their children to the bathroom. Also, I had to contend with not only the customers but the kitchen staff as well. The cooks and busboys were often undependable and surly. If I didn't treat them just right, I would wind up having to apologize to customers because their meals came late or their water glasses weren't filled. Another reason I didn't like the job was that I was always moving. Because of the constant line at the door, as soon as one group left, another would take its place. I usually had only a twenty-minute lunch break and a ten-minute break in almost nine hours of work. I think I could have put up with the job if I had been able to pause and rest more often. The last and most important reason I hated the job was my boss. She played favorites with the waiters and waitresses, giving some the best-tipping repeat customers and preferences on holidays. She would hover around during my break to make sure I didn't take a second more than the allotted time. And even when I helped out by working through a break, she never had an appreciative word but would just tell me not to be late for work the next day.

Fill in the blanks: Paragraph _____ makes its point more clearly and effectively

because _____

_____

_____ .

UNDERSTANDING COHERENCE

Paragraph B is more effective because the material is organized clearly and logically. Using emphatic order, the writer gives us a list of four reasons why the job was so bad: rude customers, unreliable kitchen staff, constant motion, and — most of all — an unfair boss. Further, the writer includes transitional words that act as signposts, making movement from one idea to the next easy to follow. The major transitions are *First of all, Also, Another reason,* and *The last and most important reason.*

While paragraph A is unified and supported, the writer does not have any clear and consistent way of organizing the material. Partly, emphatic order is used, but this is not made clear by transitions or by saving the most important reason for last. Partly, a time order is used, but it moves inconsistently from two to seven to five o'clock.

These two paragraphs lead us to the third base of effective writing: *coherence.* The supporting ideas and sentences in a composition must be organized so that they cohere or ''stick together.'' As has already been mentioned, key techniques for tying material together are a clear method of organization (such as time order or emphatic order), transitions, and other connecting words.

CHECKING FOR COHERENCE

To check a paper for coherence, ask yourself these questions:

1   Does the paper have a clear method of organization?
2   Are transitions and other connecting words used to tie together the material?

Base 4: Sentence Skills

Activity

Two versions of a paragraph are given below. Both are unified, supported, and organized, but one version communicates more clearly and effectively. Which one, and why?

Paragraph A

Falling Asleep Anywhere

[1]There are times when people are so tired that they fall asleep almost anywhere. [2]For example, there is a lot of sleeping on the bus or train on the way home from work in the evenings. [3]A man will be reading the newspaper, and seconds later it appears as if he is trying to eat it. [4]Or he will fall asleep on the shoulder of the stranger sitting next to him. [5]Another place where unplanned naps go on is the lecture hall. [6]In some classes, a student will start snoring so loudly that the professor has to ask another student to shake the sleeper awake. [7]A more embarrassing situation occurs when a student leans on one elbow and starts drifting off to sleep. [8]The weight of the head pushes the elbow off the desk, and this momentum carries the rest of the body along. [9]The student wakes up on the floor with no memory of getting there. [10]The worst time to fall asleep is when driving a car. [11]Police reports are full of accidents that occur when people lose consciousness and go off the road.[12] If the drivers are lucky, they are not seriously hurt. [13]One woman's car, for instance, went into the river. [14]She woke up in four feet of water and thought it was raining. [15]When people are really tired, nothing will stop them from falling asleep--no matter where they are.

Paragraph B

Falling Asleep Anywhere

[1]There are times when people are so tired that they fall asleep almost anywhere. [2]For example, on the bus or train on the way home from work. [3]A man will be reading the newspaper, seconds later it appears as if he is trying to eat it. [4]Or he will fall asleep on the shoulder of the stranger sitting next to him. [5]Another place where unplanned naps go on are in the lecture hall. [6]In some classes, a student will start snoring so loudly that the professor has to ask another student to shake the sleeper awake. [7]A more embarrassing situation occurs when a student leans on one elbow and starting to drift off to sleep. [8]The weight of the head push the elbow off the desk, and this momentum carries the rest of the body along. [9]The student wakes up on the floor with no memory of getting there. [10]The worst time to fall asleep is when

driving a car. ¹¹Police reports are full of accidents that occur when people conk out and go off the road. ¹²If the drivers are lucky they are not seriously hurt. ¹³One womans car, for instance went into the river. ¹⁴She woke up in four feet of water. ¹⁵And thought it was raining. ¹⁶When people are really tired, nothing will stop them from falling asleep--no matter where they are.

Fill in the blanks: Paragraph _____ makes its point more clearly and effectively because _____

_____

_____ .

UNDERSTANDING SENTENCE SKILLS

Paragraph A is more effective because it incorporates *sentence skills,* the fourth base of competent writing. See now if you can identify the ten sentence-skills mistakes in paragraph B. Do this, first of all, by going back and underlining the ten spots in paragraph B that differ in wording or punctuation from paragraph A. Then try to identify the ten sentence-skills mistakes by circling what you feel is the correct answer in each of the ten statements below.

Note: Comparing paragraph B with the correct version may help you guess correct answers even if you are not familiar with the names of certain skills.

1. In word group 2, there is a
   a. missing comma
   b. missing apostrophe
   c. sentence fragment
   d. dangling modifier

2. In word group 3, there is a
   a. run-on
   b. sentence fragment
   c. mistake in subject-verb agreement
   d. mistake involving an irregular verb

3. In word group 5, there is a
   a. sentence fragment
   b. spelling error
   c. run-on
   d. mistake in subject-verb agreement

4. In word group 7, there is a
   a. misplaced modifier
   b. dangling modifier
   c. mistake in parallelism
   d. run-on

5. In word group 8, there is a
   a. nonstandard English verb
   b. run-on
   c. comma mistake
   d. missing capital letter

6. In word group 11, there is a
   a. mistake involving an irregular verb
   b. sentence fragment
   c. slang phrase
   d. mistake in subject-verb agreement

7. In word group 12, there is a
   a. missing apostrophe
   b. missing comma
   c. mistake involving an irregular verb
   d. sentence fragment

8. In word group 13, there is a
   a. mistake in parallelism
   b. mistake involving an irregular verb
   c. missing apostrophe
   d. missing capital letter

9. In word group 13, there is a
   a. missing comma around an interrupter
   b. dangling modifier
   c. run-on
   d. cliché

10. In word group 15, there is a
    a. missing quotation mark
    b. mistake involving an irregular verb
    c. sentence fragment
    d. mistake in pronoun point of view

You should have chosen the following answers:

1. c	3. d	5. a	7. b	9. a
2. a	4. c	6. c	8. c	10. c

Part Five of this book explains these and other sentence skills. You should review all the skills carefully. Doing so will ensure that you know the most important rules of grammar, punctuation, and usage — rules needed to write clear, error-free sentences.

CHECKING FOR SENTENCE SKILLS

Sentence skills are summarized in the chart on the following page and on the inside front cover of the book.

A Summary of the Four Bases of Effective Writing

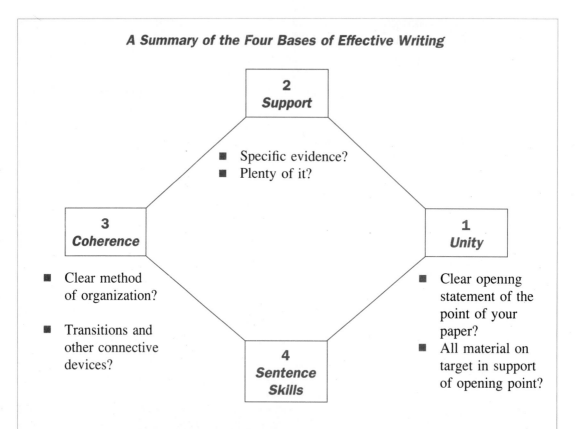

2 Support

- Specific evidence?
- Plenty of it?

3 Coherence

- Clear method of organization?
- Transitions and other connective devices?

1 Unity

- Clear opening statement of the point of your paper?
- All material on target in support of opening point?

4 Sentence Skills

- Fragments eliminated? (page 341)
- Run-ons eliminated? (358)
- Correct verb forms? (373)
- Subject and verb agreement? (390)
- Faulty parallelism and faulty modifiers eliminated? (418)
- Faulty pronouns eliminated? (399)
- Capital letters used correctly? (436)
- Punctuation marks where needed?
  - (a) Apostrophe (449)
  - (b) Quotation marks (458)
  - (c) Comma (466)
  - (d) Semicolon; colon (477)
  - (e) Hyphen; dash (478)
  - (f) Parentheses (480)
- Correct paper format? (431)
- Needless words eliminated? (517)
- Correct word choices? (513)
- Possible spelling errors checked? (489)
- Careless errors eliminated through proofreading? (32)
- Sentences varied? (521)

Practice in Using the Four Bases

You are now familiar with the four bases, or standards, of effective writing: unity, support, coherence, and sentence skills. In this closing section, you will expand and strengthen your understanding of the four bases as you work through the following activities:

1 Evaluating Scratch Outlines for Unity
2 Evaluating Paragraphs for Unity
3 Evaluating Paragraphs for Support
4 Evaluating Paragraphs for Coherence
5 Revising Paragraphs for Coherence
6 Evaluating Paragraphs for All Four Bases:
   Unity, Support, Coherence, and Sentence Skills

1 EVALUATING SCRATCH OUTLINES FOR UNITY

The best time to check a paper for unity is when it is in outline form. A scratch outline, as explained on page 22, is one of the best techniques for getting started with a paper.

Look at the following scratch outline that one student prepared and then corrected for unity:

I had a depressing weekend.
1. Hay fever bothered me
2. Had to pay seventy-seven-dollar car bill
3. Felt bad
4. Boyfriend and I had a fight
5. Did poorly in my math test today as a result
6. My mother yelled at me unfairly

Four reasons support the opening statement that the writer was depressed over the weekend. The writer crossed out "Felt bad" because it was not a specific reason for her depression. (Saying that she felt bad is only another way of saying that she was depressed.) She also crossed out the item about the day's math test because the point she is supporting is that she was depressed over the weekend.

Activity

Cross out the items that do not support the opening point in each outline. These items must be omitted in order to achieve paragraph unity.

1. The cost of raising a child keeps increasing.
   a. School taxes get higher every year.
   b. A pair of children's shoes will probably cost $200 by the year 2000.
   c. Overpopulation is a worldwide problem.
   d. Providing nutritious food is more costly because of inflated prices.
   e. Children should work at age sixteen.

2. My father's compulsive gambling hurt our family life.
   a. We were always short of money for bills.
   b. Luckily, my father didn't drink.
   c. My father ignored his children to spend time at the racetrack.
   d. Gamblers' Anonymous can help compulsive gamblers.
   e. My mother and father argued constantly.

3. There are several ways to get better mileage in your car.
   a. Check air pressure in tires regularly.
   b. Drive at the fifty-five-mile-per-hour speed limit.
   c. Orange and yellow cars are the most visible.
   d. Avoid jackrabbit starts at stop signs and traffic lights.
   e. Always have duplicate ignition and trunk keys.

4. My swimming instructor helped me overcome my terror of the water.
   a. He talked with me about my fears.
   b. I was never good at sports.
   c. He showed me how to hold my head under water and not panic.
   d. I held on to a floating board until I was confident enough to give it up.
   e. My instructor was on the swimming team at his college.

5. Fred Wilkes is the best candidate for state governor.
   a. He has fifteen years' experience in the state senate.
   b. His son is a professional football player.
   c. He has helped stop air and water pollution in the state.
   d. His opponent has been divorced.
   e. He has brought new industries and jobs to the state.

2 EVALUATING PARAGRAPHS FOR UNITY

Activity

Each of the following five paragraphs contains sentences that are off target—sentences that do not support the opening point—and so the paragraphs are not unified. In the interest of paragraph unity, such sentences must be omitted.

Cross out the irrelevant sentences and write the numbers of those sentences in the spaces provided. The number of spaces will tell you the number of irrelevant sentences in each paragraph.

1.

A Kindergarten Failure

[1]In kindergarten I experienced the fear of failure that haunts many schoolchildren. [2]My moment of panic occurred on my last day in kindergarten at Charles Foos Public School in Riverside, California. [3]My family lived in California for three years before we moved to Omaha, Nebraska, where my father was a personnel manager for Mutual of Omaha. [4]Our teacher began reading a list of names of all those students who were to line up at the door in order to visit the first-grade classroom. [5]Our teacher was a pleasant-faced woman who had resumed her career after raising her own children. [6]She called off every name but mine, and I was left sitting alone in the class while everyone else left, the teacher included. [7]I sat there in absolute horror. [8]I imagined that I was the first kid in human history who had flunked things like crayons, sandbox, and sliding board. [9]Without getting the teacher's permission, I got up and walked to the bathroom and threw up into a sink. [10]Only when I ran home in tears to my mother did I get an explanation of what had happened. [11]Since I was to go to a parochial school in the fall, I had not been taken with the other children to meet the first-grade teacher at the public school. [12]My moment of terror and shame had been only a misunderstanding.

The numbers of the irrelevant sentences: _____   _____

2.

How to Prevent Cheating

[1]Instructors should take steps to prevent students from cheating on exams. [2]To begin with, instructors should stop reusing old tests. [3]Even a test that has been used once is soon known on the student grapevine. [4]Students will check with their friends to find out, for example, what was on Dr. Thompson's biology final last term. [5]They may even manage to find a copy of the test itself, "accidentally" not turned in by a former student of Dr. Thompson's. [6]Instructors should also take some commonsense precautions at test time. [7]They should make students separate themselves--by at least one seat--during an exam, and they should watch the class closely. [8]The best place for the instructor to sit is in the rear of the room, so that a student is never sure if the instructor is looking at him or her. [9]Last of all, instructors must make it clear to students that there will be stiff penalties for cheating. [10]One of the problems with our school systems is a lack of discipline. [11]Instructors never used to give in to students' demands or put up with bad behavior, as they do today. [12]Anyone caught cheating should immediately receive a zero for the exam. [13]A person even suspected of cheating should be forced to take an alternative exam in the instructor's office. [14]Because cheating is unfair to honest students, it should not be tolerated.

The numbers of the irrelevant sentences: _____   _____

3.
A Dangerous Cook

[1]When my friend Tom sets to work in the kitchen, disaster often results. [2]Once he tried to make toasted cheese sandwiches for us by putting slices of cheese in the toaster along with the bread; he ruined the toaster. [3]Unfortunately, the toaster was a fairly new one that I had just bought for him three weeks before, on his birthday. [4]On another occasion, he had cut up some fresh beans and put them in a pot to steam. [5]I was really looking forward to the beans, for I eat nothing but canned vegetables in my dormitory. [6]I, frankly, am not much of a cook either. [7]The water in the Teflon pan steamed away while Tom was on the telephone, and both the beans and the Teflon coating in the pan were ruined. [8]Finally, another time Tom made spaghetti for us, and the noodles stuck so tightly together that we had to cut off slices with a knife and fork. [9]In addition, the meatballs were burned on the outside but almost raw inside. [10]The tomato sauce, on the other hand, turned out well. [11]For some reason, Tom is very good at making meat and vegetable sauces. [12]Because of Tom's kitchen mishaps, I never eat at his place without an Alka-Seltzer in my pocket, or without money in case we have to go out to eat.

The numbers of the irrelevant sentences:

_____   _____   _____   _____   _____

4.
Why Adults Visit Amusement Parks

[1]Adults visit amusement parks for several reasons. [2]For one thing, an amusement park is a place where it is acceptable to ''pig out'' on junk food. [3]At the park, everyone is drinking soda and eating popcorn, ice cream, or hot dogs. [4]No one seems to be on a diet, and so buying all the junk food you can eat is a guilt-free experience. [5]Parks should provide stands where healthier food, such as salads or cold chicken, would be sold. [6]Another reason people visit amusement parks is to prove themselves. [7]They want to visit the park that has the newest, scariest ride in order to say that they went on the Parachute Drop, the seven-story Elevator, the Water Chute, or the Death Slide. [8]Going on a scary ride is a way to feel courageous and adventurous without taking much of a risk. [9]Some rides, however, can be dangerous. [10]Rides that are not properly inspected or maintained have killed people all over the country. [11]A final reason people visit amusement parks is to escape from everyday pressures. [12]When people are poised at the top of a gigantic roller coaster, they are not thinking of bills, work, or personal problems. [13]A scary ride empties the mind of all worries--except making it to the bottom alive. [14]Adults at an amusement park may claim they have come for their children, but they are there for themselves as well.

The numbers of the irrelevant sentences: _____   _____   _____

5.   My Color Television

¹My color television has given me nothing but heartburn. ²I was able to buy it a little over a year ago because I had my relatives give me money for my birthday instead of a lot of clothes that wouldn't fit. ³My first dose of stomach acid came when I bought the set. ⁴I let a salesclerk fool me into buying a discontinued model. ⁵I realized this a day later, when I saw newspaper advertisements for the set at seventy-five dollars less than I had paid. ⁶The set worked so beautifully when I first got it home that I would keep it on until stations signed off for the night. ⁷Fortunately, I didn't get any channels showing all-night movies, or I would never have gotten to bed. ⁸Then I started developing a problem with the set that involved static noise. ⁹For some reason, when certain shows switched into a commercial, a loud buzz would sound for a few seconds. ¹⁰Gradually, this sound began to appear during a show, and to get rid of it, I had to click the dial to another channel and click it back. ¹¹Sometimes this technique would not work, and I had to pick up the set and shake it to remove the buzzing sound. ¹²I actually began to build up my arm muscles shaking my set; I could feel the new muscles working whenever I shot a basketball. ¹³When neither of these methods removed the static noise, I would sit popping Tums and wait for the sound to go away. ¹⁴Eventually I wound up slamming the set with my hand again, and it stopped working altogether. ¹⁵My trip to the repair shop cost me $62. ¹⁶The set is working well now, but I keep expecting more trouble.

The numbers of the irrelevant sentences: _____ _____ _____ _____

3   EVALUATING PARAGRAPHS FOR SUPPORT

Activity

The five paragraphs that follow lack sufficient supporting details. Identify the spot or spots where more specific details are needed in each paragraph.

1.   Chicken: Our Best Friend

¹Chicken is the best-selling meat today for a number of good reasons. ²First of all, its reasonable cost puts it within everyone's reach. ³Chicken is popular, too, because it can be prepared in so many different ways. ⁴It can, for example, be cooked by itself, in spaghetti sauce, or with noodles and gravy. ⁵It can be baked, boiled, broiled, or fried. ⁶Chicken is also convenient. ⁷Last and most important, chicken has a high nutritional value. ⁸Four ounces of chicken contain twenty-eight grams of protein, which is almost half the recommended daily dietary allowance.

Fill in the blanks: The first spot where supporting details are needed occurs after sentence number _____ . The second spot occurs after sentence number _____ .

2.

A Car Accident

¹I was on my way home from work when my terrible car accident took place. ²As I drove my car around the curve of the expressway exit, I saw a number of cars ahead of me, backed up because of a red light at the main road. ³I slowly came to a stop behind a dozen or more cars. ⁴In my rear-view mirror, I then noticed a car coming up behind me that did not slow down or stop. ⁵I had a horrible, helpless feeling as I realized the car would hit me. ⁶I knew there was nothing I could do to signal the driver in time, nor was there any way I could get away from the car. ⁷Minutes after the collision, I picked up my glasses, which were on the seat beside me. ⁸My lip was bleeding, and I got out a tissue to wipe it. ⁹The police arrived quickly, along with an ambulance for the driver of the car that hit me. ¹⁰My car was so damaged that it had to be towed away. ¹¹Today, eight years after the accident, I still relive the details of the experience whenever a car gets too close behind me.

Fill in the blank: The point where details are clearly needed occurs after sentence number _____ .

3.

Tips on Bringing Up Children

¹In some ways, children should be treated as mature people. ²For one thing, adults should not use baby talk with children. ³Using real words with children helps them develop language skills more quickly. ⁴Baby talk makes children feel patronized, frustrated, and confused, for they want to understand and communicate with adults by learning their speech. ⁵So animals should be called cows and dogs, not "moo-moos" and "bow-wows." ⁶Second, parents should be consistent when disciplining children. ⁷For example, if a parent tells a child, "You cannot have dessert unless you put away your toys," it is important that the parent follow through on the warning. ⁸By being consistent, parents will teach children responsibility and give them a stable center around which to grow. ⁹Finally, and most important, children should be allowed and encouraged to make simple decisions. ¹⁰Parents will thus be helping their children prepare for the complex decisions that they will have to deal with in later life.

Fill in the blank: The spot where supporting details are needed occurs after sentence number _____ .

4.

Telephone Answering Machines

¹Telephone answering machines are beginning to annoy me. ²First of all, I am so surprised when a machine answers the phone that I become tongue-tied or flustered. ³As the metallic voice says, "Please leave your message when you hear the tone," my mind goes blank. ⁴I don't like to hang up, but I know I'll sound like a fool when the owner plays back the message: "Uh, uh, Dr. Spencer, uh, I wanted to make an appointment, uh, wait a minute, for the fifth, no the second, uh, I'm not sure. . . ." ⁵Another problem I have with the machines is that they can malfunction. ⁶I sometimes call big catalog companies to place orders, and the order is taken by a recording machine. ⁷Just as I'm trying to say, "Two blouses, number B107, size 10," the machine clicks off. ⁸When I call back, another mix-up will occur. ⁹Above all, I dislike the so-called funny tapes that some people now use to answer their phones. ¹⁰One of my coworkers recently bought an answering machine and uses one of these tapes. ¹¹Answering machines seem to be spreading all over the country, but I would rather talk to a human voice anytime.

Fill in the blanks: The first spot where supporting details are needed occurs after sentence number _____. The second spot occurs after sentence number

_____ .

5.

Being on TV

¹People act a little strangely when a television camera comes their way. ²Some people behave as if a crazy puppet-master is pulling their strings. ³Their arms jerk wildly about, and they begin jumping up and down for no apparent reason. ⁴Often they accompany their body movements with loud screams, squeals, and yelps. ⁵Another group of people engage in an activity known as the cover-up. ⁶They will be calmly watching a sports game or other televised event when they realize the camera is focused on them. ⁷The camera operator can't resist zooming in for a close-up of these people. ⁸Then there are those who practice their funny faces on the unsuspecting public. ⁹They take advantage of the television time to show off their talents, hoping to get that big break that will carry them to stardom. ¹⁰Finally, there are those who pretend they are above reacting for the camera.¹¹They wipe all expression from their faces and appear to be interested in something else. ¹²Yet if the camera stays on them long enough, they will slyly check to see if they are still being watched. ¹³Everybody's behavior seems to be slightly strange in front of a TV camera.

Fill in the blanks: The first spot where supporting details are needed occurs after sentence number _____ . The second spot occurs after sentence number

_____ .

4   EVALUATING PARAGRAPHS FOR COHERENCE

Activity

Answer the questions about coherence that follow each of the two paragraphs below.

1.   Why I Bought a Handgun

[1]I bought a handgun to keep in my house for several reasons. [2]Most important, I have had a frightening experience with an obscene phone caller. [3]For several weeks, a man has called me once or twice a day, sometimes as late as three in the morning. [4]As soon as I pick up the phone, he whispers something obscene or threatens me by saying, "I'll get you." [5]I decided to buy a gun because crime is increasing in my neighborhood. [6]One neighbor's house was burglarized while she was at work; the thieves not only stole her appliances but also threw paint around her living room and slashed her furniture. [7]Not long after this incident, an elderly woman from the apartment house on the corner was mugged on her way to the supermarket. [8]The man grabbed her purse and threw her to the ground, breaking her hip. [9]I started thinking about buying a gun about a year ago when I was listening to the news one night. [10]It seemed that every news story involved violence of some kind--rapes, murders, muggings, and robberies. [11]I wondered if some of the victims in the stories would still be alive if they had been able to frighten off the criminal with a gun. [12]As time passed, I became more convinced that I should keep a gun in the house.

a. What words show emphasis in sentence 2? _____

b. What is the number of the sentence to which the transition *In addition* could be added? _____

c. In sentence 8, to whom does the pronoun *her* refer? _____

d. How many times is the key word *gun* repeated in the paragraph? _____

e. The paragraph should use emphatic order. Write a 1 before the reason that is slightly less important than the other two, a 2 before the second-most-important reason, and a 3 before the most important reason.

_____ Obscene phone caller

_____ Crime increase in neighborhood

_____ News stories about crime victims

2.   Joining a Health Club

¹You should do some investigating before you decide to join a health club. ²Make sure that the contract you sign is accurate. ³Check the agreement to be certain that the fees listed are correct, that the penalties for breaking the contract are specified, and that no hidden charges are included. ⁴As soon as you begin thinking about joining a health club, make a list of your needs and requirements. ⁵Decide if you want (or will ever use) facilities such as a swimming pool, jogging track, steam room, weight machines, racquet-ball courts, or a bar and lounge. ⁶Your requirements will determine what kind of club you should join and where you will truly get your money's worth. ⁷After you have decided what type of exercise club is best for you, visit some local clubs and check out the facilities. ⁸Make sure that the equipment is in good order, that the changing rooms and exercise areas are clean, and that there are enough instructors for everyone. ⁹Talk to some of the members to see if they are satisfied with the club and its management. ¹⁰Ask them if they have had any problems with contracts, the club's hours, or lack of equipment. ¹¹Once you have found the best club, you are ready to sign a membership contract.

a.   What is the number of the sentence to which the word *Also* could be added? _____

b.   To whom does the pronoun *them* in sentence 10 refer? _____

c.   What is a synonym for *contract* in sentence 3? _____

d.   What is the number of a sentence to which the words *For example* could be added? _____

e.   The paragraph should use time order. Put a 1 before the step that should come first, a 2 before the intermediate step, and a 3 before the final step.

_____ Make sure the contract is accurate.

_____ Make a list of your needs and requirements.

_____ Visit some local clubs to check facilities.

5   REVISING PARAGRAPHS FOR COHERENCE

The two paragraphs in this section begin with a clear point, but the supporting material that follows the point is not coherent. Read each paragraph and the comments that follow it on how to organize and connect the supporting material. Then do the activity provided in each case.

Paragraph 1

A Difficult Period

Since I arrived in the Bay Area in midsummer, I have had the most difficult period of my life. I had to look for an apartment. I found only one place that I could afford, but the landlord said I could not move in until it was painted. When I first arrived in San Francisco, my thoughts were to stay with my father and stepmother. I had to set out looking for a job so that I could afford my own place, for I soon realized that my stepmother was not at all happy having me live with them. A three-week search led to a job shampooing rugs for a housecleaning company. I painted the apartment myself, and at least that problem was ended. I was in a hurry to get settled because I was starting school at the University of San Francisco in September. A transportation problem developed because my stepmother insisted that I return my father's bike, which I was using at first to get to school. I had to rely on a bus that often arrived late, with the result that I missed some classes and was late for others. I had already had a problem with registration in early September. My counselor had made a mistake with my classes, and I had to register all over again. This meant that I was one week late for class. Now I'm riding to school with a classmate and no longer have to depend on the bus. My life is starting to order itself, but I must admit that at first I thought it was hopeless to stay here.

Comments on Paragraph 1:   The writer of this paragraph has provided a good deal of specific evidence to support the opening point. The evidence, however, needs to be organized. Before starting the paragraph, the writer should have decided to arrange the details by using time order. He or she could then have listed in a scratch outline the exact sequence of events that made for such a difficult period.

Activity 1

Here is a list of the various events described by the writer of paragraph 1. Number the events in the correct time sequence by writing a 1 in front of the first event that occurred, a 2 in front of the second event, and so on.

Since I arrived in the Bay Area in midsummer, I have had the most difficult period of my life.

___3___   I had to search for an apartment I could afford.

___2___   I had to find a job so that I could afford my own place.

___1___   My stepmother objected to my living with her and my father.

_____4_____ I had to paint the apartment before I could move in.

_____6_____ I had to find an alternative to unreliable bus transportation.

_____5_____ I had to reregister for my college courses because of a counselor's mistake. .

Your instructor may now have you rewrite the paragraph on separate paper. If so, be sure to use time signals such as *first, next, then, during, when, after,* and *now* to help guide your reader from one event to the next.

Paragraph 2

Childhood Cruelty

When I was in grade school, my classmates and I found a number of excuses for being cruel to a boy named Andy Poppovian. Sometimes Andy gave off a strong body odor, and we knew that several days had passed since he had taken a bath. Andy was very slow in speaking, as well as very careless in personal hygiene. The teacher would call on him during a math or grammar drill. He would sit there silently for so long before answering that she sometimes said, "Are you awake, Andy?" Andy had long fingernails that he never seemed to cut, with black dirt caked under them. We called him "Poppy," or we accented the first syllable in his name and mispronounced the rest of it and said to him, "How are you today, POP-o-van?" His name was funny. Other times we called him "Popeye," and we would shout at him, "Where's your spinach today, Popeye?" Andy always had sand in the corners of his eyes. When we played tag games at recess, Andy was always "it" or the first one who was caught. He was so physically slow that five guys could dance around him and he wouldn't be able to touch any of them. Even when we tried to hold a regular conversation with him about sports or a teacher, he was so slow in responding to a question that we got bored talking with him. Andy's hair was always uncombed, and it was often full of white flakes of dandruff. Only when Andy died suddenly of spinal meningitis in seventh grade did some of us begin to realize and regret our cruelty toward him.

Comments on Paragraph 2:   The writer of this paragraph provides a number of specifics that support the opening point. However, the supporting material has not been organized clearly. Before writing this paragraph, the author should have (1) decided to arrange the supporting evidence by using emphatic order and (2) listed in a scratch outline the reasons for the cruelty to Andy Poppovian and the supporting details for each reason. The writer could also have determined which reason to use in the emphatic final position of the paper.

Activity 2

Create a clear outline for paragraph 2 by filling in the scheme below. The outline is partially completed.

> When I was in grade school, my classmates and I found a number of excuses for being cruel to a boy named Andy Poppovian.

Reason 1. *Physically slow* _____

Details a. _____

         b. *Five guys could dance around him.* _____

Reason 2. _____

Details a. _____

         b. *Sand in eyes* _____

         c. _____

         d. _____

Reason 3. *Funny name* _____

Details a. _____

         b. _____

Reason 4. _____

Details a. _____

         b. *In regular conversation* _____

Your instructor may have you rewrite the paragraph on separate paper. If so, be sure to introduce each of the four reasons with transitions such as *First, Second, Another reason,* and *Finally.* You may also want to use repeated words, pronouns, and synonyms to help tie your sentences together.

6 EVALUATING PARAGRAPHS FOR ALL FOUR BASES: UNITY, SUPPORT, COHERENCE, AND SENTENCE SKILLS

Activity

In this activity, you will evaluate paragraphs in terms of all four bases: unity, support, coherence, and sentence skills. Evaluative comments follow each paragraph below. Circle the letter of the statement that best applies in each case.

1.
Ponderosa Steak House

There are a number of advantages to eating at Ponderosa Steak House. The first advantage is that the meals are moderate in price. Another reason is that the surroundings are clean, and the people are pleasant. Also, I have a variety of dinners to choose from. The last and main advantage is that I don't have to plan and prepare the meal.

a. The paragraph is not unified.
b. The paragraph is not adequately supported.
c. The paragraph is not well organized.
d. The paragraph does not show a command of sentence skills.
e. The paragraph is well written in terms of the four bases.

2.
A Frustrating Moment

A frustrating moment happened to me several days ago. When I was shopping. I had picked up a tube of crest toothpaste and a jar of noxema skin cream. After the cashier rang up the purchases, which came to $4.15. I handed her $10. Then got back my change, which was only $0.85. I told the cashier that she had made a mistake. Giving me change for $5 instead of $10. But she insist that I had only gave her $5, I became very upset and insist that she return the rest of my change. She refused to do so instead she asked me to step aside so she could wait on the next customer. I stood very rigid, trying not to lose my temper. I simply said to her, I'm not going to leave here, Miss, without my change for $10. Giving in at this point a bell was rung and the manager was summoned. After the situation was explain to him, he ask the cashier to ring off her register to check for the change. After doing so, the cashier was $5 over her sale receipts. Only then did the manager return my change and apologize for the cashier mistake.

a. The paragraph is not unified.
b. The paragraph is not adequately supported.
c. The paragraph is not well organized.
d. The paragraph does not show a command of sentence skills.
e. The paragraph is well written in terms of the four bases.

3.

Asking Girls Out

There are several reasons I have trouble asking girls to go out with me. I have asked some girls out and have been turned down. This is one reason that I can't talk to them. At one time I was very shy and quiet, and people sometimes didn't even know I was present. I can talk to girls now as friends, but as soon as I want to ask them out, I usually start to become quiet, and a little bit of shyness comes out. When I get the nerve up finally, the girl will turn me down, and I swear that I will never ask another one out again. I feel sure I will get a refusal, and I have no confidence in myself. Also, my friends mock me, though they aren't any better than I am. It can become discouraging when your friends get on you. Sometimes I just stand there and wait to hear what line the girl will use. The one they use a lot is ''We like you as a friend, Ted, and it's better that way.'' Sometimes I want to have the line put on a tape recorder, so they won't have to waste their breath on me. All my past experiences with girls have been just as bad. One girl used me to make her old boyfriend jealous. Then when she succeeded, she started going out with him again. I had a bad experience when I took a girl to the prom. I spent a lot of money on her. Two days later, she told me that she was going steady with another guy. I feel that when I meet a girl I have to be sure I can trust her. I don't want her to turn on me.

a.   The paragraph is not unified.

b.   The paragraph is not adequately supported.

c.   The paragraph is not well organized.

d.   The paragraph does not show a command of sentence skills.

e.   The paragraph is well written in terms of the four bases.

4.

A Change in My Writing

A technique in my present English class has corrected a writing problem that I've always had. In past English courses, I had major problems with commas in the wrong places, bad spelling, capitalizing the wrong words, sentence fragments, and run-on sentences. I never had any big problems with unity, support, or coherence, but the sentence skills were another matter. They were like little bugs that always appeared to infest my writing. My present instructor asked me to rewrite papers, just concentrating on sentence skills. I thought that the instructor was crazy because I didn't feel that rewriting would do any good. I soon became certain that my instructor was out of his mind, for he made me rewrite my first paper four times. It was very frustrating, for I became tired of doing the same paper over and over. I wanted to belt my instructor against the wall when I'd show him each new draft and he'd find skills mistakes and say ''Rewrite.'' Finally, my papers began to improve and the sentence skills began to fall into place. I was able to see them and correct them before turning in a paper, whereas I couldn't before. Why or how this happened I don't know, but I think that rewriting helped a lot. It took me most of the semester, but I stuck it out and the work paid off.

a. The paragraph is not unified.

b. The paragraph is not adequately supported.

c. The paragraph is not well organized.

d. The paragraph does not show a command of sentence skills.

e. The paragraph is well written in terms of the four bases.

5.   Luck and Me

    I am a very lucky young man, which has not been the case with the rest of my family. Sometimes when I get depressed, which is too frequently, it's hard to see just how lucky I am. I'm lucky that I'm living in a country that is free. I'm allowed to worship the way I want to, and that is very important to me. Without a belief in God a person cannot live with any real certainty in life. My family cares about me, though maybe not as much as I would like. My relationship with my girlfriend is a source of good fortune for me. She gives me security and that's something I need a lot. Even with these positive realities in my life, I still seem to find time for insecurity, worry, and, worst of all, depression. At times in my life I have had bouts of terrible luck. But overall, I'm a very lucky guy. I plan to further develop the positive aspects of my life and try to eliminate the negative ones.

a. The paragraph is not unified.

b. The paragraph is not adequately supported.

c. The paragraph is not well organized.

d. The paragraph does not show a command of sentence skills.

e. The paragraph is well written in terms of the four bases.

Some Suggestions on What to Do Next

1 Read "Providing Examples" (page 127) or "Narrating an Event" (page 191) in Part Two and do the writing assignments given. Then go on to the other types of paragraph development in Part Two.

2 When you have mastered the different types of paragraph development, you may want to work through "Writing the Essay" in Part Three and do one or more of the writing assignments presented.

3 Continue review of sentence skills in Part Five.

4 Begin to work on any "Special Skills" in Part Four that you may need to know.

PART TWO

PARAGRAPH DEVELOPMENT

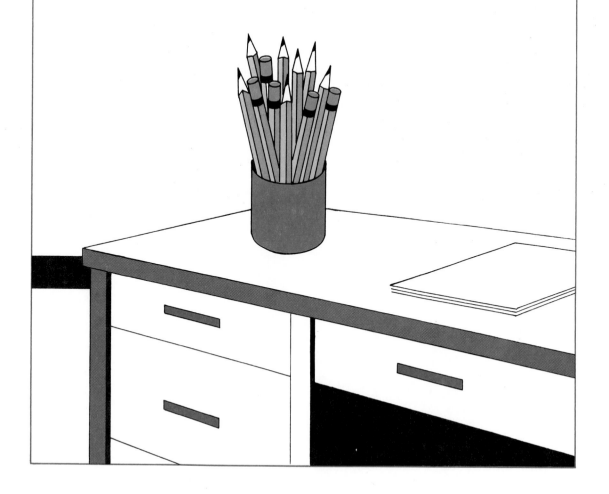

PREVIEW

Part Two introduces you to paragraph development and gives you practice in the following common types of paragraph development:

Providing Examples
Explaining a Process
Examining Cause and Effect
Comparing or Contrasting
Defining a Term
Dividing and Classifying
Describing a Scene or Person
Narrating an Event
Arguing a Position

After a brief explanation of each type of paragraph development, student paragraphs illustrating each type are presented, followed by questions about those paragraphs. The questions relate to the standards of effective writing described in Part One. You are then asked to write your own paragraph. In each case, writing assignments progress from personal-experience topics to more formal and objective topics, with the last assignment in each section requiring some simple research. At times, points or topic sentences for development are suggested, so that you can concentrate on (1) making sure your evidence is on target in support of your opening idea, (2) providing plenty of specific supporting details to back up your point, and (3) organizing your supporting material clearly. The last chapter in Part Two provides some additional assignments.

INTRODUCTION TO PARAGRAPH DEVELOPMENT

Nine Patterns of Paragraph Development

Traditionally, all writing has been divided into the following forms:

- Exposition
Examples	Comparison or contrast
Process	Definition
Cause and effect	Division and classification
- Description
- Narration
- Argumentation or persuasion

In *exposition,* the writer provides information about and explains a particular subject. The patterns of development in exposition include (1) giving examples, (2) detailing the process of doing or making something, (3) analyzing causes or effects, (4) comparing or contrasting, (5) defining a term or concept, (6) dividing something into parts or grouping it into categories. In this part of the book, each of the six patterns of exposition is presented in a separate chapter.

There are also individual chapters devoted to (7) description, (8) narration, and (9) argument. A *description* is a verbal picture of a person, place, or thing. In a *narration*, a writer tells the story of something that happened. *Argumentation* or *persuasion* is an attempt to prove a point or defend an opinion.

You will have a chance, then, to learn how nine different patterns can help organize material in your papers. Each of the nine patterns has its own internal logic and provides its own special strategies for imposing order on your ideas.

As you practice each pattern, you should keep the following two points in mind:

■ In each paragraph that you write, one pattern will predominate, but very often one or more additional patterns may also be involved. For instance, ''Good-Bye, Tony'' — a paragraph you have already read (page 5) — presents a series of causes leading to an effect: that the writer will not go out with Tony again. But the writer also presents examples to explain each of the causes (Tony was late, he was bossy, he was abrupt). And there is an element of narration, as the writer presents examples that occur from the beginning to the end of the date.

■ More important, a paragraph you write in almost any pattern will probably involve some form of argumentation. You will advance a point and then go on to support your point. To convince the reader that your thesis is valid, you may use a series of examples, or a narration, or a description, or some other pattern of organization. Among the paragraphs you will read in Part Two, one writer supports the point that a certain pet shop is depressing by providing a number of descriptive details. Another writer labels a certain experience in his life as a ''heartbreak'' and then uses a narrative to demonstrate the truth of his statement. A third writer advances the opinion that good horror movies can be easily distinguished from bad horror movies and then supplies comparative information about both to support her claim. Much of your writing, in short, will have the purpose of persuading your reader that the idea you have advanced is valid.

Writer, Purpose, and Audience

As was noted in ''Important Factors in Writing'' in Part One, the purpose of most writing is to inform, entertain, or persuade; in this book, most of your writing will involve some form of persuasion or information. The audience for your writing is primarily your instructors and sometimes other students—who are really a symbol for any general audience of educated adults. Sometimes, however, you must write for more specific audiences; therefore, it is important to develop the skills of choosing appropriate words, and adopting an appropriate tone of voice, for a particular purpose and particular readers.

This part of the book, then, includes assignments at or near the ends of chapters that ask you to write with a very specific purpose in mind and for a very specific audience. You will be asked, for example, to imagine yourself as a TV critic addressing parents at a school, as a graduate of a local high school advising a counselor there about a drug problem, as an aide at a day-care center preparing instructions for children, as an apartment tenant complaining to a landlord about neighbors, or as a client of a video dating service introducing himself or herself to potential dates. Through these and other assignments, you will learn how to adjust your style and tone of voice to a given writing situation.

Tools for Paragraph Development

USING PART TWO:
THE PROGRESSION IN EACH CHAPTER

After each type of essay development is explained, student papers illustrating that type are presented, followed by questions about the papers. The questions relate to unity, support, and coherence — the principles of effective writing explained earlier in the book. You are then asked to write your own essay. In most cases, the first assignment is fairly structured and provides a good deal of guidance for the writing process. The other assignments offer a wide choice of writing topics. The fourth assignment always requires some simple research, and the fifth assignment requires writing with a specific purpose and for a specific audience.

USING PEER REVIEW

In addition to having your instructor as an audience for your writing, you will benefit by having another student in your class as an audience. On the day a paper is due, or on a day when you are writing papers in class, your instructor may ask you to pair up with another student. That student will read your paper, and you will read his or her paper.

Ideally, read the other paper aloud while your peer listens. If that is not practical, read it in a whisper while your peer looks on. As you read, both you and your peer should look and listen for spots where the paper does not read smoothly and clearly. Check or circle the trouble spots where your reading snags.

Your peer should then read your paper, marking possible trouble spots while doing so. Then each of you should do three things.

1   Identification

On a separate sheet of paper, write at the top the title and author of the paper you have read. Underneath that, put your name as the reader of the paper.

2   Scratch Outline

"X-ray" the paper for its inner logic by making up a scratch outline. The scratch outline need be no more than twenty words or so, but it should show clearly the logical foundation on which the paper is built. It should identify and summarize the overall point of the paper and the three areas of support for the main point.

Your outline can look as follows.

Point: _____

Support:

(1) _____

(2) _____

(3) _____

For example, here is a scratch outline of the paper on a new puppy in the house on page 143:

Point: *A new puppy can have drastic effects in a house.*

Support:

(1) *Keeps family awake at night*

(2) *Destroys possessions*

(3) *Causes arguments*

3 Comments

Under the outline, write the heading "Comments." Here is what you comment on:

- Look at the spots where your reading of the paper snagged: Are words missing or misspelled? Is there a lack of parallel structure? Are there mistakes with punctuation? Is the meaning of a sentence confused? Try to figure out what the problems are and suggest ways of fixing them.

- Are there spots in the paper where you see problems with *unity,* or *support,* or *organization*? If so, offer comments. For example, you might say, "More details are needed in the first supporting paragraph," or "Some of the details in the last supporting paragraph don't really back up your point."

- Finally, make note of something you really liked about the paper, such as good use of transitions or an especially realistic or vivid specific detail.

After you have completed your evaluation of the paper, give it to your peer. Your teacher may provide you with the option of rewriting a paper in light of this feedback. Whether or not you rewrite, be sure to hand in the peer evaluation form with your paper.

USING A PERSONAL CHECKLIST

After you have completed a paper, there are three ways you should check it yourself. You should *always* do the first two checks, which take only a couple of minutes. Ideally, you should take the time to do the detailed final check as well.

1 Read the paper *out loud.* If it does not sound right — that is, if it does not read smoothly and clearly — then make the changes needed to ensure that it does.

2 Make sure you can answer clearly and concisely two basic questions: "What is the point of my essay? What are the three distinct bits of support for my point?"

3 Last, evaluate your paper in terms of the detailed checklist given on the following page. The checklist is also reproduced on the inside front cover of this book.

should

t of the Four Bases in Effective Writing

...estions below as a guide in both writing and evaluating a paper.
...n parentheses refer to the pages that explain each skill.

...nity

- ...pening statement of the point of your paper? (44–45; 53–56)
- ...rial on target in support of opening point? (93–94; 103–107

Base 2: Support

- *Specific* evidence? (46–50; 61–64; 95–96)
- Plenty of it? (51–52; 65–68; 107–109)

Base 3: Coherence

- Clear method of organization? (73–75; 82–87; 97–98; 111–114)
- Transitions and other connective devices? (75–80; 88–90)

Base 4: Sentence Skills

- Fragments eliminated? (341)
- Run-ons eliminated? (358)
- Correct verb forms? (373; 382; 390)
- Subject and verb agreement? (390)
- Faulty modifiers and faulty parallelism eliminated? (418; 422; 427)
- Faulty pronouns eliminated? (399)
- Capital letters used correctly? (436)
- Punctuation marks where needed?
  a Apostrophe (449)
  b Quotation marks (458)
  c Comma (466)
  d Colon; semicolon (477; 478)
  e Dash; hyphen (478; 479)
  f Parentheses (480)
- Correct paper format? (431)
- Needless words eliminated? (517)
- Correct word choices? (513)
- Possible spelling errors checked? (482; 489; 501)
- Careless errors removed through editing and through proofreading? (32; 539)
- Sentences varied? (521)

PROVIDING EXAMPLES

In our daily conversations, we often provide *examples* — that is, details, particulars, specific instances — to explain statements that we make. In the box below are several statements and supporting examples:

Statement	*Examples*
The A&P was crowded today.	There were at least four carts waiting at each of the checkout counters, and it took me forty-five minutes to get through a line.
The corduroy shirt I bought is poorly made.	When I washed it, the colors began to fade, one button cracked and another fell off, a shoulder seam opened, and the sleeves shrank almost two inches.
My son Peter is unreliable.	If I depend on him to turn off a pot of beans in ten minutes, the family is likely to eat burned beans. If I ask him to turn down the thermostat before he goes to bed, the heat is likely to stay on all night.

In each case, the examples help us *see for ourselves* the truth of the statement that has been made. In paragraphs, too, explanatory examples help the audience fully understand a point. Lively, specific examples also add interest to a paper.

In this chapter, you will be asked to provide a series of examples to support a topic sentence. Providing examples to support a point is one of the most common and simple methods of paragraph development. First read the paragraphs ahead; they all use examples to develop their points. Then answer the questions that follow.

PARAGRAPHS TO CONSIDER

Inconsiderate Drivers

[1]Some people are inconsiderate drivers. [2]In the city, they will at times stop right in the middle of the street while looking for a certain home or landmark. [3]If they had any consideration for the cars behind them, they would pull over to the curb first. [4]Other drivers will suddenly slow down unexpectedly at a city intersection to make a right or left turn. [5]The least they could do is use their turn signals to let those behind them know in advance of their intention. [6]On the highway, a common example of inconsiderateness is night drivers who fail to turn off their high beams, creating glare for cars approaching in the other direction. [7]Other rude highway drivers move to the second or passing lane and then stay there, making it impossible for cars behind to go around them. [8]Yet other drivers who act as if they have special privileges are those who do not wait their turn in bottleneck situations where the cars in two lanes must merge alternately into one lane. [9]Perhaps the most inconsiderate drivers are those who throw trash out their windows, creating litter that takes away some of the pleasure of driving and that must be paid for with everyone's tax dollars.

The Cruelty of Children

[1]Children can be very cruel. [2]For one thing, they start very early to use words that wound. [3]Three-year-olds in nursery school, for example, call each other "dum-dum" or "weirdo," and slightly older children use nicknames like "fatty" or "four-eyes" to tease their schoolmates. [4]Children who are just a bit older learn facts about other kids from their parents, and use those facts to make someone break down and cry. [5]Children also attack each other physically. [6]For instance, whenever a group of grade-schoolers come home from school, there is a lot of pushing, tripping, punching, and pinching. [7]An argument may end in shoving and hair-pulling. [8]But far worse than harsh words or physical violence is the emotional hurt that children can cause their classmates by their cruelty. [9]By junior high school days, for example, young teenagers start to shut out the people they do not like. [10]They ignore the kids whose looks, clothes, interests, or finances differ from their own. [11]Popular kids form groups, and the unpopular ones are left to face social isolation, loneliness, and depression. [12]Many adults think that childhood is an ideal time, but terribly cruel things can happen then.

An Egotistical Neighbor

[1]I have an egotistical neighbor named Alice. [2]If I tell Alice how beautiful I think the dress she is wearing is, she will take the time to tell me the name of the store where she bought it, the type of material that was used in making it, and the price. [3]Alice is also egotistical when it comes to her children. [4]Because they are hers, she thinks they just have to be the best children on the block. [5]I am wasting my time by trying to tell her I have seen her kids expose themselves on the street or take things from parked cars. [6]I do not think parents should praise their children too much. [7]Kids have learned how to be good at home and simply awful when they are not at home. [8]Finally, Alice is quick to describe the furnishings of her home for someone who is meeting her for the first time. [9]She tells how much she paid for the paneling in her dining room. [10]She mentions that she has two color television sets and that they were bought at an expensive furniture store. [11]She lets the person know that the stereo set in her living room cost more than a thousand dollars, and that she has such a collection of recordings that she would not be able to play them all in one week. [12]Poor Alice is so self-centered that she never realizes how boring she can be.

■ Questions

About Unity

1. Which two sentences in ''An Egotistical Neighbor'' are irrelevant to the point that Alice is egotistical? (*Write the sentence numbers here.*) __6__ __7__

About Support

2. In ''Inconsiderate Drivers,'' how many examples are given of inconsiderate drivers?

    _____ one   _____ two   _____ four   __✓__ six

3. After which sentence in ''The Cruelty of Children'' are specific details needed?

    __4__

About Coherence

4. What are the three main transition words used in ''The Cruelty of Children''?

    a. __For instance__

    b. __For one thing__

    c. __For example__

5. What are the two main transition words in "An Egotistical Neighbor"?

   a. _Also_

   b. _Finally_

6. Which two paragraphs clearly use emphatic order to organize their details, saving for last what the writers regard as their most important examples?

   _____ _up cod driver_ _____

   _____ _ego ner_ _____

WRITING AN EXAMPLES PARAGRAPH

■ Writing Assignment 1

The assignment here is to complete an unfinished paragraph (opposite page), which has as its topic sentence, "My husband Roger is a selfish person." Provide the supporting details needed to fill out the examples of Roger's selfishness. The first example has been done for you.

How to Proceed: To do this assignment, first jot down on separate paper a couple of answers for each of the following questions.

a   What specific vacations did the family go on because Roger wanted to go? Give places, length of stay, time of year. What vacations has the family never gone on (for example, to visit the wife's relatives), even though the wife wanted to?

b   What specific items has Roger bought for himself (rather than for the whole family's use) with leftover budget money?

c   What chores and duties involved in the everyday caring for the children has Roger never done?

Note: Your instructor may ask you to work with one or two other students in generating the details needed to develop the three examples in the paragraph. Each group may then be asked to read their details aloud, with the class deciding which details are the most effective for each example.

Here and in general in your writing, try to generate *more* supporting material than you need. You are then in a position to choose the most convincing details for your paper. Now take your best details, reshape them as needed, and use them to complete the paragraph about Roger.

A Selfish Person

My husband Roger is a selfish person. For one thing, he refuses to move out of the city, even though it is a bad place to raise the children. *We inherited some money when my parents died, and it might be enough for a down payment on a small house in a nearby town. But Roger says he would miss his buddies in the neighborhood.* Also, when we go on vacation, we always go where Roger wants to go.

_____

_____

_____

Another example of Roger's selfishness is that he always spends any budget money that is left over. _____

_____

_____

_____

Finally, Roger leaves all the work of caring for the children to me. _____

_____

_____

_____

_____

■ **Writing Assignment 2**

Write a paragraph about one quality of a person you know well. The person might be a member of your family, a friend, a roommate, a boss or supervisor, a neighbor, an instructor, or someone else. Listed on the next page are some descriptive words that can be applied to people. They are only suggestions; you can write about any other specific quality.

Honest	Hardworking	Jealous
Bad-tempered	Supportive	Materialistic
Ambitious	Suspicious	Sarcastic
Bigoted	Open-minded	Self-centered
Considerate	Lazy	Spineless
Argumentative	Independent	Good-humored
Softhearted	Stubborn	Cooperative
Energetic	Flirtatious	Disciplined
Patient	Irresponsible	Sentimental
Reliable	Stingy	Defensive
Generous	Trustworthy	Dishonest
Persistent	Aggressive	Insensitive
Shy	Courageous	Unpretentious
Sloppy	Compulsive	Neat

How to Proceed

a   Begin by prewriting. Make a list of examples that will support your topic sentence. For example, if you decide to write about your brother's irresponsibility, jot down several examples of times when he showed this quality. Part of your list might look like this:

Lost rent money
Forgot to return borrowed textbooks
Didn't show up for big family dinner
Left dog alone in the apartment for two days
Left my bike out in the rain
Missed conference with instructor

Another way to get started is to ask yourself questions about your topic and write down the answers. Again, if you were writing about your brother's irresponsibility, you might ask yourself questions such as these:

How has Bill been irresponsible?
What are examples of times he's shown this quality?
What happened on these occasions?
Who was involved?
What were the results of his actions?

The answers to these questions should serve as an excellent source of details for the paragraph.

b Then prepare a scratch outline made up of the strongest examples from the prewriting material you generated above. Note that as you make this outline, you should group related details together. For example, the items in the list about the irresponsible brother might be categorized as follows:

At apartment
Lost rent money
Left dog alone in apartment

At home
Missed family dinner
Left bike in rain

At school
Didn't return textbooks
Missed conference

c Next, write out your topic sentence. This first sentence should tell the name of the person you are writing about, your relationship to the person, and the specific quality you are focusing on. For example, you might write, ''Linda is a flirtatious girl I know at school,'' or ''Stubbornness is Uncle Carl's outstanding characteristic.''

Do not make the mistake of beginning with more than one quality (''I have a cousin named Alan who is softhearted and generous'') or with a general quality (''My boss is a good person''). Focus on *one specific quality*.

d Develop your examples with specific details. Remember that you don't want to *tell* us about the person; rather, you want to *show* the person to us by detailing words, actions, or both. You might want to go back and reread the examples provided in ''An Egotistical Neighbor.''

e As you are writing drafts of your paragraph, ask yourself repeatedly: ''Do my examples truly show that my subject has a certain quality?'' Your aims in this assignment are twofold: (1) to provide *truly specific* details for the quality in question and (2) to provide *enough* specific details so that you solidly support your point.

f When you are satisfied that you have provided effective examples, edit your paragraph carefully for the sentence-skills mistakes listed on the inside front cover. In addition, make sure you can answer *Yes* to the questions on unity, support, and coherence.

■ Writing Assignment 3

Write a paragraph that uses examples to develop one of the following statements or a related statement of your own.

1. _____ is a distracting place to try to study.
2. The daily life of a student is filled with conflicts.
3. Abundant evidence exists that America has become a health-conscious nation.
4. Despite modern appliances, many household chores are still drudgery.
5. One of my instructors, _____, has some good (*or* unusual) teaching techniques.
6. Wasted electricity is all around us.
7. Life in America is faster-paced than ever before.
8. Violence on television is widespread.
9. Women today are wearing some ridiculous fashions.
10. Some students here at _____ do not care about learning (*or* are overly concerned about grades).

Be sure to choose examples that truly support your point. They should be relevant facts, statistics, personal experiences, or incidents you have heard or read about. Organize your paragraph by grouping several examples that support your point. Save the most vivid, convincing, or important example for last.

■ Writing Assignment 4

Research some magazine articles or recent books to back up the topic sentence, "The diet of the average American is unhealthy." You might want to check for articles dealing with such topics as fat, fiber, fast food, junk food, caffeine, food additives, heart disease, and cancer. The *Readers' Guide to Periodical Literature* and your library card catalog (subject index) will direct you to the material you need. Try looking under such headings as *Nutrition, Food habits, Diet, Vitamins,* and *Health*. (See "Using the Library" on pages 291–307.)

■ Writing Assignment 5

Imagine that you are a television critic and are still living near the high school you attended. The principal is planning a special parents' evening and would like you to make a speech on the topic "Television and the Responsible Parent." You accept the invitation and decide to organize your talk around this thesis: "There are three television programs that represent worthwhile viewing for families." Write a one-paragraph summary of your talk.

To make an outline for your speech, think of three programs you would recommend for viewing by all members of a family. Write down the titles of the programs, and then list under each the specific features that made you choose that show.

EXPLAINING A PROCESS

Every day we perform many activities that are processes — that is, a series of steps carried out in a definite order. Many of these processes are familiar and automatic: for example, tying shoelaces, changing bed linen, using a vending machine, and starting a car. We are thus seldom aware of the sequence of steps that makes up each activity. In other cases, such as when we are asked for directions to a particular place, or when we try to read and follow the directions for a new table game, we may be painfully conscious of the whole series of steps involved in the process.

In this section, you will be asked to write a process paragraph — one that explains clearly how to do or make something. To prepare for this assignment, you should first read the student process papers below and then respond to the questions that follow.

Note: In process writing, where you are often giving instruction, the pronoun *you* can appropriately be used. Two of the model paragraphs here use *you* — as indeed does much of this book, which gives instruction on how to write effectively. As a general rule, though, do not use *you* in your writing.

PARAGRAPHS TO CONSIDER

Sneaking into the House at Night

[1]The first step I take is bringing my key along with me. [2]Obviously, I don't want to have to knock on the door at 1:30 in the morning and rouse my parents out of bed. [3]Second, I make it a point to stay out past midnight. [4]If I come in before then, my father is still up. [5]I find it hard to face his disapproving look after a night out. [6]All I need in my life right now is for him to make me feel guilty. [7]Trying to make it as a college student is as much as I'm ready to handle. [8]Next, I am careful to be very quiet upon entering the house. [9]This involves lifting the front door up slightly as I open it, so that it does not creak. [10]It also means treating the floor and steps to the second floor like a minefield, stepping carefully over the spots that squeak. [11]Finally, I stop briefly in the bathroom without turning on the light and then tiptoe to my room, put my clothes in a pile on a chair, and slip quietly into bed. [12]With my careful method of sneaking into the house at night, I have avoided some major hassles with my parents.

How to Harass an Instructor

[1]You can use a number of time-proven techniques to harass an instructor. [2]First of all, show up late for class, so that you can interrupt the beginning of the instructor's presentation. [3]Saunter in nonchalantly and try to find a seat by a friend. [4]In a normal tone of voice, speak some words of greeting to your friends as you sit down, and scrape your chair as loudly as possible while you make yourself comfortable in it. [5]Then just sit there and do anything but pay attention. [6]When the instructor sees that you are not involved in the class, he or she may pop a quick question, probably hoping to embarrass you. [7]You should then say, in a loud voice, "I DON'T KNOW THE ANSWER." [8]This declaration of ignorance will throw the instructor off guard. [9]If the instructor then asks you why you don't know the answer, say "I don't even know what page we're on" or "I thought the assignment was boring so I didn't do it." [10]Give the impression that there is no sane reason why you should be expected to know the answer. [11]After the instructor calls on someone else, get up loudly from your seat, walk to the front of the classroom, and demand to be excused for an emergency meeting in the washroom. [12]Stay at least fifteen minutes and take your time coming back. [13]If the instructor asks you where you've been when you reenter the room, simply ignore the question and go to your seat. [14]Flop into your chair, slouching back and extending your legs as far out as possible. [15]When the instructor informs you of the assignment that the class is working on, heave an exaggerated sigh and very slowly open up your book and start turning the pages. [16]As soon as he or she stops looking at you, rest your elbows on the desk, hold your pencil between the fingertips of your hands, and gaze off into space. [17]The instructor will look at you and wonder whether it wouldn't have been better to go into business instead of education.

Driving around a Traffic Circle

[1]If you want to drive around one of our country's treacherous traffic circles and live to tell about it, there are a number of steps to follow. [2]To begin, you must prepare yourself mentally as you approach the circle. [3]Do not give in to panic or begin to quiver with terror. [4]Instead, try repeating to yourself, in a calm tone of voice, "Other drivers do not want to kill me. [5]They are sane people. [6]We will all go around the circle in an orderly way." [7]Next, as you actually begin to negotiate the circle, you should practice the technique of looking in all directions at once. [8]Every second, check your rearview mirror, left mirror, right mirror, and then turn your head quickly to cover any blind spots. [9]Following this, you must deal with the flow of traffic merging onto the circle from other roads. [10]Try to maintain your speed at all costs. [11]Do not slow down to read road signs or stop to allow another car to enter the circle. [12]This will merely create a massive pileup behind you. [13]Finally, the trickiest part of traffic-circle driving is exiting from the circle. [14]At some point, you must stop going around and make a move, usually across two lanes of traffic, toward a side road. [15]You can use your directional signals, although these will have little effect on the other drivers. [16]The best way to exit is to check your rearview mirror until you spot a timid-looking driver slightly behind you in the next lane. [17]This person will slow down as you cut into his or her lane and make a successful dash for the exit. [18]Never cut across in front of a Cadillac driver smoking a thick cigar or in front of an eighteen-year-old in a Trans Am. [19]Such people will speed up, not slow down. [20]These instructions, then, should enable you to enter a traffic circle and come out in one piece on the other side.

■ **Questions**

About Unity

1. Which paragraph lacks a topic sentence?

    _____

2. Which two sentences in "Sneaking into the House at Night" should be eliminated in the interest of paragraph unity? (*Write the sentence numbers here.*) _____   _____

About Support

3. After which sentence in "How to Harass an Instructor" are supporting details needed? _____

4. Summarize the four steps in the process of driving around a traffic circle.

   a. _____

   b. _____

   c. _____

   d. _____

About Coherence

5. Do these paragraphs use time order or emphatic order?

   _____

6. List the four main transition words in "Sneaking into the House at Night."

   a. _____   c. _____

   b. _____   d. _____

WRITING A PROCESS PARAGRAPH

■ Writing Assignment 1

Choose one of the topics below to write about in a process paper.

   How to change a car or bike tire
   How to bathe a dog or cat
   How to get rid of house or garden pests such as mice, roaches, or wasps
   How to fall asleep (if you need to and can't)
   How to play a simple game like checkers, tic-tac-toe, or an easy card game
   How to load a van
   How to learn a song
   How to live on a limited budget
   How to shorten a skirt or pants
   How to plant a garden
   How to take care of plants
   How to fix a leaky faucet, a clogged drain, or the like
   How to build a campfire
   How to make your house look lived in when you are away
   How to study for an important exam
   How to paint a ceiling
   How to conduct a yard or garage sale
   How to wash dishes efficiently, clean a bathroom, do laundry, or the like

How to Proceed

a Begin by prewriting. Freewrite for ten minutes on the topic you have chosen. Do not worry about spelling, grammar, organization, or other matters of correct form. Just write whatever comes into your head regarding the topic. Keep writing for more than ten minutes if added details about the topic occur to you. This freewriting will give you a base of raw material that you can draw on in the next phase of your work on the paragraph. After freewriting, you should have a sense of whether there is enough material available for you to write a process paragraph about the topic. If so, continue as explained below. If not, choose another topic and freewrite about *it* for ten minutes.

b Write a clear, direct topic sentence about the process you are going to describe. In your topic sentence, you can (1) say that it is important for your audience to know about the process ("Knowing how to study effectively for a major exam can mean the difference between passing and failing a course"), or (2) state your opinion of the process ("My technique for building a campfire is almost foolproof").

c List all the steps you can think of that may be part of the process. Don't worry, at this point, about how each step fits or whether certain steps overlap. Here, for example, is the list prepared by the author of "Sneaking into the House at Night":

Quiet on stairs	Lift up front door
Come in after Dad's asleep	Late dances on Saturday night
House is freezing at night	Don't turn on bathroom light
Bring key	Avoid squeaky spots on floor
Know which steps to avoid	Get into bed quietly

d Number your items in time order; strike out items that do not fit in the list; add others that come to mind. Thus:

~~Quiet on stairs~~
2 Come in after Dad's asleep
~~House is freezing at night~~
1 Bring key
5 Know which steps to avoid
3 Lift up front door
~~Late dances on Saturday night~~
6 Don't turn on bathroom light
4 Avoid squeaky spots on floor
8 Get into bed quietly
7 *Undress quietly*

e Use your list as a guide to write the first rough draft of your paper. As you write, try to think of additional details that will support your opening sentence. Do not expect to finish your paper in one draft. You should, in fact, be ready to write a series of lists and drafts as you work toward the goals of unity, support, and coherence.

f Be sure that the point of view in your paragraph is consistent. For example, if you begin to write "How *I* got rid of mice" (first person), do not switch suddenly to "*You* must buy the right traps" (second person). Write your paragraph either from the first-person point of view (*I-we*) *or* from the second-person point of view (*you*). As noted at the beginning of this chapter, do not hesitate to use the second-person *you* point of view. A process paragraph in which you give instructions is one of the few situations in formal writing where the second-person *you* is acceptable.

g Be sure to use some transitions such as *first, next, also, then, after, now, during,* and *finally* so that your paper moves smoothly and clearly from one step in the process to the next.

h While working on your paper, refer to the checklist on the inside front cover to make sure you can answer *Yes* to the questions about unity, support, and coherence. Also, refer to the checklist when you edit the next-to-final draft of your paper for sentence-skills mistakes, including spelling.

■ Writing Assignment 2

For this assignment, you will be working with more general topics than those in Writing Assignment 1. You will find, in many cases, that you must invent your own steps in a particular process. You will also have to make decisions about how many steps to include and what order to place them in.

How to break a bad habit such as smoking, overeating, or excess drinking
How to improve a course you have taken
How to make someone you know happy
How to go about meeting people
How to discipline a child
How to improve the place where you work
How to show appreciation to others
How to make someone forgive you
How to con an instructor
How to make yourself depressed
How to get over a broken relationship
How to procrastinate
How to flirt

■ **Writing Assignment 3**

Everyone is an expert at something. Write a process paragraph on some skill that you can perform very well. The skill might be, for example, ''refereeing a game,'' ''fishing for perch,'' ''playing third base,'' ''putting up a tent,'' ''making an ice cream soda,'' ''becoming a long-distance runner,'' or ''fine-tuning a car engine.'' Write from the point of view that ''This is how _____ should be done.''

■ **Writing Assignment 4**

Write a process paragraph on how to succeed at a job interview. Here are some ways to research this topic.

■ Consult material that is available in your college placement center.

■ Look in your college bookstore or any paperback bookstore for ''how to'' books on getting a job; many such books will contain a chapter on doing well in a job interview.

■ In your library, look up relevant books under a subject heading such as *Career Planning, Vocational Guidance,* and *Occupations.*

■ Browse through the magazine rack in your library or the magazine section of any newspaper store; articles on this common topic frequently appear in popular magazines.

■ Look in a recent *Readers' Guide to Periodical Literature* in your library under the heading *Interviewing* for articles on the topic. (See ''Using the Library'' on pages 291–307 for information on the *Readers' Guide.*)

Condense the material you have found into three, four, or five basic steps. Choose the steps, tips, and pointers that seem most important to you or that recur most often in the material. Remember that you are reading only to obtain background information for your paper. Do not copy material or repeat someone else's words or phrases in your own work.

■ **Writing Assignment 5**

Option 1: You have a part-time job helping out in a day-care center. The director, who is pleased with your work and wants to give you more responsibility, has assigned you to be in charge of a group activity (for example, an exercise session, an alphabet lesson, or a valentine-making project). But before you actually begin the activity, the director wants to see a summary of how you would go about it. What advance preparation would be needed, and what exactly would you be doing throughout the time of the project? Write a paragraph explaining the steps you would follow in conducting the activity.

Option 2:   Alternatively, write an explanation you might give to one of the children on how to do a simple classroom task — serving juice and cookies, getting ready for nap time, watering a plant, putting toys or other classroom materials away, or any other task you choose. Explain each step of the task in a way that a child would understand.

EXAMINING
CAUSE AND EFFECT

What caused Pat to drop out of school? Why are soap operas so popular? Why does our football team do so poorly each year? How has retirement affected Dad? What effects does divorce have on children? Every day we ask questions similar to these and look for answers. We realize that many actions do not occur without causes, and we realize also that a given action can have a series of effects — for good or bad. By examining the causes or effects of an action, we seek to understand and explain things that happen in our lives.

In this section, you will be asked to do some detective work by examining the causes of something or the effects of something. First read the three paragraphs that follow and answer the questions about them. All three paragraphs support their opening points by explaining a series of causes or a series of effects.

PARAGRAPHS TO CONSIDER

New Puppy in the House

[1]Buying a new puppy can have drastic effects on a quiet household. [2]For one thing, the puppy keeps the entire family awake for at least two solid weeks. [3]Every night when the puppy is placed in its box, it begins to howl, yip, and whine. [4]Even after the lights go out and the house quiets down, the puppy continues to moan. [5]Since it is impossible to sleep while listening to a heartbroken, trembling "Woo-wooo," the family soon begins to suffer the effects of loss of sleep. [6]Everyone becomes hostile, short-tempered, depressed, and irritable. [7]A second effect is that the puppy tortures the family by destroying its material possessions. [8]Every day something different is damaged. [9]Family members find chewed belts and shoes, gnawed table legs, and leaking sofa cushions. [10]In addition, the puppy usually ruins the wall-to-wall carpeting and makes the house smell like a public restroom at a big-city bus station. [11]Worst of all, though, the puppy causes family arguments. [12]Parents argue with children about who is supposed to feed and walk the dog. [13]Children argue among themselves about whose turn it is to play with the puppy. [14]Everyone argues about whose idea it was to get the puppy in the first place. [15]These continual arguments, along with the effects of sleeplessness and the loss of valued possessions, seriously disrupt a household. [16]Only when the puppy gets a bit older will the house be peaceful again.

My Car Accident

[1]Several factors caused my recent car accident. [2]First of all, because a heavy snow and freezing rain had fallen the day before, the road that I was driving on was hazardous. [3]The road had been plowed but was dangerously icy in spots where dense clusters of trees kept the early morning sun from hitting the road. [4]Second, despite the slick patches, I was stupidly going along at about fifty miles an hour instead of driving more cautiously. [5]I have a daredevil streak in my nature and sometimes feel I want to become a stock-car racer after I finish school, rather than an accountant as my parents want me to be. [6]Another contributing factor to my accident was a dirty green Chevy van that suddenly pulled onto the road from a small intersecting street about fifty yards ahead of me. [7]The road was a sheet of ice at that point, but I was forced to apply my brake and also swing my car into the next lane. [8]Unfortunately, the fourth and final cause of my accident now presented itself. [9]The rear of my Honda Civic was heavy because I had a barbell set in the backseat. [10]I was selling the fairly new weight-lifting set to someone at school, since the weights had failed to build up my muscles immediately and I had gotten tired of practicing with them. [11]The result of all the weight in the rear was that after I passed the van, my car spun completely around on the slick road. [12]For a few horrifying, helpless moments, I was sliding down the highway backwards at fifty miles an hour, with no control whatsoever over the car. [13]Then, abruptly, I slid off the road, thumping into a high plowed snowbank. [14]I felt stunned for a moment but then also relieved. [15]I saw a telephone pole about six feet to the right of me and realized that my accident could have been really disastrous.

Why I Stopped Smoking

[1]For one thing, I realized that my cigarette smoke bothered others, particularly my wife and children, irritating their eyes and causing them to cough and sneeze. [2]Also, cigarettes are a messy habit. [3]Our house was littered with ashtrays piled high with butts, matchsticks, and ashes, and the children were always knocking them over. [4]Cigarettes are expensive, and I estimated that the carton a week that I was smoking cost me about $950 a year. [5]Another reason I stopped was that the message about cigarettes being harmful to health finally got through to me. [6]A heavy smoker I know from work is in Eagleville Hospital now with lung cancer. [7]Cigarettes were also inconvenient. [8]When I would smoke, I would have to drink something to wet down my dry throat, and that meant I had to keep going to the bathroom all the time. [9]I sometimes seemed to spend whole weekends doing nothing but smoking, drinking, and going to the bathroom. [10]Most of all, I resolved to stop smoking because I felt exploited. [11]I hated the thought of wealthy, greed-filled corporations making money off my sweat and blood. [12]The rich may keep getting richer, but--at least as regards to cigarettes--with no thanks to me.

■ Questions

About Unity

1. Which two sentences in "My Car Accident" are not on target in support of the opening idea and so should be omitted? (*Write the sentence numbers here.*) _____ _____

2. Which paragraph lacks a topic sentence?

_____

About Support

3. How many causes are given to support the opening idea in "My Car Accident"?

_____one   _____two   _____three   _____four

In "Why I Stopped Smoking"?

____one   ____two   ____three   ____four   ____five   ____six

4. How many effects of bringing a new puppy into the home are given in "New Puppy in the House"?

_____one   _____two   _____three   _____four

About Coherence

5. What are the five major transition words used in "Why I Stopped Smoking"?

a. _____   c. _____   e. _____

b. _____   d. _____

6. What words signal the most important effect in "New Puppy in the House"?

_____

Activity 1

Complete the following outline of "Why I Stopped Smoking." The effect is that the author stopped smoking; the causes are what make up the paragraph. Summarize each in a few words. The first cause and details are given for you as an example.

Point

There are a number of reasons why I stopped smoking.

1. Reason: <u>*Bothered others*</u>

   Details: <u>*Eye irritation, coughing, sneezing*</u>

2. Reason: _____

   Details: _____

3. Reason: _____

   Details: _____

4. Reason: _____

   Details: _____

5. Reason: _____

   Details: _____

6. Reason: _____

   Details: _____

Activity 2

This exercise will help you tell the difference between *reasons* that back up a point and the supporting *details* that go with each one of the reasons. The scrambled list below contains both reasons and supporting details. Complete the outline following the list by writing the reasons in the lettered blanks (a, b, c, d) and the appropriate supporting details in the numbered blanks (1, 2, 3). Arrange the reasons in what *you feel* is their order of importance. Also, summarize the reasons and details in a few words rather than writing them out completely.

Point

There are a number of reasons why people enjoy eating at Burger Village.

Reasons and Details

An order is ready no more than three minutes or so after it is placed.

A hostess is present in the dining room to help parents with children.

The workers wear clean uniforms, and their hands are clean.

There are french fries in two sizes.

The hostess hands out moistened cloths to wash the children after they eat.

The waiting line moves quickly.

Customers can order hamburgers or fish, chicken, or ham sandwiches.

The hostess helps with the children's coats and gets a highchair for the baby.

The place is clean.

There are several flavors of milk shakes and several kinds of soft drinks, as well as coffee and hot chocolate.

Someone is always sweeping the floor, collecting trays, and wiping off tables.

The kitchen area is all clean and polished stainless steel.

The service is fast and convenient.

The hostess gives the children small cups to drink from and special hats.

There are a variety of items on the menu.

Orders come packaged in bags or boxes for easy carrying out.

Outline

a. _____

   (1) _____

   (2) _____

   (3) _____

b. _____

   (1) _____

   (2) _____

   (3) _____

c. _____

   (1) _____

   (2) _____

   (3) _____

d. _____

   (1) _____

   (2) _____

   (3) _____

WRITING A CAUSE-EFFECT PARAGRAPH

■ Writing Assignment 1

Listed below are topic sentences and brief outlines for three cause or effect paragraphs. Choose one of them to develop into a paragraph.

Option 1

Topic sentence: There are several reasons why some high school graduates are unable to read.
a. Failure of parents (cause)
b. Failure of schools (cause)
c. Failure of students themselves (cause)

Option 2

Topic sentence: Attending college has changed my personality in positive ways.
a. More confident (effect)
b. More knowledgeable (effect)
c. More assertive (effect)

Option 3

Topic sentence: Living with roommates (*or* family) makes attending college difficult.
a. Late night hours (cause)
b. More temptations to cut class (cause)
c. More distractions from studying (cause)

How to Proceed

a Begin by prewriting. On separate paper, make a list of details that might go under each of the supporting points. Provide more details than you can actually use. Here, for example, are some of the details generated by the writer of ''New Puppy in the House'' while working on the paragraph:

Whines and moans
Arguments about walking dog
Arguments about feeding dog
Purchase collar, leash, food
Chewed belts and shoes
Arguments about playing with dog
Loss of sleep

Visits to vet
Short tempers
Accidents on carpet
Chewed cushions and tables

b Decide which details you will use to develop the paragraph. Also, number the details in the order in which you will present them. Here is how the writer of ''New Puppy in the House'' made decisions about details:

2 Whines and moans
6 Arguments about walking dog
6 Arguments about feeding dog
~~Purchase collar, leash, food~~
4 Chewed belts and shoes
6 Arguments about playing with dog
1 Loss of sleep
~~Visits to vet~~
3 Short tempers
5 Accidents on carpet
4 Chewed cushions and tables

Notice that the writer has put the same number in front of certain details that go together. For example, there is a ''4'' in front of ''Chewed belts and shoes'' and also in front of ''Chewed cushions and tables.''

c As you are working on your paper, keep checking your material to make sure it is unified, supported, and coherent.

d Finally, edit the next-to-final draft of your paper for sentence-skills mistakes, including spelling.

■ Writing Assignment 2

Below are ten topic sentences for a cause or an effect paper. In scratch outline form on separate paper, provide brief supporting points for five of the ten statements.

List the Causes

1. There are several reasons so many accidents occur on _____ (*name a local road, traffic circle, or intersection*).

2. _____ is (*or* is not) a good instructor (*or* employer), for several reasons.

3. _____ is a sport that cannot be appreciated on television.

4. _____ is the most difficult course I have ever taken.

5. For several reasons, many students live at home while going to school.

List the Effects

6. Watching too much TV can have a bad effect on students.

7. When I heard the news that _____, I was affected in various ways.

8. Conflicts between parents can have harmful effects on a child.

9. Breaking my bad habit of _____ has changed my life (*or* would change my life).

10. My fear of _____ has affected my everyday life.

Decide which of your outlines would be most promising to develop into a paragraph. Make sure that your causes or effects are logical ones that truly support the point in the topic sentence. Then follow the directions on ''How to Proceed'' in Writing Assignment 1.

■ Writing Assignment 3

Most of us criticize others readily, but we find it more difficult to give compliments. For this assignment, write a one-paragraph letter praising someone. The letter may be to a person you know (parent, relative, friend); to a public figure (actor, politician, leader, sports star, and so on); or to a company or organization (for example, the people who manufactured a product you own, a newspaper, a TV network, or a government agency).

To start, make a list of reasons why you admire the person or organization. Here are examples of reasons for praising an automobile manufacturer:

My car's dependability
Prompt action on a complaint
Well-thought-out design
Friendly dealer service

Here are reasons for admiring a parent:

Sacrifices you made
Patience with me
Your sense of humor
Your fairness
Your encouragement

Then follow the suggestions on ''How to Proceed'' in Writing Assignment 1.

■ Writing Assignment 4

Investigate the reasons behind a current news event. For example, you may want to investigate the causes of:

A labor strike

A military action by our or some other government

A Supreme Court decision or some other court decision

A murder or some other act of violence

A particular tax increase

A natural disaster, a traffic accident, a plane crash, or some other catastrophe

Research the reasons for the event by reading current newspapers (especially big-city dailies that are covering the story in detail), by reading weekly news magazines (such as *Time, Newsweek,* and *U.S. News & World Report*), by watching television news shows and specials, or by asking authorities on the subject (an instructor in law, history, or political science, for example).

Decide on the three or four major causes of the event, and write a paragraph detailing those causes.

■ Writing Assignment 5

Option 1: Assume that there has been an alarming increase in drug abuse among the students at the high school you attended. What might be the causes of this increase? Spend some time thinking about the several possible causes. Then, as a concerned member of the local community, write a letter to the high school guidance counselor explaining the reasons for the increased drug abuse. Your purpose in the letter is to provide helpful information that the counselor may be able to use in dealing with the problem.

Option 2: Your roommate has been complaining that it's impossible to succeed in Mr. X's class because the class is too stressful. You volunteer to attend the class and see for yourself. Afterward, you decide to write a letter to the instructor, calling attention to the stressful conditions in the class and suggesting concrete ways that he or she could deal with these conditions. Write this letter, dealing with the causes and effects of stress in the class.

COMPARING
OR
CONTRASTING

Comparison and contrast are two thought processes we constantly perform in everyday life. When we *compare* two things, we show how they are similar; when we *contrast* two things, we show how they are different. We might compare or contrast two brand-name products (for example, Levi's versus Wrangler jeans), two television shows, two cars, two instructors, two jobs, two friends, or two courses of action we could take in a given situation. The purpose of comparing or contrasting is to understand each of the two things more clearly and, at times, to make judgments about them.

In this section, you will be asked to write a paper of comparison or contrast. First, however, you must learn the two common methods of developing a comparison or contrast paragraph. Read the two paragraphs that follow and try to explain the difference in the two methods of development.

PARAGRAPHS TO CONSIDER

My Senior Prom

[1]My senior prom was nothing like what I had expected it to be. [2]From the start of my senior year, I had pictured getting dressed in a blue gown that my aunt would make and that would cost two hundred dollars in any store. [3]No one else would have a gown as attractive as mine. [4]I imagined my boyfriend coming to the door with a lovely blue corsage, and I pictured myself happily inhaling its perfume all evening long. [5]I saw us setting off for the evening in his brother's 1992 Lincoln Continental. [6]We would make a flourish as we swept in and out of a series of parties before the prom. [7]Our evening would be capped by a delicious shrimp dinner at the prom and by dancing closely together into the early morning hours. [8]The prom was held on May 16, 1992, at the Pony Club on the Black Horse Pike. [9]However, because of sickness in

her family, my aunt had no time to finish my gown and I had to buy an ugly pink one at the last minute for eighty dollars. [10] My corsage of yellow carnations looked terrible on my pink gown, and I do not remember its having any scent. [11] My boyfriend's brother was out of town, and I stepped outside to the stripped-down Chevy that he used at the races on weekends. [12] We went to one party where I drank a lot of wine that made me sleepy and upset my stomach. [13] After we arrived at the prom, I did not have much more to eat than a roll and some celery sticks. [14] Worst of all, we left early without dancing because my boyfriend and I had had a fight several days before and at the time we did not really want to be with each other.

Electronic versus Electric Typewriters

[1] The electronic typewriter I use in my job as a secretary is a great improvement over the electric model I use at home. [2] First of all, the electronic Brother C-60 is more convenient to use than my ten-year-old Royal. [3] The electronic typewriter has a one-step correction key. [4] If I make mistakes on a page, all I have to do is hit the correction key. [5] The typewriter instantly removes a mistyped letter, a word, or even an entire line. [6] In contrast, my electric typewriter has no correction key. [7] If I spot an error, I have to paint it with correction fluid and wait for the fluid to dry before correcting the mistake. [8] This is very time-consuming, especially when there are a lot of corrections to be made. [9] Second, the electronic typewriter is fast and does not jam. [10] When I type at top speed, the machine remembers all the letters I hit and keeps printing them even if I go a bit ahead of the machine's typing speed. [11] On the other hand, the keys of my electric will sometimes jam together if I type too fast. [12] Then I have to reach in and separate the metal bars before I can type again. [13] Finally, the electronic typewriter is a much quieter machine. [14] The sound of typing on an electronic is a gentle, low-pitched series of taps. [15] The electric, however, makes a loud, clattery noise as I type. [16] The sound seems to echo through the room and continue in my head even after I stop typing. [17] The convenience, speed, and quiet of the electronic Brother I use at work has spoiled me for my old-fashioned Royal at home. [18] An electronic typewriter simply outclasses everything that has come before.

Complete this comment: The difference in the methods of contrast in the two paragraphs is _____

_____

_____

_____ .

Compare your answer with the following explanation of the two methods of development used in comparison or contrast paragraphs.

METHODS OF DEVELOPMENT

There are two common methods of development in a comparison or contrast paper. Details can be presented in a *one-side-at-a-time* format or in a *point-by-point* format. Each format is illustrated below.

One Side at a Time

Look at the outline of "My Senior Prom":

    a. Expectations (first half of paper)
       (1) Gown (expensive, blue)
       (2) Corsage (lovely, fragrant blue)
       (3) Car (Lincoln Continental)
       (4) Partying (much)
       (5) Dinner (shrimp)
       (6) Dancing (all night)
    b. Reality (second half of paper)
       (1) Gown (cheap, pink)
       (2) Corsage (wrong color, no scent)
       (3) Car (stripped-down Chevy)
       (4) Partying (little)
       (5) Dinner (roll and celery sticks)
       (6) Dancing (didn't because of quarrel)

The first half of the paragraph explains fully one side of the contrast; the second half of the paragraph deals entirely with the other side. In using this method, be sure to follow the same order of points of contrast (or comparison) for each side.

Point by Point

Now look at the outline of "Electronic versus Electric Typewriters":

    a. Convenience
       (1) Electronic
       (2) Electric
    b. Speed
       (1) Electronic
       (2) Electric
    c. Noise level
       (1) Electronic
       (2) Electric

The outline shows how the two kinds of typewriters are contrasted point by point. First, the writer compares the convenience of the electronic, which has a correction key, with the lack of a correction key on the electric; next, the writer compares the speed of the electronic with the slowness of the electric; finally, the writer compares the quietness of the electronic with the noise of the electric.

When you begin a comparison or contrast paper, you should decide right away whether you are going to use the one-side-at-a-time format or the point-by-point format. An outline is an essential step in helping you decide which format will be more workable for your topic.

Activity 1

Complete the partial outlines provided for the two paragraphs that follow.

1.   How My Parents' Divorce Changed Me

In the three years since my parents' divorce, I have changed from a spoiled brat to a reasonably normal college student. Before the divorce, I expected my mother to wait on me. She did my laundry, cooked and cleaned up after meals, and even straightened up my room. My only response was to complain if the meat was too well done or if the sweater I wanted to wear was not clean. In addition, I expected money for anything I wanted. Whether it was an expensive bowling ball or a new school jacket, I expected Mom to hand over the money. If she refused, I would get it from Dad. However, he left when I was fifteen, and things changed. When Mom got a full-time job to support us, I was the one with the free time to do housework. Now, I did the laundry, started the dinner, and cleaned not only my own room but the whole house. Fortunately, Mom was tolerant. She did not even complain when my first laundry project left us with streaky blue underwear. Also, I no longer asked her for money, since I knew there was none to spare. Instead, I got a part-time job on weekends to earn my own spending money. Today I have my own car that I am paying for, and I am putting myself through college. Things have been hard sometimes, but I am glad not to be that spoiled kid any more.

Topic sentence: In the three years since my parents' divorce, I have changed from a spoiled brat to a reasonably normal college student.

a. Before the divorce

    (1) _____

    (2) _____

b. After the divorce

    (1) _____

    (2) he/she got her own job to earn their own spending money.

Complete the following statement: Paragraph 1 uses a _____ method of development.

2. Good and Bad Horror Movies

A good horror movie is easily distinguished from a bad one. A good horror movie, first of all, has both male and female victims. Both sexes suffer terrible fates at the hands of monsters and maniacs. Therefore, everyone in the audience has a chance to identify with the victim. Bad horror movies, on the other hand, tend to concentrate on women, especially half-dressed ones. These movies are obviously prejudiced against half the human race. Second, a good horror movie inspires compassion for its characters. For example, the audience will feel sympathy for the Wolfman's victims and also for the Wolfman, who is shown to be a sad victim of fate. In contrast, a bad horror movie encourages feelings of aggression and violence in viewers. For instance, in the <u>Halloween</u> films, the murder scenes use the murderer's point of view. The effect is that the audience stalks the victims along with the killer and feels the same thrill he does. Finally, every good horror movie has a sense of humor. In <u>Dracula</u>, the Count says meaningfully at dinner, "I don't drink wine" as he stares at a young woman's juicy neck. Humor provides relief from the horror and makes the characters more human. A bad horror movie, though, is humorless and boring. One murder is piled on top of another, and the characters are just cardboard figures. Bad horror movies may provide cheap thrills, but the good ones touch our emotions and live forever.

Topic sentence: A good horror movie is easily distinguished from a bad one.

a. Kinds of victims

    (1) _____

    (2) _____

b. Effect on audience

    (1) _____

    (2) _____

c. Tone

    (1) _____

    (2) _____

Complete the following statement: Paragraph 2 uses a _____ method of development.

Activity 2

Write the number 1 beside the point that all the other scrambled sentences in the list below support. Then number the rest of the sentences in a logical order. To do this, you will have to decide whether the sentences should be arranged according to a *one-side-at-a-time* order or *point-by-point* order.

A Change in Attitude

6 — Eventually I could not find a dress or pair of slacks in my wardrobe that I could wear while still continuing to breathe.

10 — In the evening when I got hungry, I made myself tomato, lettuce, and onion salad with vinegar dressing.

14 — I have kicked my chocolate habit, saved my clothes wardrobe, and enabled myself to breathe again.

3 — I also could seldom resist driving over to the nearby Snack Shack in the evening to get a large chocolate shake.

9 — For dessert at lunch I had an orange or other fruit.

2 — I gobbled chocolate bars during breaks at work, had chocolate cake for dessert at lunch, and ate chocolate-covered butter creams in the evening.

7 — At this point I began using willpower to control my urge for chocolate.

12 — As a result, the twelve pounds that I didn't want dropped off steadily and eventually disappeared.

4 — When the pounds began to multiply steadily, I tried to console myself.

13 — I have been able to lose weight by changing my attitude about chocolate.

5 — I said, "Well, that's only three pounds; I can lose that next week."

11 — Also, I made myself go and step on the bathroom scale whenever I got the urge to drive over to the Snack Shack.

1 — There was a time when I made chocolate a big part of my daily diet.

8 — Instead of eating chocolate bars on my break, I munched celery and carrots.

Complete the following statement: The sentences can be organized using _____ order.

ADDITIONAL PARAGRAPHS TO CONSIDER

Read these additional paragraphs of comparison or contrast and then answer the questions that follow.

My Broken Dream

¹When I became a police officer in my town, the job was not as I had dreamed it would be. ²I began to dream about being a police officer at about age ten. ³I could picture myself wearing a handsome blue uniform and having an impressive-looking badge on my chest. ⁴I could also picture myself driving a powerful patrol car through town and seeing everyone stare at me with envy. ⁵But most of all, I dreamed of wearing a gun and using all the equipment that "TV cops" use. ⁶I just knew everyone would be proud of me. ⁷I could almost hear the guys on the block saying, "Boy, Steve made it big. ⁸Did you hear he's a cop?" ⁹I dreamed of leading an exciting life, solving big crimes, and meeting lots of people. ¹⁰I just knew that if I became a cop everyone in town would look up to me. ¹¹However, when I actually did become a police officer, I soon found out that it was not as I had dreamed it would be. ¹²My first disappointment came when I was sworn in and handed a well-used, baggy uniform. ¹³My disappointment continued when I was given a badge that looked like something pulled out of a Cracker Jack box. ¹⁴I was assigned a beat-up old junker and told that it would be my patrol car. ¹⁵It had a striking resemblance to a car that had lost a demolition derby at a stock-car raceway. ¹⁶Disappointment seemed to continue. ¹⁷I soon found out that I was not the envy of all my friends. ¹⁸When I drove through town, they acted as if they had not seen me. ¹⁹I was told I was crazy doing this kind of job by people I thought would look up to me. ²⁰My job was not as exciting as I had dreamed it would be either. ²¹Instead of solving robberies and murders every day, I found that I spent a great deal of time comforting a local resident because a neighborhood dog had watered his favorite bush.

Two Views on Toys

¹There is a vast difference between children and adults where presents are concerned. ²First, there is the matter of taste. ³Adults pride themselves on taste, while children ignore the matter of taste in favor of things that are fun. ⁴Adults, especially grandparents, pick out tasteful toys that go unused, while children love the cheap playthings advertised on television. ⁵Then, of course, there is the matter of money. ⁶The new games on the market today are a case in point. ⁷Have you ever tried to lure a child away from some expensive game in order to get him or her to play with an old-fashioned game or toy? ⁸Finally, there is a difference between an adult's and a child's idea of what is educational. ⁹Adults, filled with memories of their own childhoods, tend to be fond of the written word. ¹⁰Today's children, on the other hand, concentrate on anything electronic. ¹¹These things mean much more to them

than to adults. ¹²Next holiday season, examine the toys that adults choose for children. ¹³Then look at the toys the children prefer. ¹⁴You will see the difference.

Mike and Helen

¹Like his wife Helen, Mike has a good sense of humor. ²Also, they are both short, dark-haired, and slightly pudgy. ³Both Mike and Helen can be charming when they want to be, and they seem to handle small crises in a calm, cool way. ⁴A problem such as an overflowing washer, a stalled car, or a sick child is not a cause for panic; they seem to take such events in stride. ⁵Unlike Helen, though, Mike tends to be disorganized. ⁶He is late for appointments and unable to keep important documents--bank records, receipts, and insurance papers--where he can find them. ⁷And unlike Helen, Mike tends to hold a grudge. ⁸He is slow to forget a cruel remark, a careless joke, or an unfriendly slight. ⁹Also, Mike enjoys swimming, camping, and tennis, unlike Helen, who is an indoors type.

■ Questions

About Unity

1. Which paragraph lacks a topic sentence?

2. Which paragraph has a topic sentence that is too broad?

About Support

3. Which paragraph contains virtually no specific details?

4. Which paragraph do you feel offers the most effective details?

About Coherence

5. What method of development (one side at a time or point by point) is used in "My Broken Dream"?

In "Two Views on Toys"?

6. Which paragraph offers specific details, but lacks a clear, consistent method of development?

WRITING A COMPARISON OR CONTRAST PARAGRAPH

■ Writing Assignment 1

Below are topic sentences and supporting points for three contrast paragraphs. Choose one of the three to develop into a paragraph.

Option 1

Topic sentence: I abused my body when I was twenty, but I treat myself much differently at the age of thirty.

a.   At twenty, I was a heavy smoker. . . .
     Now, instead of smoking, I . . .
b.   At twenty, I had a highly irregular diet. . . .
     Today, on the other hand, I eat . . .
c.   Finally, at twenty, I never exercised. . . .
     Today, I work out regularly. . . .

Option 2

Topic sentence: Dating is still much easier for boys than it is for girls.

a.   A boy can take the initial step of asking for a date without seeming too aggressive. . . .
     In contrast, a girl . . .
b.   The boy is usually in charge of deciding where to go and what to do. . . .
     The girl, on the other hand, has to live with the choices. . . .
c.   At the end of the night, the boy knows the moves he is planning to make, if any. . . .
     But the girl waits nervously to see what will happen. . . .

Option 3

Topic sentence: My sociology instructor teaches a class quite differently from my psychology instructor.

a.   For one thing, Ms. _____ demands many hours of homework each week. . . .

     In contrast, Mr. _____ does not believe in much work outside class. . . .

b. In addition, Ms. _____ gives difficult tests—with no study aids. . . .

Mr. _____'s tests, on the other hand, are easy. . . .

c. Finally, Ms. _____ keeps every class strictly on the subject of the day. . . .

But Mr. _____ will wander off onto any topic that snags his interest. . . .

How to Proceed

a Begin by prewriting. To develop some supporting details for the paragraph, freewrite for five minutes on the topic sentence you have chosen.

b Then add to the material you have written by asking yourself questions. If you were writing about how you treat your body differently at thirty, for example, you might ask yourself:

How many cigarettes did I smoke at twenty?

What have I substituted for cigarettes?

What kinds of food did I eat?

How regularly did I eat?

What foods do I eat today?

Why didn't I exercise at twenty?

What made me decide to start exercising?

What kind of exercises do I do?

How often do I exercise?

Write down whatever answers occur to you for these and other questions. As with the freewriting, do not worry at this stage about writing correctly. Instead, concentrate on getting down all the information you can think of that supports each point.

c Now go through all the material you have accumulated. Perhaps some of the details you have written down may help you think of even better details that would fit. If so, write them down.

d As you work on the drafts of your paper, use words such as *in contrast, but, on the other hand,* and *however* to tie together your material.

e Be sure to edit the next-to-final draft of your paper for sentence-skills mistakes, including spelling.

■ Writing Assignment 2

Write a comparison or contrast paragraph on one of the twenty topics below.

Two holidays	Two jobs
Two instructors	Two characters in the same
Two children	movie or TV show
Two kinds of eaters	Two commercials
Two drivers	Two methods of studying
Two coworkers	Two cartoon strips
Two members of a team	Two cars
(or two teams)	Two friends
Two singers or groups	Two crises
Two animals	Two employees
Two parties	Two magazines

How to Proceed

a You must begin by making two decisions: (1) what your topic will be and (2) whether you are going to do a comparison or a contrast paper. Many times, students choose to do essays centered on the differences between two things. For example, you might write about how a math instructor you have in college differs from a math teacher you had in high school. You might discuss important differences between two coworkers or between two of your friends. You might contrast a factory job you had packing vegetables with a white-collar job you had as a salesperson in a clothing store.

b After you choose a tentative topic, write a simple topic sentence expressing it. Then see what kind of support you can generate for that topic. For instance, if you plan to contrast two cars, see if you can think of and jot down three distinct ways they differ. In other words, prepare a scratch outline. An outline is an excellent prewriting technique to use when doing any paragraph; it is almost indispensable when planning a comparison or contrast paragraph. For a model, look back at the outlines given on pages 26–31.

Keep in mind that this planning stage is probably the *single most important phase* of work you will do on your paper. Without clear planning, you are not likely to write an effective paragraph.

c After you have decided on a topic and the main lines of support, you must decide whether to use a one-side-at-a-time or a point-by-point method of development. Both methods are explained and illustrated in this chapter.

d Now, freewrite for ten minutes on the topic you have chosen. Do not worry about punctuation, spelling, or other matters relating to correct form. Just get as many details as you can onto the page. You want a base of raw material that you can add to and select from as you now work on the first draft of your paper. After you do a first draft, try to put it aside for a day or at least several hours. You will then be ready to return with a fresh perspective on the material and build upon what you have already done.

e Use transition words like *first, in addition, also, in contrast, another difference, on the other hand, but, however,* and *most important* to link points together.

f As you continue working on your paper, refer to the checklist on the inside front cover. Make sure that you can answer *Yes* to the questions about unity, support, and coherence.

g Finally, use the checklist on the inside front cover to edit the next-to-final draft of your paper for sentence-skills mistakes, including spelling.

■ Writing Assignment 3

Write a contrast paragraph on one of the fifteen topics below.

Neighborhood stores versus a shopping mall
Driving on an expressway versus driving on country roads
People versus *Us* (or any other two popular magazines)
Camping in a tent versus camping in a recreational vehicle
Working parents versus stay-at-home parents
Last year's fashions versus this year's
Used car versus new car
Records versus tapes
PG-rated movies versus R-rated movies
News in a newspaper versus news on television
Yesterday's toys versus today's
Fresh food versus canned or frozen food
Winning locker room after a game versus losing one
Ad on television versus ad (for the same product) in a magazine
Amateur sport versus professional sport

Follow the directions on ''How to Proceed'' given in Writing Assignment 2.

■ Writing Assignment 4

Write a summary of a television show in which two contrasting points of view are presented. The show might be *Phil Donahue* or any other program which presents two different viewpoints on a subject. As you listen, take notes. You may want to organize your notes in two columns so that you can compare viewpoints at a glance.

In your topic sentence, include the date and subject of the show. For example:

Two opposing views on gay teachers were presented on the *Phil Donahue* show on December 14, 1992.

Limit the support you will cover in your paragraph to three or four major areas. After you have generated your support, decide which method of organization you will use and prepare a brief outline of the paragraph. Be sure to use transition words and to edit the next-to-final draft carefully.

■ Writing Assignment 5

You are living in an apartment building in which new tenants are making life unpleasant for you. Write a letter of complaint to your landlord comparing and contrasting life before and after the tenants arrived. You might want to focus on one or more of the following:

Noise level
Trash
Safety hazards
Parking situation

DEFINING
A TERM

In talking with other people, we at times offer informal definitions to explain just what we mean by particular terms. Suppose, for example, we say to a friend, ''Ted is an anxious person.'' We might then expand on our idea of ''anxious'' by saying, ''He's always worrying about the future. Yesterday he was talking about how many bills he'll probably have this year. Then he was worrying about what he would ever do if he got laid off.'' In a written definition, we make clear in a more complete and formal way our own personal understanding of a term. Such a definition typically starts with one meaning of a term. The meaning is then illustrated with a series of examples or a story.

In this section, you will be asked to write a paragraph in which you define a term. The three student papers below are all examples of definition paragraphs. Read them and then answer the questions that follow.

PARAGRAPHS TO CONSIDER

Luck

[1]Luck is putting $1.75 into a vending machine and getting the money back with your snacks. [2]It is a teacher's decision to give a retest on a test where you first scored thirty. [3]Luck refers to moments of good fortune that happen in everyday life. [4]It is not going to the dentist for two years and then going and finding out that you do not have any cavities. [5]It is calling up a plumber to fix a leak on a day when the plumber has no other work to do. [6]Luck is finding a used car for sale at a good price at exactly the time when your car rolls its last mile. [7]It is driving into a traffic bottleneck and choosing the lane that winds up moving most rapidly. [8]Luck is being late for work on a day when your boss arrives later than you do. [9]It is having a new checkout aisle at the supermarket open up just as your cart arrives. [10]The best kind of luck is winning a new color TV set on a chance for which you paid only a quarter.

Disillusionment

¹Disillusionment is the feeling of having one of our most cherished beliefs stolen from us. ²I learned about disillusionment firsthand the day Mr. Keller, our eighth-grade teacher, handed out the grades on our class biology projects. ³I had worked hard to assemble what I thought was the best insect collection any school had ever seen. ⁴For weeks, I had set up homemade traps around our house, in the woods, and in vacant lots. ⁵At night, I would stretch a white sheet between two trees, shine a lantern on it, and collect the night-flying insects that gathered there. ⁶With my own money, I had bought killing jars, insect pins, gummed labels, and display boxes. ⁷I carefully arranged related insects together, with labels listing each scientific name and the place and date of capture. ⁸Slowly and painfully, I wrote and typed the report that accompanied my project at the school science fair. ⁹In contrast, my friend Eddie did almost nothing for his project. ¹⁰He had his father, a doctor, build an impressive maze complete with live rats and a sign that read, ''You are the trainer.'' ¹¹A person could lift a little plastic door, send a rat running through the maze, and then hit a button to release a pellet of rat food as a reward. ¹²This exhibit turned out to be the most popular one at the fair. ¹³I felt sure that our teacher would know that Eddie could not have built it, and I was certain that my hard work would be recognized and rewarded. ¹⁴Then the grades were finally handed out, and I was crushed. ¹⁵Eddie had gotten an A plus, but my grade was a B. ¹⁶I suddenly realized that honesty and hard work don't always pay off in the end. ¹⁷The idea that life is not fair, that sometimes it pays to cheat, hit me with such force that I felt sick. ¹⁸I will never forget that moment.

A Mickey Mouse Course

¹A Mickey Mouse course is any college course that is so easy that even Mickey or Minnie Mouse could achieve an A grade. ²A student who is taking a heavy schedule, or who does not want four or five especially difficult courses, will try to sandwich in a Mickey Mouse course. ³A student can find out about such a course by consulting other students, since word of a genuine Mickey Mouse course spreads like wildfire. ⁴Or a student can study the college master schedule for telltale course titles like The Art of Pressing Wild Flowers, History of the Comic Book, or Watching Television Creatively. ⁵In an advanced course such as microbiology, though, a student had better be prepared to spend a good deal of time during the semester on that course. ⁶Students in a Mickey Mouse course can attend the classes while half-asleep, hung-over, or wearing stereo earphones or a blindfold; they will still pass. ⁷The course exams (if there are any) would not challenge a five-year-old. ⁸The course lectures usually consist of information that anyone with

common sense knows anyway. ^{9}Attendance may be required, but participation or involvement in the class is not. ^{10}The main requirement for passing is that a student's body is there, warming a seat in the classroom. ^{11}There are no difficult labs or special projects, and term papers are never mentioned. ^{12}Once safely registered for such a course, all the students have to do is sit back and watch the credits accumulate on their transcripts.

■ **Questions**

About Unity

1. Which paragraph places its topic sentence within the paragraph rather than, more appropriately, at the beginning?

   _____

2. Which sentence in ''A Mickey Mouse Course'' should be omitted in the interest of paragraph unity? (*Write the sentence number here.*) _____

About Support

3. Which two paragraphs develop their definitions through a series of short examples?

   _____

   _____

4. Which paragraph develops its definition through a single extended example?

   _____

About Coherence

5. Which paragraph uses emphatic order, saving its best detail for last?

   _____

6. Which paragraph uses time order to organize its details?

   _____

WRITING A DEFINITION PARAGRAPH

■ **Writing Assignment 1**

Following are a topic sentence and three supporting points for a paragraph that defines the term *TV addict*. Using separate paper, plan out and write the secondary supporting details and closing sentence needed to complete the paragraph. Refer to the suggestions on ''How to Proceed'' that follow.

Topic sentence: Television addicts are people who will watch all the programs they can, for as long as they can, rather than do anything else.

   a.   TV addicts, first of all, will watch anything on the tube, no matter how bad it is. . . .
   b.   In addition, addicts watch TV more hours than normal people do. . . .
   c.   Finally, addicts feel that TV is more important than any other activities or events that might be going on. . . .

How to Proceed

a   Begin by prewriting. Prepare examples for each of the three qualities of a TV addict. For each quality, you should have at least two or three sentences that provide either an extended example or shorter examples of this quality in action.

b   To generate these details, ask yourself the following questions:

What are some examples of terrible shows that I (or people I know) watch just because the television is turned on?
What are some examples of how frequently I (or people I know) watch TV?
What are some other activities or events that I (or people I know) give up in order to watch TV?

Write down quickly whatever answers occur to you. Do not worry about writing correct sentences; just concentrate on getting down all the details about television addicts that you can think of.

c   Draw from and add to this material as you work on the paragraph. Make sure that your paragraph is unified, supported, and coherent.

d   Finally, edit the next-to-final draft of your paper for sentence-skills mistakes, including spelling.

■ **Writing Assignment 2**

Write an essay that defines one of the following terms. Each term refers to a certain kind of person.

Bigmouth	Clown	Good example
Charmer	Jellyfish	Hypocrite
Loser	Leader	Perfectionist
Lazybones	Nerd	Pack rat
Kibitzer	Good neighbor	Hard worker
Con artist	Optimist	Apple-polisher
Fair-weather friend	Pessimist	Fusspot
Team player		

How to Proceed

a To write a topic sentence for your definition paragraph, your first step should be to *place the term in a class or category.* Then *describe what you feel are the special features that distinguish your term from all the other members of its class.*

In the sample topic sentences below, underline the class, or category, that the term belongs to and double-underline the distinguishing details of that class, or category. One is done for you as an example.

A klutz is a <u>person</u> who <u>stumbles through life.</u>

A worrywart is a person who sees danger everywhere.

The class clown is a student who gets attention in the wrong way.

A clotheshorse is a person who needs new clothes to be happy.

b In order to develop your definition, use one of the following methods:

Examples. Give several examples that support your topic sentence.

Extended example. Use one longer example to support your topic sentence.

Contrast. Support your topic sentence by showing what your term is *not.* For instance, you may want to define a ''fair-weather friend'' by contrasting his or her actions with those of a true friend.

c Once you have created a topic sentence and decided how to develop your paragraph, write a scratch outline. This step is especially important if you are using a contrast method of development.

d Be sure you touch the four bases of unity, support, coherence, and sentence skills in your writing.

■ Writing Assignment 3

Write an essay that defines one of the abstract terms below.

Persistence	Responsibility	Fear
Rebellion	Insecurity	Arrogance
Sense of humor	Assertiveness	Conscience
Escape	Jealousy	Class
Danger	Nostalgia	Innocence
Curiosity	Gentleness	Freedom
Common sense	Depression	Violence
Family	Obsession	Shyness
Practicality	Self-control	

As a guide in writing your paper, use the suggestions on ''How to Proceed'' in Writing Assignment 2. Remember to place your term in a class, or category, and to describe what *you* feel are the distinguishing features of that term. Three examples follow.

Laziness is a quality that doesn't deserve its bad reputation.

Jealousy is the emotion that hurts the most.

Persistence is the quality of not giving up even during rough times.

■ Writing Assignment 4

Since it affects almost all of us, *stress* is an important word to define. In order to write a definition paragraph on stress, do some research in popular magazines, paperbacks, and textbooks.

How to Proceed

a   Look up *stress* in the index of any psychology or sociology textbook.

b   Using a recent issue of the *Readers' Guide to Periodical Literature* in your library, locate recent magazine articles listed under the heading *Stress*.

c   Also in your library, look up books in the card catalog under the subject heading *Stress*. You may also want to check your public library, which will probably carry more popular paperbacks dealing with stress than your college library.

d   Check your college bookstore, or any paperback bookstore, for recent books about stress; many of these will be in the category of ''self-help'' books.

e You should organize this definition paragraph in one of the ways listed below:

- Use a series of examples (see page 127) of stress.
- Use division and classification (see page 172) to describe several types of stress.
- Use narrative (see page 191); for example, create a hypothetical person and show how stress affects this person in a typical morning or day.

Hints: Do not simply write a series of general, abstract sentences that repeat and reword your definition. If you concentrate on providing specific support, you will avoid the common trap of getting lost in a maze of generalities.

Make sure your paper is firmly anchored on the bases of unity, support, coherence, and good sentence skills. Edit the next-to-final draft of the paragraph carefully.

■ Writing Assignment 5

Option 1: At the place where you work, one employee has just quit, creating a new job opening. Since you have been working there for a while, your boss has asked you to write a job description of the position. That description, which is really a detailed definition of the job, will be sent to employment services. These services will be responsible for interviewing candidates. Choose any position you know about, and write a job description for it. First give the purpose of the job, and then list its duties and responsibilities. Finally, give the qualifications for the position.

Option 2: Alternatively, imagine that a new worker has been hired, and your boss has asked you to explain ''team spirit'' to him or her. The purpose of your explanation will be to give the newcomer an idea of the teamwork that is expected in this workplace. Write a paragraph that defines in detail what your boss means by *team spirit*. Use examples or one extended example to illustrate your general statements.

DIVIDING
AND
CLASSIFYING

If you were doing the laundry, you would probably begin by separating the clothing into piles. You might put all the whites in one pile and all the colors in another. You might put all cottons in one pile, polyesters in another, silks in a third, and so on. Or you might divide and classify the laundry not according to color or fabrics but on the basis of use. You might put bath towels in one pile, bed sheets in another, personal garments in a third, and so on. Sorting clothes in various ways is just one small example of how we spend a great deal of time organizing our environment in one manner or another.

In this section, you will be asked to write a paragraph in which you divide or classify a subject according to a single principle. To prepare for this assignment, first read the division and classification paragraphs below and then work through the questions and the activity that follow.

PARAGRAPHS TO CONSIDER

Automobile Drivers

[1]One type of automobile driver is the slowpoke. [2]A woman who is a slowpoke, for instance, will drive forty miles per hour in a fifty-five-mile zone. [3]She will slow down and start signaling for a left-hand turn three blocks before making it. [4]Or her car will slow down while she is in avid conversation with other people in the car, or while she puzzles over street signs to get her bearings, or while she looks at displays and sale signs in shop windows, or as she struggles to open the wrapping of her Burger King Whopper. [5]A second type is the high-speed driver. [6]A man who is a high-speed driver, for

example, will limit his speed only when he suspects that the state police or radar traps are nearby. [7]The state police must develop a system to ensure that they begin to catch this kind of driver. [8]He typically speeds past cars on the left and right sides, weaving in and around them sharply, and he closely tailgates a car that holds him up until it shifts to another lane. [9]He races to get through yellow or just-red lights at highway and city intersections, and he speeds down city streets, oblivious to the possibility that children may run out from between parked cars or that people may open a car door. [10]The final type is the sensible-speed driver who, road conditions being normal, maintains the posted speed limits and drives at a consistent and moderate rate. [11]If these drivers do change their rate, they do so because they are driving defensively. [12]They are speeding up to pass the driver in front, who is creeping along to look at the pumpkins on display at a roadside stand. [13]Or they are slowing down to allow the speed demon who has tried passing to get back in lane and out of the path of an oncoming truck.

Studying for a Test

[1]The time a student spends studying for a test can be divided into three distinct phases. [2]Phase 1, often called the "no problem" phase, runs from the day the test is announced to approximately forty-eight hours before the dreaded exam is passed out. [3]During phase 1, the student is carefree, smiling, and kind to helpless animals and small children. [4]When asked by classmates if he or she has studied for the test yet, the reply will be an assured "No problem." [5]During phase 1, no actual studying takes place. [6]Phase 2 is entered two days before the test. [7]For example, if the test is scheduled for 9 A.M. Friday, phase 2 begins at 9 A.M. Wednesday. [8]During phase 2, again, no actual studying takes place. [9]Phase 3, the final phase, is entered twelve hours before "zero hour." [10]This is the acute phase, characterized by sweaty palms, nervous twitches, and confused mental patterns. [11]For a test at nine o'clock on Friday morning, a student begins exhibiting these symptoms at approximately nine o'clock on Thursday night. [12]Phase 3 is also termed the "shock" phase, since the student is shocked to discover the imminent nature of the exam and the amount of material to be studied. [13]During this phase, the student will probably be unable to sleep and will mumble meaningless phrases like "$a^2 + c^2$." [14]This phase will not end until the exam is over. [15]If the cram session has worked, the student will fall gratefully asleep. [16]On waking up, he or she will be ready to go through the whole cycle again with the next test.

The Dangers of Tools

[1]Tools can be divided into three categories according to how badly people can injure themselves with them. [2]The first group of tools causes dark-purple bruises to appear on the user's feet, fingers, or arms. [3]Hammers are famous for this, as millions of cartoons and comic strips have shown. [4]Mallets and crowbars, too, can go a bit off target and thud onto a bit of exposed flesh. [5]But first-class bruises can also be caused by clamps, pliers, vise grips, and wrenches. [6]In a split second, any one of these tools can lash out and badly squeeze a stray finger. [7]Later, the victim will see blue-black blood forming under the fingernail. [8]The second type of tool usually attacks by cutting or tearing. [9]Saws seem to enjoy cutting through human flesh as well as wood. [10]Keeping a hand too close to the saw, or using a pair of knees as a sawhorse, will help the saw satisfy its urge. [11]Planes, chisels, and screwdrivers also cut into skin. [12]And the utility knife--the kind with a razor blade projecting from a metal handle--probably cuts more people than it does linoleum. [13]The most dangerous tools, however, are the mutilators, the ones that send people directly to the emergency room. [14]People were definitely not made to handle monster tools like chain saws, table saws, power drills, and power hammers. [15]Newspapers are filled with stories of shocking accidents people have had with power tools. [16]In summary, if people are not careful, tools can definitely be hazardous to their health.

■ Questions

About Unity

1. Which paragraph lacks a topic sentence? _____

2. Which sentence in ''Automobile Drivers'' should be omitted in the interest of paragraph unity? (*Write the sentence number here.*) _____

About Support

3. Which of the three phases in ''Studying for a Test'' lacks specific details?
   <u>phases 2</u>

4. After which sentence in ''The Dangers of Tools'' are supporting details needed? _____

About Coherence

5. Which paragraph uses time order to organize its details?

   _____

6. Which paragraph uses emphatic order to organize its details?

   _____

7. What words in the emphatic-order paragraph signal the most important detail?

   _____

Activity

This activity will sharpen your sense of the classifying process. In each of the following ten groups, cross out the one item that has not been classified on the same basis as the other three. Also, indicate in the space provided the single principle of classification used for the three items. Note the examples.

Examples   Water
   a.   Cold
   b.   ~~Lake~~
   c.   Hot
   d.   Lukewarm
   Unifying principle:

   Temperature

Household pests
   a.   ~~Mice~~
   b.   Ants
   c.   Roaches
   d.   Flies
   Unifying principle:

   Insects

1.   Eyes
     a.   Blue
     b.   Nearsighted
     c.   Brown
     d.   Hazel
     Unifying principle:

     _____

2.   Mattresses
     a.   Double
     b.   Twin
     c.   Queen
     d.   Firm
     Unifying principle:

     _____

3.   Zoo animals
     a.   Flamingo
     b.   Peacock
     c.   Polar bear
     d.   Ostrich
     Unifying principle:

     _____

4.   Vacation
     a.   Summer
     b.   Holiday
     c.   Seashore
     d.   Weekend
     Unifying principle:

     _____

5.   College classes
     a.   Enjoy
     b.   Dislike
     c.   Tolerate
     d.   Morning
     Unifying principle:

     _____

6.   Wallets
     a.   Leather
     b.   Plastic
     c.   Stolen
     d.   Fabric
     Unifying principle:

     _____

7. Newspaper
    a. Wrapping garbage
    b. Editorials
    c. Making paper planes
    d. Covering floor while painting
    Unifying principle:

    _____

8. Students
    a. Freshman
    b. Transfer
    c. Junior
    d. Sophomore
    Unifying principle:

    _____

9. Exercise
    a. Running
    b. Swimming
    c. Gymnastics
    d. Fatigue
    Unifying principle:

    _____

10. Leftovers
    a. Cold chicken
    b. Feed to dog
    c. Reheat
    d. Use in a stew
    Unifying principle:

    _____

WRITING A DIVISION AND CLASSIFICATION PARAGRAPH

■ Writing Assignment 1

Below are four possible division and classification writing assignments, along with possible divisions. Choose *one* of them to develop into a paragraph.

Option 1
Supermarket shoppers
a. Slow, careful shoppers
b. Average shoppers
c. Rushed, hurried shoppers

Option 2
Eaters
a. Super-conservative eaters
b. Typical eaters
c. Adventurous eaters

Option 3
Methods of housekeeping
a. Never clean
b. Clean regularly
c. Clean constantly

Option 4
Attitudes toward money
a. Tightfisted
b. Sometimes splurge
c. Spendthrift

How to Proceed

a Begin by prewriting. To develop some ideas for the paragraph, freewrite for five or ten minutes on your topic.

b Add to the material you have written by asking yourself questions. If you are writing about supermarket shoppers, for example, you might ask:

How do the three kinds of shoppers pick out the items they want?

How many aisles will each type of shopper visit?

Which shoppers bring lists, calculators, coupons, and so on?

How much time does it take each type of shopper to finish shopping?

Write down whatever answers occur to you for these and other questions. As with freewriting, do not worry at this stage about writing correctly. Instead, concentrate on getting down all the information you can think of that supports each of the three points.

c Now go through all the material you have accumulated. Perhaps some of the details you have written down may help you think of even better details that would fit. If so, write them down. Then make decisions about the exact information you will use to support each point. Number the details 1, 2, 3, and so on, in the order you will present them.

d As you work on the drafts of your paragraph, make sure that it touches the bases of unity, support, and coherence.

e Finally, edit the next-to-final draft of your paper for sentence-skills mistakes, including spelling.

■ Writing Assignment 2

Write a division and classification essay on one of the following subjects:

Instructors	Drivers
Sports fans	Mothers or fathers
Eating places	Women's or men's magazines
Attitudes toward life	Presents
Commercials	Neighbors
Employers	Rock, pop, or country singers
Jobs	Amusement parks or rides
Bars	Guests or company
Family get-togethers	Ways to get an A (or F) in a course
Shoes	Car accessories

How to Proceed

a   The first step in writing a division and classification paragraph is to divide your tentative topic into three reasonably complete parts. *Always use a single principle of division when you form your three parts.* For example, if your topic was "Automobile Drivers" and you divided them into slow, moderate, and fast drivers, your single basis for division would be "rate of speed." It would be illogical, then, to have as a fourth type "teenage drivers" (the basis of such a division would be "age") or "female drivers" (the basis of such a division would be "sex"). You could probably classify automobile drivers on the basis of age or sex or another division, for almost any subject can be analyzed in more than one way. What is important, however, is that in any single paper you choose only one basis for division and stick to it. Be consistent.

In "Studying for a Test," the writer divides the process of studying into three time phases: from the time the test is announced to forty-eight hours before the test; the day and a half before the test; the final twelve hours before the test. In "The Dangers of Tools," the single basis for dividing tools into three categories is the kind of injury each type inflicts: bruises, cuts, and major injuries.

b   To ensure a clear three-part division in your own paragraph, fill in the outline below before starting your paper and make sure you can answer *Yes* to the questions that follow. You should expect to do a fair amount of thinking before coming up with a logical plan for your paper.

Topic: _____

Three-part division of the topic:

(1)   _____

(2)   _____

(3)   _____

Is there a single basis of division for the three parts? _____

Is the division reasonably complete? _____

c   Refer to the checklist of the four bases on the inside front cover while writing the drafts of your paper. Make sure you can answer *Yes* to the questions about unity, support, coherence, and sentence skills. Also, use the checklist when you edit the next-to-final draft of your paper for sentence-skills mistakes, including spelling.

■ Writing Assignment 3

There are many ways you could classify the students around you. Pick out one of your courses and write a division and classification paragraph on the students in that class. You might want to categorize the students according to one of the principles of division below:

Attitude toward the class

Participation in the class

Method of taking notes in class

Method of taking a test in class

Punctuality

Attendance

Level of confidence

Performance during oral reports, speeches, presentations, lab sessions

Of course, you may use any other principle of division that seems appropriate. Follow the steps listed in "How to Proceed" for Writing Assignment 2.

■ Writing Assignment 4

Write a review of a restaurant by analyzing a particular restaurant's (1) food, (2) service, and (3) atmosphere. In order to do the assignment, you should visit a restaurant, take a notebook with you, and write down observations about such elements as:

Quantity of food you receive

Taste of the food

Temperature of the food

Freshness of the ingredients

How the food is presented (garnishes, dishes, and so on)

Attitude of the servers

Efficiency of the servers

Decor

Level of cleanliness

Noise level and music, if any

You may want to write about details other than the ones listed above. Just be sure they fit into one of your three categories of food, service, or atmosphere.

For your topic sentence, rate the restaurant by giving it from one to five stars, on the basis of your overall impression. Include the restaurant's name and location in your topic sentence. Here are some examples:

Guido's, a center-city Italian restaurant, deserves three stars.

The McDonald's on Route 70 merits a four-star rating.

The Circle Diner in Westfield barely earns a one-star rating.

■ **Writing Assignment 5**

You are teaching a class in safe driving at a high school, and part of today's lecture is about types of drivers to avoid. For this part of your presentation, write a paragraph dividing the category "unsafe drivers" into three or more types according to their driving habits. For each type, include both a description and suggestions to your students on how to avoid an accident with this type of driver.

DESCRIBING
A SCENE
OR PERSON

When you describe something or someone, you give your readers a picture in words. To make this "word picture" as vivid and real as possible, you must observe and record specific details that appeal to your readers' senses (sight, hearing, taste, smell, and touch). More than any other type of writing, a descriptive paragraph needs sharp, colorful details.

Here is a description in which only the sense of sight is used:

A rug covers the living-room floor.

In contrast, here is a description rich in sense impressions:

A thick, reddish-brown shag rug is laid wall to wall across the living-room floor. The long, curled fibers of the shag seem to whisper as you walk through them in your bare feet, and when you squeeze your toes into the deep covering, the soft fibers push back at you with a spongy resiliency.

Sense impressions include sight (*thick, reddish-brown shag rug; laid wall to wall; walk through them in your bare feet; squeeze your toes into the deep covering; push back*), hearing (*whisper*), and touch (*bare feet, soft fibers, spongy resiliency*). The sharp, vivid images provided by the sensory details give us a clear picture of the rug and enable us to share in the writer's experience.

In this section, you will be asked to describe a person, place, or thing for your readers through the use of words rich in sensory details. To help you prepare for the assignment, first read the three paragraphs ahead and then answer the questions that follow.

PARAGRAPHS TO CONSIDER

An Athlete's Room

[1]As I entered the bright, cheerful space, with its beige walls and practical, flat-pile carpet, I noticed a closet to my right with the door open. [2]On the shelf above the bunched-together clothes were a red baseball cap, a fielder's glove, and a battered brown gym bag. [3]Turning from the closet, I noticed a single bed with its wooden headboard against the far wall. [4]The bedspread was a brown, orange, and beige print of basketball, football, and baseball scenes. [5]A lamp shaped like a baseball and a copy of Sports Illustrated were on the top of a nightstand to the left of the bed. [6]A sports schedule and several yellowing newspaper clippings were tacked to the cork bulletin board on the wall above the nightstand. [7]A desk with a bookcase top stood against the left wall. [8]I walked toward it to examine it more closely. [9]As I ran my fingers over the items on the dusty shelves, I noticed some tarnished medals and faded ribbons for track accomplishments. [10]These lay next to a heavy gold trophy that read, "MVP: Pinewood Varsity Basketball." [11]I accidentally tipped an autograph-covered, slightly deflated basketball off one shelf, and the ball bounced with dull thuds across the width of the room. [12]Next to the desk was a window with brightly printed curtains that matched the bedspread. [13]Between the window and the left corner stood a dresser with one drawer half open, revealing a tangle of odd sweat socks and a few stretched-out T shirts emblazoned with team insignias. [14]As I turned to leave the room, I carefully picked my way around scattered pairs of worn-out sneakers.

A Depressing Place

[1]The pet shop in the mall is a depressing place. [2]A display window attracts passers-by who stare at the prisoners penned inside. [3]In the right-hand side of the window, two puppies press their forepaws against the glass and attempt to lick the human hands that press from the outside. [4]A cardboard barrier separates the dogs from several black-and-white kittens piled together in the opposite end of the window. [5]Inside the shop, rows of wire cages line one wall from top to bottom. [6]At first, it is hard to tell whether a bird, hamster, gerbil, cat, or dog is locked inside each cage. [7]Only an occasional movement or clawing, shuffling sound tells visitors that living creatures are inside. [8]Running down the center of the store is a line of large wooden perches that look like coat racks. [9]When customers pass by, the parrots and mynas chained to these perches flutter their clipped wings in a useless attempt to escape. [10]At the end of this center aisle is a large plastic

tub of dirty, stagnant-looking water containing a few motionless turtles. [11]The shelves against the left-hand wall are packed with all kinds of pet-related items. [12]The smell inside the entire shop is an unpleasant mixture of strong chemical deodorizers, urine-soaked newspapers, and musty sawdust. [13]Because so many animals are crammed together, the normally pleasant, slightly milky smell of the puppies and kittens is sour and strong. [14]The droppings inside the uncleaned birdcages give off a dry, stinging odor. [15]Visitors hurry out of the shop, anxious to feel fresh air and sunlight. [16]The animals stay on.

Karla

[1]Karla, my brother's new girlfriend, is a catlike creature. [2]Her face, with its wide forehead, sharp cheekbones, and narrow, pointed chin, resembles a triangle. [3]Karla's skin is a soft, velvety brown. [4]Her large brown eyes slant upward at the corners, and she emphasizes their angle with a sweep of maroon eye shadow. [5]Karla's habit of looking sidelong out of the tail of her eye makes her look cautious, as if she were expecting something to sneak up on her. [6]Her nose is small and flat. [7]The sharply outlined depression under it leads the observer's eye to a pair of red-tinted lips. [8]With their slight upward tilt at the corners, Karla's lips make her seem self-satisfied and secretly pleased. [9]One reason Karla may be happy is that she recently was asked to be in a local beauty contest. [10]Karla's face is framed by a smooth layer of brown hair that always looks just combed. [11]Her long neck and slim body are perfectly in proportion with her face. [12]Karla manages to look elegant and sleek no matter how she is standing or sitting, for her body seems to be made up of graceful angles. [13]Her slender hands are tipped with long, polished nails. [14]Her narrow feet are long, too, but they appear delicate even in flat-soled running shoes. [15]Somehow, Karla would look perfect in a cat's jeweled collar.

■ Questions

About Unity

1. Which paragraph lacks a topic sentence?

   _____

2. Which sentence in the paragraph on Karla should be omitted in the interest of paragraph unity? (*Write the sentence number here.*) _____

About Support

3. Label as *sight, touch, hearing,* or *smell* all the sensory details in the following sentences taken from the three paragraphs. The first one is done for you as an example.

   touch   sight   sight

   a. I accidentally tipped an autograph-covered, slightly deflated basketball off

   sight   hearing

   one shelf, and the ball bounced with dull thuds across the width of the

   sight

   room.

   b. Because so many animals are crammed together, the normally pleasant, slightly milky smell of the puppies and kittens is sour and strong.

   c. Her slender hands are tipped with long, polished nails.

   d. As I ran my fingers over the items on the dusty shelves, I noticed some tarnished medals and faded ribbons for track accomplishments.

4. After which sentence in "A Depressing Place" are specific details needed?

   _____

About Coherence

5. Spatial signals (*above, next to, to the right,* and so on) are often used to help organize details in descriptive paragraphs. List four space signals that appear in "An Athlete's Room":

   _____

   _____

   _____

   _____

6. The writer of "Karla" organizes the details by observing Karla in an orderly way. Which of Karla's features is described first? _____ Which is described last? _____ Check the method of spatial organization that best describes the paragraph:

   _____ Interior to exterior

   _____ Near to far

   _____ Top to bottom

WRITING A DESCRIPTIVE PARAGRAPH

■ Writing Assignment 1

Write a paragraph describing a special kind of room. Use as your topic sentence "I could tell by looking at the room that a _____ lived there." There are many kinds of people who could be the focus for such a paragraph. You can select any one of the following, or think of some other type of person.

Photographer	Music lover	Carpenter
Cook	TV addict	Baby
Student	Camper	Cat or dog lover
Musician	Grandparent	World traveler
Hunter	Cheerleader	Drug addict
Slob	Football player	Little boy or girl
Outdoors person	Actor	Alcoholic
Instructor	Prostitute	Roller skater

How to Proceed

a   Begin by prewriting. After choosing a topic, spend a few minutes making sure it will work. Prepare a list of all the details you can think of that support the topic. For example, the writer of "An Athlete's Room" made this list:

Sports trophy
Autographed basketball
Sports Illustrated
Baseball lamp
Sports schedule
Medals and ribbons
Sports print on bedspread, curtains
Sweat socks, T shirts
Baseball cap
Baseball glove
Gym bag
Sports clippings

If you don't have enough details, then choose another type of person, and check your new choice with a list of details before committing yourself to the topic.

b   As you work on the paragraph, you should keep in mind all four bases of effective writing.

Base 1: Unity. Everything in the paragraph should support your point. For example, if you are writing about an athlete's room, all the details should serve to show that the person who lives in the room is an athlete. Other details should be omitted. Then, after your paragraph is finished, imagine omitting the key word in your topic sentence. Your details alone should make it clear to the reader what word should fit in that empty space.

Base 2: Support. Description depends on the use of *specific* rather than *general* descriptive words. For example:

General	Specific
Old sports trophies	Tarnished medals and faded ribbons for track accomplishments
Ugly turtle tub	Large plastic tub of dirty, stagnant-looking water containing a few motionless turtles
Unpleasant smell	Unpleasant mixture of strong chemical deodorizers, urine-soaked newspapers, and musty sawdust
Nice skin	Soft, velvety brown skin

Remember that you want your readers to experience the room vividly as they read. Your words should be as detailed as a clear photograph and should give your readers a real feel for the room as well. Use as many senses as possible in describing the room. Chiefly you will use sight, but to an extent you may be able to use touch, hearing, and smell as well.

Base 3: Coherence. Organize your descriptive paragraph by using spatial order. Spatial order means that you move from right to left or from larger items to smaller ones, just as a visitor's eye might move around a room. For instance, the writer of ''An Athlete's Room'' presents an orderly description in which the eye moves from right to left around the room. Here are transition words that will help you connect your sentences as you describe the room:

to the left	across from	on the opposite side
to the right	above	nearby
next to	below	

Such transitions will help prevent you—and your reader—from getting lost as the description proceeds.

Base 4: Sentence skills. In the later drafts of your paper, edit carefully for sentence-skills mistakes. Refer to the checklist of such skills on the inside front cover of the book.

■ Writing Assignment 2

Write a paragraph about a particular place that you can observe carefully or that you already know well. It might be one of the following or some other place:

Student lounge area	Hair salon
Car showroom	Doctor's or dentist's office
Gymnasium	Classroom
Fast-food restaurant	Bank
Inside of a car	Dressing room
Ladies' or men's room	Attic
Movie theater	Street market
Auto repair garage	Place where you work
Record shop	Porch

How to Proceed

a Remember that, like all paragraphs, a descriptive paper must have an opening point. This point, or topic sentence, should state a dominant impression about the place you are describing. State the place you want to describe and the dominant impression you want to make in a single short sentence. The sentence can be refined later. For now, you just want to find and express a workable topic. You might write, for example, a sentence like one of the following:

The student lounge was hectic.

The record shop was noisy.

The car's interior was very clean.

The dressing room in the department store was stifling.

The dentist's office was soothing.

The movie theater was freezing.

The gymnasium was tense.

The attic was gloomy.

The men's room was classy.

The office where I work was strangely quiet.

b Now make a list of all the details you can think of that support the general impression. For example, the writer of "A Depressing Place" made the list shown on the next page:

A Depressing Place
Puppies behind glass
Unpleasant smell
Chained birds
Rows of cages
Dirty tub of turtles
Stuffy atmosphere
Kittens in window
Sounds of caged animals
Droppings and urine on newspapers

c   Organize your paper by using any one or a combination of the following methods.

In terms of physical order: That is, move from left to right, or far to near, or in some other consistent order

In terms of size: That is, begin with large features or objects and work down to smaller ones

In terms of a special order: Use a special order appropriate to the subject.

For instance, the writer of "A Depressing Place" organizes the paper in terms of physical order (from one side of the pet shop to the center to the other side).

d   Use as many senses as possible in describing a scene. Chiefly you will use sight, but to some extent you may be able to use touch, hearing, smell, and perhaps even taste as well. Remember that it is through the richness of your sense impressions that the reader will gain a picture of the scene.

e   As you are working on the drafts of your paper, refer to the checklist on the inside front cover. Make sure you can answer *Yes* to the questions about unity, support, coherence, and sentence skills.

■ Writing Assignment 3

Write a paragraph describing a person. Decide on a dominant impression you have about the person, and use only those details that will add to that impression. Here are some examples of people you might want to write about.

TV or movie personality	Coworker
Instructor	Clergyman
Employer	Police officer
Child	Store owner or manager
Older person	Bartender
Close friend	Joker
Enemy	Neighbor

Before you begin, you may want to look carefully at the paragraph on Karla given earlier in this chapter and at ''How to Proceed'' in Writing Assignment 2.

Here are some possible topic sentences. Your instructor may let you develop one of these or may require you to write your own.

Kate gives the impression of being permanently nervous.

The old man was as faded and brittle as a dying leaf.

The child was a cherubic little figure.

Our high school principal resembled a cartoon drawing.

The young woman seemed to belong to another era.

Our neighbor is a fussy person.

The rock singer seemed to be plugged in to some special kind of energy source.

The drug addict looked as lifeless as a corpse.

My friend Mike is a slow, deliberate person.

The owner of that grocery store seems burdened with troubles.

■ Writing Assignment 4

Visit a place you have never been to before and write a paragraph describing it. You may want to go to:

A restaurant

A classroom, a laboratory, an office, a workroom, or some other room in your school

A kind of store you ordinarily don't visit: for example, a certain men's or women's clothing store, hardware store, toy store, record shop, gun shop, or sports shop

A bus terminal, train station, or airport

A church or synagogue

A park, vacant lot, or street corner

You may want to jot down details about the place while you are there or very soon after you leave. Again, decide on a dominant impression you have about the place, and use only those details which will add to that impression. Follow the notes on ''How to Proceed'' for Writing Assignment 2.

■ **Writing Assignment 5**

Option 1: You have just subscribed to a video dating service. Clients of this service are required to make a three-minute presentation, which will be recorded on videotape. In this presentation, clients describe the kind of person they would like to date. Write a one-paragraph description for your video presentation. Begin by brainstorming for a few minutes on what your "ideal date" would be like. Then arrange the details you come up with into some or all of the following categories:

■ *Character and personality*   (Are his or her attitudes important to you? Do you prefer someone who's quiet or someone who's outgoing?)

■ *Interests*   (Should your date have some of the same interests as you? If so, which ones?)

■ *Personal habits*   (Do you care, for instance, if your date is a nonsmoker?)

■ *Physical qualities*   (How might your ideal date look and dress?)

Option 2: Alternatively, write a similar presentation in which you describe *yourself.* Your aim is to present yourself as honestly as possible, so that interested members of the dating service will get a good sense of what you are like.

NARRATING
AN EVENT

At times we make a statement clear by relating in detail something that has happened to us. In the story we tell, we present the details in the order in which they happened. A person might say, for example, "I was embarrassed yesterday," and then go on to illustrate the statement with the following narrative:

> I was hurrying across campus to get to a class. It had rained heavily all morning, so I was hopscotching my way around puddles in the pathway. I called to two friends ahead to wait for me, and right before I caught up to them, I came to a large puddle that covered the entire path. I had to make a quick choice of either stepping into the puddle or trying to jump over it. I jumped, wanting to seem cool since my friends were watching, but didn't clear the puddle. Water splashed everywhere, drenching my shoe, sock, and pants cuff, and also spraying the pants of my friends as well. "Well done, Dave!" they said. I felt the more embarrassed because I had tried to look so casual.

The speaker's details have made his moment of embarrassment vivid and real for us, and we can see and understand just why he felt as he did.

In this section, you will be asked to tell a story that illustrates some point. The paragraphs below all present narrative experiences that support a point. Read them and then answer the questions that follow.

PARAGRAPHS TO CONSIDER

Heartbreak

[1]Bonnie and I had gotten engaged in August, just before she left for college at Penn State. [2]A week before Thanksgiving, I drove up to see her as a surprise. [3]When I knocked on the door of her dorm room, she was indeed surprised but not in a pleasant way. [4]She introduced me to her roommate, who looked uncomfortable and quickly left. [5]I asked Bonnie how classes were going, and at the same time I tugged on the sleeve of my heavy sweater in order to pull it off. [6]As I was pulling it off, a large poster caught my eye. [7]It was decorated with paper flowers and yellow ribbon, and it said, "Bonnie and Blake." [8]"What's going on?" I said. [9]I stood there stunned and then felt an anger that grew rapidly. [10]"Who is Blake?" I asked. [11]Bonnie laughed nervously and said, "What do you want to hear about--my classes or Blake?" [12]I don't really remember what she then told me, except that Blake was a sophomore math major. [13]I felt a terrible pain in the pit of my stomach, and I wanted to rest my head on someone's shoulder and cry. [14]I wanted to tear down the sign and run out, but I did nothing. [15]Clumsily I pulled on my sweater again. [16]My knees felt weak, and I barely had control of my body. [17]I opened the room door, and suddenly more than anything I wanted to slam the door shut so hard that the dorm walls would collapse. [18]Instead, I managed to close the door quietly. [19]I walked away understanding what was meant by a broken heart.

A Childhood Disappointment

[1]The time I almost won a car when I was ten years old was probably the most disappointing moment of my childhood. [2]One hot summer afternoon I was wandering around a local department store, waiting for my mother to finish shopping. [3]Near the toy department, I was attracted to a crowd of people gathered around a bright blue car that was on display in the main aisle. [4]A sign indicated that the car was the first prize in a sweepstakes celebrating the store's tenth anniversary. [5]The sign also said that a person did not have to buy anything to fill out an entry form. [6]White entry cards and shiny yellow pencils were scattered on a card table nearby, and the table was just low enough for me to write on, so I filled out a card. [7]Then, feeling very much like an adult, I slipped my card into the slot of a heavy blue wooden box that rested on another table nearby. [8]I then proceeded to the toy department, completely forgetting about the car. [9]However, about a month later, just as I was walking into the house from my first day back at school, the telephone rang. [10]When my mother answered it, a man asked to speak to a Michael Winchester. [11]My mother said, "There's a Michael Williams here, but not a Michael Winchester." [12]He asked, "Is this 862-9715 at 29 Williams Street?" [13]My mother said, "That's the right number, but this is 29 Winchester Street." [14]She then asked him, "What is this all about?" and he explained to her about the sweepstakes contest. [15]My mother then called me

to ask if I had ever filled out an application for a sweepstakes drawing. [16]I said that I had, and she told me to get on the phone. [17]The man by this time had realized that I had filled in my first name and street name on the line where my full name was to be. [18]He told me I could not qualify for the prize because I had filled out the application incorrectly. [19]For the rest of the day, I cried whenever I thought of how close I had come to winning the car. [20]I am probably fated for the rest of my life to think of the "almost" prize whenever I fill out any kind of contest form.

A Frustrating Job

[1]Working as a baby-sitter was the most frustrating job I ever had. [2]I discovered this fact when my sister asked me to stay with her two sons for the evening. [3]I figured I would get them dinner, let them watch a little TV, and then put them to bed early. [4]The rest of the night I planned to watch TV and collect an easy twenty dollars. [5]It turned out to be anything but easy. [6]First, right before we were about to sit down for a pizza dinner, Rickie let the parakeet out of its cage. [7]This bird is really intelligent and can repeat almost any phrase. [8]The dog started chasing it around the house, so I decided to catch it before the dog did. [9]Rickie and Jeff volunteered to help, following at my heels. [10]We had the bird cornered by the fireplace when Rickie jumped for it and knocked over the hamster cage. [11]Then the bird escaped again, and the hamsters began scurrying around their cage like crazy creatures. [12]The dog had disappeared by this point, so I decided to clean up the hamsters' cage and try to calm them down. [13]While I was doing this, Rickie and Jeff caught the parakeet and put it back in its cage. [14]It was time to return to the kitchen and eat cold pizza. [15]But upon entering the kitchen, I discovered why the dog had lost interest in the bird chase. [16]What was left of the pizza was lying on the floor, and tomato sauce was dripping from the dog's chin. [17]I cleaned up the mess and then served chicken noodle soup and ice cream to the boys. [18]Only at nine o'clock did I get the kids to bed. [19]I then returned downstairs to find that the dog had thrown up pizza on the living-room rug. [20]When I finished cleaning the rug, my sister returned. [21]I took the twenty dollars and told her that she should get someone else next time.

■ Questions

About Unity

1. Which paragraph lacks a topic sentence?

    _____

    Write a topic sentence for the paragraph:

    _____

    _____

    _____

2. Which sentence in "A Frustrating Job" should be omitted in the interest of paragraph unity? (*Write the sentence number here.*) _____

About Support

3. What is for you the best (most real and vivid) detail or image in the paragraph "Heartbreak"?

_____

_____

What is the best detail or image in "A Childhood Disappointment"?

_____

_____

What is the best detail or image in "A Frustrating Job"?

_____

_____

4. Which two paragraphs provide details in the form of the actual words used by the participants?

_____

_____

About Coherence

5. Do the three paragraphs use time order or emphatic order to organize details?

_____

6. What are four transition words used in "A Frustrating Job"?

   a.   _____

   b.   _____

   c.   _____

   d.   _____

WRITING A NARRATIVE PARAGRAPH

■ Writing Assignment 1

Write an essay about an experience in which a certain emotion was predominant. The emotion might be fear, pride, satisfaction, embarrassment, or any of the following:

Frustration	Sympathy	Shyness
Love	Bitterness	Disappointment
Sadness	Violence	Happiness
Terror	Surprise	Jealousy
Shock	Nostalgia	Anger
Relief	Loss	Hate
Envy	Silliness	Nervousness

The experience should be limited in time. Note that the three paragraphs presented in this chapter all detail experiences that occurred within relatively short periods. One writer describes a heartbreaking surprise he received the day he visited his girlfriend; another describes the disappointing loss of a prize; the third describes a frustrating night of baby-sitting.

A good way to re-create an event is to include some dialog, as the writers of two of the three paragraphs in this chapter have done. Repeating what you have said or what you have heard someone else say helps make the situation come alive. First, though, be sure to check the section on quotation marks on pages 458–465.

How to Proceed

a Begin by prewriting. Think of an experience or event in your life in which you felt a certain emotion strongly. Then spend ten minutes freewriting about the experience. Do not worry at this point about such matters as spelling or grammar or putting things in the right order; instead, just try to get down all the details you can think of that seem related to the experience.

b This preliminary writing will help you decide whether your topic is promising enough to develop further. If it is not, choose another emotion. If it is, do two things:

■ First, write your topic sentence, underlining the emotion you will focus on. For example, ''My first day in kindergarten was one of the <u>scariest</u> days of my life.''

■ Second, make up a list of all the details involved in the experience. Then arrange these details in time order.

c Using the list as a guide, prepare a rough draft of your paper. Use time signals such as *first, then, next, after, while, during,* and *finally* to help connect details as you move from the beginning to the middle to the end of your narrative.

d As you work on the drafts of your paper, refer to the checklist on the inside front cover to make sure that you can answer *Yes* to the questions about unity, support, and coherence. Also use the checklist to edit the next-to-final draft of your paper for sentence-skills mistakes, including spelling.

■ Writing Assignment 2

Write a paper that shows, through some experience you have had, the truth *or* falsity of a popular belief. You might write about any one of the following statements or some other popular saying.

Every person has a price.

Haste makes waste.

Don't count your chickens before they're hatched.

A bird in the hand is worth two in the bush.

It isn't what you know, it's who you know.

Borrowing can get you into trouble.

What you don't know won't hurt you.

Keeping a promise is easier said than done.

You never really know people until you see them in an emergency.

If you don't help yourself, nobody will.

An ounce of prevention is worth a pound of cure.

Hope for the best but expect the worst.

Never give advice to a friend.

You get what you pay for.

A stitch in time saves nine.

A fool and his money are soon parted.

There is an exception to every rule.

Nice guys finish last.

Begin your narrative paragraph with a topic sentence that expresses your agreement or disagreement with a popular saying. For example, ''My sister learned recently that 'Keeping a promise is easier said than done.' '' Or '' 'Never give advice to a friend' is not always good advice, as I learned after helping a friend reunite with her boyfriend.''

Refer to the suggestions about "How to Proceed" on pages 195–196 when doing your paper. Remember that the purpose of your story is to *support* your topic sentence. Feel free to select carefully from and even add to your experience so that the details truly support the point of your story.

■ Writing Assignment 3

Write an account of a memorable personal experience. Make sure that your story has a point, expressed in the first sentence of the paper. If necessary, tailor your narrative to fit your purpose. Use time order to organize your details (*first* this happened; *then* this; *after* that, this; *next,* this; and so on). Concentrate on providing as many specific details as possible so that the reader can really share your experience. Try to make it as vivid for the reader as it was for you when you first experienced it.

You might want to use one of the topics below, or a topic of your own choosing. Regardless, remember that your story must illustrate or support a point stated in the first sentence of your paper.

The first time you felt grown-up

A major decision

A moment you knew you were happy

The occasion of your best or worst date

A time you took a foolish risk

An argument you will never forget

An incident that changed your life

A time when you did or did not do the right thing

Your best or worst holiday, birthday, or other special occasion

A time you learned a lesson or taught one to someone else

An occasion of triumph in sports or some other event

You may want to refer to the suggestions on "How to Proceed" in Writing Assignment 1.

■ Writing Assignment 4

For this assignment, you will re-create a real-life scene you have witnessed. Here are some places where you will find interesting interactions among people:

The traffic or small-claims court in your area

The dinner table at your or someone else's home

A waiting line at a supermarket, unemployment office, ticket counter, movie theater, or cafeteria

A doctor's office

An audience at a movie or sports event

A classroom

A restaurant

A student lounge

In your topic sentence, name the place where the incident happened and make some point about the incident. Here are some possibilities:

I witnessed a heartwarming incident at Burger King yesterday.

Two fans at last week's baseball game got into a hilarious argument.

The scene at our family dinner table Monday was one of complete confusion.

A painful dispute went on in Atlantic County small-claims court yesterday.

In your paragraph, describe the characters you watched: who they were, their facial expressions, their tones of voice, and so on. Use dialog to re-create what they said to each other. As you write the paragraph, keep in mind that your goal is to "photograph" the incident in writing. You want to achieve with words a sharp, clear re-creation of a limited subject.

■ Writing Assignment 5

Imagine that a younger brother or sister, or a young friend, has to make a difficult decision of some kind. Perhaps he or she must decide how to go about preparing for a job interview, whether or not to get help with a difficult class, or what to do about a coworker who is taking money from the cash register. Write a narration from your own experience (or that of someone you know) that will teach a younger person something about the decision he or she must make. In your paragraph, include a comment or two about the lesson your story teaches. You may narrate an experience about any problem young people face, including any of those already mentioned or those listed below.

Should he or she save a little from a weekly paycheck?

Should he or she live at home or move to an apartment with some friends?

How should he or she deal with a group of friends who are involved with drugs, stealing, or both?

ARGUING
A POSITION

Most of us know someone who enjoys a good argument. Such a person usually challenges any sweeping statement we might make. ''Why do you say that?'' he or she will ask. ''Give your reasons.'' Our questioner then listens carefully as we cite our reasons, waiting to see if we really do have solid evidence to support our point of view. Such a questioner may make us feel a bit nervous, but we may also feel grateful to him or her for helping us think through our opinions.

The ability to advance sound and compelling arguments is an important skill in everyday life. We can use persuasion to get an extension on a term paper, obtain a favor from a friend, or convince an employer that we are the right person for a job. Understanding persuasion based on clear, logical reasoning can also help us see through the sometimes faulty arguments advanced by advertisers, editors, politicians, and others who try to bring us over to their side.

In this section, you will be asked to argue a position and defend it with a series of solid reasons. You are in a general way doing the same thing — making a point and then supporting it—with all the paragraphs in the book. The difference here is that, in a more direct and formal manner, you will advance a point about which you feel strongly and seek to convince others to agree with you.

PARAGRAPHS TO CONSIDER

Let's Ban Proms

[1]While many students regard proms as peak events in high school life, I believe that high school proms should be banned. [2]One reason is that even before the prom takes place, it causes problems. [3]Teenagers are separated into "the ones who were asked" and "the ones who weren't." [4]Being one of those who weren't asked can be heartbreaking to a sensitive young person. [5]Another pre-prom problem is money. [6]The price of the various items needed can add up quickly to a lot of money. [7]The prom itself can be unpleasant and frustrating, too. [8]At the beginning of the evening, the girls enviously compare dresses while the boys sweat nervously inside their rented suits. [9]During the dance, the couples who have gotten together only to go to the prom have split up into miserable singles. [10]When the prom draws to a close, the popular teenagers drive off happily to other parties while the less-popular ones head home, as usual. [11]Perhaps the main reason proms should be banned, however, is the drinking and driving that go on after the prom is over. [12]Teenagers pile into their cars on their way to "after-proms" and pull out the bottles and cans stashed under the seat. [13]By the time the big night is finally over, at 4 or 5 A.M., students are trying to weave home without encountering the police or a roadside tree. [14]Some of them do not make it, and prom night turns into tragedy. [15]For all these reasons, proms have no place in our schools.

A Terrible Vacation

[1]Despite much advertising to the contrary, taking a cruise is a terrible way to spend a vacation. [2]For one thing, there is too much food. [3]You are force-fed seven times a day: breakfast, midmorning snack, lunch, afternoon punch and pastries, dinner, the midnight buffet, and the 1:30 A.M. pizza in the disco. [4]Also, the waiters will not take "no" for an answer when they bring the food to your table. [5]They think "no" means "give me a medium-sized portion." [6]Another problem with a cruise is that there is too little genuine exercise. [7]The swimming pool is the size of a large bathtub. [8]Three strokes bring you to the other side. [9]And if you want to jog around the deck, you will have to dodge flying Ping-Pong balls and leapfrog over the shuffleboard players as you go. [10]Finally, the shipboard activities are boring. [11]The big event of the afternoon is bingo, and at night, for excitement, there is the Kentucky Derby, complete with little wooden horses and a social director throwing dice to see which one wins. [12]Many people are opposed to gambling anyway, and these games can be offensive to them. [13]If you try to start a conversation on deck with one of the other passengers, you will find that most of them have sent their minds on vacation along with their bodies.

[14]All they are interested in is what kind of suntan lotion you are using or what you think will be served at the next meal. [15]You will soon give up and join them in the chief activity on board--staring at the ocean. [16]So the next time you look through your vacation folders, pick the mountains, the seashore--anything but a cruise ship. [17]This way, your vacation will expand your mind and your muscles but not your waistline.

Living Alone

[1]Living alone is quite an experience. [2]People who live alone, for one thing, have to learn to do all kinds of tasks by themselves. [3]They must learn--even if they have had no experience--to change fuses, put up curtains and shades, temporarily dam an overflowing toilet, cook a meal, and defrost a refrigerator. [4]When there are no fathers, husbands, mothers, or wives to depend on, a person can't fall back on the excuse, ''I don't know how to do that.'' [5]Those who live alone also need the strength to deal with people. [6]Alone, singles must face noisy neighbors, unresponsive landlords, dishonest repair people, and aggressive bill collectors. [7]Because there are no buffers between themselves and the outside world, people living alone have to handle every visitor--friendly or unfriendly--alone. [8]Finally, singles need a large dose of courage to cope with occasional panic and unavoidable loneliness. [9]That weird thump in the night is even more terrifying when there is no one in the next bed or the next room. [10]Frightening weather or unexpected bad news is doubly bad when the worry can't be shared. [11]Even when life is going well, little moments of sudden loneliness can send shivers through the heart. [12]Struggling through such bad times taps into reserves of courage that people may not have known they possessed. [13]Facing everyday tasks, confronting all types of people, and handling panic and loneliness can shape singles into brave, resourceful, and more independent people.

■ Questions

About Unity

1. The topic sentence in ''Living Alone'' is too broad. Circle the topic sentence below that states accurately what the paragraph is about.
   a. Living alone takes courage.
   b. Living alone can create feelings of loneliness.
   c. Living alone should be avoided.
2. Which sentence in ''A Terrible Vacation'' should be eliminated in the interest of paragraph unity? (*Write the sentence number here.*) _____

About Support

3. How many reasons are given to support the topic sentence in each paragraph?

   a. In "A Terrible Vacation"   ____ one ____ two ____ three ____ four

   b. In "Let's Ban Proms"   ____ one ____ two ____ three ____ four

   c. In "Living Alone"   ____ one ____ two ____ three ____ four

4. After which sentence in "Let's Ban Proms" are more specific details needed?

   _____

About Coherence

5. Which paragraph uses a combination of time and emphatic order to organize its details?

   _____

6. What are the three main transition words in "Living Alone"?

   a. _____   b. _____   c. _____

Activity

Complete the outline below of "A Terrible Vacation." Summarize in a few words the primary and secondary supporting material that fits under the topic sentence. Two items have been done for you as examples.

Topic sentence: Despite much advertising to the contrary, taking a cruise is a terrible way to spend a vacation.

   a. _____

       (1) _____

       (2) _____

   b. _____

       (1) _____

       (2) *Little room for jogging* _____

   c. _____

       (1) _____

       (2) *Dull conversations with other passengers* ____

       (3) _____

WRITING AN ARGUMENT PARAGRAPH

■ Writing Assignment 1

On separate paper, make up brief outlines for any four of the eight statements that follow. Note the example. Make sure that you have three separate and distinct reasons for each statement.

Example   Large cities should outlaw passenger cars.
  a.   Cut down on smog and pollution
  b.   Cut down on noise
  c.   Create more room for pedestrians

1. Condoms should (*or* should not) be made available in schools.
2. _____ (*name a specific sports team*) should win its league championship.
3. Television is one of the best (*or* worst) inventions of this century.
4. _____ are the best (*or* worst) pets.
5. All cigarette and alcohol advertising should be banned.
6. Teenagers make poor parents.
7. _____ is one public figure today who can be considered a hero.
8. This college needs a better _____ (cafeteria *or* library *or* student center *or* grading policy *or* attendance policy).

How to Proceed

a   Decide, perhaps through discussion with your instructor or classmates, which of your outlines would be most promising to develop into a paragraph. Make sure that your supporting points are logical ones that actually back up your topic sentence. Ask yourself in each case, "Does this item truly support my topic sentence?"

b   Now do some prewriting. Prepare a list of all the details you can think of that might support your point. To begin with, prepare more details than you can actually use. Here, for example, are details generated by the writer of "Let's Ban Proms" while working on the paragraph:

Car accidents (most important)	Waste of school money
Drinking after prom	Going with someone you don't like
Competition over dates	License to stay out all night
Preparation for prom cuts into school hours	Separates popular from unpopular
	Expenses
Rejection of not being asked	Parents' interference

c   Decide which details you will use to develop your paragraph. Also, number the details in the order in which you will present them. (You may also want to make an outline of your paragraph at this point.) Because emphatic order (most important reason last) is the most effective way to organize an argument paragraph, be sure to save your most powerful reason for last. Here is how the writer of "Let's Ban Proms" made decisions about details:

   8   Car accidents (most important)
   7   Drinking after prom
   3   Competition over dates
   ~~Preparation for prom cuts into school hours~~
   1   Rejection of not being asked
   ~~Waste of school money~~
   4   Going with someone you don't like
   6   License to stay out all night
   5   Separates popular from unpopular
   2   Expenses
   ~~Parents' interference~~

d   Develop each reason with specific details. For example, in "Let's Ban Proms," notice how the writer explains how unpleasant the prom can be by describing boys "who sweat nervously" and one-night-only dates splitting up into "miserable singles." The writer also expands the idea of after-prom drinking by describing the "bottles and cans stashed under the seat" and the teenagers "trying to weave home."

e   As you write, imagine that your audience is a jury that will ultimately render a verdict on your argument. Have you presented a convincing case? If *you* were on the jury, would you be favorably impressed with this argument?

f   As you are working on the drafts of your paper, keep the four bases of unity, support, coherence, and sentence skills in mind.

g   Finally, edit the next-to-final draft of your paper for sentence-skills mistakes, including spelling.

■ Writing Assignment 2

Write a paragraph that uses reasons to develop a point of some kind. You may advance and defend a point of your own about which you feel strongly, or you could support any one of the following statements:

1. Junk food should be banned from school cafeterias.
2. Being young is better than being old.
3. Being old is better than being young.
4. Fall can be seen as the saddest season.
5. Many college instructors know their subjects, but some are poor teachers.

6. _____ is a sport that should be banned.

7. _____ is a subject that should be taught in every school.

8. Athletes at schools with national reputations in sports should be paid for their work.

9. _____ is the one material possession that is indispensable in everyday life.

10. A college diploma is (*or* is not) essential for an ambitious person.

Use the suggestions in ''How to Proceed'' on pages 203–204 as a guide in writing your paragraph.

■ Writing Assignment 3

Write a paragraph in which you take a stand on one of the controversial subjects below. As a lead-in to this writing project, your instructor might give the class a chance to ''stand up for what they believe in.'' One side of the front of the room should be designated *strong agreement* and the other side *strong disagreement,* with the space between for varying intermediate degrees of agreement or disagreement. As the class stands in front of the room, the instructor will read one value statement at a time from the list below, and students will move to the appropriate spot depending on their degree of agreement or disagreement. Some time will be allowed for students, first, to discuss with those near them the reasons they are standing where they are, and second, to state to those on the other end of the scale the reasons for their position.

1. Students should not be required to attend high school.
2. Prostitution should be legalized.
3. Homosexuals and lesbians should not be allowed to teach in schools.
4. The death penalty should exist for certain crimes.
5. Abortion should be legal.
6. Federal prisons should be coed, and prisoners should be allowed to marry.
7. Parents of girls under eighteen should be informed if their daughters receive birth-control aids.
8. The government should set up centers where sick or aged persons can go voluntarily to commit suicide.
9. Any woman on welfare who has more than two illegitimate children should be sterilized.
10. Parents should never hit their children.

Begin your paragraph by writing a sentence that expresses your attitude toward one of these value statements. For example, ''I feel that prostitution should be legalized.''

Outline the reason or reasons you hold the opinion that you do. Your support may be based on your own experience, the experience of someone you know, or logic. For example, an outline of a paragraph based on one student's logic proceeded as follows:

I feel that prostitution should be legalized for the following reasons:

1. Prostitutes would then have to pay their fair share of income tax.
2. Government health centers would administer regular checkups and thus help prevent the spread of venereal disease.
3. Prostitutes would be able to work openly and independently and would not be subject to exploitation by others.
4. Most of all, prostitutes would no longer be so much regarded as social outcasts--an attitude that is psychologically damaging to those who may already have emotional problems.

Another outline, based on experience, proceeded as follows:

I do not feel that prostitution should be legalized, because of a woman I know who was once a prostitute.

1. The attention Linda received as a prostitute prevented her from seeing and working on her personal problems.
2. She became embittered toward all men, whom she always suspected of wanting to exploit her.
3. She developed a negative self-image and felt that no one could love her.

Use your outline as the basis for writing a paragraph. Be sure to refer to the suggestions on ''How to Proceed'' on pages 203–204.

■ Writing Assignment 4

Write a paper in which you use research findings to help support one of the following statements.

Wearing seat belts in automobiles should be mandatory.

Many people do not need vitamin pills.

Disposable cans and bottles should be banned.

Everyone should own a pet.

Mandatory retirement ages should be abolished.

Cigarettes should be illegal.

Penalties against drunken drivers should be sharply increased.

Advertising should not be permitted on Saturday morning cartoon shows.

Research the topic you have chosen in one or more of the following ways:

■ Look up the topic in the subject section of your library card catalog. (You may want to review "Using the Library" on page 294.) Possible subject headings for these statements might be *Automobile safety, Vitamin pills, Pollution, Pets, Retirement, Smoking, Alcohol,* and *Advertising.* Select the books listed under your heading that seem likely to give you relevant information about your topic. Then find the books in the library stacks.

■ Look up the topic in recent issues of *Readers' Guide to Periodical Literature.* (You may want to review "Using the Library.") Try the same headings suggested above. Select the articles listed under your heading that appear most likely to provide information on your topic. Then see if you can find some of these articles in the periodicals section of your library.

■ You may also want to check a paperback bookstore. Ask a salesperson to help you find recent books on your topic.

Reading material on your topic will help you think about that topic. See if you can organize your paper in the form of three reasons that support the topic. Put these reasons into a scratch outline, and use it as a guide in writing your paragraph. Here is an example:

Wearing seat belts should be mandatory.

1. Seat belts are now comfortable and easy to use. . . .
2. A seat-belt law would be easy to enforce. . . .
3. Seat belts would save lives. . . .

Note that statistics (on how many lives could be saved) would support the last reason. Do not hesitate to cite studies and other data in a limited way; they make your argument more objective and compelling.

■ Writing Assignment 5

You have finally met Mr. or Ms. Right—but your parents don't approve of him or her. Specifically, they are against your doing one of the following:

Continuing to see this person
Going steady
Moving in together
Getting married at the end of the school year

Write a letter to your parents explaining in a fully detailed way why you have made your choice. Do your best to convince them that it is a good choice.

ADDITIONAL PARAGRAPH ASSIGNMENTS

This section contains a variety of paragraph writing assignments. The earlier assignments are especially suited for writing practice at the beginning of a course; the later ones can be used to measure progress at the end of the course. In general, more detailed instructions are provided with the earlier assignments; fewer guidelines appear for the later ones, so that writers must make more individual decisions about exactly how to proceed. In short, the section provides a wide range of writing assignments. Many choices are possible, depending on the needs and interests of students and the purposes of the instructor.

■ 1 Best or Worst Experience

Your instructor may pass out slips of paper and ask you to write, in the middle of the slip, your name; in the top left-hand corner, the best or worst job (or chore) you have ever had; in the top right-hand corner, the best or worst instructor you have ever had; in the lower left-hand corner, the best or worst place you have ever eaten in; in the lower right-hand corner, the best or worst thing that has happened to you in the past week. The instructor may also participate by writing on the board. Here is one student's paper.

Baby-sitting for my sister	*B. O. Sullivan (Tenth-grade history teacher)*
	Gail Battaglia	
Fourth Street Diner	*Trying to register*

You should then get together with any person in the room whom you do not know, exchange papers, and talk for a bit about what you wrote. Then the two of you should join another pair, with members of the resulting group of four doing two things:

- Mastering the first names of all the members of the group, so that, if asked, they could introduce the instructor to everyone in the group.
- Giving a "mini" speech to the group in which they talk with *as much specific detail as possible* about any one of the four responses on their slips of paper. During or after this speech, other members of the group should ask questions to get as full a sense as possible of why the experience described was a "best" or "worst" one.

Finally, you should write a paper about any one of the best or worst experiences. The main purpose in writing this paper is to provide plenty of specific details that *show clearly* why your choice was a "best" or "worst" one. The papers on pages 97–98 are examples of students' responses to this assignment.

■ 2 Writing Up an Interview

Interview someone in the class. Take notes as you ask the person a series of questions.

How to Proceed

a Begin by asking a series of factual questions about the person. You might ask such questions as:

Where is the person from? Where does he or she live now?

Does the person have brothers or sisters? Does the person live with other people, or alone?

What kinds of jobs (if any) has the person had? Where does he or she work now?

What are the person's school or career plans? What courses is he or she taking?

What are the person's favorite leisure activities?

Work at getting specific details rather than general ones. You do not want your introduction to include lines such as "Regina graduated from high school and worked for a year." You want to state specific places and dates: "Regina graduated from DeWitt Clinton High School in the Bronx in 1991. Within a week of graduation, she had gotten a job as a typist for a branch of the Allstate

Insurance Company located in Queens.'' Or if you are writing about a person's favorite activities, you do not want to simply say, ''Regina enjoys watching TV in her few spare hours.'' Instead, go on and add details such as ''Her favorite shows are *60 Minutes, Cheers,* and *L.A. Law.*

b   Then ask a series of questions about the person's attitudes and thoughts on various matters. You might ask the person's feelings about his or her

Writing ability

Parents

Boss (if any)

Courses

Past schooling

Strengths and talents

Areas for self-improvement

You might also ask what things make the person angry or sad or happy, and why.

c   After collecting all this information, use it in two paragraphs. Begin your introduction to the person with a line like ''This is a short introduction to _____. Here is some factual information about him (her).'' Then begin your second paragraph with the line, ''Now let's take a brief look at some of _____'s attitudes and beliefs.''

■ 3   Keeping a Journal

Keep a journal for one week, or for whatever time period your instructor indicates. At some point during each day—perhaps right before going to bed—write for ten minutes or more about some of the specific happenings, thoughts, and feelings of your day. You do not have to prepare what to write or be in the mood or worry about making mistakes; just write down whatever words come out. As a minimum, you should complete at least one page in each writing session.

Keeping a journal will help you develop the habits of thinking on paper and writing in terms of specific details. Also, the journal can serve as a sourcebook of ideas for possible papers.

A sample journal entry was given on page 17. It includes general ideas that the writer might develop into paragraphs; for example:

Working at a department store means that you have to deal with some irritating customers.

Certain preparations are advisable before you quit a job.

See if you can construct another general point from this journal entry that might be the basis for a detailed and interesting paragraph. Write the point in the space below.

■ 4 Writing a Dialog

Make up and write a *realistic* dialog between two or more people. Don't have your characters talk like cardboard figures; have them talk the way people would in real life. Also, make sure their voices are consistent. (Do not have them suddenly talk out of character.)

The dialog should deal with a lifelike situation. It may, for example, be a discussion or argument of some kind between two friends or acquaintances, a husband and wife, a parent and child, a brother and sister, a boyfriend and girl-friend, a clerk and customer, or other people. The conversation may or may not lead to a decision or action of some kind.

When writing dialog, enclose your characters' exact words within quotation marks. (You should first review the material on quotation marks on pages 458–465.) Also, include brief descriptions of whether your characters smile, sit down or stand up, or make other facial gestures or movements during the conversation. And be sure to include a title for your dialog. The example that follows can serve as a guide.

A Supermarket Conversation

The supermarket checker rang up the total and said to the young woman in line, "That'll be $43.61."

The young woman fumbled with her pocketbook and then said in an embarrassed voice, "I don't think I have more than $40. How much did you say it was again?"

"It's $43.61," the checker said in a sharp, impatient tone.

As the young woman searched her pocketbook for the dollars she needed, the checker said loudly, "If you don't have enough money, you'll have to put something back."

A middle-aged man behind the young woman spoke up. "Look, Ma'am, I'll lend you a couple of dollars."

"No, I couldn't do that," said the young woman. "If . . . I don't think I need those sodas," she said hesitantly.

"Look, lady, make up your mind. You're holding up the line," the checker snapped.

The man turned to the checker and said coldly, "Why don't you try being a little more courteous to people? If we weren't here buying things, you'd be out of a job."

■ 5 Annoyances in Everyday Life

Make up a list of things that bother you in everyday life. One student's list of "pet peeves" included the following items:

> Drivers that suddenly slow down to turn without having signaled
> The cold floor in my bathroom on a winter morning
> Not having cable television to watch football and basketball games
> The small napkin holder in my parents' home that is always running out of napkins
> Not being able to fall asleep at night when I know I have to get up at 6:30 the next morning

Suggestions on How to Proceed

a Brainstorm a list of everyday annoyances by asking yourself questions: "What annoys me at home (or about my kitchen, bathroom, closets, and so on)?" "What annoys me about getting to school?" "What annoys me at school or work?" "What annoys me while I am driving or shopping?" You will probably be able to think of other questions.

b Decide which annoyances seem most promising to develop. Which are the most interesting or important? Which ones can be developed with specific, vivid details? Cross out the items you will not use. Next, number the annoyances you have listed in the order in which you will present them. You may want to group related items together (all the ones that are connected with shopping, for instance). Be sure to end with the item that annoys you the most.

c Now write a rough draft of the paragraph. Begin with a topic sentence that makes clear what your paragraph is about. Concentrate on providing plenty of details about each of the annoyances you are describing.

d In a second or third draft, add signal words (such as *one, also, another,* and *last*) to set off each annoyance.

e Use the checklist on the inside back cover to edit your paper for sentence-skills mistakes, including spelling.

■ 6 Getting Comfortable

Getting comfortable is a quiet pleasure in life that we all share. Write a paper about the special way you make yourself comfortable, providing plenty of specific details so that the reader can really see and understand your method. Use transition words such as *first, next, then, in addition, also, finally,* and so on to guide readers through your paper. Transitions act like signposts on an unfamiliar route—they prevent your readers from getting lost.

A student paragraph on getting comfortable ("How I Relax") is on page 73.

■ 7 A Special Person

Write in detail about a person who provided help at an important moment in your life. State in the first sentence who the person is and the person's relationship to you (friend, father, cousin, etc.). For example, "My grandmother gave me a lot of direction during the difficult time when my parents were getting divorced." Then show through specific examples (the person's words and actions) why he or she was so special for you.

■ 8 A Favorite Childhood Place

Describe a favorite childhood place that made you feel secure, safe, private, or in a world of your own. Here are some possibilities:

A closet

Under a piece of furniture

A grandparent's room

A basement or attic

The woods

A shed or barn

A tree

A bunk bed

Begin with a topic sentence something like this: "_____ was a place that made me feel _____ when I was a child." Keep the point of your topic sentence in mind as you describe this place. Include only details that will support the idea that your place was one of *security, safety, privacy,* or the like.

■ 9 Expressing Uniqueness

Write a paragraph providing examples of one quality or habit that helps make you unique. One student's response to this assignment appears below.

Floor-Cleaning Freak

The one habit that makes me unique is that I am a floor-cleaning freak. I use my Dustbuster to snap up crumbs seconds after they fall. When a rubber heel mark appears on my vinyl floor, I run for the steel wool. As I work in my kitchen preparing meals, I constantly scan the tiles, looking for spots where some liquid has been spilled or for a crumb that has somehow miraculously escaped my vision. After I scrub and wax my floors, I stand to one side of the room and try to catch the light in such a way as to reveal spots that have gone unwaxed. As I travel from one room to the other, my experienced eye is faithfully searching for lint that may have invaded my domain since my last passing. If I discover an offender, I discreetly tuck it into my pocket. The amount of lint I have gathered in the course of the day is the ultimate test of how diligently I am performing my task. I give my vacuum cleaner quite a workout, and I spend an excessive amount on replacement bags. My expenses for floor-cleaners and wax are alarmingly high, but somehow this does not stop me. Where my floors are concerned, money is not a consideration!

■ 10 Making It through the Day

Write about techniques you use to make it through a day of school or work. These may include:

Caffeine

A system of rewards

Humor

Food

Fantasizing

You might organize the paragraph by using time order. Show how you turn to your supports at various times during the day in order to cope with fatigue or boredom. For example, in the morning you might use coffee (with its dose of caffeine) to get started. Later in the day, you would go on to use other supports.

■ 11 Life without TV

Imagine that all the televisions of America go blank, starting tonight. What would you and your family do on a typical night without television? You may want to write about

What each individual would be doing

What the family could do together

Problems the lack of TV would cause

Benefits of family life without TV

Choose any of these approaches, or some other single approach, in writing about your family life without TV.

■ 12 Ten Topics

Write a paper on one of the following topics. Begin with a clear, direct sentence that states exactly what your paper will be about. For example, if you choose the first topic, your opening sentence might be, ''There were several delightful childhood games I played that occupied many of my summer days.'' An opening sentence for the second topic might be, ''The work I had to do to secure my high school diploma is one of the special accomplishments of my life.'' Be sure to follow your opening sentence with plenty of specific supporting details that develop your topic.

A way you had fun as a child

A special accomplishment

A favorite holiday and why

Some problems a family member or friend is having

A superstition or fear

A disagreement you have had with someone

A debt you have repaid or have yet to repay

The sickest you've ever been

How your parents (or you and a special person in your life) met

Your father's or mother's attitude toward you

■ 13 Ten More Topics

Write a paper on one of the topics below. Follow the instructions given for assignment 12.

A wish or dream you have or had

Everyday pleasures

Ways you were punished by your parents as a child

Ways you were rewarded by your parents as a child

A difficult moment in your life

An experience you or someone you know has had with drugs

Your weaknesses as a student

Your strengths as a student

A time a prayer was answered

Something you would like to change

■ 14   Fifteen Topics

Write a paper on one of the following topics:

Crime	Music	Books
Lies	Exercise	Transportation
Television	Debt	Exhaustion
Plants	Parking meters	Telephone
Comic books	Hunger	Drugs

Suggestions on How to Proceed

a   You might begin by writing several statements about your general topic. For example, suppose that you choose to do a paper on the subject ''Neighborhood.'' Here are some statements you might write:

My neighborhood is fairly rural.

The neighborhood where I grew up was unique.

Many city neighborhoods have problems with crime.

My new neighborhood has no playgrounds for the children.

A neighborhood's appearance reflects the people who live in it.

Everyone in my neighborhood seems to cut the grass almost daily.

My neighborhood became a community when it was faced with a hurricane last summer.

My neighborhood is a noisy place.

b   Choose (or revise) one of the statements that you could go on to develop in a paragraph. You should not select a narrow statement like ''My new neighborhood has no playgrounds for the children,'' for it is a simple factual sentence that needs no support. Nor should you begin with a point such as ''Many city neighborhoods have problems with crime,'' which is too broad for you to develop adequately in a single paragraph. (See also the information on topic sentences on pages 53–56.)

c   After you have chosen a promising sentence, make a scratch outline of supporting details that will develop the point of that sentence. For example, one student provided the following outline:

My neighborhood is a noisy place.
1. Businesses
   a. Tavern with loud music
   b. Twenty-four-hour drive-in burger restaurant

    2. Children
        a. Skating and biking while carrying loud radios
        b. Street games
    3. Traffic
        a. Truck route nearby
        b. Horn-blowing during frequent delays at intersection

d   While writing your paper, use the checklist on the inside front cover to make sure you can answer *Yes* to the questions about unity, support, and coherence. Also, refer to the checklist when you edit the next-to-final draft of your paper for sentence-skills mistakes, including spelling.

■ 15   Fifteen More Topics

Write a paper on one of the topics below. Follow the instructions given for assignment 14.

Comics	Tryouts
Babies	Pens
Vacation	Hospital
Red tape	Parties
Dependability	Criticism
Illness	Success
Failure	Wisdom teeth
Home	

PART THREE
ESSAY DEVELOPMENT

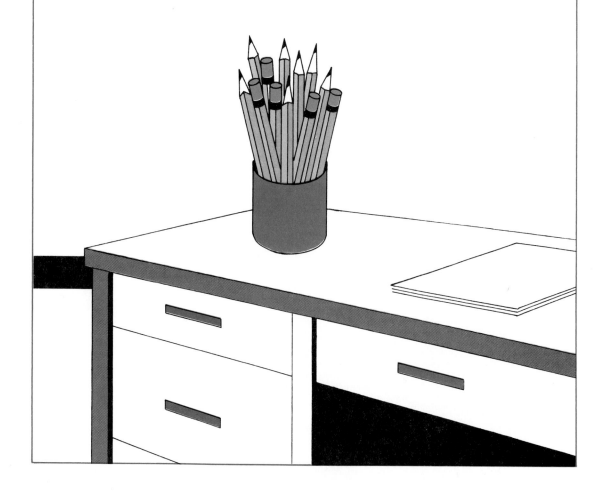

PREVIEW

Part Three moves from the single-paragraph paper to the several-paragraph essay. The differences between a paragraph and an essay are explained and then illustrated with a paragraph that has been expanded into an essay. You are shown how to begin an essay, how to tie its supporting paragraphs together, and how to conclude it. Three student essays are presented, along with questions to increase your understanding of the essay form. Finally, directions on how to plan an essay are followed by a series of essay writing assignments.

WRITING
THE ESSAY

What Is an Essay?

DIFFERENCES BETWEEN AN ESSAY AND A PARAGRAPH

An essay is simply a paper of several paragraphs, rather than one paragraph, that supports a single point. In an essay, subjects can and should be treated more fully than they would be in a single-paragraph paper.

The main idea or point developed in an essay is called the *thesis statement* (rather than, as in a paragraph, the *topic sentence)*. The thesis statement appears in the introductory paragraph, and it is then developed in the supporting paragraphs that follow. A short concluding paragraph closes the essay.

THE FORM OF AN ESSAY

The diagram on the next page shows the form of an essay.

Introductory Paragraph

Introduction
Thesis sentence
Plan of development:
Points 1, 2, 3

The *introduction* attracts the reader's interest.
The *thesis sentence* states the main idea advanced in the paper.
The *plan of development* is a list of the points that support the thesis. The points are presented in the order in which they will be developed in the paper.

First Supporting Paragraph

Topic sentence (point 1)
Specific evidence

The *topic sentence* advances the first supporting point for the thesis, and the *specific evidence* in the rest of the paragraph develops that first point.

Second Supporting Paragraph

Topic sentence (point 2)
Specific evidence

The *topic sentence* advances the second supporting point for the thesis, and the *specific evidence* in the rest of the paragraph develops that second point.

Third Supporting Paragraph

Topic sentence (point 2)
Specific evidence

The *topic sentence* advances the third supporting point for the thesis, and the *specific evidence* in the rest of the paragraph develops that third point.

Concluding Paragraph

Summary, conclusion,
or both

A *summary* is a brief restatement of the thesis and its main points. A *conclusion* is a final thought or two stemming from the subject of the paper.

A MODEL ESSAY

Gene, the writer of the paragraph on working in an apple plant (page 37), later decided to develop his subject more fully. Here is the essay that resulted.

My Job in an Apple Plant

Introductory paragraph

[1] In the course of working my way through school, I have taken many jobs I would rather forget. [2] I have spent nine hours a day lifting heavy automobile and truck batteries off the end of an assembly belt. [3] I have risked the loss of eyes and fingers working a punch press in a textile factory. [4] I have served as a ward aide in a mental hospital, helping care for brain-damaged men who would break into violent fits at unexpected moments. [5] But none of these jobs was as dreadful as my job in an apple plant. [6] The work was physically hard; the pay was poor; and, most of all, the working conditions were dismal.

First supporting paragraph

[7] First of all, the job made enormous demands on my strength and energy. [8] For ten hours a night, I took cartons that rolled down a metal track and stacked them onto wooden skids in a tractor trailer. [9] Each carton contained twelve heavy cans or bottles of apple juice. [10] A carton shot down the track about every fifteen seconds. [11] I once figured out that I was lifting an average of twelve tons of apple juice every night. [12] When a truck was almost filled, I or my partner had to drag fourteen bulky wooden skids into the empty trailer nearby and then set up added sections of the heavy metal track so that we could start routing cartons to the back of the empty van. [13] While one of us did that, the other performed the stacking work of two men.

Second supporting paragraph

[14] I would not have minded the difficulty of the work so much if the pay had not been so poor. [15] I was paid the minimum wage of that time, two dollars an hour, plus the minimum of a quarter extra for working the night shift. [16] Because of the low salary, I felt compelled to get as much overtime pay as possible. [17] Everything over eight hours a night was time-and-a-half, so I typically worked twelve hours a night. [18] On Friday I would sometimes work straight through until Saturday at noon--eighteen hours. [19] I averaged over sixty hours a week but did not take home much more than $100.

Third supporting paragraph

[20] But even more than the low pay, what upset me about my apple plant job was the working conditions. [21] Our humorless supervisor cared only about his production record for each night and tried to keep the assembly line moving at a breakneck pace. [22] During work I was limited to two ten-minute breaks and an unpaid half hour for lunch. [23] Most of my time was spent outside on the truck loading dock in near-zero-degree temperatures. [24] The steel floors of the trucks were like ice; the quickly penetrating cold made my feet feel like stone. [25] I had no shared interests with the man I loaded cartons with, and so I had to work without companionship on the job. [26] And after the production line shut down and most people left, I had to spend two hours alone scrubbing clean the apple vats, which were coated with a sticky residue.

Concluding paragraph

[27] I stayed on the job for five months, all the while hating the difficulty of the work, the poor money, and the conditions under which I worked. [28] By the time I quit, I was determined never to do such degrading work again.

Important Points about the Essay

INTRODUCTORY PARAGRAPH

An introductory paragraph has certain purposes or functions and can be constructed using various methods.

Purposes of the Introduction

An introductory paragraph should do three things:

1   Attract the reader's *interest*. Using one of the suggested methods of introduction described below can help draw the reader into your paper.

2   Present a *thesis sentence*—a clear, direct statement of the central idea that you will develop in your paper. The thesis statement, like a topic sentence, should have a key word or words that reflect your attitude about the subject. For example, in the essay on the apple plant job, the key word is *dreadful*.

3   Indicate a *plan of development*—a preview of the major points that will support your thesis statement, listed in the order in which they will be presented. In some cases, the thesis statement and plan of development may appear in the same sentence. In some cases, also, the plan of development may be omitted.

Activity

1.   In "My Job in an Apple Plant," which sentences are used to attract the reader's interest?

_____ Sentences 1 to 3   _____1 to 4   _____1 to 5

2.   The thesis in "My Job in an Apple Plant" is presented in

_____ Sentence 4   _____ Sentence 5   _____ Sentence 6

3.   The thesis is followed by a plan of development.

_____ Yes   _____ No

4.   Which words in the plan of development announce the three major supporting points in the essay? Write them below.

a.   _____

b.   _____

c.   _____

Common Methods of Introduction

Here are some common methods of introduction. Use any one method, or a combination of methods, to introduce your subject in an interesting way.

1 *Broad statement.* Begin with a broad, general statement of your topic and narrow it down to your thesis statement. Broad, general statements ease the reader into your thesis statement by providing a background for it. In "My Job in an Apple Plant," Gene writes generally on the topic of his worst jobs and then narrows down to a specific worst job.

2 *Contrast.* Start with an idea or situation that is the opposite of the one you will develop. This approach works because your readers will be surprised, and then intrigued, by the contrast between the opening idea and the thesis that follows it. Here is an example of a "contrast" introduction:

> When I was a girl, I never argued with my parents about differences between their attitudes and mine. My father would deliver his judgment on an issue, and that was usually the end of the matter. Discussion seldom changed his mind, and disagreement was not tolerated. But the situation is different with today's parents and children. My husband and I have to contend with radical differences between what our children think about a given situation and what we think about it. We have had disagreements with all three of our daughters, Stephanie, Diana, and Gisel.

3 *"Relevance."* Explain the importance of your topic. If you can convince your readers that the subject applies to them in some way, or is something they should know more about, they will want to continue reading. The introductory paragraph of "Sports-Crazy America" (page 229) provides an example of a "relevance" introduction.

4 *Anecdote.* Use an incident or brief story. Stories are naturally interesting. They appeal to a reader's curiosity. In your introduction, an anecdote will grab the reader's attention right away. The story should be brief and should be related to your central idea. The incident in the story can be something that happened to you, something that you have heard about, or something that you have read about in a newspaper or magazine. Here is an example of a paragraph that begins with a story:

> The husky man pushes open the door of the bedroom and grins as he pulls out a .38 revolver. An elderly man wearing thin pajamas looks at him and whimpers. In a feeble effort at escape, the old man slides out of his bed and moves to the door of the room. The husky man, still grinning, blocks his way. With the face of a small, frightened animal, the old man looks up and whispers, "Oh God, please don't hurt me." The grinning man then fires four times. The television movie cuts now to a soap commercial, but the little boy

who has been watching the set has begun to cry. Such scenes of direct violence on television must surely be harmful to children for a number of psychological reasons.

5   *Questions.* Ask your readers one or more questions. These questions catch the readers' interest and make them want to read on. Here is an example of a paragraph that begins with questions:

> What would happen if we were totally honest with ourselves? Would we be able to stand the pain of our own self-deception? Would the complete truth be too much for us to bear? Such questions will probably never be answered, for in everyday life we protect ourselves from the onslaught of too much reality. All of us cultivate defense mechanisms that prevent us from seeing and hearing and feeling too much. Included among such defense mechanisms are rationalization, reaction formation, and substitution.

Note, however, that the thesis itself must not be a question.

6   *Quotation.* A quotation can be something you have read in a book or an article. It can also be something that you have heard: a popular saying or proverb (''Never give advice to a friend''); a current or recent advertising slogan (''Reach out and touch someone''); a favorite expression used by your friends or family (''My father always says . . .''). Using a quotation in your introductory paragraph lets you add someone else's voice to your own. Here is an example of a paragraph that begins with a quotation:

> ''Evil,'' wrote Martin Buber, ''is lack of direction.'' In my school days as a fatherless boy, with a mother too confused by her own life to really care for me, I strayed down a number of dangerous paths. Before my eighteenth birthday, I had been a car thief, a burglar, and a drug seller.

SUPPORTING PARAGRAPHS

Most essays have three supporting points, developed in three separate paragraphs. (Some essays will have two supporting points; others, four or more.) Each of the supporting paragraphs should begin with a topic sentence that states the point to be detailed in that paragraph. Just as the thesis provides a focus for the entire essay, the topic sentence provides a focus for each supporting paragraph.

Activity

1. What is the topic sentence for the first supporting paragraph of ''My Job in an Apple Plant''? (*Write the sentence number here.*) _____

2. What is the topic sentence for the second supporting paragraph? _____

3. What is the topic sentence for the third supporting paragraph? _____

TRANSITIONAL SENTENCES

In paragraphs, transitions and other connective devices (pages 75–80) are used to help link sentences. Similarly, in an essay *transitional sentences* are used to help tie the supporting paragraphs together. Such transitional sentences usually occur near the end of one paragraph or the beginning of the next.

In "My Job in an Apple Plant," the first transitional sentence is:

> I would not have minded the difficulty of the work so much if the pay had not been so poor.

In this sentence, the key word *difficulty* reminds us of the point of the first supporting paragraph, while *pay* tells us the point to be developed in the second supporting paragraph.

Activity

Here is the other transitional sentence in "My Job in an Apple Plant":

> But even more than the low pay, what upset me about my apple plant job were the working conditions.

Complete the following statement: In the sentence above, the key words _____ echo the point of the second supporting paragraph, and the key words _____ announce the topic of the third supporting paragraph.

CONCLUDING PARAGRAPH

The concluding paragraph often summarizes the essay by briefly restating the thesis and, at times, the main supporting points of the essay. Also, the conclusion brings the paper to a natural and graceful end, sometimes leaving the reader with a final thought on the subject.

Activity

1. Which sentence in the concluding paragraph of "My Job in an Apple Plant" restates the thesis and supporting points of the essay? _____

2. Which sentence contains the concluding thought of the essay? _____

Essays to Consider

Read the three student essays below and then answer the questions that follow.

Giving Up a Baby

[1]As I awoke, I overheard a nurse say, "It's a lovely baby boy. [2]How could a mother give him up?" [3]"Be quiet," another voice said. [4]"She's going to wake up soon." [5]Then I heard the baby cry, but I never heard him again. [6]Three years ago, I gave up my child to two strangers, people who wanted a baby but could not have one. [7]I was in pain over my decision, and I can still hear the voices of people who said I was selfish or crazy. [8]But the reasons I gave up my child were important ones, at least to me.

[9]I gave up my baby, first of all, because I was very young. [10]I was only seventeen, and I was unmarried. [11]Because I was so young, I did not yet feel the desire to have and raise a baby. [12]I knew that I would be a child raising a child and that, when I had to stay home to care for the baby, I would resent the loss of my freedom. [13]I might also blame the baby for that loss. [14]In addition, I had not had the experiences in life that would make me a responsible, giving parent. [15]What could I teach my child, when I barely knew what life was all about myself?

[16]Besides my age, another factor in my decision was the problems my parents would have. [17]I had dropped out of high school before graduation, and I did not have a job or even the chance of a job, at least for a while. [18]My parents would have to support my child and me, possibly for years. [19]My mom and dad had already struggled to raise their family and were not well off financially. [20]I knew I could not burden them with an unemployed teenager and her baby. [21]Even if I eventually got a job, my parents would have to help raise my child. [22]They would have to be full-time baby-sitters while I tried to make a life of my own. [23]Because my parents are good people, they would have done all this for me. [24]But I felt I could not ask for such a big sacrifice from them.

[25]The most important factor in my decision was, I suppose, a selfish one. [26]I was worried about my own future. [27]I didn't want to marry the baby's father. [28]I realized during the time I was pregnant that we didn't love each other. [29]My future as an unmarried mother with no education or skills would certainly have been limited. [30]I would be struggling to survive, and I would have to give up for years my dreams of getting a job and my own car and apartment. [31]It is hard to admit, but I also considered the fact that, with a baby, I would not have the social life most young people have. [32]I would not be able to stay out late, go to parties, or feel carefree and irresponsible, for I would always have an enormous responsibility waiting for me at home. [33]With a baby, the future looked limited and insecure.

[34]In summary, thinking about my age, my responsibility to my parents, and my own future made me decide to give up my baby. [35]As I look back today at my decision, I know that it was the right one for me at the time.

Sports-Crazy America

[1]Almost all Americans are involved with sports in some way. [2]They may play basketball or volleyball or go swimming or skiing. [3]They may watch football or basketball games on the high school, college, or professional level. [4]Sports may seem like an innocent pleasure, but it is important to look under the surface. [5]In reality, sports have reached a point where they play Thesis too large a part in daily life. [6]They take up too much media time, play too large a role in the raising of children, and give too much power and prestige to athletes.

[7]The overemphasis on sports can be seen most obviously in the vast media coverage of athletic events. [8]It seems as if every bowl game play-off, tournament, trial, bout, race, meet, or match is shown on one television channel or another. [9]On Saturday and Sunday, a check of TV Guide will show almost forty sports programs on UHF and VHF alone, and many more on cable stations. [10]In addition, sports makes up about 30 percent of local news at six and eleven, and network news shows often devote several minutes of world news to major American sports events. [11]Radio offers a full roster of games and a wide assortment of sports talk shows. [12]Furthermore, many daily papers such as USA Today are devoting more and more space to sports coverage, often in an attempt to improve circulation. [13]The paper with the biggest sports section is the one people will buy.

[14]The way we raise and educate our children also illustrates our sports mania. [15]As early as six or seven, kids are placed in little leagues, often to play under screaming coaches and pressuring parents. [16]Later, in high school, students who are singled out by the school and by the community are not those who are best academically but those who are best athletically. [17]And college sometimes seems to be more about sports than about learning. [18]America may be the only country in the world where people often think of their colleges as teams first and schools second. [19]The names Penn State, Notre Dame, and Southern Cal mean ''sports'' to the public.

[20]Our sports craziness is especially evident in the prestige given to athletes in America. [21]For one thing, we reward them with enormous salaries. [22]In 1990, for example, baseball players averaged $350,000 a year; the average annual salary in America is $18,000. [23]Besides their huge salaries, athletes receive the awe, admiration, and sometimes the votes of the public. [24]Kids look up to a Michael Jordan or a Roger Clemens as a true hero, while adults wear the jerseys and jackets of their favorite teams. [25]Ex-players become senators and congressmen. [26]And an athlete like Monica Seles or Jim Kelly needs to make only one commercial for advertisers to see the sales of a product boom.

[27]Americans are truly mad about sports. [28]Perhaps we like to see the competitiveness we experience in our daily lives acted out on playing fields. [29]Perhaps we need heroes who can achieve clear-cut victories in the space of only an hour or two. [30]Whatever the reason, the sports scene in this country is more popular than ever.

An Interpretation of <u>Lord of the Flies</u>

[1]Modern history has shown us the evil that exists in human beings. [2]Assassinations are common, governments use torture to discourage dissent, and six million Jews were exterminated during World War II. [3]In <u>Lord of the Flies</u>, William Golding describes a group of schoolboys shipwrecked on an island with no authority figures to control their behavior. [4]One of the boys soon yields to dark forces within himself, and his corruption symbolizes the evil in all of us. [5]First, Jack Merridew kills a living creature; then, he rebels against the group leader; and finally, he seizes power and sets up his own murderous society.

[6]The first stage in Jack's downfall is his killing of a living creature. [7]In Chapter 1, Jack aims at a pig but is unable to kill. [8]His upraised arm pauses "because of the enormity of the knife descending and cutting into living flesh, because of the unbearable blood," and the pig escapes. [9]Three chapters later, however, Jack leads some boys on a successful hunt. [10]He returns triumphantly with a freshly killed pig and reports excitedly to the others, "I cut the pig's throat." [11]Yet Jack twitches as he says this, and he wipes his bloody hands on his shorts as if eager to remove the stains. [12]There is still some civilization left in him.

[13]After the initial act of killing the pig, Jack's refusal to cooperate with Ralph shows us that this civilized part is rapidly disappearing. [14]With no adults around, Ralph has made some rules. [15]One is that a signal fire must be kept burning. [16]But Jack tempts the boys watching the fire to go hunting, and the fire goes out. [17]Another rule is that at a meeting, only the person holding a special seashell has the right to speak. [18]In Chapter 5, another boy is speaking when Jack rudely tells him to shut up. [19]Ralph accuses Jack of breaking the rules. [20]Jack shouts: "Bollocks to the rules! We're strong--we hunt! If there's a beast, we'll hunt it down! We'll close in and beat and beat and beat--!" [21]He gives a "wild whoop" and leaps off the platform, throwing the meeting into chaos. [22]Jack is now much more savage than civilized.

[23]The most obvious proof of Jack's corruption comes in Chapter 8, when he establishes his own murderous society. [24]Insisting that Ralph is not a "proper chief" because he does not hunt, Jack asks for a new election. [25]After he again loses, Jack announces, "I'm going off by myself. . . . Anyone who wants to hunt when I do can come too." [26]Eventually, nearly all the boys join Jack's "tribe." [27]Following his example, they paint their faces like savages, sacrifice to "the beast," brutally murder two of their schoolmates, and nearly succeed in killing Ralph as well. [28]Jack has now become completely savage--and so have the others.

[29]Through Jack Merridew, then, Golding shows how easily moral laws can be forgotten. [30]Freed from grown-ups and their rules, Jack learns to kill living things, defy authority, and lead a tribe of murdering savages. [31]Jack's example is a frightening reminder of humanity's potential for evil. [32]The "beast" the boys try to hunt and kill is actually within every human being.

■ **Questions**

1. In which essay does the thesis statement appear in the last sentence of the introductory paragraph?

   _____

2. In the essay on *Lord of the Flies,* which sentence of the introductory paragraph

   contains the plan of development? _____

3. Which method of introduction is used in ''Giving Up a Baby''?
   a. General to narrow   c. Incident or story
   b. Stating importance of topic   d. Questions

4. Complete the following brief outline of ''Giving Up a Baby'':
   I gave up my baby for three reasons:

   a. _____

   b. _____

   c. _____

5. Which *two* essays use a transitional sentence between the first and second supporting paragraphs?

   _____

   _____

6. *Complete the following statement:* Emphatic order is shown in the last support-ing paragraph of ''Giving Up a Baby'' with the words *most important factor;* the last supporting paragraph of ''Sports-Crazy America '' with the words

   _____; in the last supporting paragraph of ''An Interpretation

   of *Lord of the Flies''* with the words _____.

7. Which essay uses time order as well as emphatic order to organize its three

   supporting paragraphs? _____

8. List four major transitions used in the supporting paragraphs of ''An Interpreta-tion of *Lord of the Flies.''*

   a. _____   c. _____

   b. _____   d. _____

9. Which *two* essays include a sentence in the concluding paragraph that summa-rizes the three supporting points?

   _____

   _____

10. Which essay includes two final thoughts in its concluding paragraph?

   _____

Planning the Essay

OUTLINING THE ESSAY

When you write an essay, advance planning is crucial for success. You should plan your essay by outlining in two ways:

1 Prepare a scratch outline. This should consist of a short statement of the thesis followed by the main supporting points for the thesis. Here is Gene's scratch outline for his essay on the apple plant:

Working at an apple plant was my worst job.
1. Hard work
2. Poor pay
3. Bad working conditions

Do not underestimate the value of this initial outline—or the work involved in achieving it. Be prepared to do a good deal of plain hard thinking at this first and most important stage of your paper.

2 Prepare a more detailed outline. The outline form that follows will serve as a guide. Your instructor may ask you to submit a copy of this form either before you actually write an essay or along with your finished essay.

FORM FOR PLANNING AN ESSAY

To write an effective essay, use a form like the one that follows.

Introduction

Opening remarks

Thesis statement _____

_____

Plan of development

Body

Topic sentence 1 _____

_____

Specific supporting evidence

Topic sentence 2 _____

_____

Specific supporting evidence

Topic sentence 3 _____

_____

Specific supporting evidence

Conclusion

Summary, closing remarks, or both

Essay Writing Assignments

Hints: Keep the following points in mind when writing an essay on any of the topics below.

1   Your first step must be to plan your essay. Prepare both a scratch outline and a more detailed outline, as explained on the preceding pages.

2   While writing your essay, use the checklist below to make sure your essay touches all four bases of effective writing.

Base 1: Unity

_____ Clearly stated thesis in the introductory paragraph of your paper

_____ All the supporting paragraphs on target in backing up your thesis

Base 2: Support

_____ Three separate supporting points for your thesis

_____ *Specific* evidence for each of the three supporting points

_____ *Plenty* of specific evidence for each supporting point

Base 3: Coherence

_____ Clear method of organization

_____ Transitions and other connecting words

_____ Effective introduction and conclusion

Base 4: Sentence Skills

_____ Clear, error-free sentences (use the checklist on the inside front cover of this book)

■ 1   Your House or Apartment

Write an essay on the advantages *or* disadvantages (not both) of the house or apartment where you live. In your introductory paragraph, describe briefly the place you plan to write about. End the paragraph with your thesis statement and a plan of development. Here are some suggestions for thesis statements:

> The best features of my apartment are its large windows, roomy closets, and great location.

The drawbacks of my house are its unreliable oil burner, tiny kitchen, and old-fashioned bathroom.

An inquisitive landlord, sloppy neighbors, and platoons of cockroaches came along with our rented house.

My apartment has several advantages, including friendly neighbors, lots of storage space, and a good security system.

■ 2 A Big Mistake

Write an essay about the biggest mistake you made within the past year. Describe the mistake and show how its effects have convinced you that it was the wrong thing to do. For instance, if you write about "taking on a full-time job while going to school" as your biggest mistake, show the problems it caused. (You might discuss such matters as low grades, constant exhaustion, and poor performance at work, for example.)

To get started, make a list of all the things you did last year that, with hindsight, now seem to be mistakes. Then pick out the action that has had the most serious consequences for you. Make a brief outline to guide you as you write, as in the examples below.

> Thesis: Separating from my husband was the worst mistake I made last year.
> 1. Children have suffered
> 2. Financial troubles
> 3. Loneliness
>
> Thesis: Buying a used car to commute to school was the worst mistake of last year.
> 1. Unreliable--late for class or missed class
> 2. Expenses for insurance, repairs
> 3. Led to an accident

■ 3 A Valued Possession

Write an essay about a valued material possession. Here are some suggestions:

Car	Appliance
Portable radio	Cassette deck
TV set	Photograph album
Piece of furniture	Piece of clothing
Piece of jewelry	Stereo system (car or home)
Camera	Piece of hobby equipment

In your introductory paragraph, describe the possession: tell what it is, when and where you got it, and how long you have owned it. Your thesis statement should center on the idea that there are several reasons this possession is so important to you. In each of your supporting paragraphs, provide details to back up one of the reasons.

For example, here is a brief outline of an essay written about a leather jacket:

1. It is comfortable.
2. It wears well.
3. It makes me look and feel good.

■ 4   Summarizing a Selection

Write an essay in which you summarize three of the study skills described on pages 253–266. Summarizing involves condensing material by highlighting main points and key supporting details. You can eliminate minor details and most examples given in the original material. You should avoid using the exact language in the original material; put the ideas into your own words.

The introductory paragraph of the essay and suggested topic sentences for the supporting paragraphs are provided below. In addition to developing the supporting paragraphs, you should write a brief conclusion for the essay.

Introductory Paragraph

Using Study Skills

Why do some students in a college class receive A grades, while others get D's and F's? Are some people just naturally smarter? Are other students doomed to failure? Motivation--willingness to do the work--is a factor in good grades. But the main difference between successful and unsuccessful students is that the ones who do well have mastered the specific skills needed to handle college work. Fortunately, these skills can be learned by anyone. Doing well in college depends on knowing how to . . . *[Complete this sentence with the three study skills you decide to write about.]*

Suggested Topic Sentences for the Supporting Paragraphs (Choose Any Three)

Time control is one aid to success as a student. . . .

Another aid is the use of memory techniques. . . .

Knowing how to concentrate is another essential skill. . . .

Studying a textbook effectively is another key to success. . . .

Perhaps the most crucial step of all is effective classroom note-taking. . . .

■ 5 How Study Skills Help

You may already be practicing some of the study skills described on pages 253–266. If so, write an essay on how study skills are helping you to succeed in school. Your thesis might be, ''Study skills are helping me to succeed in college.'' You could organize the essay by describing, in separate paragraphs, how three different study skills have improved your work. Your topic sentences might be similar to these:

> First of all, time control has helped me to make the best use of my time.
>
> In addition, taking good notes in class has enabled me to do well in discussions and on tests.
>
> Finally, I can study a textbook effectively now.

Alternatively, begin applying some of the techniques and be prepared to write an essay at a later time on how the study skills helped you become a better student. Or you might want to write about three study techniques of your own that have helped you succeed in your studies.

■ 6 Single Life

Write an essay on the advantages or drawbacks of single life. To get started, make a list of all the advantages and drawbacks you can think of for single life. Advantages might include:

> Fewer expenses
> Fewer responsibilities
> More personal freedom
> More opportunities to move or travel

Drawbacks might include:

> Parental disapproval
> Being alone at social events
> No companion for shopping, movies, and so on
> Sadness at holiday time

After you make up two lists, select the thesis for which you feel you have more supporting material. Then organize your material into a scratch outline. Be sure to include an introduction, a clear topic sentence for each supporting paragraph, and a conclusion.

Alternatively, write an essay on the advantages or drawbacks of married life. Follow the directions given above.

■ 7 Influences on Your Writing

Are you as good a writer as you want to be? Write an essay analyzing the reasons you have become a good writer or explaining why you are not as good as you'd like to be. Begin by considering some factors that may have influenced your level of writing ability.

> *Your family background:* Did you see people writing at home? Did your parents respect and value the ability to write?
>
> *Your school experience:* Did you have good writing teachers? Did you have a history of failure or success with writing? Was writing fun, or was it a chore? Did your school emphasize writing?
>
> *Social influences:* How did your school friends do at writing? What were your friends' attitudes toward writing? What feelings about writing did you pick up from TV or the movies?

You might want to organize your essay by describing the three greatest influences on your writing skill (or lack of writing skill). Show how each of these has contributed to the present state of your writing.

■ 8 A Major Decision

All of us come to various crossroads in our lives — times when we must make an important decision about which course of action to follow. Think about a major decision you had to make (or one you are planning to make). Then write an essay on the reasons for your decision. In your introduction, describe the decision you have reached. Each of the body paragraphs that follow should fully explain one of the reasons for your decision. Here are some examples of major decisions that often confront people:

Enrolling in or dropping out of college

Accepting or quitting a job

Getting married or divorced

Breaking up with a boyfriend or girlfriend

Having a baby

Moving away from home

Student papers on this topic include the essay on page 228 and the paragraphs on pages 47–48.

■ **9   Reviewing a TV Show or Movie**

Write an essay about a television show or movie you have seen very recently. The thesis of your essay will be that the show (or movie) has both good and bad features. (If you are writing about a TV series, be sure that you evaluate only one episode.)

In your first supporting paragraph, briefly summarize the show or movie. Don't get bogged down in small details here; just describe briefly the major characters and give the highlights of the action.

In your second supporting paragraph, explain what you feel are the best features of the show or movie. Listed below are some examples of good features you might write about:

A suspenseful, ingenious, or realistic plot

Good acting

Good scenery or special effects

A surprise ending

Good music

Believable characters

In your third supporting paragraph, explain what you feel are the worst features of the show or movie. Here are some possibilities:

Farfetched, confusing, or dull plot

Poor special effects

Bad acting

Cardboard characters

Unrealistic dialog

Remember to cover only a few features in each paragraph; do not try to include everything.

■ **10   Good Qualities**

We are often quick to point out a person's flaws, saying, for example, ''That teacher is conceited,'' ''My boss has no patience,'' or ''My sister is lazy.'' We are usually equally hard on ourselves; we constantly analyze our own faults. We rarely, though, spend as much time thinking about another person's, or our own, good qualities. Write an essay on the good qualities of a particular person. The person might be a teacher, a job supervisor, a friend, a relative, some other person you know well, or even yourself.

In your introductory paragraph, give some brief background information about the person you are describing. And include in your thesis statement a plan of development that names the three qualities you will write about. Here are several suggestions:

Patience, fairness, and kindness are my boss's best qualities.

My boyfriend is hardworking, ambitious, and determined.

Our psychology instructor has a good sense of humor, a strong sense of justice, and a genuine interest in his students.

When planning your paper, you may find it helpful to look at the positive qualities included in the list on page 132.

■ 11   Your High School

Imagine that you are an outside consultant called in as a neutral observer to examine the high school you attended. After your visit, you must send the school board a five-paragraph letter in which you describe the most striking features (good, bad, or a combination of both) of the school and the evidence for each of these features.

In order to write the letter, you may want to think about the following features of your high school:

Attitude of the teachers, student body, or administration
Condition of the buildings, classrooms, recreational areas, and so on
Curriculum
How classes are conducted
Extracurricular activities
Crowded or uncrowded conditions

Be sure to include an introduction, a clear topic sentence for each supporting paragraph, and a conclusion.

■ 12   Reacting to a Reading Selection

Read the following article about a boy who committed suicide. (The story is true, but the names of the persons involved have been changed for the sake of privacy.) Then do the activities that follow the article.

A Suicide at Twelve: "Why, Steve?"*
By Richard E. Meyer

He lived to be almost thirteen. Walnut eyes. Brown thatch. Boy Scout. Altar boy. He grew up in white, middle-class America. He played football, and he played baseball. His mother, father, two brothers, and sister loved him. On the fourth day of the eleventh month of his twelfth year, a sunny afternoon in suburban Cincinnati, he walked down his favorite trail in the woods behind his house, climbed a tree, knotted a rope, and hanged himself.

Why, Steve?

In the past year, at least 210 others as young as Steve Dailey killed themselves in the United States. Reported suicides among the very young have more than doubled in twenty years. Even adjusted for population growth, the rate has climbed. The story of Steve Dailey, all-American boy, is an American tragedy: a story about the good life and the possibilities it offers for hidden pressure, subtle loneliness, quiet frustration--and unanswered questions.

Why, Steve?

Steven Dailey was born July 30, 1961, in the Cincinnati suburb of Clifton. One month after his first birthday, his parents, Sue and Charles Dailey, presented him with a brother, Mike. The two boys would become good friends. When Steve was two or a little older, Grandpa Rafton, in charge of the tailors at MacGregor, the sporting goods company which made uniforms for the Cincinnati Reds, presented Steve and Mike with baseball uniforms of their own, cut in the Reds' own patterns from the Reds' own cloth. Steve's had pitcher Jimmy O'Toole's old number, 31, sewn on the back.

Almost from the day he was married, Charles Dailey worked with Boy Scouts, first as an assistant scoutmaster for a year, then as a scoutmaster for five. When Steve and Mike were still toddlers, he took them along to Scout meetings. One night, he told a meeting of Scout parents: "You know, these boys are growing up awfully fast. If you're ever going to get to know your sons, you better get to know them now--because soon they're going to be at an age when you can't really get to know them."

In the second grade, Steve entered St. Catherine's School, in the parish where the Daileys had moved in the suburb of Westwood. His father became a volunteer football coach in St. Catherine's growing athletic program. Steve Dailey was big enough to play second level, or "pony," football. But he got paired in practice against a youngster everybody called Mugsy. "After Mugsy kind of tore him up a few times, he decided that maybe he ought to play 'bandits' a year and kind of find out what it's all about first," his father remembers. "Bandits" are the beginners. "That kinda bugged the devil out

* Richard E. Meyer, "A Suicide at Twelve: 'Why, Steve?' " Reprinted by permission of Associated Press.

of me,'' Steve's father says. Charles Dailey thinks he probably told his son he was disappointed. ''But Steve says, 'Well, I just don't want to play ''pony'' ball. I'm just not good enough.' And it was probably a good choice on his part. But that was at the stage when I really wanted him to be the best football player in the world, you know. And I wanted him to be better.''

Steve preferred quieter pursuits. He started a stamp collection. At seven, he caught his first fish--a little bluegill he tugged from the lake at Houston Woods State Park on a camping trip with his family. In 1969, when he was eight years old, Steve joined the Cub Scouts. He advanced to Webelos, where he met Dan Carella, who would become his assistant scoutmaster. Just before becoming a full-fledged Boy Scout, Steve was given Cub Scouting's highest award, the Arrow of Light.

Steve was graduated from the ''bandits'' after a year of learning the fundamentals of football. He played ''pony'' football for two years. But he was a large boy, and he found himself paired off against Mugsy again. Charles Dailey resigned himself: ''Steve didn't mind getting knocked down, getting blocked out and all that kind of stuff; but he just did not have the--what?--the killer instinct.''

In school, Steve got B's and C's. He received his First Communion, was confirmed, and learned how to serve Mass. He was a faithful altar boy who kept his serving appointments on holidays and vacations. But he wasn't above draining the last few drops of altar wine--or clowning with the incense in the vestry. By 1972, when he was eleven, Steve was well on his way toward his most important goal: to become an Eagle Scout. He worked at it steadily. By now his father was a Scout commissioner. He went along with Steve and his troop on most of their hikes and campouts. And he counseled Steve on five of the dozen merit badges he earned.

''Steve went after the merit badges that took a little more brains and thought,'' says Dan Carella. ''He was sensitive--not a rough kid. He wasn't a real loner, but he wasn't outgoing as much as some of the other kids. He liked to be with the older boys and the grown-ups. But there were a lot of older boys and younger ones, and he was in-between. That's one of the reasons he had no real close buddies. I can't really remember ever seeing him with any close buddy.''

At home, Steve and Mike started a beer-can collection. Steve learned to play chess. He read Hardy Boy mysteries. He got a new ten-speed bicycle for Christmas. And he went on a month-long camping trip to California with the whole family: Mike, sister Kay, and a new Dailey, his smallest brother, Jamie. Everybody visited Disneyland.

Back home, Sue and Charles Dailey noticed something--Mike was always outside playing baseball with the kids in the neighborhood. Steve preferred being alone. He worked on Scout projects or watched color television. His father thought it was because the other kids made up street rules for their game--and Steve insisted on playing by the correct rules.

By now Steve's father was athletic director at St. Catherine's. Steve worked long hours at fund-raising for the Dads' Club, which sponsored the parish teams. He took over the popcorn concession at basketball tournaments. "He'd get upset when I'd suggest he take a break and try to get some other kid to replace him so he could go watch the games," says Jay Deakin, past president of the Dads' Club.

During the 1972–1973 school year, Steve played "pee-wee" football, one level above "pony." So did Mugsy. "Steve always fought him off, but he'd get beat all the time," his father says. "There'd be nights when Steve'd say, 'Oh, he really wiped me out!' "

"It didn't frighten Steve to get hit," says coach Ray Bertran. "However, some boys, they go out and they look to hit the other kid. He wasn't that way. In 'pee-wee' I guess he was the biggest kid, but he just wasn't that aggressive."

Steve wasn't on the starting team. But one October evening, he came home from practice smiling.

"What happened?" asked his father.

"Boy, I really wiped him out tonight. I really got him."

Steve meant Mugsy. It was probably the only time that ever happened, Charles Dailey says.

"Steve never missed a Scout meeting. He added up the requirements to become an Eagle, allotted himself so much time to accomplish each, and put himself on a rigid schedule.

"Steve was really good at Scouting," says Charles Dailey, "and I really had a lot of pride in that."

Steve set his heart on a trip to the Philmont Scout Ranch in New Mexico, and started working at Scout projects to earn his way. He planned to work at a Scout car wash. And he never missed a Scout paper drive.

But he didn't go in for Scout roughhousing or free-for-alls. "Steve had sort of soft feelings," remembers Carella. "He was a very personal boy. He stuck up for the guys who were being picked on. During the district camporee in Mt. Airy Forest, there were a couple of kids who--well, they weren't mamma's boys, but they just didn't know how to handle themselves. A lot of the boys preferred to tent with other kids. But Steve said, 'Well, I'll go with them.' "

Last fall, Steve's father told him he had to play a fall sport. "I was thinking in terms of football," says Charles Dailey. But St. Catherine's had started soccer. Steve said he'd rather play that.

"He was aggressive on the soccer team," says football coach George Kugler. Dan Carella describes him as "a good soccer player." But soccer was not the prestige sport at St. Catherine's.

"Football at St. Catherine's is king," says Bill Coffey, a history teacher.

Carella discounts any attempt by Charles Dailey to pressure his son to play football. But he adds: "There probably was some pressure in the situation. His father is athletic director. The situation says, 'Hey, how come you're not playing, Steve?' "

To make enough money to go to the Philmont Scout Ranch, Steve wanted a summer job. His father arranged for the job with a company that shares space in the Cardinal Engineering Building, where he works as a civil engineer. But he didn't tell Steve about the arrangement. Better, he thought, to have Steve ask--and think he got the job himself. Lump in his throat, Steve accompanied his father to the Cardinal Building one morning in April. Fortified by a cup of chocolate milk and a donut from a bakery along the way, Steve marched in--and came out with a job making catalogs and cleaning up for 75 cents an hour starting when school let out this summer.

"Boy, he was a king then," Charles Dailey remembers.

Steve also got a newspaper route, with the weekly <u>Press</u>, which circulates in Westwood. That money would go toward Philmont. And he made arrangements with Aunt Beth McGinnis to mow her lawn for two dollars whenever it needed it. That would go toward Philmont, too.

In his own way, Steve Dailey was shy. When his mother went from room to room at night to check on her brood, she always got a kiss from Mike, a kiss from Kay, and a kiss from Jamie. But never from Steve. Kissing embarrassed him. His mother always thought: "Well, he'll come to me when he's ready."

On Mother's Day, Sunday, May 12, her children brought Sue Dailey breakfast in bed. Steve presented her with a terrarium he'd made in Scouts. And he put his arms around her and kissed her. By now, though, school wasn't going entirely well. Steve wasn't doing his homework for language arts. That was Margaret Linahan's class. And Steve was getting a D.

"In content subjects, like science and social studies, I suppose he could take his own path. But in English grammar there is only one way to go," Mrs. Linahan smiles. "As long as I'm your teacher." She told Sue and Charles Dailey their son's grades were falling.

"Hey, is something bothering you?" Steve's father asked him.

"No," Steve said.

"Hey, you know, if you fail anything you're going to be grounded in the yard the whole summer."

In Bill Coffey's history class, Steve slipped from an A to a B or B-plus. Coffey was one of his favorites. He, in turn, appreciated Steve's sense of humor. "In the last few weeks, he didn't talk as much," Coffey remembers. "He didn't participate. And his dry wit was no longer as present."

Steve paid a sentimental visit to Sister Marie Russell's fourth-grade classroom. "I wondered why he was not with his class," she recalls, eyes puzzled behind her glasses. "Why was he wandering in the hall? And why was he by himself? You'd think a twelve-year-old would be with the boys."

Though Steve was never what Bill Coffey calls "Joe Popularity," he was well-liked--and he was good friends with Ted Hutchinson, for instance, and Rick Flannery. But Charles Dailey was unaware that Steve had any close friends. He never went to any of his friends' houses to play--and never invited any of them to his house to play.

With spring came baseball, and a peak of activity in the Dailey household. "We would have to fix a pot of stew, where you could keep heating it up when people would come in and out, or chili, or something like that," Sue remembers. Her husband says: "Sometimes we'd just eat, and then the person that wasn't here, he'd have to warm up the stuff that was left."

Steve played on an intermediate-ability team. He was a starter. But manager Ray Kendrick says, "I'm not sure he really liked sports, at least not baseball. He wasn't that enthusiastic about it. . . ."

But now Charles Dailey headed in his spare time an athletic organization at St. Catherine's that totaled 110 coaches, almost all of them fathers who had volunteered. Four football teams . . . fourteen baseball teams . . . ten basketball teams . . . track . . . soccer . . . softball . . . volleyball . . . kickball. The parish sports budget totaled $11,491.

Steve's father says, "This year I think he wanted to play soccer again. But I told him that there wasn't any way, because in high school, well, he's just not going to be a soccer man . . . because he's plain too big, and never was real fast. I still had the hopes that this year he would finally find out, with the size and all on his side, that he would become more aggressive."

Steve Dailey, twelve years old, stood 5 feet, 5 inches, and weighed 140 pounds.

"Steve, you ready for football?" coach George Kugler asked him. "You ought to play. Get some fundamentals. You're gonna be a big kid. You can make tackle."

Ted Hutchinson remembers Steve Dailey saying: "My dad wants me to play football, but I'd rather play soccer."

Two weeks before the end of school, Bill Coffey asked him: "Steve, you gonna play football?"

"Yeah, I guess I have to," he shrugged. "My Dad wants me to lose 10 pounds because of the weight limit."

Steve put himself on another schedule, this time with weights. Across the top of a piece of notebook paper he marked places for the dates of each day until fall. Beneath that, he charted sit-ups, bench presses, snatches, lifts, push-ups, windmills, jumping jacks; he measured an oval in his backyard with a tape and started running laps.

Affectionately, Charles Dailey teased him about being a "big lop"--the nickname he'd been given when he'd grown to be 6 feet 2 inches, as a young man. But Mike told him Steve didn't like it. And he stopped.

On Sunday, May 26, Steve helped haul stones and build a form for the concrete foundation to a utility shed-workshop his father was putting up behind the house. He hurt his back and missed school on Monday. He missed baseball practice, too. And that was the second time--the first had been a short while before when he'd had to stay at home with Kay and Jamie while his mother took Mike to the doctor.

"Then he didn't show up for one of our games," says Ray Kendrick, the baseball manager. Charles Dailey thinks it was a make-up game. "I started someone else in his place," Kendrick says.

That week, smiling, Steve told Sister Marie Russell about his summer job. But he didn't dress up for Roaring Twenties Day in Bill Coffey's history class. Ted Hutchinson remembers: "He just sat there." And that week, Coffey remembers, he discussed Japanese hara-kiri in class. He recalls no reaction from Steve.

On Saturday, June 1, Steve's father took him to a Scout show. He bought Steve a souvenir patch. That evening Steve worked on his personal management merit badge, for which he drew up a budget. It set a fixed amount aside each month for the trip to Philmont. After dinner, he tried to show his family photo slides of Philmont, but the projector bulb blew out.

On Sunday, June 2, Steve helped clean the family camper for a Scout canoe trip the coming weekend. He wire-brushed the rust from its wheels and painted them white.

On Monday, June 3, he rode his bicycle in front of his house and hit a hole in the pavement. It pitched him over the handlebars. A neighbor was sure he'd been hurt, but he got up, looked around to be certain nobody had seen him and got back on his bike. One of its pedals was bent.

On Tuesday, June 4, two days before the end of school, Sue Dailey volunteered to staple the PTA bulletin together at St. Catherine's. She met Steve in the hall on his way to history. He called out, "Hi, Mom."

Margaret Linahan kept him after school to finish an assignment. When he got done, he found that Aunt Beth had already left. She was to have picked him up and taken him to her house so he could earn more Philmont money mowing the lawn. But she had biscuits in the oven and couldn't wait.

Steve walked home. He called his father at work: "I just want to tell you that Monday I wrecked my bike."

"Oh? Did you get hurt?"

"Yeah, I hurt my hand, and you know, it's pretty sore. I think I might have broken it."

Charles Dailey didn't think it was all that bad, or his son would have mentioned it before. He and Steve talked about the bicycle. Steve's father remembers saying, without raising his voice: "We'll take a look at it, and if you broke it that means you're going to have to pay for it."

"You know, I can't play ball, so I don't want to go to practice," Steve said.

"Well, you know, I think you ought to go, because you've missed here a few times, and if you're going to be part of the team you've got to go to practices, too."

"Well, I'm not gonna take my glove."

"I think that you ought to take the glove and all and just go on up."

Steve handed the telephone receiver to his mother, and she hung it up.

Steve walked out the back door. He had tears in his eyes. He went to the garage, found the rope, carried it down the trail to a dead tree in the woods. His father found his body the next morning. The baseball glove was nearby.

The terrible ifs accumulated:

Dan Carella: "If he'd come to me. . . ."

Sister Marie Russell: "Oh, if only I would have known, I would have gone out of my way to get him and really talk to him. . . ."

Margaret Linahan: "If I wouldn't have kept him after school. . . ."

Beth McGinnis: "If I would have just waited for him. . . ."

Sue Dailey: "If I'd have only said he didn't have to go to baseball practice. . . ."

Charles Dailey: "If I had gone back there [to the woods that night], he might have been able to keep his weight off the rope for a period of time, or something like that, and, you know, you could have helped him. . . ."

Steve's father says a police officer friend told him the rope wasn't tied, but only looped, around the tree limb. He believes his son didn't intend to die--but that the rope had held accidentally.

"Yet I don't question the fact that he got the rope and he went back there and he had tied the rope around his neck. You know, I just can't believe that Steve would really do that. Except that he had to have done it, I guess."

Other police officers and county coroner Philip Holman determined that the fastening around the tree limb was secure enough to rule out an accident. They declared Steve a suicide.

"Not infrequently, suicides are caused by intense anger or frustration," says Dr. Fedor Hagenauer, a pediatric psychiatrist at the University of Cincinnati. "Because this anger or frustration is addressed at people who are very important, children have a lot of guilty feelings about them. And then, because of the guilty feelings, and because the anger or frustration has to come out in some way, they might try to take it out on themselves . . . even with a token gesture, or going through the motions . . . maybe with a fantasy that they'll be rescued at the last minute . . . and they'll do it thinking, 'Everybody will see how unhappy I am and they'll learn and give in to what I'm unhappy about.' " It would have been impossible, he said, to predict Steve's fate.

The Rev. James Conway, who celebrated Requiem Mass at St. Catherine's, doesn't think Steve was morally responsible for his death.

During the Mass, Boy Scouts presented gifts to God symbolizing Steve's life. At Mike's suggestion, one was a soccer ball.

Questions on Assignment 12

Your instructor may have you discuss the following questions with a small group of your classmates. If so, appoint one member to keep brief notes on how the group responds to each question, for a general class discussion may follow. Whether you work in a small group or answer the questions individually, be sure to provide *specific evidence from the text* to back up all your points and ideas.

1. What kind of boy was Steve? What words (for example, *quiet, active, bright, lazy, shy*) would you use to describe his personality? Be sure to provide details from the text to support the qualities that you name.

2. a. In what ways was Steve a highly organized person?
    b. Why was he so organized, do you think?

3. Why did Steve feel pressured to play football rather than soccer (which he preferred)?

4. a. What kind of person does Margaret Linahan, Steve's English teacher, seem to be?
    b. Do you think that she was a good teacher for Steve?

5. a. What kind of person does Steve's mother, Sue Dailey, seem to be?
    b. Do you think that she was a good mother for Steve?

6. a. What are some examples of how Charles Dailey was sensitive to his son?
    b. What are some examples of how Charles Dailey was insensitive to his son?

7. Why do you think Steve's father believes his death was accidental?

8. Do you think that Charles Dailey's love for his son was destructive?

9. Why do you think Steve killed himself at the point when he did?

10. Why does the writer end his piece with the detail of the soccer ball?

Writing Activity for Assignment 12

Option 1: Imagine that you work for an adoption agency, and that Steve's father and mother have come to the agency to adopt a boy. After an investigation in which you learn all the details described in the article, would you deny or grant Steve's parents the right to adopt a boy? To make your decision, you will have to think carefully about and interpret the information about Steve's parents that is presented in the article.

Write a several-paragraph letter to your superior in which you recommend or do not recommend that they be granted the right of adoption. Include a brief introductory paragraph stating your recommendation (*thesis*) and summarizing briefly your reasons for the recommendation (*plan of development*). Then write a supporting paragraph for each of your reasons. Refer to specific words and actions of Steve's parents in developing the reasons for your recommendation.

Option 2: Write a several-paragraph essay in which you provide evidence from the text to support either one of the following points:

Steve's father and other adults in his life were insensitive to him and his needs.

Steve's father and other adults in his life should not be blamed for his suicide.

PART FOUR

SPECIAL SKILLS

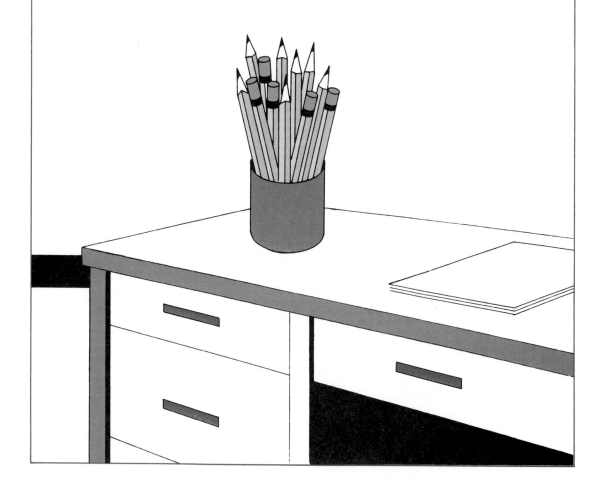

PREVIEW

Part Four presents skills that will help you in a variety of study and writing situations. "Developing Key Study Skills" explains five skills that will make you a better student in all your courses. In "Writing an Exam Essay," you will learn how to prepare for and take the essay exams that are among the most common types of writing students do in college. Similarly, "Writing a Summary," "Writing a Report," and "Writing a Résumé and Job Application Letter" provide explanations and models for three practical kinds of writing that you may need to know. Finally, "Using the Library" and "Writing a Research Paper" will prepare you for writing projects that require library research.

DEVELOPING KEY STUDY SKILLS

Following are brief guides to several important reading and study skills needed for success in school. A short comprehension test follows each guide. The skills include class note-taking, time control, concentrating, studying a textbook, and memory training.

TAKING EFFECTIVE CLASSROOM NOTES

One of the most important single things you can do to perform well in a college course is to take effective class notes. The following hints should help you become a better note-taker.

Attend Class

First, attend class faithfully. Your alternatives — reading the text or someone else's notes, or both — cannot substitute for the experience of hearing ideas in person as someone presents them to you. Also, in class lectures and discussions your instructor typically presents and develops the main ideas and facts of the course — the ones you will be expected to know on exams.

Use Abbreviations

Another valuable hint is to make use of abbreviations while taking notes. Using abbreviations saves time when you are trying to get down a great deal of information. Abbreviate terms that recur frequently in a lecture, and put a key to your abbreviations at the top of your notes. For example, in a sociology class, *eth* could stand for *ethnocentrism;* in a psychology class, *STM* could stand for *short-term memory*. (When a lecture is over, you may want to go back and write out the terms you have abbreviated.) In addition, abbreviate words that often recur in any lecture. For instance, use *e* or *ex* or *X* for *example; def* for *definition; info* for *information,* + for *and*, and so on. If you use the same abbreviations all the time, you will soon develop a kind of personal shorthand that makes taking notes much easier.

Be Alert for Signals

A third hint when taking notes is to be on the lookout for signals of importance. For one thing, write down *whatever your instructor puts on the board*. If he or she takes the time to put material on the board, it is probably important, and the chances are good that it will come up later on exams.

Always write down *definitions* and *enumerations*. Enumerations are lists of items. They are signaled in such ways as: ''The four steps in the process are . . .''; ''There were three reasons for . . .''; ''The two effects were . . .''; ''Five characteristics of . . .''; and so on. Always number (1, 2, 3, etc.) such enumerations in your notes. They will help you understand relationships among ideas and organize the material of the lecture.

Watch for *emphasis words*—words your instructor may use to indicate that something is important. Examples of such words are ''This is an important reason . . .''; ''A point that will keep coming up later . . .''; ''The chief cause was . . .''; ''The basic idea here is . . .''; and so on. Always write down the important statements announced by these and other emphasis words.

If your instructor *repeats* a point, you can assume it is important. You might put an *R* for *repeated* in the margin, so that later you will know that your instructor has stressed it.

Be sure to write down the instructor's *examples* and mark them with an *e, ex,* or *X*. The examples help you understand abstract points. If you do not write them down, you are likely to forget them later when they are needed to help make sense of an idea.

Also, be sure to write down the *connections between ideas*. Too many students merely copy the terms the instructor puts on the board. They forget that, as time passes, the details that serve as connecting bridges between ideas quickly fade. You should, then, write down the relationships and connections in class. That way you'll have them to help tie together your notes later on.

Review Your Notes

Review your notes as soon as possible after class. You must make them as clear as possible while they are fresh in your mind. A day later may be too late, because forgetting sets in very quickly. Make sure that punctuation is clear, that all words are readable and correctly spelled, and that unfinished sentences are completed (or at least marked off so that you can check your notes with another student's). Add clarifying or connecting comments wherever necessary. Make sure important ideas are clearly marked. Improve the organization if necessary, so that you can see at a glance main points and relationships among them.

Keep a Written Record

Finally, try in general to get down a written record of each class. You must do this because forgetting begins almost immediately. Studies have shown that within two weeks you are likely to have forgotten 80 percent or more of what you have heard. And in four weeks you are lucky if 5 percent remains! The significance of this is so crucial that it bears repeating: to guard against the relentlessness of forgetting, it is absolutely essential that you write down what you hear in class. Later on you can concentrate on working to understand fully and to remember the ideas that have been presented in class. And the more complete your notes are at this time of study, the more you are likely to learn.

Activity

Check your understanding by answering the following questions.

1. *True or false?* _____ Forgetting will begin several days after you have attended a class.

2. How would you abbreviate the term *traditional authority* during a fast-moving sociology lecture?

   _____

3. Often the most important single step you can take to perform well in a course is to
   a. sit where the instructor can see you and listen carefully.
   b. write down definitions and examples.
   c. be there and take effective notes.
   d. continue taking notes during discussion periods or at the end of a class.

4. Which of the following should be taken down in your notes?
   a. Enumerations
   b. Definitions
   c. Material written on the board
   d. All of the above

5. When you review your notes after a class, you should
   a. memorize them.
   b. clarify them.
   c. type them.
   d. reduce them.

CONTROLLING YOUR TIME

Success in college depends on time control. Time control means that you deliberately organize and plan your time, instead of letting it drift by. Planning means that you should never be faced with a night-before-the-test "cram" session or an overdue term paper. There are three steps involved in time control.

Prepare a Monthly Calendar

First, you should prepare a large monthly calendar. Buy a calendar with a large white block around each date, or make one yourself. At the beginning of the college semester, circle important dates on this calendar. Circle the days on which tests are scheduled; circle the days on which papers are due. This calendar can also be used to schedule study plans. You can jot down your plans for each day at the beginning of the week. An alternative method would be to make plans for each day the night before. On Tuesday night, for example, you might write down "Read Chapter 5 in psychology" in the Wednesday block. Be sure to hang this calendar in a place where you will see it every day — your kitchen, your bedroom, even your bathroom!

Prepare a Weekly Study Schedule

The *second step* in time control is to have a weekly study schedule for the semester. To prepare this schedule, make up a chart that covers all the days of the week and all the waking hours in each day. On the opposite page is part of one student's schedule.

On your schedule, mark in all the fixed hours in each day — hours for meals, classes, job (if any), and travel time. Next, mark in the time blocks that you can *realistically* use for study each day. Depending on the number of courses you are taking and the demands of the courses, you may want to block off five, ten, or even twenty or more hours of study time a week. Keep in mind that you should not block off time for study that you do not truly intend to use for study. Otherwise, your schedule will be a meaningless gimmick. Also, remember that you should allow time for "rest and relaxation" in your schedule. You will be happiest, and you will be able to accomplish the most, when you have time for both work and play.

Sample Weekly Study Schedule

	Monday	Tuesday	Wednesday	Thursday	Friday	Saturday	Sunday
6:00 A.M.							
7:00	B	B	B	B	B		
8:00	Math	STUDY	Math	STUDY	Math		
9:00	STUDY	Biology	STUDY	Biology	STUDY	Job	
10:00	Psychology		Psychology		Psychology		
11:00	STUDY	English	English	English			
12:00	L		L		L		

Make Up a "To Do" List

The *third step* in time control is to make a daily or weekly "to do" list. This may be the most valuable time control method you ever use. On this list, you write down the things you need to do for the following day or the following week. If you choose to write a weekly list, do it on Sunday night. If you choose to write a daily list, do it the night before.

You may use a three- by five-inch notepad or a small spiral-bound notebook for this list. Carry the list around with you during the day. Always concentrate on doing first the most important items on your list. To make the best use of your time, mark high-priority items with an asterisk and give them precedence over low-priority items. For instance, you may find yourself wondering what to do after dinner on Thursday evening. Among the items on your list are "Clean inside of car" and "Review chapter for math quiz." It is obviously more important for you to review your notes at this point; you can clean the car some other time. As you complete items on your "to do" list, cross them out. Do not worry about unfinished items. They can be rescheduled. You will still be accomplishing a great deal and making more effective use of your time.

Here is part of one student's daily "to do" list:

To Do   Tuesday

*1.   Review biology notes before class
*2.   Proofread English paper due today
* 3.   See Dick about game on Friday*
*4.   Gas for car
* 5.   Read next chapter of psychology text*

Activity

Check your understanding by answering the following questions.

1. *True or false?* _____ Do not list personal chores on your "to do" list.

2. *True or false?* _____ On a large monthly calendar, you should circle the dates when tests are scheduled and papers are due.

3. When you make up a "to do" list, you should
    a. schedule one-hour blocks of study time.
    b. mark down deadlines.
    c. decide on priorities.
    d. hang it on the wall.

4. On your weekly schedule, you should block out time periods for
    a. commuting time.
    b. study time.
    c. rest and relaxation.
    d. all of the above.

5. Circle the highest-priority item on the following "to do" list:
    a. Buy new shirt
    b. Study notes for tomorrow's lab quiz
    c. Wax car
    d. Preview next chapter for Friday's sociology class

DEVELOPING CONCENTRATION

Lack of concentration happens to everyone. But if you are a student, you cannot afford to let it happen too often. Nor can you merely say, "This time, I'm going to keep my mind on my work," and have it magically happen. A much better method is to figure out what in particular is interfering with your concentration — and then to do something about it.

Causes of Lack of Concentration

What causes lack of concentration? It can happen, first of all, because of physical factors. If you are sick, hungry, thirsty, or sleepy, you will not be able to concentrate. Or the factors could be mental, not physical. You could be suffering from indecision: other things keep popping into your head, and you do not know what to do first. You could be distracted by daydreaming or worrying about a personal problem. Finally, the cause of your failure to concentrate could be external. Your study environment could be the problem if it is too warm or too cold, too noisy, or not properly lighted. If your study spot lacks the right equipment — yellow pads, pens and pencils, or a dictionary, for instance — distraction may result. Any of these factors can take your mind off the assignment.

How to Concentrate

Concentration allows you to focus on just one thought or bit of information at a time, eliminating everything else. The first step in achieving a high level of concentration is having the right attitude. If you want to absorb something, you'll do so, no matter what else is going on — as anyone glued to a television set in the middle of a family argument proves. Take a positive attitude — here's a chance to learn something new. Or take a realistic attitude — the job simply has to be done today. And forget about how you felt or performed in the past. You are making a fresh start.

Next, if you want to concentrate, create a good atmosphere. Choose a room or area where you can work without being disturbed. Find a desk or table and clear it off. Sit in a straight chair. On your desk or table, place a good light source, such as a gooseneck fluorescent lamp. Then, add the study equipment you'll need: textbooks and notebooks, dictionary, pens, pencils, paper, a clock, calendar, calculator, and so on. Keep this special area only for studying; when you sit down there, the atmosphere will say "Work!"

The third step toward concentration is planning study time. Make up a weekly schedule and block out regular study periods that you plan to stick to. Schedule at least one hour of study time for each hour of class time. Organized, preplanned study time can aid concentration — you know exactly when you will study, and you're sure that there won't be any conflicts then with other activities.

Finally, study with a goal in mind. Do not just read the history chapter; read it to discover the specific causes of World War II. Work quickly. Have a clock in sight or put your watch next to your book, and try to cover each page in a designated period of time. Plan to take a break after a certain number of pages; keep the break brief, however. Remember, you want to get the job done so that you can go on to the next item on your schedule. If you do feel your concentration slipping as you study, put a pencil mark in the margin each time you are distracted. The mark will remind you to concentrate.

What to Avoid: A Few Warnings

When you are trying to concentrate, watch out for several pitfalls. For one thing, do not try to study when you are physically uncomfortable. If you are too warm, open a window. If you are hungry, eat. Make sure you get enough sleep. And make a real effort to exercise regularly. Your mind will concentrate better when your body is in shape.

In addition, do not allow distractions to disturb you. Some people claim that noise does not bother them when they are trying to study. In general, though, it helps to avoid noisy places such as a college cafeteria or a lounge, and to keep the radio and television turned off. Do not waste brain power blocking out sounds that intrude on your concentration. Should a distracting thought get in your way, write it down and go back to your studying. You can always deal with the idea later; once it's written down, it can't be lost.

Last of all, don't be a passive learner. When you read, do not just sit quietly and let your eyes roam across the page. Get involved with the assignment. Have a pen or pencil in hand and use it to underline, check, and star material. Write key words and questions in the margins. Turn the chapter headings and subheadings into questions and look for the answers as you read. You might want to close your eyes after you have read several paragraphs and see if you can restate what you have learned. If you cannot, go back, reread, and try again. If nothing else works, read aloud. You'll hear the words as well as see them — and the more senses you involve in studying, the easier the process will be.

Activity

Check your understanding by answering the following questions.

1. *True or false?* _____ Lack of concentration could be caused by hunger.
2. The first step in achieving a high level of concentration is
   a. buying a fluorescent light.
   b. solving a personal problem.
   c. studying with a goal in mind.
   d. having the right attitude.

3. *True or false?* _____ Concentrating means reading a textbook selection as slowly as you can.
4. Being an active learner means
   a. underlining material.
   b. turning headings into questions.
   c. trying to recall what you read.
   d. all of the above.

5. A good study area should have
   a. soothing music.
   b. a straight chair.
   c. a telephone.
   d. exercise equipment.

STUDYING TEXTBOOKS SYSTEMATICALLY

In many college courses, success means being able to read and study a textbook skillfully. For many students, unfortunately, textbooks are heavy going. After an hour or two of study, the textbook material is as formless and as hard to understand as ever. But there is a way to attack even the most difficult textbook and make sense of it. Use a sequence in which you preview a chapter, mark it, take notes on it, and then study the notes.

Previewing

Previewing a selection is an important first step to understanding. Taking the time to preview a section or chapter can give you a bird's-eye view of the way the material is organized. You will have a sense of where you are beginning, what you will cover, and where you will end.

There are several steps in previewing a selection. First, study the title. The title is the shortest possible summary of a selection and will often tell you the limits of the material you will cover. For example, the title "FDR and the Supreme Court" tells you to expect a discussion of President Roosevelt's dealing with the Court. You know that you will probably not encounter any material dealing with FDR's foreign policies or personal life. Next, read over quickly the first and last paragraphs of the selection; these may contain important introductions to, and summaries of, the main ideas. Then examine briefly the headings and subheadings in the selection. Together, the headings and subheadings are a minioutline of what you are reading. Headings are often main ideas or important concepts in capsule form; subheadings are breakdowns of ideas within main areas. Finally, read the first sentence of some paragraphs, look for words set off in **boldface** or *italics,* and look at pictures or diagrams. After you have previewed a selection in this way, you should have a good general sense of the material to be read.

Textbook Marking

You should mark a textbook selection at the same time that you read it through carefully. Use a felt-tip highlighter to shade material that seems important, or use a regular ballpoint pen and put a symbol in the margin next to the material: a star, a check, or *NB* (for *nota bene,* a Latin phrase meaning "note well"). What to mark is not as mysterious as some students believe. You should try to find main ideas by looking for the following clues: definitions and examples, enumerations, and emphasis words.

1 ***Definitions and examples:*** Definitions are often among the most important ideas in a selection. They are particularly significant in introductory courses in almost any subject area, where much of your learning involves mastering the specialized vocabulary of that subject. In a sense, you are learning the "language" of psychology or business or whatever the subject might be.

   Most definitions are abstract, and so they usually are followed by one or more examples to help clarify their meaning. Always mark off definitions and at least one example that makes a definition clear to you. In a psychology text, for example, we are told that "rationalization is an attempt to reduce anxiety by deciding that you have not really been frustrated." Several examples follow, among them: "A young man, frustrated because he was rejected when he asked for a date, convinces himself that the woman is not very attractive and is much less interesting than he had supposed."

2 ***Enumerations:*** Enumerations are lists of items (causes, reasons, types, and so on) that are numbered 1, 2, 3, . . . or that could easily be numbered in an outline. They are often signaled by addition words. Many of the paragraphs

in this book use words like *First of all, Another, In addition,* and *Finally* to signal items in a series. Other textbooks also use this very common and effective organizational method.

3   ***Emphasis words:*** Emphasis words tell you that an idea is important. Common emphasis words include phrases such as *a major event, a key feature, the chief factor, important to note, above all,* and *most of all.* Here is an example: "The most significant contemporary use of marketing is its application to nonbusiness areas, such as political parties."

Textbook Note-Taking

Next, you should take notes. Go through the chapter a second time, rereading the most important parts. Try to write down the main ideas in a simple outline form. For example, in taking notes on a psychology selection, under a main heading, "Defense Mechanisms," you might write down the subheadings "Definition" and "Kinds." Below the second subheading you would number and describe each kind and give an example of each.

> *Defense Mechanisms*
>
> a.   *Definition: unconscious attempts to reduce anxiety*
> b.   *Kinds:*
>   (1)   *Rationalization: An attempt to reduce anxiety by deciding that you have not really been frustrated*
>   *Example: A man turned down for a date decides that the woman was not worth going out with anyway*
>   (2)   *Projection: Projecting onto other people motives or thoughts of one's own*
>   *Example: A wife who wants to have an affair accuses her husband of having one*

Studying Your Text Notes

To study your notes, use the method of repeated self-testing. For example, look at the subheading "Kinds" under "Defense Mechanisms" and say to yourself, "What are the kinds of defense mechanisms?" When you can recite them, then say to yourself, "What is rationalization?" "What is an example of rationalization?" Then ask yourself, "What is projection?" "What is an example of projection?" After you learn each section, review it, and then go on to the next section.

Do not simply read your notes; keep looking away and seeing if you can recite them to yourself. This self-testing is the key to effective learning.

In conclusion, remember this sequence in order to deal with a textbook: previewing, marking, taking notes, studying the notes. Approaching a textbook in this methodical way will give you very positive results. You will no longer feel bogged down in a swamp of words, unable to figure out what you are supposed to know. Instead, you will understand exactly what you have to do, and how to go about doing it.

Activity

Check your understanding by answering the following questions.

1. *True or false?* _____ Previewing a textbook means marking off definitions and examples.
2. A textbook writer might let you know an idea is important by using
   a. stars.
   b. emphasis signals.
   c. illustrations.
   d. addition signals.
3. Previewing a textbook selection includes
   a. reading the title.
   b. looking at diagrams.
   c. examining subheadings.
   d. all of the above.
4. *True or false?* _____ Headings are usually set off in larger type or different-colored ink.
5. Enumerations are
   a. definitions.
   b. outlines.
   c. lists of items.
   d. examples.

BUILDING A MORE POWERFUL MEMORY

Two Basic Steps in Remembering: Organization and Self-Testing

Effective studying and remembering require, first, that you organize the material to be learned. Organization means preparing study notes made up of headings and subheadings, definitions and examples, enumerations, and other important points. The very process of organizing material and condensing it to the main points will help you understand and remember it.

After you have organized your subject material, memorize it through repeated self-testing. Look at the first item in your notes; then look away and try to repeat it to yourself. When you can, look at the next item; then look away and try to repeat it. When you can repeat the second item, go back without looking at your notes and try to repeat both the first and second items. This constant review is at the heart of self-testing. After you can repeat the first two items without looking at your notes, go on to the third item. When you learn it, try to repeat all three items without looking at your paper. After you learn each new item, go back and test yourself on all the previous items you have studied.

Six Aids to Memorization

1 *Intend to remember:*   The first aid to memory is intending to remember. This bit of advice appears to be so obvious that many people overlook its value. But if you have made the decision to remember something, and you then work at mastering it, you will remember. Anyone can have a bear-trap memory by working at it; no one is born with a naturally poor memory.

2 *Overlearn:*   Overlearning is a second memory aid. If you study a subject beyond the time needed for perfect recall, you will increase the length of time that you will remember it. The method of repeated self-testing uses the principle of overlearning; you can also apply the principle by going over several times a lesson you have already learned perfectly.

3 *Space memory work:*   Spacing memory work over several sessions, rather than a single long one, is the third aid. Just as with physical exercise, five two-hour sessions spaced over several days are more helpful than ten hours all at once. The spaced sessions allow material time to ''sink in'' (psychologists would say ''transfer from short-term to long-term memory''). Spacing the sessions also helps you ''lock in'' material you have studied in the first session but have begun to forget. Studies show that forgetting occurs most rapidly soon after learning ends — but that review within a day or two afterward counters much memory loss.

4 *Study before bedtime:*   A fourth aid in memorizing material is studying just before going to bed. Do not watch a late movie or allow any other interference between studying and sleep. Then be sure to review the material immediately when you get up in the morning. Set your clock a half hour earlier than usual so that you will have time to do this. The review in the morning will help ''lock in'' the material that you have studied the night before and that your mind has worked over during the night.

5 *Use key words as ''hooks'':*   A fifth helpful tool is using key words in an outline as ''hooks.'' Reduce your outline of a passage to a few key words and memorize those words. The key words you master will then serve as ''hooks'' that will help you pull back entire ideas into your memory.

6 ***Use memory formulas:*** Sixth and finally, another tool is using memory formulas to help you recall points under a main idea, items in a list, steps in a procedure, or other things arranged in a series. For example, you might remember the four methods used in behavior therapy (extinction, reinforcement, desensitization, and imitation) by writing down the first letter in each word (*e r d i*) and remembering the letters by forming an easily recalled catchphrase (''Ellen's rolling dice inside'') or rearranging them to form an easily recalled word (*r i d e*). The letters serve as hooks that help you pull in words that are often themselves hooks for entire ideas.

Activity

Check your understanding by answering the following questions.

1. The two basic steps in remembering are organizing the material to be learned and
    a. studying before bedtime.
    b. testing yourself repeatedly.
    c. preparing study notes.
    d. memorizing definitions.

2. *True or false?* _____ Material is best studied in a single long session rather than spaced out over several sessions.

3. *True or false?* _____ Some people are born with naturally poor memories.

4. Overlearning means
    a. studying beyond the time needed for perfect recall.
    b. transferring facts to the short-term memory.
    c. using memory formulas.
    d. studying to the point of diminishing returns.

5. The first aid to memory is
    a. creating ''hooks.''
    b. self-testing.
    c. intending to remember.
    d. using study notes.

WRITING
AN
EXAM ESSAY

Examination essays are among the most common types of writing that you will do in college. They include one or more questions to which you must respond in detail, writing your answers in a clear, well-organized manner. This section describes five basic steps needed to prepare adequately for an essay test and to take the test. It is assumed, however, that you are already doing two essential things: first, attending class regularly and taking notes on what happens in class; second, reading your textbook and other assignments and taking notes on them. If you are not consistently going to class, reading your text, and taking notes in both cases, perhaps you should be asking yourself and talking with others (counselors, instructors, friends) about your feelings on being in school.

To write an effective exam essay, follow these five steps:

Step 1: Try to anticipate the probable questions on the exam.

Step 2: Prepare and memorize an informal outline answer for each question.

Step 3: Read exam directions and questions carefully, and budget your time.

Step 4: Prepare a brief outline before answering an essay question.

Step 5: Write a clear, well-organized essay.

Each step will be explained and illustrated on the pages that follow.

STEP 1: ANTICIPATE PROBABLE QUESTIONS

Anticipating probable questions is not as hard as you might think. Because exam time is limited, the instructor can give you only several questions to answer. He or she will — reasonably enough — focus on questions dealing with the most important parts of the subject. You can probably guess most of them as you make up a list of ten or more likely questions.

Your class notes are one key. What topics and ideas did your instructor spend a good deal of time on? Similarly, in your textbook, what ideas are emphasized? Usually the keys to the most important ideas are headings and subheadings, definitions and examples, enumerations, and ideas marked by emphasis signals. Also, take advantage of any study guides that may have been given out, any questions you may have gotten on quizzes, and any reviews that your instructor may have provided in class. You should, then, be able to determine the most important areas of the subject and make up questions that cover them.

STEP 2: PREPARE AND MEMORIZE AN INFORMAL OUTLINE ANSWER FOR EACH QUESTION

Write out each question you have made up and, under it, list the main points that need to be discussed. Put important supporting information in parentheses after each main point. You now have an informal outline that you can memorize.

An Illustration of Step 2

One class was given a day to prepare for an essay exam. The students were told that the essay item would be "Describe six aids to memory." One student, Teri, made up the following outline answer.

Six Memory Aids

1. *(Intend) to remember (personal decision is crucial)*
2. *(Overlearn) (helps you remember longer)*
3. *(Space) memory work (several sessions rather than one long session)*
4. *Time before (bed) as a study period (no interference; review the next morning)*
5. *Use (key) words as "hooks" (help pull into memory whole ideas)*
6. *Memory (formulas) (letters serve as hooks to help you remember items in a series)*

IOSBKF (I often see Bill kicking footballs.)

Activity

See whether you can complete the following explanation of what Teri has done in preparing for the essay item.

First, Teri wrote down the heading and numbered the different items under it.

Also, in parentheses beside each point she added _____ . Next, she circled the key word for each hint and wrote down the first

_____ of each key word below her outline. She then used the first letter in each key word to make up a catchphrase she could remember

easily. Finally, she _____ herself repeatedly until she could recall the key words the letters stood for and the main points the key words represented.

STEP 3: LOOK AT THE EXAM CAREFULLY AND BUDGET YOUR TIME

First, read all test directions carefully. This seems like an obvious point, but people are often so anxious at the start of a test that they fail to read the instructions, and so they never understand clearly and completely just what they must do.

Second, read all the essay questions carefully, first noting the *direction word* or *words* in each question that tell you just what to do. For example, *enumerate,* or *list,* means ''number 1, 2, 3, and so on''; *illustrate* means ''explain by giving examples''; *compare* means ''give similarities''; *contrast* means ''give differences''; *summarize* means ''offer a condensed account of the main ideas.''

Finally, budget your time, depending on the point value of each question and its difficulty for you. Write in the test margin the approximate time you give yourself to answer each question. This way you will not end the exam with too little time to respond to a question that you know you can answer.

An Illustration of Step 3

When Teri received the exam, she circled the direction word *Describe,* which she knew meant ''Explain in detail.'' She also jotted a ''30'' in the margin when the teacher said that students would have a half hour to write their answers.

STEP 4: PREPARE A BRIEF OUTLINE
BEFORE ANSWERING AN ESSAY QUESTION

Too many students make the mistake of anxiously and blindly starting to write. Instead, you should first jot down the main points you want to discuss in an answer. (Use the margin of the exam or a separate piece of scratch paper.) Then decide how you will order the points in your essay. Write *1* in front of the first point, *2* beside the second, and so on. You now have an informal outline to guide you as you write your essay answer.

If there is a question on an exam which is similar to the questions you anticipated and outlined at home, quickly write down the catchphrase that calls back the content of your outline. Below the catchphrase, write the key words represented by each letter in it. The key words, in turn, will remind you of the ideas they represent. If you have prepared properly, this step will take only a minute or so, and you will have before you the guide you need to write a focused, supported, organized answer.

An Illustration of Step 4

Teri first recited her catchphrase to herself, using it as a guide to write down her memory formula (IOSBKF) at the bottom of the test sheet. The letters in the formula helped her call into memory the key words in her study outline, and she wrote down those words. Then in parentheses beside each word she added the supporting points she remembered.

By investing only a couple of minutes, Teri was able to reconstruct her study outline and give herself a clear and solid guide to use in writing her answer.

STEP 5: WRITE A CLEAR,
WELL-ORGANIZED ESSAY

If you have followed the suggestions to this point, you have done all the preliminary work needed to write an effective essay. Be sure not to wreck your chances of getting a good grade by writing carelessly. Instead, as you prepare your response, keep in mind the principles of good writing: unity, support, coherence, and clear, error-free sentences.

First, start your essay with a sentence that clearly states what your paper will be about. Then make sure that everything in your paper relates to your opening statement.

Second — although you must obviously take time limitations into account — provide as much support as possible for each of your main points.

Third, use transitions to guide your reader through your answer. Words such as *first, next, then, however,* and *finally* make it easy for the reader to follow your train of thought.

Last, leave time to edit your essay for sentence-skills mistakes you may have made while you concentrated on writing your answer. Look for words omitted, miswritten, or misspelled (if it is possible, bring a dictionary with you); for awkward phrasing or misplaced punctuation marks; for whatever else may prevent the reader from understanding your thought. Cross out any mistakes and make your corrections neatly above the errors. If you want to change or add to some point, insert an asterisk at the appropriate spot, put another asterisk at the bottom of the page, and add the corrected or additional material there.

An Illustration of Step 5

Read through Teri's answer, reproduced here, and then do the activity that follows.

	There are six aids to memory. One aid is to intend to remember.
	An important part of success is making the ~~decison~~ *decision* that you are truly
	going to remember something. A second aid is overlearning. ~~Studing~~
	Studying a lesson several more times after you know it will help you
	remember it longer. Also, space study over several sessions rather
	than one long one.* Another aid is to study *just* just before bedtime and
	not watch TV or do anything else after study. The mind ~~ad~~ absorbs
	the material during the night. ~~The most~~ You should then review the
	material the next morning. A fifth memory aid is *to* use key words as
	hooks. For example, the key word "bed" helped me remember the
	entire idea about the value of studying right before sleep. A final aid
	to memory is memory formulas. I used the catchphrase "I often see
	Bill kicking footballs" and IOSBKF to recall the first letters of the six
	memory aids.
	* You need time in between for material to "sink in."

Activity

The following sentences comment on Teri's essay. Fill in the missing word or words in each case.

1. Teri begins with a sentence that clearly states what her paper

   _____. Always begin with such a clear signal!

2. The six transitions that Teri used to guide her reader, and herself, through the six points of her answer are:

   _____   _____   _____

   _____   _____   _____

3. Notice the various _____ that Teri made when writing and proofreading her paper. She neatly crossed out miswritten or unwanted words; she used her _____ after she had finished her essay to correct a misspelled word; she used insertion signs ($_\wedge$) to add omitted words; and she used an asterisk (*) to add an omitted detail.

PRACTICE IN WRITING AN EXAM ESSAY

Activity 1

Check your understanding of the chapter by answering the following questions.

1. When you receive an exam essay in class, you should first
   a. tackle the easiest question.
   b. write a brief outline of each question.
   c. read all test directions carefully.
   d. memorize a catchphrase.

2. *True or false?* _____ Because of time limits, you do not have to worry about the four bases of good writing on an exam essay.

3. *True or false?* _____ It is possible to anticipate essay exam questions.

4. Clues to possible essay exam questions are
   a. enumerations.
   b. ideas marked by emphasis signals.
   c. reviews in class.
   d. all of the above.

5. A catchphrase can help you
   a. recall an informal outline prepared for a test question.
   b. proofread your essay for sentence-skills mistakes.
   c. budget your time on an essay.
   d. use transitions to guide the reader through your answer.

Activity 2

As the assignment for your next class, your instructor may ask you to study pages 253–255 so that you will be ready to write an in-class paper in response to the statement, ''Describe five hints that will make you a better note-taker.''

You should do well on this paper if you follow the five basic steps in writing an effective exam essay that have been presented in this chapter.

Activity 3

1. Make up five questions you might be expected to answer on an essay exam in a social science or physical science course (sociology, psychology, biology, or some other).
2. Then, for each of the five questions, make up an outline answer comparable to the one on memory training.
3. Finally, write a full essay answer, in complete sentences, for one of the questions. Your outline will serve as your guide.

Be sure to begin your essay with a statement that makes the direction of your answer clear. For example, ''The six different kinds of defense mechanisms are defined and illustrated below.'' If you are explaining in detail the different causes of, reasons for, or characteristics of something, you may want to develop each point in a separate paragraph. For instance, if you were answering a question in sociology about the primary functions of the family unit, you could, after starting with a statement that ''There are three primary functions of the family unit,'' go on to develop and describe each function in a separate paragraph.

You will turn in the essay answer to your English instructor, who will evaluate it using the standards for effective writing applied to your other written assignments.

WRITING
A SUMMARY

At some point in a course, your instructor may ask you to write a summary of a book, an article, a TV show, or the like. In a *summary* (also referred to as a *précis* or an *abstract*), you reduce material in an original work to its main ideas and key supporting points. Unlike an outline, however, a summary does not use symbols such as I, A, 1, 2, and so on to indicate the relationships among parts of the original material.

A summary may be a word, a phrase, several sentences, or one or more paragraphs in length. The length of the summary you prepare will depend on your instructor's expectations and the length of the original work. Most often, you will be asked to write a summary of one or more paragraphs.

Writing a summary brings together a number of important reading, study, and writing skills. To condense the original matter, you must preview, read, evaluate, organize, and perhaps outline the material. Summarizing, then, can be a real aid to understanding; you must "get inside" the material and realize fully what is being said before you can reduce its meaning to a few words.

HOW TO SUMMARIZE AN ARTICLE

To write a summary of an article, follow the steps described below. If the assigned material is a TV show or film, adapt the suggestions accordingly.

1 Take a few minutes to preview the work. You can preview an article in a magazine by taking a quick look at the following.
   a *Title:* The title often summarizes what the article is about. Think about the title for a minute and how it may condense the meaning of an article.
   b *Subtitle:* A subtitle, if given, is a short summary appearing under or next to the title. For example, in a *Newsweek* article titled "Splitting Up the Family," the following caption appeared: "The courts are changing rules of divorce and child custody—and often making things worse." The subtitle, the caption, or any other words in large print under or next to the title often provide a quick insight into the meaning of an article.

c   *First and last several paragraphs:* In the opening paragraphs, the author may introduce the subject and state the purpose of the article. In the closing paragraphs, the writer may present conclusions or a summary. In either case, you get a quick overview of what the entire article is about.

d   *Other items:* Note any headings or subheadings that appear in the article. They often provide clues to the main points and give an immediate sense of the content of each section. Look carefully at any pictures, charts, or diagrams that accompany the article. Page space in a magazine or journal is limited, and such visual aids are generally used only when they help illustrate important points in the article. Note any words or phrases set off in *italic type* or **boldface type;** such words have probably been emphasized because they deal with important points.

2   Quickly read the article for a general understanding the first time through. Do not slow down or turn back. Mark off what seem to be main ideas and key supporting points. Pay special attention to all the items noted in the preview. Look for major enumerations (lists of items), since these often indicate key ideas. Also, try to identify important information by turning headings and subheadings into questions and by reading to find the answers to the questions.

3   Go back and reread more carefully the areas you have identified as most important. Also, focus on other key points you may have missed in your first reading.

4   Keep the following items in mind when working on the rough drafts of your summary.

a   Express the author's ideas in your own words. Do not imitate or stay too close to the style of the original work.

b   Do not write an overly detailed summary. Remember that the purpose of a summary is to reduce the original material to its main ideas and essential supporting points. A paragraph summary should be between 150 and 200 words in length.

c   Do not begin your sentences with expressions like ''the author says''; equally important, do not introduce your own opinions into the summary with comments like ''another good point made by the author.'' Instead, concentrate on presenting the author's main ideas directly and briefly.

d   Preserve the balance and proportion of the original work. If the original devoted 70 percent of its space to one idea and only 30 percent to another, your summary should reflect that emphasis.

e   As you work on the summary, pay attention to the principles of effective writing (unity, support, coherence, and clear, error-free sentences) explained in Part One.

f   Write the final draft of the summary.

A MODEL SUMMARY

Here is a model summary of a magazine article.

> In ''Breaking the Divorce Cycle'' (Newsweek, January 13, 1992), Barbara Kantrowitz reports on the first generation of children who have seen widespread divorce and who are now adults. Still carrying wounds from their childhood traumas, they often have difficulties with their own love relationships. Reports show that they are more likely than children of intact families to fear commitment and to have troubled relationships and broken marriages. Social scientists aren't sure whether these adjustment problems stem from the unhappy parental relationship before the divorce, the divorce experience itself, the economic decline following the divorce, or life in a single-parent family. In addition, there seems to be no way to predict how well children of divorce will do as adults. Some factors, however, have been identified as having a positive influence. One is having had a parent with a strong positive attitude. Another is having had parents who continued to communicate with each other about their children and to share parenting decisions. For whatever reasons, some adult children of divorce have managed to overcome their childhood pain and somehow learn from their parents' mistakes. These are the adults who have ended up with the independence and maturity to build a strong marriage of their own.

HOW TO SUMMARIZE A BOOK

To write a summary of a book, first preview the book by taking a quick look at the following:

1   *Title:* The title is often the shortest possible summary of what a book is about. Think about the title for a minute and how it may summarize the whole meaning of the work.

2   *Table of contents:* The contents will tell you the number of chapters in the book and the subject of each chapter. Use the contents to get a general sense of how the book is organized. You should also note the number of pages in each chapter. If thirty pages are devoted to one episode or idea and an average of fifteen pages to other episodes or ideas in the book, you should probably give more space in your summary to the contents of the longer chapter.

3   *Preface:* Here you will probably find the author's reasons for writing the book. Also, the preface may summarize the main ideas developed in the book and may describe briefly how the book is organized.

4 *First and last chapters:* In these chapters, the author may preview and review important ideas and themes developed in the book.

5 *Other items:* Note the way the author has used headings and subheadings to organize the information in the book. Check the opening and closing paragraphs of each chapter to see if they contain introductions or summaries. Look quickly at charts, diagrams, and pictures in the book, since they are probably there to illustrate key points. Note any special features (index, glossary, appendices) that may appear at the end of the book.

After previewing the book, use steps 2 through 4 on page 275 as a guide in preparing a summary.

PRACTICE IN WRITING A SUMMARY

Activity 1

Check your understanding of the chapter by answering the following questions.

1. Key ideas in an article may be signaled by the
   a. table of contents.
   b. title and subtitle.
   c. preface.
   d. index.

2. After previewing an article, you should
   a. read the article.
   b. take notes on the material.
   c. write a first draft of your summary.
   d. quote from the material.

3. *True or false?* _____ Use expressions such as ''the author says'' when writing your summary.

4. The table of contents in a book will give you an idea of
   a. how much space in the book is devoted to a particular idea.
   b. why the author wrote the book.
   c. the conclusions the author has reached.
   d. all of the above.

5. *True or false?* _____ A summary should include the key supporting points of a selection.

Activity 2

Use the guidelines in this chapter to write a one-paragraph summary of one of the following sections of the chapter "Developing Key Study Skills":

Controlling Your Time
Developing Concentration
Studying Textbooks Systematically
Building a More Powerful Memory

Your summary should be no fewer than 100 words and no more than 125 words. A summary of one section is given below as a model.

A Summary of "Taking Effective Classroom Notes"

The following hints can help you take better notes in your classes. First, attend class faithfully. Copying another student's notes cannot substitute for hearing the ideas in person. Another important hint is to develop a system of abbreviations. Using abbreviations for often-used words and special terms will speed note-taking. Putting a key to abbreviations at the top of the page will prevent confusion. In addition, be alert for signals of importance: anything written on the board, definitions, enumerations, emphasis words, and repeated ideas. Writing down examples and connections between ideas is also essential. Next, review your notes after class to clarify and expand them. Finally, take notes in every class you attend. Otherwise, you are sure to forget most of what you hear.

Activity 3

Write a one-paragraph summary of an article of interest to you in a weekly or monthly magazine. Be sure to use the guidelines in this chapter when reading the article and preparing the summary. Hand in a copy of the article along with your summary.

Activity 4

Watch a television show. Then prepare a summary of the show. In your first sentence, give basic information about the show by using a format such as the following: "The January 31, 1993, broadcast of CBS's *60 Minutes* examined. . . ."

WRITING
A REPORT

Each semester, you will probably be asked by at least one instructor to read a book or an article (or watch a TV show or a film) and to write a paper recording your response to the material. In these reports, or reaction papers, your instructor will most likely expect you to do two things: summarize the material and detail your reaction to it. The following pages explain both parts of a report.

PART 1 OF A REPORT:
A SUMMARY OF THE WORK

To develop the first part of a report, do the following:

1 Identify the author and title of the work and include in parentheses the publisher and publication date. An example follows on page 281. For magazines, give the date of publication.

2 Write an informative summary of the material. Condense the content of the work by highlighting its main points and key supporting points. (See pages 274–278 for a complete discussion of summarizing techniques.) Use direct quotations from the work to illustrate important ideas.

   Do not discuss any single aspect of the work in great detail while neglecting to mention other equally important points. Summarize the material so that the reader gets a general sense of *all* key aspects of the original work. Also, keep the summary objective and factual. Do not include in the first part of the paper your personal reaction to the work; your subjective impression will form the basis of the second part of the paper.

PART 2 OF A REPORT:
YOUR REACTION TO THE WORK

To develop the second part of a report, do the following:

1   Focus on any or all of the following questions. (Check with your instructor to see if he or she wants you to emphasize specific points.)

    a   How is the assigned work related to ideas and concerns discussed in the course for which you are preparing the paper? For example, what points made in the course textbook, class discussions, or lectures are treated more fully in the work?

    b   How is the work related to problems in our present-day world?

    c   How is the material related to your life, experiences, feelings, and ideas? For instance, what emotions did the work arouse in you? Did the work increase your understanding of a particular issue? Did it change your perspective in any way?

2   Evaluate the merit of the work: the importance of its points, its accuracy, completeness, organization, and so on. You should also indicate here whether or not you would recommend the work to others, and why.

POINTS TO KEEP IN MIND
WHEN WRITING A REPORT

Here are some important matters to consider as you prepare a report.

1   Apply the four basic standards of effective writing (unity, support, coherence, and clear, error-free sentences).

    a   Make sure each major paragraph presents and then develops a single main point. For example, in the model report in the next section, the first paragraph summarizes the book, and the three paragraphs that follow detail three separate reactions of the student writer. The student then closes the report with a short concluding paragraph.

    b   Support your general points or attitudes with specific reasons and details. Statements such as ''I agreed with many ideas in this article'' or ''I found the book very interesting'' are meaningless without specific evidence that shows why you feel as you do. Look at the model report closely to see the way the main point or topic sentence of each paragraph is developed by specific supporting evidence.

  c Organize the material in the paper. Follow this basic *plan of organization:* a summary of one or more paragraphs, a reaction of two or more paragraphs, and a conclusion. Also, use *transitions* to make the relationships among ideas in the paper clear.

  d Edit the paper carefully for errors in grammar, mechanics, punctuation, word use, and spelling.

2 Document quotations from all works by placing the page number in parentheses after the quoted material. Look at the example in the first paragraph of the model report. You may use quotations in the summary and reaction parts of the paper, but do not rely on them too much. Use them only to emphasize key ideas.

A MODEL REPORT

Here is a report written by a student in an introductory psychology course. Look at the paper closely to see how it follows the guidelines for report writing described in this chapter.

A Report on <u>Man's Search for Meaning</u>

**PART 1:
SUMMARY
Topic
sentence
for
summary
paragraph**

Dr. Viktor Frankl's book <u>Man's Search for Meaning</u> (New York: Washington Square Press, 1985) is both an autobiographical account of his years as a prisoner in Nazi concentration camps and a presentation of his ideas about the meaning of life. The three years of deprivation and suffering he spent at Auschwitz and other Nazi camps led to the development of his theory of Logotherapy, which, very briefly, states that the primary force in human beings is "a striving to find a meaning in one's life" (154). Without a meaning in life, Frankl feels, we experience emptiness and loneliness that lead to apathy and despair. This need for meaning was demonstrated to Frankl time and again with both himself and other prisoners who were faced with the horrors of camp existence. Frankl was able to sustain himself partly through the love he felt for his wife. In a moment of spiritual insight, he realized that this love was stronger and more meaningful than death, and would be a real and sustaining force within him even if he knew his wife was dead. Frankl's comrades also had reasons to live that gave them strength. One had a child waiting for him; another was a scientist who was working on a series of books that needed to be finished. Finally, both Frankl and his friends found meaning through their decision to accept and bear their fate with courage. He says that the words of Dostoevski came frequently to mind: "There is one thing that I dread: not to be worthy of my suffering." When Frankl's prison experience was over and he returned to his profession of psychiatry, he found that his theory of meaning held true not only for the prisoners but with all people. He has since had great success in working with patients by helping them locate in their own lives meanings of love, work, and suffering.

PART 2: REACTION Topic sentence for first reaction paragraph

One of my reactions to the book was the relationship I saw between the "Capos" and ideas about anxiety, standards, and aggression discussed in our psychology class. The Capos were prisoners who acted as trustees, and Frankl says they acted more cruelly toward the prisoners than the guards or the SS men. Several psychological factors help explain this cruelty. The Capos must have been suppressing intense anxiety about "selling themselves out" to the Nazis in return for small favors. Frankl and other prisoners must have been a constant reminder to the Capos of the courage and integrity they themselves lacked. When our behavior and values are threatened by someone else acting in a different way, one way we may react is with anger and aggression. The Capos are an extreme example of how, if the situation is right and we will not risk our "civilized image," we may be capable of great cruelty to those whose actions threaten our standards.

Topic sentence for second reaction paragraph

I think that Frankl's idea that meaning is the most important force in human beings helps explain some of the disorder and discontent in the world today. Many people are unhappy because they are caught in jobs where they have no responsibility and creativity; their work lacks meaning. Many are also unhappy because our culture seems to stress sexual technique in social relationships rather than human caring. People buy popular books that may help them become better partners in bed, but that may not make them more sensitive to each other's human needs. Where there is no real care, there is no meaning. To hide the inner emptiness that results from impersonal work and sex, people busy themselves with the accumulation of material things. With television sets, stereos, cars, expensive clothes, and the like, they try to forget that their lives lack true meaning. Instead of working or going to school to get a meaningful job, or trying to be decent human beings and friends with the people around them, or accepting their troubles with quiet determination and dignity, they frantically buy, buy, buy to distract themselves from the lack of real meaning within.

Topic sentence for third reaction paragraph

I have also found that Frankl's idea that suffering can have meaning helps me understand the behavior of people I know. I have a friend named Jim who was always poor and did not have much of a family--only a stepmother who never cared for him as much as for her own children. What Jim did have, though, was determination. He worked two jobs to save money to go to school, and then he worked and went to school at the same time. The fact that his life was hard seemed to make him bear down all the more. On the other hand, I can think of a man in my neighborhood who for all the years I've known him has done nothing with his life. He spends whole days standing on his front porch, or on the sidewalk, or on the street corner, smoking and looking at cars go by. He is a burned-out case. Somewhere in the past his problems must have become too much for him, and he gave up. He could have found meaning in his life by deciding to fight his troubles like Jim, but he didn't, and now he is a sad shadow of a man. Without determination and the desire to face his hardships, he lost his chance to make his life meaningful.

Concluding
paragraph
 In conclusion, I would strongly recommend Frankl's book to persons who care about why they are alive, and who want to truly think about the purpose and meaning of their lives.

PRACTICE IN WRITING A REPORT

Activity 1

Check your understanding of the chapter by answering the following questions.

1. *True or false?* _____ Each major paragraph in a report should develop a single main point.
2. A summary of a work should
   a. be opinionated.
   b. focus on one key idea.
   c. be objective and factual.
   d. show how the work relates to your life.

3. *True or false?* _____ Quotations should be documented by placing the page number in parentheses after the quoted material.
4. The two basic parts of a report are
   a. a summary and supporting details.
   b. a summary and a reaction.
   c. a plan of organization and quotations.
   d. a reaction and a final draft.
5. A reaction to a work could include how the work relates to
   a. a certain course.
   b. current problems.
   c. your own life.
   d. all of the above.

Activity 2

Read a magazine article that interests you. Then write a report on the article. Include a one-paragraph summary, a reaction of one or more paragraphs, and a brief concluding paragraph. You may, if you like, quote briefly from the article. Be sure to enclose in quotation marks the words that you take from the article and put the page number in parentheses at the end of the quoted material.

Activity 3

Read a book suggested by your instructor. Then write a report on the book. Include a summary of at least one paragraph, a reaction of two or more paragraphs, and a brief concluding paragraph. Also, make sure that each major paragraph in your report develops a single main point. You may quote some sentences from the book, but they should be only a small part of your report. Follow the directions in Activity 2 above when you quote material.

WRITING
A RÉSUMÉ
AND
JOB APPLICATION
LETTER

When applying for a job through the mail, you should ordinarily send (1) a résumé and (2) a job application letter.

RÉSUMÉ

A résumé is a summary of your personal background and your qualifications for a job. It helps a potential employer see at a glance whether you are suited for a job opening. A sample job résumé is shown on the following page.

Points to Note about the Résumé

1 Your résumé, along with your letter of application, is your introduction to a potential employer. First impressions count; thus it is extremely important that you make the résumé neat!

    a If possible, type the résumé on good-quality letter (8½- by 11-inch) paper. Keep at least a 1-inch margin on all sides.

    b Edit and proofread very carefully for sentence-skills and spelling mistakes. A potential employer may regard such mistakes as signs of carelessness in your character. You might even want to get someone else to proofread the résumé for you.

    c If possible, be brief and to the point: use only one page.

    d Use a format like that of the model résumé. (See also the variations described below.) Balance your résumé on the page so that you have at least a 1-inch margin on all sides.

    e Note that you should start with your most recent education or job experience and work backward in time.

ROBERTA LYONS
715 Baltic Avenue
Atlantic City, New Jersey 08401
609-821-8715

Education

1990–Present   Atlantic Community College
Mays Landing, New Jersey 08330

By June I will have completed my Associate of Science degree, including twenty-four hours of courses in my major, Hotel-Motel Management. In addition, I will have taken twelve hours of courses in accounting. Presently I have a B+ average in my major and a B in my other courses.

1986–1990   Pleasantville High School
Pleasantville, New Jersey 08310

Job Experience

1990–Present   Checker and Assistant to the Manager, A & P Supermarket, Atlantic City, New Jersey 08401

During holiday periods in the two years I have worked at the A & P, I have served as an assistant to the manager by supervising the preparation of party trays.

1989–1990   Clerk, Seven-Eleven Food Store, Absecon, New Jersey 08300

My job also included some bookkeeping and stock inventory.

Special Training

As part of one of my courses, Seminar in Hospitality Management, I have helped plan and supervise the preparation of luncheons for several college affairs. I have also privately catered several baptisms and weddings for relatives and friends.

References

My references are available upon request from the Placement Office at Atlantic Community College, Mays Landing, New Jersey 08330.

2 Your résumé should point up strengths, not weaknesses.

    a Do not include "Special Training" if you have had none. Do not refer to your grade-point average if it is a low C.

    b On the other hand, include a main heading like "Extracurricular Activities" if the activities or awards seem relevant. For example, if Roberta Lyons had been a member of the Management Club or vice president of the Hospitality Committee in high school or college, she should have mentioned these facts.

    c If you have no work experience related to the job for which you are applying, then list the jobs you have had. Any job that shows a period of responsible employment may favorably impress a potential employer.

3 You can list the names of your references directly on the résumé. Be sure to get the permission of people you cite before listing their names. You can also give the address of a placement office file that holds references, as shown on the model report. Or you can simply say that you will provide references on request.

JOB APPLICATION LETTER

The purpose of the letter of application that goes with your résumé is to introduce yourself briefly and to try to make an employer interested in you. You should include only the high points of the information about yourself in the résumé.

Shown on the next page is the letter of application that Roberta Lyons sent with her résumé.

Points to Note about the Job Application Letter

1 Your letter should do the following:

    a In the first paragraph, state that you are an applicant for the job and note the source through which you learned about the job.

        Here is how Roberta's letter might have opened if her source had been a newspaper: "I would like to apply for the position of Assistant Banquet Manager which was advertised in yesterday's *Philadelphia Inquirer*."

        Sometimes an ad will list only a box number (such as Y-140) to reply to. Your inside address should then be:

Y-140
Philadelphia Inquirer
Philadelphia, Pennsylvania 19101

Dear Sir or Madam:

715 Baltic Avenue
Atlantic City, New Jersey 08401
March 3, 1992

Mr. Philip McGuire
Banquet Manager
Sheraton Hotel
1725 John F. Kennedy Boulevard
Philadelphia, Pennsylvania 19103

Dear Mr. McGuire:

I learned through the Placement Office at Atlantic Community College of the Assistant Banquet Manager position at the Sheraton Hotel. I would like to be considered as a candidate for the job.

I will graduate from Atlantic Community College in June with an Associate of Science degree in Hotel-Motel Management. The degree will include twenty-four hours of courses related to the hospitality industry and twelve hours of accounting. Also, you will see from my résumé that I have work-related experience as well, including two years as an assistant to the manager at an A & P Supermarket.

If my qualifications and experience are of interest to you, I will be happy to come for an interview at your convenience. I think you will find me an eager, motivated, and able person, truly interested in working in the hospitality field.

Sincerely yours,

Roberta Lyons

Roberta Lyons

b In the second paragraph, state briefly your qualifications for the job and refer the reader to your résumé.

c In the last paragraph, state your willingness to come for an interview. If you can be available for an interview at only certain times, indicate this.

2 As with the résumé, neatness is crucial. Follow the same hints for the letter that you did for the résumé:

a Type the letter on good-quality letter (8½- by 11-inch) paper.

b Proofread very carefully for sentence-skills and spelling mistakes. Use the checklist of sentence skills on the inside front cover.

c Be brief and to the point: use no more than one page.

d Use a format like the model letter. Keep at least a 1-inch margin on all sides.

e Use punctuation and spelling in the model letter as a guide. For example:

Skip two spaces between the inside address and the salutation (''Dear Mr. McGuire:'').

Use a colon after the salutation.

Sign your name at the bottom, in addition to typing it.

PRACTICE IN WRITING A RÉSUMÉ AND JOB APPLICATION LETTER

Activity 1

Check your understanding of the chapter by answering the following questions.

1. *True or false?* _____ A job application letter should tell an employer about your family and hobbies.

2. A résumé should always list
   a. names of references.
   b. previous jobs.
   c. height and weight.
   d. your grades.

3. *True or false?* _____ A résumé should be brief: one page if possible.

4. On a résumé, you should
   a. admit your weak areas.
   b. list your most recent job last.
   c. highlight your strengths.
   d. do all of the above.

5. The purpose of a job application letter is to
   a. try to make an employer interested in you.
   b. repeat all the information in your résumé.
   c. take the place of an interview.
   d. do all of the above.

Activity 2

Clip a job listing from a newspaper or copy a job description posted in your school placement office. The job should be one that you feel you are presently qualified for or that you would be qualified for in the future.

Write a résumé and a letter of application for the job. Use the models already considered as guides. Your instructor may let you invent some of the specific supporting material needed to make you appear a strong candidate for the job.

Use the checklist of the four bases on the inside front cover as a guide in your writing. Be sure to edit and proofread both the résumé and the application letter carefully, and give them to someone else to check as well. Each should be nothing less than perfect, because a prospective employer will use them as a first indication of whether you are a good candidate for a job.

USING
THE
LIBRARY

This chapter provides the basic information you need to use your college library with confidence. It also describes the basic steps you should follow in researching a topic.

Most students seem to know that libraries provide study space, typing facilities, and copying machines. They are also usually aware that a library has a reading area, which contains recent copies of magazines and newspapers. But the true heart of a library consists of the following: a *main desk,* a *book file, book stacks,* a *magazine file,* and a *magazine storage area.* Each of these will be discussed on the pages that follow.

PARTS OF THE LIBRARY

Main Desk

The main desk is usually located in a central spot. Check with the main desk to see if there is a brochure that describes the layout and services of the library. You might also ask if the library staff provides tours of the library. If not, explore your library to find each of the areas described below.

Activity

Make up a floor plan of your college library. Label the main desk, book file, book stacks, magazine file, and magazine storage area.

Book File

The book file will be your starting point for almost any research project. The book file is a list of all the books in the library. It may be an actual card catalog: a file of cards alphabetically arranged in drawers. Increasingly, however, the book file is computerized, and it appears on a number of computer terminals located at different spots in the library.

Finding a Book — Author, Title, and Subject: Whether you use an actual file of cards or a computer terminal, it is important for you to know that there are three ways to look up a book. You can look it up according to *author, title,* or *subject.*

For example, suppose you want to see if the library has *Savage Inequalities,* by Jonathan Kozol. You could check for the book in any of three ways:

1 You could go to the *Titles* section of the book file and look it up there under *S*. Note that you always look up a book under the first word in the title, excluding the word *A, An,* or *The*.

2 You could go to the *Authors* section of the book file and look it up there under *K*. An author is always listed under his or her last name. Here is the author entry in a card catalog for Kozol's book *Savage Inequalities*:

LC4091
K69

Kozol, Jonathan
Savage inequalities: children in America's schools/Jonathan Kozol
— New York: Crown Publishers, 1991.
262 pp.
Includes notes on sources.

1. Socially handicapped children — Education — United States.   2. Children of minorities — Education — United States.   3. Segregation in education — United States.
4. Education, Urban — Social aspects — United States.

---

3 Or, since you know the subjects that the book deals with — in this case, ''education of children'' — you could go to the *Subjects* section of the book file and look it up under *E*.

Generally, if you are looking for a particular book, it is easier to use the *Authors* or the *Titles* section of the book file. But if you hope to find other books about your topic, then the *Subjects* section is where you should look. There you can get a list of all the books in the library that deal with, say, the education of children. The *Subjects* section will also give you specific subject headings within your topic, making it easier for you to find books about a limited area.

■ In the catalog card shown on the preceding page, how many subject headings are listed under which you might find specific kinds of books about the education of children? _____

Using a Computerized Book File: Recently I visited a local library that had just been computerized. The card catalog was gone, and in its place was a table with ten computer terminals. I approached a terminal and looked, a bit uneasily, at the instructions placed nearby. The instructions turned out to be very simple. They told me that if I wanted to look up the author of a book, I should type "A = " on the keyboard in front of the terminal and then the name of the author. I typed "A = Kozol Jonathan," and then (following the directions) I hit the keyboard Enter/Return key.

In seconds a new screen appeared showing me a numbered list of eight books by Jonathan Kozol, one of which was *Savage Inequalities.* This title was numbered "8" on the list, and at the bottom of the screen was a direction to type the number of the title I wanted more information about. So I typed the number "8" and hit the Enter/Return key. I then got the following screen:

AUTHOR:	Kozol, Jonathan
TITLE:	Savage Inequalities
PUBLISHER:	Crown, 1991
SUBJECTS:	1. Socially handicapped children — Education — United States. 2. Children of minorities — Education — United States. 3. Segregation in education — United States. 4. Education, Urban — Social aspects — United States.

Call Number	Material	Location	Status
371.967Koz	Book	Cherry Hill	Available

I was very impressed. The terminal was easier and quicker to use than a card catalog. The screen gave me the basic information I needed to know about the book, including where to find it. In addition, the screen told me that the book was "Available" on the shelves. (A display card nearby explained that if the book was not on the shelves, the message under "Status" would be "Out on loan.") I noticed other options. If the book was not on the shelves at the Cherry Hill location of the library, I would be told if it was available at other libraries nearby, by means of interlibrary loan.

The computer gave me two other choices. I could type "T = " plus a name to look up the title of a book. Or I could type "S = " plus the subject to get the names of any books that the library had dealing with the subject.

294 PART FOUR: SPECIAL SKILLS

Using Subject Headings to Research a Topic: Whether your library has a card catalog or a computer terminal, it is the *Subjects* section that will be extremely valuable to you when you are researching a topic. If you have a general topic, the *Subjects* section will help you find books on that general topic and will suggest more limited topics.

For example, after I typed "S = Education of children," a screen came up showing me eighteen different titles. In addition, the screen listed a number of subtopics under which I could find related books about the topic. One of these subtopics was "Socially handicapped children — Education — United States." When I typed in this heading, I found Kozol's book and similar books on socially handicapped children.

The *Subjects* section of the book file, then, can be extremely helpful when you are researching a topic. There are two points to remember: (1) Start researching a topic by using the *Subjects* section of the book file. (2) Use the subtopics and related topics suggested by the book file to help you narrow your topic. Chances are, you will be asked to do a paper of about five to fifteen pages. You do not want to choose a topic so broad that it could be covered only by an entire book or more. Instead, you want to come up with a limited topic that can be adequately supported in a relatively short paper.

Activity

Part A: Answer the following questions about the card catalog.

1. Is your library's book file an actual file of cards in drawers, or is the book file on computer terminals?

    _____

2. What are the three ways of looking up a book in the library?

    a. _____

    b. _____

    c. _____

3. Which section of the book file will help you research and limit a topic?

    _____

Part B: Use your library book file to answer the following questions.

1. What is the title of one book by Alice Walker?

   _____

2. What is the title of one book by George Will?

   _____

3. Who is the author of *The Making of the President?* (Remember to look up the title under *Making,* not *The.*)

   _____

4. Who is the author of *A Separate Peace?* _____

5. List two books dealing with the subject of marriage, and note their authors.

   a. _____

   b. _____

6. List two books dealing with the subject of AIDS, and note their authors.

   a. _____

   b. _____

7. Look up a book titled *The Road Less Travelled* or *Passages* or *The American Way of Death* and give the following information:

   a. Author _____

   b. Publisher _____

   c. Date of publication _____

   d. Call number _____

   e. Subject headings: _____

8. Look up a book written by Barbara Tuchman or Russell Baker or Bruce Catton and give the following information:

   a. Title _____

   b. Publisher _____

   c. Date of publication _____

   d. Call number _____

   e. Subject headings: _____

Book Stacks

The book stacks are the library shelves where books are arranged according to their call numbers. The call number, as distinctive as a social security number, always appears in a book file for any book. It is also printed on the spine of the book.

If your library has open stacks (ones that you are permitted to enter), follow these steps to find a book. Suppose you are looking for *Savage Inequalities,* which has the call number LC4091.K69 in the Library of Congress system. (Libraries using the Dewey Decimal system have call letters made up entirely of numbers rather than letters and numbers. However, you use the same basic method to locate a book.) First you go to the section of the stacks that holds the *L*'s. After you locate the *L*'s, you look for the *LC*'s. After that, you look for *LC4091*. Finally, you look for *LC4091/K69*, and you have the book.

If your library has closed stacks (ones you are not permitted to enter), you will have to write down title, author, and call number on a slip of paper. (Such paper will be available near the card catalog or computer terminals.) You'll then give the slip to a library staff person, who will locate the book and bring it to you.

Activity

Use the book stacks to answer one of the following set of questions. Choose the questions that relate to the system of classifying books (Library of Congress or Dewey Decimal) used by your library.

Option 1: Library of Congress System (Letters and Numbers)

1. Books in the BF21 to BF833 area refer to
   a. philosophy.
   b. sociology.
   c. psychology.
   d. history.

2. Books in the HV580 to HV5840 area deal with what type of social problem?
   a. drugs
   b. suicide
   c. white-collar crime
   d. domestic violence

3. Books in the PR4740 to PR4757 area deal with
   a. James Joyce.
   b. Jane Austen.
   c. George Eliot.
   d. Thomas Hardy.

4. What aspect of the environment is dealt with in the area between TD196 and TD763?
    a. air
    b. water
    c. soil
    d. vegetation

Option 2: Dewey Decimal System (Numbers)

1. Books in the 320 area deal with
    a. self-help.
    b. divorce.
    c. science.
    d. politics.
2. Books in the 546 to 547 area deal with
    a. biology.
    b. chemistry.
    c. physics.
    d. anthropology.
3. Books in the 636 area deal with
    a. animals.
    b. computers.
    c. marketing.
    d. senior citizens.
4. Books in the 709 area deal with
    a. camping.
    b. science fiction.
    c. art.
    d. poetry.

Magazine File

The magazine file is also known as the *periodicals* file. *Periodicals* (from the word *periodic*, which means "at regular periods") are magazines, journals, and newspapers. In this chapter, the word *magazine* stands for any periodical.

The magazine file often contains recent information about a given subject, or very specialized information about a subject, which may not be available in a book. It is important, then, to check magazines as well as books when you are doing research.

Just as you use the book file to find books on your subject, you use the magazine file to find articles on your subject in magazines and other publications. There are two files in particular that should help: *Readers' Guide to Periodical Literature* and the *Magazine Index*.

Readers' Guide to Periodical Literature: The familiar green volumes of the *Readers' Guide,* found in just about every library, list articles published in almost two hundred popular magazines, such as *Newsweek, Health, People, Ebony, Redbook,* and *Popular Science.* Articles are listed alphabetically under both subject and author. For example, if you wanted to learn the names of articles published on the subject of child abuse within a certain time span, you would look under the heading "Child abuse." An extract from the *Readers' Guide* is shown on the opposite page; here is a typical entry:

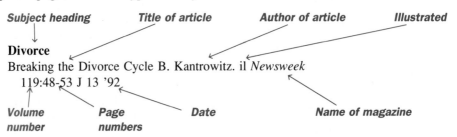

Note the sequence in which information is given about the article:

1 Subject heading.
2 Title of the article. In some cases, there will be bracketed words after the title that help make clear just what the article is about.
3 Author (if it is a signed article). The author's first name is always abbreviated.
4 Whether the article has a bibliography (*bibl*) or is illustrated with pictures (*il*). Other abbreviations are shown in the front of the *Readers' Guide.*
5 Name of the magazine. A short title like *Time* is not abbreviated, but longer titles are. For example, the magazine *Popular Science* is abbreviated *Pop Sci.* Refer to the list of magazines in the front of the index to identify abbreviations.
6 Volume number of the magazine (preceding the colon).
7 Page numbers on which the article appears (after the colon).
8 Date when the article appeared. Dates are abbreviated: for example, *Mr* stands for *March, Ag* for *August, N* for *November.* Other abbreviations are shown in the front of the *Readers' Guide.*

The *Readers' Guide* is published in monthly supplements. At the end of a year, a volume is published covering the entire year. You will see in your library large green volumes that say, for instance, *Readers' Guide 1989* or *Readers' Guide 1991.* You will also see the small monthly supplements for the current year.

The *Readers' Guide* is also now available in a much more useful form, on a computer. I was amazed to see at my local library that I could now sit down at a terminal and quickly search for an article on almost any subject published in the last seven years. Searching on the computer was much easier than having to go through seven or so different paper volumes of the *Readers' Guide.*

An Excerpt from Readers' Guide

EXERCISE
See also
Aerobics
Bodybuilding
Boxercise
Exercising equipment
Eye exercises
Flo-Motion
Gymnastics
Health clubs
Leg exercises
Periodization (Athletic training)
Running
Sports
Stretching exercises
Tape recordings — Exercise use
Walking
Water exercises
Weight lifting
Yoga
Abdominal showman. R. Brody. il *Men's Health* 6:24-5 Je '91
Build your back through your belly. G. Gutfeld. il *Prevention (Emmaus, Pa.)* 43:58-64 Jl '91
Exercises fit for a princess [exercise routine used by Princess Diana] L. Webb. il pors *Good Housekeeping* 213:92+ S '91
Fitness. See issues of *Men's Health*
Get fit fast [40 minute workouts] D. Bensimhon. il *Men's Health* 6:66-71 Ag '91
Growing fitter. S. Levin. il *New Choices for the Best Years* 31:42-4 Je '91
I want a flat stomach, fast! [special section] il *Glamour* 89:146-51 Je '91
Making love better with exercise [research by Phillip Whitten] J. S. Chou. il *McCall's* 118:20 S '91
Rear view: exercises and tips to improve your bottom line. il *Redbook* 177:16 S '91
The rewards of total transformation; ed. by Stephanie Ebbert. S. Miller. il pors *Prevention (Emmaus, Pa.)* 43:87-8+ Jl '91
Shape up! [cover story; special section] R. Sutton. il *American Health* 10:37-51 Je '91
Summer slimmers: take off 6 inches in 60 days. S. Lally. il *Prevention (Emmaus, Pa.)* 43:33-41+ Jl '91
Tuning up your body for summer fitness & fun. il *Ebony* 46:118+ Jl '91

When should you exercise? M. J. Schnatter. il *McCall's* 118:14 Jl '91
Workout wonders right at home. '*Teen* 35:49 Je '91

Physiological effects
The athlete's dilemma [metabolic ceiling] J. M. Diamond. il *Discover* 12:78-83 Ag '91
Cruise control [antidizzines or seasickness exercises] il *Prevention (Emmaus, Pa.)* 43:10+ Jl '91
Daily exercise fights hypertension, clots. K. Fackelmann. *Science News* 139:342 Je 1 '91
Effects of exercise on the human immune system. L. D. Caren. bibl f il *BioScience* 41:410-15 Je '91
Exercise! P. D. Wood. il *World Health* p25-7 My/Je '91
Exercise is hip [reducing risk of hip fractures; research by Annlia Paganini-Hill] il *Prevention (Emmaus, Pa.)* 43:12 Je '91
Pump down high blood pressure. G. Gutfeld. il *Prevention (Emmaus, Pa.)* 43:60-4+ Je '91
Putting an end to sore muscles. S. Y. Lee. il *McCall's* 118:18 Ag '91
Regular exercise cuts diabetes risk. K. Fackelmann. *Science News* 140:36 Jl 20 '91
Twofold path to saving aging bones [calcium and exercise; research by Miriam E. Nelson] *Science News* 139:367 Je 8 '91

Psychological aspects
A bridge too far: confessions of a fitness agnostic. R. Lipsyte. il *American Health* 10:26-7 Je '91
The burnout factor. S. Levin. il *Women's Sports & Fitness* 13:12-13 Jl/Ag '91
Don't be an exercise dropout. S. Browder. il *Reader's Digest* 139:168-9+ Ag '91
The never-ending workout [exercise addiction] *Mademoiselle* 97:88+ Je '91
EXERCISE CLUBS *See* Health clubs
EXERCISE STRESS TESTS *See* Physical fitness — Testing
EXERCISE TESTING *See* Physical fitness — Testing
EXERCISES, MILITARY *See* Military maneuvers
EXERCISING EQUIPMENT
Working in [home gym] S. Omelianuk. il *Gentlemen's Quarterly* 61:183+ S '91

Magazine Index: The *Magazine Index* is an automated system that lists articles in about four hundred general-interest magazines. Given a choice, you should always use this system rather than the *Readers' Guide*: it lists articles from twice as many sources as the *Readers' Guide* and is both fast and easy to use.

You sit in front of what looks like a large television screen that has already been loaded with a microfilmed index. By pushing the first of two buttons, you quickly advance the film forward from *A* to *B* to *C,* and so on. By pushing the other button, you move in the opposite direction. It really is as simple as that! The entries on the screen look just like the entries in the *Readers' Guide.* You'll note that the most recent articles on a topic are given first. This machine is an excellent research tool that is finding its way into more and more libraries.

Activity 1

At this point in the chapter, you now know the two basic steps in researching a topic in the library. What are the steps?

1. _____

2. _____

Activity 2

Use the excerpt from the *Readers' Guide* on page 299 to answer the following questions.

1. Who is the author of an article titled "Don't Be an Exercise Dropout"?

    _____

2. What is the title of an article by S. Levin?

    _____

3. How many articles are listed that deal with the psychological aspects of exercise?

    _____

4. In what issue of *McCall's* is there an article about exercise?

    _____

5. On what pages of *World Health* is the article "Exercise!"?

    _____

Activity 3

1. Look up a recent article on divorce in the *Readers' Guide* or the *Magazine Index* and fill in the following information:

    a. Article title _____

    _____

    b. Author (if given) _____

    c. Name of magazine _____

    d. Pages _____

    e. Date _____

2. Look up a recent article on date rape in the *Readers' Guide* or the *Magazine Index* and fill in the following information:

    a. Article title _____

    _____

    b. Author (if given) _____

    c. Name of magazine _____

    d. Pages _____

    e. Date _____

Specialized Indexes: Once you know how to use the *Readers' Guide,* you will find it easy to use some of the more specialized indexes in most libraries. Here are some helpful ones:

■ *New York Times Index.* This is an index to articles published in *The New York Times.* After you look up a subject, you'll find a list of articles published on that topic, with a short summary of each article.

■ *Business Periodicals Index.* The articles here are from over three hundred publications that generally treat a subject in more detail than it would receive in the popular magazines indexed in the *Readers' Guide.* At the same time, the articles are usually not too technical or too hard to read.

■ *Social Sciences Index.* This is an index to articles published by journals in the areas of anthropology, environmental science, psychology, and sociology. Your instructors in these areas may expect you to consult this index while doing a research project on one of these subjects.

■   Other specialized indexes that your library may have include the following:

Art Index
Applied Science and Technology Index
Biological and Agricultural Index
Book Review Digest
Education Index
General Science Index
Humanities Index
Nursing Index
Religious Periodical Literature Index

Depending on the subject area you are researching, you may want to consult the appropriate index listed above. Note that some libraries have most of these indexes on a computer, as well as *Readers' Guide*.

Activity

1.   Check the magazine area in your library. (It might be known as the *periodicals* area.) Place a check by each of the indexes that it includes:

_____ *Readers' Guide*

_____ *Business Periodicals Index*

_____ *Magazine Index*

_____ *Social Sciences Index*

_____ *New York Times Index*

2.   Are any of these indexes available on a computer as well as in paperbound

volumes? _____ If so, which ones? _____

3.   What are two other indexes in this area of your library besides the five mentioned above?

_____

_____

A Note on Other Reference Materials: Every library has a reference area, often close to the place where the *Readers' Guide* is located, in which other reference materials can be found. Such general resource materials include dictionaries, encyclopedias, atlases, yearbooks, almanacs, a subject guide to books in print (this can help in locating books on a particular subject), anthologies of quotations, and other items.

You may also find in the reference area a series of filing cabinets called the *pamphlet file*. This will consist of a series of file cabinets full of pamphlets, booklets, and newsletters on a multitude of topics. One entire file drawer, for example, may include all the pamphlets and the like for subjects that start with *A*. I looked in the *A* drawer of the pamphlet file in my library and found lots of booklets about subjects like abortion, adoption, and animal rights, along with many other topics starting with *A*. On top of these filing cabinets may be a booklet titled ''Pamphlet File Subject Headings''; it will quickly tell you if the file includes material on your subject of interest.

Activity

1. What is one encyclopedia that your library has?

   _____

2. What unabridged dictionary does your library have?

   _____

3. Where is your library's pamphlet file located?

   _____

4. Is there a booklet or small file that tells you what subject headings are included in the pamphlet file? _____

   Where it is? _____

Magazine Storage Area

Near your library's *Readers' Guide* or *Magazine Index,* you'll probably notice slips of paper. Here, for instance, is a copy of the slip used in my local library:

PERIODICAL REQUEST

Name of magazine _____

Date of magazine _____

(For your reference: Title and pages of article:)

_____

As you locate each magazine and journal article that you would like to look at, fill out a slip. When you are done, take the slips to a library staff person working nearby. Don't hesitate to do this: helping you obtain the articles you want is part of his or her job.

Here's what will probably happen next:

- If a magazine that you want is a very recent one, it may be on the open shelves in the library. The staff member will tell you, and you can go find it yourself.
- If the magazine you want is up to a year old or so, it may be kept in a closed area. The staff person will go find it and bring it to you.
- Sometimes you'll ask to see an article in a magazine that the library does not carry. You'll then have to plan to use other articles, or go to a larger library. However, most college libraries and large county libraries should have what you need.
- In many cases, especially with older issues, the magazine will be on microfilm or microfiche. (*Microfilm* is a roll of film on which articles have been reproduced in greatly reduced size; *microfiche* is the same thing but on easily handled sheets of film rather than on a roll.) The staff person will bring you the film or fiche and *at your request* will then show you how to load this material onto a microfiche or microfilm machine nearby.

Faced with learning how to use a new machine, many people are intimidated and nervous. I know I was. What is important is that you ask for as much help as you need. Have the staff person demonstrate how the machine is used and then watch you as you use it. (Remember that this person is being paid by the library to help you learn how to use the resources in the library, including the machine.) The machine may seem complex at first, but in fact most of the time it turns out to be easy to use. Don't be afraid to insist that the person give you as much time as you need to learn to use the machine.

After you are sure you can use the machine to look up any article, check to see if the machine will make a copy of the article. Many will. Make sure you have some change to cover the copying fee, and then go back to the staff person and ask him or her to show you how to use the print option on the machine. You'll be amazed at how quickly and easily you can get a printed copy of almost any article you want.

Activity

1. Use the *Readers' Guide* or the *Magazine Index* to find an article on divorce that was published in the last three months. Write down the name of the magazine and the date on a slip of paper and give it to a library staff person.

   Is the article available in the actual magazine? _____ If so, is it on an open shelf or is it in a closed area where a staff person must bring it to you?

   _____

2. Use the *Readers' Guide* or the *Magazine Index* to find an article on divorce that was published more than one year ago. Write down the name of the magazine and the date on a slip of paper and give it to a library staff person. Is the article available in the actual magazine or is it on microfiche or microfilm?

———————————————————————————————

3. Place a check if your library has:

———— Microfiche machine ———— with a print option

———— Microfilm machine ———— with a print option

A Summary of Library Areas

You now know about the five areas of the library that will be most useful to you in doing research:

1 *Main desk.*
2 *Book file.* In particular, you can use the *Subjects* section of the card file to find the names of books on your subject, as well as suggestions about other subject headings under which you might find books. It is by exploring your general subject in books and then in magazine articles that you will gradually be able to decide on a subject limited enough to cover in your research paper.
3 *Book stacks.* Here you will get the books themselves.
4 *Magazine files and indexes.* Once again, you can use the *Subjects* sections of these files to get the names of magazine and journal articles on your subject.
5 *Magazine storage area.* Here you will get the articles themselves.

PRACTICE IN USING THE LIBRARY

Activity

Use your library to research a subject that interests you. Select one of the topics listed below and on the following page, or (with your teacher's permission) a topic of your own choice:

Sexual harassment	Pro-life movement
State lotteries	Health insurance reform
Greenhouse effect	Drinking water pollution
Nursing home costs	Problems of retirement
Pro-choice movement	Cremation

Capital punishment	Organ donation
Prenatal care	Child abuse
Acid rain	Voucher system in schools
Animal rights movement	Food poisoning (salmonella)
Noise control	Alzheimer's disease
Drug treatment programs for adolescents	Persian Gulf war
Fertility drugs	Sports betting
Witchcraft in the 1990s	National health insurance
New treatments for AIDS	Ethical aspects of hunting
Magic Johnson and AIDS	Euthanasia
Video display terminals — health aspects	Recent consumer frauds
Hazardous substances in the home	Stress reduction in the workplace
Airbags	Sex in television
Gambling and youth	Everyday addictions
Nongraded schools	Toxic waste disposal
Forecasting earthquakes	Self-help groups
New aid for the handicapped	Telephone crimes
New remedies for allergies	Date rape
Censorship in the 1990s	Heroes for the 1990s
New prison reforms	Steroids
Drug treatment programs	Surrogate mothers
Sudden infant death syndrome	Vegetarianism
New treatments for insomnia	Safe sex

Research the topic first through the *Subjects* section of the book file and then through the *Subjects* section of one or more magazine files and indexes.

On a separate sheet of paper, provide the following information:

1. Topic
2. Three books that either cover the topic directly or at least touch on the topic in some way. Include these items:

   Author

   Title

   Place of publication

   Publisher

   Date of publication

3. Three articles on the topic published in 1990 or later from the *Readers' Guide* or the *Magazine Index*. Include these items:

    Title of article
    Author (if given)
    Title of magazine
    Date
    Page(s)

4. Three articles on the topic published in 1990 or later from other indexes (such as the *New York Times Index, Business Periodicals Index, Social Sciences Index,* or *Humanities Index*). Include these items:

    Title of article
    Author (if given)
    Title of magazine
    Date
    Page(s)

5. Finally, include a photocopy of one of the three articles. Note whether the source of the copy was the article on paper, on microfiche, or on microfilm.

WRITING A RESEARCH PAPER

The process of writing a research paper can be divided into six steps:

1. Select a topic that you can readily research.
2. Limit your topic and make the purpose of your paper clear.
3. Gather information on your limited topic.
4. Plan your paper and take notes on your limited topic.
5. Write the paper.
6. Use an acceptable format and method of documentation.

This chapter explains and illustrates each of these steps and then provides a model research paper.

STEP 1: SELECT A TOPIC THAT YOU CAN READILY RESEARCH

First of all, go to the *Subjects* section of your library book file (as described on page 294) and see whether there are at least three books on your general topic. For example, if you initially choose the topic "marriage contracts," see if you can find at least three books on such contracts. Make sure that the books are actually available on the library shelves.

Next, go to the *Magazine Index* or the *Readers' Guide* (see pages 298–300), and try to find five or more articles on your subject.

If both books and articles are at hand, pursue your topic. Otherwise, you may have to choose another topic. You cannot write a paper on a topic for which research materials are not readily available.

STEP 2: LIMIT YOUR TOPIC AND
MAKE THE PURPOSE OF YOUR PAPER CLEAR

A research paper should develop a *limited* topic. It should be narrow and deep rather than broad and shallow. Therefore, as you read through books and articles on your general topic, look for ways to limit the topic.

For instance, in reading through materials on the general topic "adoption," you might decide to limit your topic to the problems that single people have in adopting a child. The general topic "drug abuse" might be narrowed to successful drug treatment programs for adolescents. After doing some reading on the general problem of "overpopulation," you might decide to limit your paper to the birth-control policies of the Chinese government. The broad subject "death" could be reduced to unfair pricing practices in funeral homes; "divorce" might be limited to its most damaging effects on the children of dicorced parents; "stress in everyday life" could be narrowed to methods of reducing stress in the workplace.

The subject headings in the book file and the magazine file will give you helpful ideas about how to limit your subject. For example, under the subject heading "Adoption" in the *book file* at one library were several related headings, such as "Intercountry adoption" and "Interracial adoption." In addition, there was a list of eighteen books, with several of the titles suggesting limited directions for research: the tendency toward adopting older children; problems faced by the adopted child; problems faced by foster parents. Under the subject heading "Adoption" in the *magazine file* at the same library were subheadings and titles of many articles which suggested additional limited directions that a research paper might explore: corrupt practices in adoption; the increase in mixed-race adoptions; ways to find a child for adoption. The point is that *subject headings and related headings, as well as book and article titles, may be of great help to you in narrowing your topic.* Take advantage of them.

Do not expect to limit your topic and make your purpose clear all at once. You may have to do quite a bit of reading as you work out the limited focus of your paper. Note that many research papers have one of two general purposes. Your purpose might be to *make and defend a point* of some kind. (For example, your purpose in a paper might be to provide evidence that gambling should be legalized.) Or, depending on the course and the instructor, your purpose might simply be to *present information* about a particular subject. (For instance, you might be asked to do a paper that describes the latest scientific findings about what happens when we dream.)

STEP 3: GATHER INFORMATION ON YOUR LIMITED TOPIC

After you have a good sense of your limited topic, you can begin gathering information that is relevant to it.

A helpful way to proceed is to sign out the books you need from your library — or to use the copier in your library to duplicate the pages you need from those books. In addition, make copies of all the articles you need from magazines, journals, or other reference materials. Remember that, as described in ''Using the Library'' on page 304, you should even be able to make copies of articles on microfiche or microfilm.

In other words, take the steps needed to get all your key source materials together in one place. You can then sit and work on these materials in a quiet, unhurried way in your home or some other place of study.

STEP 4: PLAN YOUR PAPER AND TAKE NOTES ON YOUR LIMITED TOPIC

Preparing a Scratch Outline

As you carefully read through the material you have gathered, think constantly about the specific content and organization of your paper. Begin making decisions about exactly what information you will present and how you will arrange it. Prepare a scratch outline of your paper that shows both its thesis and the areas of support for the thesis. It may help to try to plan at least three areas of support.

Thesis: _____

Support: (1) _____

(2) _____

(3) _____

Here, for example, is the brief outline that one student, Toni Grant, prepared for a paper on premarital agreements:

Thesis: Prenuptial agreements have a number of advantages.
Support: (1) Economic advantages
                (2) Legal advantages
                (3) Lifestyle advantages

Note-Taking

With a tentative outline in mind, you can begin taking notes on the information that you expect to include in your paper. Write your notes on four- by six-inch or five- by eight-inch cards or on sheets of loose-leaf paper. The notes you take should be in the form of *direct quotations, summaries in your own words,* or both. (At times you may also *paraphrase* — use an approximately equal number of your own words in place of someone else's words. Since most research involves condensing information, you will summarize much more than you will paraphrase.)

A *direct quotation* must be written *exactly* as it appears in the original work. But as long as you don't change the meaning, you may omit words from a quotation if they are not relevant to your point. Show such an omission with three spaced periods known as an *ellipsis* in place of the deleted words:

Original Passage

In creating prenuptial agreements, people are attempting to formulate the rules of their own marriages and, if it comes to it, their own divorces, at least to a point. Thus, even though it may not be romantic, it is often practical to be clear up front.

Direct Quotation with Ellipsis

"In creating prenuptial agreements, people are attempting to formulate the rules of their own marriages and, if it comes to it, their own divorces. . . . It is often practical to be clear up front."

(Note that there are four dots in the above example, with the first dot indicating the period at the end of the sentence.)

In a *summary,* you condense the original material by expressing it in your own words. Summaries may be written as lists, as brief paragraphs, or both. On the following page is one of Toni Grant's summary note cards:

A Summary Note Card

Legal advantages

A premarital contract has three major legal advantages. First, it allows couples to create a contract that is realistic in today's world, in contrast to outdated traditional contracts. Second, it can ensure equal rights for both partners. Third, it gives couples freedom and privacy to structure their relationship as they see fit.

Weitzman 227–228

Keep in mind the following points about your research notes:

- Write on only one side of each card or sheet of paper.
- Write only one kind of information, from one source, on any one card or sheet. For example, the sample card above has information on only one idea (premarital contracts) from one source (Weitzman).
- Include at the top of each card or sheet a heading that summarizes its content. This will help you organize the different kinds of information you gather.
- Identify the source and page number at the bottom.

Whether you quote or summarize, be sure to record the exact source and page from which you take each piece of information. In a research paper, you must document all information that is not common knowledge or a matter of historical record. For example, the birth and death dates of Martin Luther King are established facts and do not need documenting. On the other hand, the number of adoptions granted to single people in 1991 is a specialized fact that should be documented. As you read several sources on a subject, you will develop a sense of what authors regard as generally shared or common information about a subject and what is more specialized information that must be documented.

If you do not document specialized information or ideas that are not your own, you will be stealing (the formal term is *plagiarizing*—using someone else's work as your own work). A good deal of the material in research writing, it can usually be assumed, will need to be documented.

STEP 5: WRITE THE PAPER

After you have finished your reading and note-taking, you should have a fairly clear idea of the plan of your paper. Make a *final outline* and use it as a guide to write your first full draft. If your instructor requires an outline as part of your paper, you should prepare either a *topic outline*, which contains your thesis plus words and phrases, or a *sentence outline,* which contains all complete sentences. A topic outline is shown in the model paper on page 319. You will note that roman numerals are used for first-level headings, capital letters for second-level headings, and numbers for third-level headings.

In your *introductory paragraph*, include a thesis statement expressing the purpose of your paper and indicate the plan of development that you will follow. The section on writing an introductory paragraph for the essay (pages 225–226) is appropriate for the introductory section of the research paper as well. Note also the opening paragraph in the model research paper on page 320.

As you move from introduction to *main body* to *summary, conclusion,* or *both,* strive for unity, support, and coherence so that your paper will be clear and effective. Repeatedly ask, ''Does each of my supporting paragraphs develop the thesis of my paper?'' Use the checklist on page 126 and the inside front cover to make sure that your paper touches all four bases of effective writing.

STEP 6: USE AN ACCEPTABLE FORMAT AND METHOD OF DOCUMENTATION

Format

The model paper on pages 318–324 shows an acceptable format for a research paper. Comments and directions are set in small print in the margins of each page; be sure to note these.

Documentation of Sources

You must tell the reader the sources (books, articles, and so on) of the borrowed material in your paper. Whether you quote directly or summarize ideas in your own words, you must acknowledge your sources. In the past, you may have used footnotes and a bibliography to cite your sources. Now, you will learn a simplified documentation style used by the Modern Language Association. This easy-to-learn style resembles the documentation used in the social sciences and natural sciences.

Citations within a Paper: When citing a source, you must mention the author's name and the relevant page number. The author's name may be given either in the sentence you are writing or in parentheses following the sentence. Here are two examples:

Emily Card writes, "People who discuss money management issues before they marry probably find it easier to develop a money management plan that squares with the value system of both partners and ensures that each has a full say in the family finances" (72).

One expert writes, "People who discuss money management issues before they marry probably find it easier to develop a money management plan that squares with the value system of both partners and ensures that each has a full say in the family finances" (Card 72).

There are several points to note about citations within the paper:

- When the author's name is provided within the parentheses, only his or her last name is given.
- There is no punctuation between the author's name and the page number.
- The parenthetical citation is placed after the borrowed material but before the period at the end of the sentence.
- If you are using more than one work by the same author, include a shortened version of the title within the parenthetical citation. For example, suppose you are using several articles by Emily Card and you include the quotation above, which is from her article "Before — or Even After — You Say, 'I Do.' " Your citation within the text would be:

(Card, "Before," 72)

Citations at the End of a Paper: Your paper should end with a list of "Works Cited" which includes all the sources actually used in the paper. (Don't list any other sources, no matter how many you have read.) Look at "Works Cited" in the model research paper (page 324) and note the following points:

- The list is organized alphabetically according to the authors' last names. Entries are not numbered.
- If no author is given for a source, use a shortened version of the title of the article. For example, a citation on page 2 of the paper is a single word ("Coming"); it refers to the unsigned article titled "Coming to Terms before the Wedding" that appeared in *U.S. News & World Report*.
- When more than one work by the same author or authors is listed, three hyphens followed by a period should be substituted for the author's or authors' names after the first entry.

- Entries are double-spaced, with no extra space between entries.
- After the first line of each entry, there is an indentation for each additional line in the entry.
- Use the abbreviation *qtd. in* when citing a quotation found in someone else's article. For example, a quotation from Gail J. Koff on page 1 of the paper is from an article not by Koff but by Sue Berkman. The citation is therefore handled as follows:

Attorney Gail J. Koff observes that older newlyweds "are bringing into the marriage assets they've accumulated while single and working" (qtd. in Berkman 243).

Model Entries for a List of "Works Cited": Model entries for "Works Cited" are given below. Use these entries as a guide when you prepare your own list.

Book by One Author

Kozol, Jonathan. Rachel and Her Children: Homeless Families in America. New York: Crown, 1986.

Note that the author's name is reversed. Always give the complete title, including any subtitle. Use a colon to separate a title from a subtitle.

Two or More Entries by the Same Author

---. Savage Inequalities: Children in America's Schools. New York: Crown, 1991.

If you cite two or more entries by the same author (in the example above, a second book by Jonathan Kozol is cited), do not repeat the author's name. Instead, begin with a line made up of three hyphens followed by a period. Then give the remaining information as usual. Arrange works by the same author alphabetically by title. The words *A, An,* and *The* are ignored when alphabetizing by title.

Book by Two or More Authors

Burke, Chris, and Jo Beth McDaniel. A Special Kind of Hero. New York: Doubleday, 1991.

For a book with two or more authors, give all the authors' names but reverse only the first name.

Magazine Article

Aimes, Katrine. ''Practicing the Safest Sex of All.'' <u>Newsweek</u> 20 Dec. 1992: 52.

Write the date of the issue as follows: day, month (abbreviated in most cases to three letters), and year. The final number or numbers refer to the page or pages of the issue on which the article appears.

Newspaper Article

Lopez, Steve. ''Real Jobs for Real People.'' <u>The Philadelphia Inquirer</u>. 15 Dec. 1992, sec. B:1.

The final letter and number refer to section B, page 1.

Editorial

''The Excuse Maker.'' Editorial. <u>The New York Times</u> 14 Jan. 1992, sec. A:1.

List an editorial as you would any signed or unsigned article, but indicate the nature of the piece by adding *Editorial* or *Letter* after the article's title.

Encyclopedia Article

Foulkes, David, and Rosalind D. Cartwright. ''Sleep and Dreams.'' <u>Encyclopaedia Britannica</u>. 1989 ed.

Selection in an Edited Collection

McClane, Kenneth A. ''A Death in the Family.'' <u>Bearing Witness</u>. Ed. Henry Louis Gates, Jr. New York: Pantheon, 1991.

Revised or Later Edition

Quinn, Virginia Nichols. <u>Applying Psychology</u>. 2d ed. New York: McGraw-Hill, 1990.

Note: The abbreviations *Rev. ed., 2d ed., 3d ed.,* and so on, are placed right after the title.

Chapter or Section in a Book by One Author

Lewis, Michael. "Leave Home without It: The Absurdity of the American Express Card." The Money Culture. New York: Norton, 1991. 11-20.

Pamphlet

Pets in Nursing Homes. California Veterinary Medical Association, 1990.

Television Program

"World's Biggest Shopping Spree." Narr. Leslie Stahl. Prod. Howard L. Rosenberg. 60 Minutes. CBS 19 Jan. 1992.

Film

Grand Canyon. Dir. Laurence Kasdan. 20th Century Fox, 1991.

Recording

Cole, Natalie. "Straighten Up and Fly Right." Unforgettable. Elektra Records CD 960149-2.

Personal Interview

Peterson, Dr. May. Personal interview. 19 Nov. 1992.

Activity

On a separate sheet of paper, convert the information in each of the following references into the correct form for a list of "Works Cited." Use the appropriate model on pages 315–317 as a guide.

1. A book by Philip Slater called *A Dream Deferred* and published in Boston by Beacon Press in 1991.
2. An article by Christine Gorman titled "Why Are Men and Women Different?" on pages 42–51 of the January 20, 1992, issue of *Time*.
3. An article by Nanci Hellmich titled "Casting Doubt on Seafood Quality" on page 1D of the January 16, 1992, issue of *USA Today*.
4. A book by Diane E. Papalia and Sally Wendkos Olds titled *Human Development* and published in a fifth edition by McGraw-Hill in New York in 1992.
5. An article by Roger Rosenblatt titled "Self-Reliance in Hard Times" on page 25 of the February 1992 issue of *Life*.

The title should be centered and about one-third down the page.

Your name, the title of your course, your instructor's name, and the date should all be double-spaced and centered.

PRENUPTIAL AGREEMENTS

by

Toni Grant

English 101
Professor Davidson

25 January 1993

*Your instructor may or may not require that
you do an outline for your paper.*

OUTLINE

Thesis: Prenuptial agreements reduce marital problems by getting couples to deal with important economic, legal, and lifestyle decisions before they marry.

I.   Introduction
II. Advantages of prenuptial agreements
    A. Economic advantages
        1. Avoiding misunderstandings about money
        2. Conforming to the value system of both partners
        3. Protecting economic interests of both partners
        4. Dealing with today's more economically complex marriages
        5. Increasing trust between the partners
    B. Legal advantages
        1. Preventing legal problems
        2. Giving the couple, not the state, legal control
    C. Lifestyle advantages
        1. Clarifying the couple's goals and desires
        2. Clarifying the partners' expectations for each other
III. How well courts honor prenuptial agreements
    A. Lifestyle segments of agreements
    B. Other segments of agreements
IV. How to create a valid prenuptial agreement
V.   Conclusion

Double-space between lines of the text. Leave about a 1-inch margin all the way around the page.

Thesis, with plan of development.

Prenuptial agreements are no longer only for the very rich. In fact, it is estimated that from 1978 to 1988, the number of prenuptial agreements tripled in the United States (Quint 132). Recognized in all 50 states, prenuptial agreements allow couples to set up their marriages as they want to instead of in the way the state orders. Also, because writing a prenuptial agreement requires a great deal of communication, a couple who work together on an agreement give their marriage a sound start. A prenuptial agreement helps a couple examine their expectations regarding marriage, each other, and themselves (Gupta 31). Indeed, prenuptial agreements reduce marital problems by helping couples make important economic, legal, and lifestyle decisions before they marry.

Parenthetic citations, with author and page number but no comma.

Including economic issues in a prenuptial agreement has several advantages. Discussing their personal economic views helps couples to create a mutually satisfying financial framework for their marriage. Many misunderstandings about money can be resolved with clear ground rules (Mahar 131). In <u>Ms.</u> magazine, Emily Card writes, ''People who discuss money management issues before they marry probably find it easier to develop a money management plan that squares with the value system of both partners and ensures that each has a full say in the family finances'' (72). And, as attorneys point out, a prenuptial agreement safeguards the economic interests of both partners because, according to the law, the agreements must be ''fair when made'' (Kuntz 176).

The abbreviation qtd. *means quoted.*

The increase in prenuptial agreements reflects the economic needs of today's more complex marriages. People marrying for the first time are older than ever before. Attorney Gail J. Koff observes that older newlyweds ''are bringing into the marriage assets they've accumulated while single and working'' (qtd. in Berkman 243). For those with money, property, or even debts, a prenuptial agreement can clarify ownership. San Francisco attorney Rosario Billingsly has her clients list their assets and liabilities. Her clients then state whether each is to be treated as separate property, as joint property, or as joint property with conditions

Grant 2

(Malveaux 15). Prenuptial agreements are also useful for today's many dual-income couples. An agreement can state how a couple will share in the assets they build together during the marriage and how those holdings will be divided should the marriage fail ("Coming"). The marriage agreement may even go so far as to specify which partner's income will be used to pay particular household expenses (Berkman 243).

Many couples feel that talking about money and property implies a distrust of each other. But in fact, working together openly on economic matters helps couples develop a more trusting relationship. In their book Don't Get Married until You Read This, attorneys David Saltman and Harry Schaffner write:

Direct quotations of four lines or more are indented from the left margin. Quotation marks are not used.

> Most agreements are quite fair to both spouses, particularly where one spouse has a far larger estate than the other. Lawyers are cautious in preparing the agreement to make sure that it is legally enforceable. They strive for the fairness that they know will help the agreement withstand the tests that may occur in a court of law. The very process of discussing these matters, assets, and income, helps the relationship mature toward the good trusting relationship that every marriage must have to survive (41-42).

Writing a prenuptial agreement not only helps couples make important economic decisions; it also gets them to deal with serious legal issues. In doing so, the partners gain some control over their destiny and may avoid legal problems (Koff 15). Among the legal issues generally discussed are division of property, inheritance rights, spousal support, and business interests (Sack 63-64). A prenuptial agreement overrides the way the law treats property, income, and responsibility at the end of the marriage (through divorce or death). As New York attorney Jacalyn F. Barnett points out, with a prenuptial agreement "you arrange your affairs the way you want them, instead of how the state decrees" (qtd. in Vreeland 21). In a Forbes magazine article, William G. Flanagan and David Stix put it this way: "More and more couples are taking

Grant 3

The spaced periods (ellipses) show that material has been omitted from the quotation.

charge of their own . . . affairs rather than risk having a judge or state law decide their fates. The law and judges' opinions can be bizarre'' (117).

Prenuptial agreements also allow couples to make lifestyle decisions before the wedding. This permits them to anticipate and deal with possible sources of friction in their daily lives. According to sociologist Lenore Weitzman, author of The Marriage Contract, discussing lifestyle issues gives couples the freedom and privacy to order their personal relationship as they wish (227-228). She goes on to say that a prenuptial agreement ''provides a positive alternative--the option of creating a personally tailored structure to facilitate . . . goals and desires'' (229).

Lifestyle issues in a prenuptial agreement include the respective rights, duties, and obligations of each partner. These then become the standard of conduct for the couple's personal relationship. Agreements may cover, for example, living arrangements, responsibility for household tasks, sexual arrangements, personal behavior, relations with family and friends, plans for having or not having children, raising children, religion, and medical care (Weitzman 269-282). Some prenuptial agreements even determine whose career will come first, how long vacations will be, how free time will be spent, and whose surname will be used (Weitzman 86-90). Attorney Gail J. Koff sees these provisions as serving ''a valuable purpose. They are thought provoking and even work to avoid future acrimony in a marriage'' (91).

For the most part, however, the courts are unwilling to enforce the lifestyle segments of prenuptial agreements. The aims of these lifestyle ''contracts'' are primarily social (Zwack 6). Attorney Steven Sack, in his book The Complete Legal Guide to Marriage, Divorce, Custody, and Living Together, states ''such provisions may carry only moral rather than legal weight'' (75). One exception is that in some states, the agreement to raise a child in a particular religion has been upheld by the courts (Zwack 83).

Grant 4

Most courts will uphold the other elements of prenuptial agreements provided that these elements are fair and have been entered into freely by both parties. Recently, in fact, courts have taken the position that the agreements "promote marital stability by defining the expectations and responsibilities of each spouse" (Koff 93).

To be considered valid by the courts, however, prenuptial agreements must be carefully prepared. Thus couples must follow established legal guidelines. First, the agreement must be prepared well in advance of the wedding. Gary Skoloff, chairman of the family law section of the American Bar Association, claims that a prenuptial agreement signed at the last minute under pressure will probably not be upheld (Flanagan and Stix 118). Second, each partner should be represented by his or her own lawyer to ensure that each of them understands and is protected by the agreement (Roha 64). Third, both sides must provide full and accurate written lists of the current value of their assets and liabilities. "For a prenuptial agreement to stand up in court, there has to be full disclosure," says Michael Albano, a vice-president of the American Academy of Matrimonial Lawyers (qtd. in Flanagan and Stix 120). Fourth, the prenuptial agreement must not look as if it is promoting divorce. Both parties should enter the agreement assuming that the marriage will last (Koff 93). Finally, the prenuptial agreement must be a written document signed by both partners before a notary public (Zwack 17-18).

The seventeenth-century English playwright William Congreve observed, "Married in haste, we may repent at leisure." Couples can avoid such repentence with a prenuptial agreement, which allows them to clearly state their unique economic, legal, and lifestyle goals and obligations.

Grant 5

Works Cited

Berkman, Sue. "Those Pre-Marriage Contracts: A Guide for the Bride." Good Housekeeping June 1989:243.

Card, Emily. "Before—or Even After—You Say, 'I Do.' " Ms. June 1984:72-73.

"Coming to Terms before the Wedding." U.S. News & World Report 18 Apr. 1988:85.

Flanagan, William G., and David Stix. "Share and Share Unalike." Forbes 10 June 1991:116-120.

Gupta, Udayan. "Prenuptial Pacts (or, Look before You Leap)." Black Enterprise Apr. 1984:31.

Koff, Gail J. Love and the Law. New York: Simon & Schuster, 1989.

Kuntz, Mary. "Planning for the Worst." Forbes 6 Oct. 1986:173, 176.

Mahar, Margaret. "Striking a Deal before Tying the Knot." Money Jan. 1984: 131-132, 134, 136.

Malveaux, Julianne. "Love Insurance." Essence Feb. 1986:15, 124-127.

Quint, Barbara Gilder. "Prenuptial Agreements." Glamour Nov. 1989:132, 134.

Roha, Ronaleen. "Spelling It Out in a Contract." Changing Times Apr. 1987:64.

Sack, Steven Mitchell. The Complete Legal Guide to Marriage, Divorce, Custody, and Living Together. New York: McGraw-Hill, 1987.

Saltman, David, and Harry Schaffner. Don't Get Married until You Read This. New York: Barron's, 1989.

Vreeland, Leslie N. "From Wedlock to Deadlock? How to Negotiate a Prenup." Money Apr. 1990:21.

Weitzman, Lenore J. The Marriage Contract. New York: Free Press, 1981.

Zwack, Joseph P. Premarital Agreements. New York: Harper & Row, 1987.

Works cited should be double-spaced. The second and following lines of an entry should be indented five spaces. Titles of books, magazines, and the like should be underlined.

PART FIVE

SENTENCE SKILLS

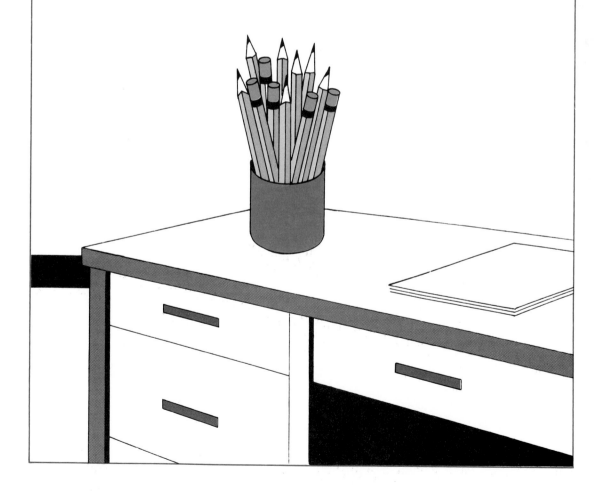

PREVIEW

As explained in Part One, there are four steps, or bases, in effective writing. Part Five is concerned with the *fourth step: the ability to write clear, error-free sentences.* First a diagnostic test is given so that you can check your present understanding of important sentence skills. Then the skills themselves appear under the general headings ''Grammar,'' ''Mechanics,'' ''Punctuation,'' and ''Word Use,'' ending with a chapter on sentence variety which helps develop your sense of the various options and methods available for composing sentences. Next come mastery tests and then editing tests that reinforce many basic writing skills and give you work in editing and proofreading. Finally, an achievement test helps you measure your improvement in important sentence skills.

SENTENCE-SKILLS DIAGNOSTIC TEST

PART 1

This test will help you check your knowledge of important sentence skills. Certain parts of the following word groups are underlined. Write an X in the answer space if you think a mistake appears at the underlined part. Write a C in the answer space if you think the underlined part is correct.

    A series of headings (''Sentence Fragments,'' ''Run-Ons,'' and so on) will give you clues to the mistakes to look for. However, you do not have to understand the label to find a mistake. What you are checking is your own sense of effective written English.

Sentence Fragments

_____ 1. <u>Because Tom had eaten and drunk too much</u>. He had to leave the party early. His stomach was like a volcano ready to erupt.

_____ 2. <u>After I slid my aching bones into the hot water of the tub, I realized there was no soap</u>. I didn't want to get out again.

_____ 3. I spent two hours on the phone yesterday. <u>Trying to find a garage to repair my car</u>. Eventually I had to have the car towed to a garage in another town.

_____ 4. <u>Sweating under his heavy load</u>. Brian staggered up the stairs to his apartment. He felt as though his legs were crumbling beneath him.

_____ 5. <u>I love to eat and cook Italian food, especially lasagna and ravioli</u>. I make everything from scratch.

_____ 6. One of my greatest joys in life is eating desserts. <u>Such as blueberry cheesecake or vanilla cream puffs</u>. Almond fudge cake makes me want to dance.

Run-Ons

_____   7. He decided to stop <u>smoking, for</u> he didn't want to die of lung cancer.

_____   8. The window shade snapped up like a <u>gunshot her</u> cat leaped four feet off the floor.

_____   9. Billy is the meanest little kid on his <u>block, he</u> eats only the heads of animal crackers.

_____   10. He knew he had flunked the driver's <u>exam, he</u> ran over a stop sign.

_____   11. My first boyfriend was five years <u>old. We</u> met everyday in the playground sandbox.

_____   12. Lola wanted to go disco <u>dancing, Tony</u> preferred going to a movie.

Standard English Verbs

_____   13. Jed <u>tows</u> away cars for a living and is ashamed of his job.

_____   14. You <u>snored</u> like a chain saw last night.

_____   15. When I was about to finish work last night, a man <u>walk</u> into the restaurant and ordered two dozen hamburgers.

_____   16. Charlotte <u>react</u> badly whenever she gets caught in a traffic jam.

Irregular Verbs

_____   17. I <u>knowed</u> her from somewhere, but I couldn't remember just where.

_____   18. I had <u>eaten</u> so much food at the buffet dinner that I went into the bathroom just to loosen my belt.

_____   19. When the mud slide started, the whole neighborhood <u>began</u> going downhill.

_____   20. Juan has <u>rode</u> the bus to school for two years while saving for a car.

Subject-Verb Agreement

_____   21. There <u>is</u> long lines at the checkout counter.

_____   22. The camping blanket <u>have</u> to be washed.

_____   23. One of the crooked politicians <u>was</u> jailed for a month.

_____   24. The cockroaches behind my stove <u>gets</u> high on Raid.

Consistent Verb Tense

————— 25. I played my stereo and watched television before I <u>decides</u> to do some homework.

————— 26. The first thing Jerry does everyday is weigh himself. The scale <u>informs</u> him what kind of meals he can eat that day.

————— 27. Sandy eats a nutritional breakfast, <u>skips</u> lunch, and then enjoys a big dinner.

————— 28. His parents stayed together for his sake; only after he <u>graduates</u> from college were they divorced.

Pronoun Agreement, Reference, and Point of View

————— 29. I work at a clothes shop where <u>you</u> do not get paid for all the holidays I should.

————— 30. I enjoy movies like *The Return of the Vampire* that frighten <u>me</u>.

————— 31. A student should write <u>their</u> own papers.

————— 32. Persons camping in those woods should watch <u>their</u> step because of wild dogs.

————— 33. Angry at striking out, Tony hurled the baseball bat at the fence and broke <u>it</u>.

————— 34. I love Parmesan cheese, but <u>it</u> does not always agree with me.

Pronoun Types

————— 35. Rick and <u>me</u> would get along better if he left town.

————— 36. No one is a better cook than <u>she</u>.

Adjectives and Adverbs

————— 37. Bonnie ran <u>quick</u> up the steps, taking them two at a time.

————— 38. Larry is <u>more better</u> than I am at darts.

Misplaced Modifiers

————— 39. He swatted the wasp that stung him <u>with a newspaper</u>.

————— 40. Charlotte returned the hamburger <u>that was spoiled</u> to the supermarket.

————— 41. My aunt once met Jerry Lewis at a benefit, <u>whom she found to be a very engaging person</u>.

————— 42. I adopted a dog from a junkyard <u>which is very close to my heart</u>.

Dangling Modifiers

_____ 43. Going to work, Clyde saw a three-car accident.

_____ 44. Flunking out of school, my parents demanded that I get a job.

_____ 45. While I was waiting for the bus, rain began to fall.

_____ 46. Braking the car suddenly, the shopping bags tumbled onto the floor.

Faulty Parallelism

_____ 47. Bill enjoys hunting for rabbits, socializing with friends, and to read the funnies.

_____ 48. Lola likes to wear soft sweaters, to eat exotic foods, and to bathe in Calgon bath oil.

_____ 49. When I saw my roommate with my girlfriend, I felt worried, angry, and embarrassment as well.

_____ 50. Frances enjoys shopping for new clothes, trying different cosmetics, and reading beauty magazines.

Capital Letters

_____ 51. Clyde uses certs mints after smoking his pipe or a cigar.

_____ 52. During july, Frank's company works a four-day week.

_____ 53. I asked my dad, ''When's Uncle Bill getting his toupee?''

_____ 54. On Summer days I like to sit in the backyard and sunbathe.

Apostrophe

_____ 55. The Wolfman's bite is worse than his bark.

_____ 56. Clydes quick hands reached out to break his son's fall.

_____ 57. I'll be with you shortly if youll just wait a minute.

_____ 58. You shouldn't drink any more if you're hoping to get home safely.

Quotation Marks

_____ 59. Mark Twain once said, ''The more I know about human beings, the more I like my dog.''

_____ 60. Say something tender to me, ''whispered Tony to Lola.''

_____ 61. ''I hate that commercial, he muttered.''

_____ 62. ''If you don't leave soon,'' he warned, ''you'll be late for work.''

Comma

_____ 63. Bill relaxes by reading <u>Donald Duck Archie and Bugs Bunny</u> comic books.

_____ 64. Although I have a black belt in <u>karate</u> I decided to go easy on the demented bully who had kicked sand in my face.

_____ 65. Dracula, who had a way with <u>women,</u> is his favorite movie hero.

_____ 66. We could always tell when our instructor felt <u>disorganized for</u> his shirt would not be tucked into his pants.

_____ 67. <u>You, my man</u>, are going to get yours.

_____ 68. His father <u>shouted</u> ''Why don't you go out and get a job?''

Commonly Confused Words

_____ 69. Some stores will accept your credit cards but not <u>you're</u> money.

_____ 70. That issue is <u>to</u> hot for any politician to handle.

_____ 71. <u>They're</u> planning to trade in their old car.

_____ 72. <u>Its</u> important to get this job done properly.

_____ 73. You should have the brakes on your car replaced <u>write</u> away.

_____ 74. <u>Who's</u> the culprit who left the paint can on the table?

Effective Word Use

_____ 75. I <u>comprehended her statement.</u>

_____ 76. The movie was a <u>real bomb</u>, and so we left early.

_____ 77. The victims of the car accident were shaken but <u>none the worse for wear.</u>

_____ 78. Anne is <u>of the opinion that</u> the death penalty should be abolished.

Answers are on page 561.

PART 2 (OPTIONAL)

Do the following at your instructor's request. This second part of the test will provide more detailed information about skills you need to know. On separate paper, number and correct all the items you have marked with an X. For example, suppose you had marked the word groups on the following page with an X. (Note that these examples are not taken from the test.)

4. If football games disappeared entirely from television. I would not even miss them. Other people in my family would perish.

7. The kitten suddenly saw her reflection in the mirror, she jumped back in surprise.

15. The tree in my cousins front yard always sheds its leaves two weeks before others on the street.

29. When we go out to a restaurant we always order something we would not cook for ourselves.

Here is how you should write your corrections on a separate sheet of paper.

4. television, I
7. mirror, and
15. cousin's
29. restaurant, we

There are over forty corrections to make in all.

SUBJECTS AND VERBS

The basic building blocks of English sentences are subjects and verbs. Understanding them is an important first step toward mastering a number of sentence skills.

Every sentence has a subject and a verb. Who or what the sentence speaks about is called the subject; what the sentence says about the subject is called the verb.

> The children laughed.
> Several branches fell.
> Most students passed the test.
> That man is a crook.

A SIMPLE WAY TO FIND A SUBJECT

To find a subject, ask *who* or *what* the sentence is about. As shown below, your answer is the subject.

> *Who* is the first sentence about? Children
> *What* is the second sentence about? Several branches
> *Who* is the third sentence about? Most students
> *Who* is the fourth sentence about? That man

A SIMPLE WAY TO FIND A VERB

To find a verb, ask what the sentence *says about* the subject. As shown below, your answer is the verb.

What does the first sentence *say about* the children? They <u>laughed</u>.
What does the second sentence *say about* the branches? They <u>fell</u>.
What does the third sentence *say about* the students? They <u>passed</u>.
What does the fourth sentence *say about* that man? He <u>is</u> (a crook).

A second way to find the verb is to put *I, you, he, she, it,* or *they* in front of the word you think is a verb. If the result makes sense, you have a verb. For example, you could put *they* in front of *laughed* in the first sentence above, with the result, *they laughed,* making sense. Therefore you know that *laughed* is a verb. You could use *they* or *he* to test the other verbs as well.

Finally, it helps to remember that most verbs show action. In the sentences already considered, the three action verbs are *laughed, fell,* and *passed.* Certain other verbs, known as *linking verbs,* do not show action. They do, however, give information about the subject. In ''That man is a crook,'' the linking verb *is* tells us that the man is a crook. Other common linking verbs include *am, are, was, were, feel, appear, look, become,* and *seem.*

Activity

In each of the following sentences, draw one line under the subject and two lines under the verb.

1. The heavy purse cut into my shoulder.
2. Small stones pinged onto the windshield.
3. The test directions confused the students.
4. Cotton shirts feel softer than polyester ones.
5. The fog rolled into the cemetery.
6. Sparrows live in the eaves of my porch.
7. A green bottle fly stung her on the ankle.
8. Every other night, garbage trucks rumble down my street on their way to the river.
9. The family played badminton and volleyball, in addition to a game of softball, at the picnic.
10. With their fingers, the children drew pictures on the steamed window.

MORE ABOUT SUBJECTS AND VERBS

1 A pronoun (a word like *he, she, it, we, you,* or *they* used in place of a noun) can serve as the subject of a sentence. For example:

He seems like a lonely person.
They both like to gamble.

Without a surrounding context (so that we know who *He* or *They* refers to), the sentences may not seem clear, but they *are* complete.

2 A sentence may have more than one verb, more than one subject, or several subjects and verbs:

My heart skipped and pounded.
The radio and tape player were stolen from the car.
Dave and Ellen prepared the report together and presented it to the class.

3 The subject of a sentence never appears within a prepositional phrase. A prepositional phrase is simply a group of words that begins with a preposition. Following is a list of common prepositions:

about	before	by	inside	over
above	behind	during	into	through
across	below	except	of	to
among	beneath	for	off	toward
around	beside	from	on	under
at	between	in	onto	with

Cross out prepositional phrases when looking for the subject of a sentence.

Under my pillow I found a quarter left by the Tooth Fairy.
One of the yellow lights at the school crossing began flashing.
The funny pages of the newspaper disappeared.
In spite of my efforts, Bob dropped out of school.
During a rainstorm, I sat in my car reading magazines.

4   Many verbs consist of more than one word. Here, for example, are some of the many forms of the verb *smile*.

smile	smiled	should smile
smiles	were smiling	will be smiling
does smile	have smiled	can smile
is smiling	had smiled	could be smiling
are smiling	had been smiling	must have smiled

Notes

a   Words like *not, just, never, only,* and *always* are not part of the verb although they may appear within the verb.

Larry did not finish the paper before class.
The road was just completed only last week.

b   No verb preceded by *to* is ever the verb of a sentence.

My car suddenly began to sputter on the freeway.
I swerved to avoid a squirrel on the road.

c   No *-ing* word by itself is ever the verb of a sentence. (It may be part of the verb, but it must have a helping verb in front of it.)

They leaving early for the game. (not a sentence, because the verb is not complete)
They are leaving early for the game. (a sentence)

Activity

Draw a single line under the subjects and a double line under the verbs in the following sentences. Be sure to include all parts of the verb.

1. A burning odor from the wood saw filled the room.
2. At first, sticks of gum always feel powdery on your tongue.
3. Vampires and werewolves are repelled by garlic.

4. Three people in the long bank line looked impatiently at their watches.

5. The driving rain had pasted wet leaves all over the car.

6. She has decided to buy a condominium.

7. The trees in the mall were glittering with tiny white lights.

8. The puppies slipped and tumbled on the vinyl kitchen floor.

9. Tony and Lola ate at Pizza Hut and then went to a movie.

10. We have not met our new neighbors in the apartment building.

■ **Review Test**

Draw a single line under subjects and a double line under verbs. Crossing out prepositional phrases may help to find the subjects.

1. A cloud of fruit flies hovered over the bananas.

2. Candle wax dripped onto the table and hardened into pools.

3. Nick and Fran are both excellent Frisbee players.

4. The leaves of my dying rubber plant resembled limp brown rags.

5. During the first week of vacation, Ken slept until noon every day.

6. They have just decided to go on a diet together.

7. Psychology and word processing are my favorite subjects.

8. The sofa in the living room has not been cleaned for over a year.

9. The water stains on her suede shoes did not disappear with brushing.

10. Fred stayed in bed too long and, as a result, arrived late for work.

SENTENCE
SENSE

WHAT IS SENTENCE SENSE?

As a speaker of English, you already possess the most important of all sentence skills. You have *sentence sense*—an instinctive feel for where a sentence begins, where it ends, and how it can be developed. You learned sentence sense automatically and naturally, as part of learning the English language, and you have practiced it through all the years that you have been speaking English. It is as much a part of you as your ability to speak and understand English is a part of you.

Sentence sense can help you recognize and avoid fragments and run-ons, two of the most common and serious sentence-skills mistakes in written English. Sentence sense will also help you to place commas, spot awkward and unclear phrasings, and add variety to your sentences.

You may ask, ''If I already have this 'sentence sense,' why do I still make mistakes in punctuating sentences?'' One answer could be that your past school experiences in writing were unrewarding or unpleasant. English may have been a series of dry writing topics and heavy doses of ''correct'' grammar and usage, or it may have given no attention at all to sentence skills. For any of these reasons, or perhaps for other reasons, the instinctive sentence skills you practice while *speaking* may turn off when you start *writing*. The very act of picking up a pen may shut down your whole natural system of language abilities and skills.

TURNING ON YOUR SENTENCE SENSE

Chances are, you don't *read a paper aloud* after you write it, or you don't do the next best thing: read it "aloud" in your head. But reading aloud is essential to turn on the natural language system within you. By reading aloud, you will be able to hear the points where your sentences begin and end. In addition, you will be able to pick up any trouble spots where your thoughts are not communicated clearly and well.

The activities that follow will help you turn on and rediscover the enormous language power within you. You will be able to see how your built-in sentence sense can guide your writing just as it does your speaking.

Activity

Each selection that follows lacks basic sentence punctuation. There is no period to mark the end of one sentence and no capital letter to mark the start of the next. Read each selection aloud (or in your head) so that you "hear" where each sentence begins and ends. Your voice will tend to drop and come to a pause at the point of each sentence break. Put a light slash mark (/) at every point where you hear a break. Then go back and read over the selection a second time. If you are now sure of each place where a split occurs, insert a period and change the first small letter after it to a capital. Minor pauses are often marked in English by commas; these are already inserted. Part of the first selection is done for you as an example.

1. I take my dog for a walk on Saturdays in the big park by /the lake I do this very early in the morning before children come to the park. /that way I can let my dog run freely he jumps out the minute I open the car door and soon sees the first innocent squirrel then he is off like a shot and doesn't stop running for at least half an hour.

2. Lola hates huge tractor trailers that sometimes tailgate her Honda Civic the enormous smoke-belching machines seem ready to swallow her small car she shakes her fist at the drivers, and she rips out a lot of angry words recently she had a very satisfying dream she broke into an army supply depot and stole a bazooka she then became the first person in history to murder a truck

3. When I sit down to write, my mind is blank all I can think of is my name, which seems to me the most boring name in the world often I get sleepy and tell myself I should take a short nap other times I start daydreaming about things I want to buy sometimes I decide I should make a telephone call to someone I know the piece of paper in front of me is usually still blank when I leave to watch my favorite television show

4. One of the biggest regrets of my life is that I never told my father I loved him I resented the fact that he had never been able to say the words ''I love you'' to his children even during the long period of my father's illness, I remained silent and unforgiving then one morning he was dead, with my words left unspoken a guilt I shall never forget tore a hole in my heart I determined not to hold in my feelings with my daughters they know they are loved, because I both show and tell them this all people, no matter who they are, want to be told that they are loved

5. Two days ago, Greg killed several flying ants in his bedroom he also sprayed a column of ants that was forming a colony along the kitchen baseboard yesterday, he picked the evening newspaper off the porch and two black army ants scurried onto his hand this morning, he found an ant crawling on a lollipop he had left in his shirt pocket if any more insects appear, he is going to call Orkin Pest Control he feels like the victim in a Hitchcock movie called *The Ants* he is half afraid to sleep tonight he imagines the darkness will be full of tiny squirming things waiting to crawl all over him

SUMMARY: USING SENTENCE SENSE

You probably did well in locating the end stops in these selections — proving to yourself that you *do* have sentence sense. This instinctive sense will help you deal with sentence fragments and run-ons, perhaps the two most common sentence-skills mistakes.

Remember the importance of *reading your paper aloud*. By doing so, you turn on all the natural language skills that come from a lifetime of speaking English. The same sentence sense that helps you communicate effectively in speaking will help you communicate effectively in writing.

SENTENCE FRAGMENTS

Introductory Project

Every sentence must have a subject and a verb and must express a complete thought. A word group that lacks a subject or a verb and that does not express a complete thought is a fragment.

Listed below are a number of fragments and sentences. See if you can complete the statement that explains each fragment.

1. People. *Fragment*

    People gossip. *Sentence*

    "People" is a fragment because, while it has a subject (*People*), it lacks a _____ (*gossip*) and so does not express a complete thought.

2. Wrestles. *Fragment*

    Tony wrestles. *Sentence*

    "Wrestles" is a fragment because, while it has a verb (*Wrestles*), it lacks a _____ (*Tony*) and so does not express a complete thought.

3. Drinking more than anyone else at the bar. *Fragment*

    Grandmother was drinking more than anyone else at the bar. *Sentence*

    "Drinking more than anyone else at the bar" is a fragment because it lacks a _____ (*Grandmother*) and also part of the _____ (*was*) and so does not express a complete thought.

4. When Lola turned eighteen. *Fragment*

    When Lola turned eighteen, she got her own apartment. *Sentence*

    "When Lola turned eighteen" is a fragment because we want to know *what happened when* Lola turned eighteen. The word group does not follow through and _____ _____

Answers are on page 562.

WHAT ARE SENTENCE FRAGMENTS?

Every sentence must have a subject and a verb and must express a complete thought. A word group that lacks a subject or a verb and that does not express a complete thought is a *fragment*. The most common types of fragments are:

1 Dependent-word fragments
2 *-ing* and *to* fragments
3 Added-detail fragments
4 Missing-subject fragments

Once you understand the specific kind or kinds of fragments that you may write, you should be able to eliminate them from your writing. The following pages explain all four fragment types.

DEPENDENT-WORD FRAGMENTS

Some word groups that begin with a dependent word are fragments. Here is a list of common dependent words:

Dependent Words		
after	if, even if	when, whenever
although, though	in order that	where, wherever
as	since	whether
because	that, so that	which, whichever
before	unless	while
even though	until	who, whoever
how	what, whatever	whose

Whenever you start a sentence with one of these words, you must be careful that a fragment does not result.

The word group beginning with the dependent word *After* in the example below is a fragment.

After I learned the price of new cars. I decided to keep my old Buick.

A *dependent statement*—one starting with a dependent word like *After*—cannot stand alone. It depends on another statement to complete the thought. "After I learned the price of new cars" is a dependent statement. It leaves us hanging. We expect in the same sentence to find out *what happened after* the writer learned the price of new cars. When a writer does not follow through and complete a thought, a fragment results.

To correct the fragment, simply follow through and complete the thought:

After I learned the price of new cars, I decided to keep my old Buick.

Remember, then, that *dependent statements by themselves are fragments.* They must be attached to a statement that makes sense standing alone.

Here are two other examples of dependent-word fragments:

My daughter refused to stop smoking. Unless I quit also.
Bill asked for a loan. Which he promised to pay back in two weeks.

"Unless I quit also" is a fragment; it does not make sense standing by itself. We want to know in the same statement *what would not happen unless* the writer quit also. The writer must complete the thought. Likewise, "Which he promised to pay back in two weeks" is not in itself a complete thought. We want to know in the same statement what *which* refers to.

Correcting a Dependent-Word Fragment

In most cases you can correct a dependent-word fragment by attaching it to the sentence that comes after it or the sentence that comes before it:

After I learned the price of new cars, I decided to keep my old Buick.
(The fragment has been attached to the sentence that comes after it.)
My daughter refused to quit smoking unless I quit also.
(The fragment has been attached to the sentence that comes before it.)
Bill asked for a loan which he promised to pay back in two weeks.
(The fragment has been attached to the sentence that comes before it.)

Another way of correcting a dependent-word fragment is simply to eliminate the dependent word by rewriting the sentence:

I learned the price of new cars and decided to keep my old Buick.
She wanted me to quit also.
He promised to pay it back in two weeks.

Do not use this method of correction too frequently, however, for it may cut down on interest and variety in your writing style.

Notes

1   Use a comma if a dependent-word group comes at the beginning of a sentence (see also page 468):

After I learned the price of new cars, I decided to keep my old Buick.

However, do not generally use a comma if the dependent-word group comes at the end of a sentence:

My daughter refused to stop smoking unless I quit also.
Bill asked for a loan which he promised to pay back in two weeks.

2   Sometimes the dependent words *who, that, which,* or *where* appear not at the very start, but near the start, of a word group. A fragment often results:

The town council decided to put more lights on South Street. <u>A place where several people have been mugged.</u>

''A place where several people have been mugged'' is not in itself a complete thought. We want to know in the same statement *where the place was* that several people were mugged. The fragment can be corrected by attaching it to the sentence that comes before it:

The town council decided to put more lights on South Street, a place where several people have been mugged.

Activity 1

Turn each of the following dependent-word groups into a sentence by adding a complete thought. Put a comma after the dependent-word group if a dependent word starts the sentence.

Examples   Although I arrived in class late
Although I arrived in class late, I still did well on the test.

The little boy who plays with our daughter
The little boy who plays with our daughter just came down with

German measles.

1. Because the weather is bad

    _____

    _____

2. If I lend you twenty dollars

    _____

    _____

3. The car that we bought

    _____

    _____

4. Since I was tired

    _____

    _____

5. Before the instructor entered the room

    _____

    _____

Activity 2

Underline the dependent-word fragment or fragments in each selection. Then correct each fragment by attaching it to the sentence that comes before or the sentence that comes after—whichever sounds more natural. Put a comma after the dependent-word group if it starts the sentence.

1. Whenever our front and back doors are open. The air current causes the back door to slam shut. The noise makes everyone in the house jump.

    _____

    _____

2. Bill always turns on the radio in the morning to hear the news. He wants to be sure that World War III has not started. Before he gets on with his day.

    _____

    _____

    _____

3. Since the line at the Motor Vehicle Bureau crawls at a snail's pace. Fred waited two hours there. When there was only one person left in front of him. The office closed for the day.

_____

_____

_____

4. My dog ran in joyous circles on the wide beach. Until she found a dead fish. Before I had a chance to drag her away. She began sniffing and nudging the smelly remains.

_____

_____

_____

5. When the air conditioner broke down. The temperature was over ninety degrees. I then found an old fan. Which turned out to be broken also.

_____

_____

_____

-ING AND *TO* FRAGMENTS

When an *-ing* word appears at or near the start of a word group, a fragment may result. Such fragments often lack a subject and part of the verb. Underline the word groups in the selections below that contain *-ing* words. Each is a fragment.

Selection 1

I spent almost two hours on the phone yesterday. Trying to find a garage to repair my car. Eventually I had to have it towed to a garage in another town.

Selection 2

Maggie was at first very happy with the blue sports car she had bought for only five hundred dollars. Not realizing until a week later that the car averaged seven miles per gallon of gas.

Selection 3

He looked forward to the study period at school. It being the only time he could sit unbothered and dream about his future. He imagined himself as a lawyer with lots of money and women to spend it on.

People sometimes write *-ing* fragments because they think the subject in one sentence will work for the next word group as well. Thus, in the first selection, the writer thinks that the subject *I* in the opening sentence will also serve as the subject for "Trying to find a garage to repair my car." But the subject must actually be *in* the sentence.

Correcting *-ing* Fragments

1   Attach the *-ing* fragment to the sentence that comes before it or the sentence that comes after it, whichever makes sense. Selection 1 could read: "I spent two hours on the phone yesterday, trying to find a garage to repair my car."

2   Add a subject and change the *-ing* verb part to the correct form of the verb. Selection 2 could read: "She realized only a week later that the car averaged seven miles per gallon of gas."

3   Change *being* to the correct form of the verb *be* (*am, are, is, was, were*). Selection 3 could read: "It was the only time he could sit unbothered and dream about his future."

Correcting *to* Fragments

When *to* appears at or near the start of a word group, a fragment sometimes results:

I plan on working overtime. To get this job finished. Otherwise, my boss may get angry at me.

The second word group is a fragment and can be corrected by adding it to the preceding sentence:

I plan on working overtime to get this job finished.

Activity 1

Underline the *-ing* fragment in each of the selections that follow. Then make it a sentence by rewriting it, using the method described in parentheses.

Example   A thunderstorm was brewing. A sudden breeze shot through the windows. Driving the stuffiness out of the room.
(Add the fragment to the preceding sentence.)

A sudden breeze shot through the windows, driving the stuffiness

out of the room.

(In the example, a comma is used to set off "driving the stuffiness out of the room," which is extra material placed at the end of the sentence.)

1. Sweating under his heavy load. Brian staggered up the stairs to his apartment. He felt as though his legs were crumbling beneath him.
    (Add the fragment to the sentence that comes after it.)

    _____

    _____

2. He works 10 hours a day. Then going to class for 2½ hours. It is no wonder he writes sentence fragments.
    (Correct the fragment by adding the subject *he* and changing *going* to the proper form of the verb, *goes.*)

    _____

    _____

3. Charlotte loved the movie *Gone with the Wind,* but Clyde hated it. His chief objection being that it lasted four hours.
    (Correct the fragment by changing *being* to the proper verb form, *was.*)

    _____

    _____

Activity 2

Underline the *-ing* or *to* fragment or fragments in each selection. Then rewrite each selection, correcting the fragments by using one of the three methods of correction described on page 347.

1. A mysterious package arrived on my porch yesterday. Bearing no return address. I half expected to find a bomb inside.

    _____

    _____

2. Jack bundled up and went outside on the bitterly cold day. To saw wood for his fireplace. He returned half frozen with only two logs.

    _____

    _____

3. Looking tired and drawn. The little girl's parents sat in the waiting room. The operation would be over in a few minutes.

    _____

    _____

4. Sighing with resignation. Jill switched on her television set. She knew that the picture would be snowy and crackling with static. Her house being in a weak reception area.

_____

_____

_____

_____

5. Jabbing the ice with a screwdriver. Bill attempted to speed up the defrosting process in his freezer. However, he used too much force. The result being a freezer compartment riddled with holes.

_____

_____

_____

_____

ADDED-DETAIL FRAGMENTS

Added-detail fragments lack a subject and a verb. They often begin with one of the following words:

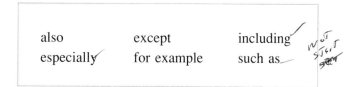

also	except	including
especially	for example	such as

See if you can locate and underline the one added-detail fragment in each of the selections that follow:

Selection 1

I love to cook and eat Italian food. Especially spaghetti and lasagna. I make everything from scratch.

Selection 2

The class often starts late. For example, yesterday at quarter after nine instead of at nine sharp. Today the class started at five after nine.

Selection 3

He failed a number of courses before he earned his degree. Among them, English I, Economics, and General Biology.

People often write added-detail fragments for much the same reason they write *-ing* fragments. They think the subject and verb in one sentence will serve for the next word group as well. But the subject and verb must be in *each* word group.

Correcting Added-Detail Fragments

1   Attach the fragment to the complete thought that precedes it. Selection 1 could read: "I love to cook and eat Italian food, especially spaghetti and lasagna."

2   Add a subject and a verb to the fragment to make it a complete sentence. Selection 2 could read: "The class often starts late. For example, yesterday it began at quarter after nine instead of at nine sharp."

3   Change words as necessary to make the fragment part of the preceding sentence. Selection 3 could read: "Among the courses he failed before he earned his degree were English I, Economics, and General Biology."

Activity 1

Underline the fragment in each of the selections below. Then make it a sentence by rewriting it, using the method described in parentheses.

Example   I am always short of pocket money. Especially for everyday items like magazines and sodas. Luckily my friends often have change.
(Add the fragment to the preceding sentence.)

I am always short of pocket money, especially for everyday items

like magazines and sodas.

1.   Nina is trying hard for a promotion. For example, through night classes and a Dale Carnegie course. She is also working overtime for no pay.
(Correct the fragment by adding the subject and verb *she is taking*.)

_____

_____

2.   I could feel Bill's anger building. Like a land mine ready to explode. I was silent because I didn't want to be the one to set it off.
(Add the fragment to the preceding sentence.)

_____

_____

3. We went on vacation without several essential items. Among other things, our sneakers and sweat jackets.
    (Correct the fragment by adding the subject and verb *we forgot*.)

    _____

    _____

Activity 2

Underline the added-detail fragment in each selection. Then rewrite that part of the selection needed to correct the fragment. Use one of the three methods of correction described above.

1. It's always hard for me to get up for work. Especially on Mondays after a holiday weekend. However, I always wake up early on free days.

    _____

    _____

2. Tony has enormous endurance. For example, the ability to run five miles in the morning and then play basketball all afternoon.

    _____

    _____

3. A counselor gives you a chance to talk about your problems. Whether with your family or the boss at work. You learn how to cope better with life.

    _____

    _____

4. Fred and Martha do most of their shopping through mail-order catalogs. Especially the Sears and J. C. Penney catalogs.

    _____

    _____

5. One of my greatest joys in life is eating desserts. Such as cherry cheesecake and vanilla cream puffs. Almond fudge cake makes me want to dance.

    _____

    _____

MISSING-SUBJECT FRAGMENTS

In each selection below, underline the word group in which the subject is missing.

Selection 1

The truck skidded on the rain-slick highway. But missed a telephone pole on the side of the road.

Selection 2

Michelle tried each of the appetizers on the table. And then found that, when the dinner arrived, her appetite had gone.

People write missing-subject fragments because they think the subject in one sentence will apply to the next word group as well. But the subject, as well as the verb, must be in each word group to make it a sentence.

Correcting Missing-Subject Fragments

1   Attach the fragment to the preceding sentence. Selection 1 could read: ''The truck skidded on the rain-slick highway but missed a telephone pole on the side of the road.''

2   Add a subject (which can often be a pronoun standing for the subject in the preceding sentence). Selection 2 could read: ''She then found that, when the dinner arrived, her appetite had gone.''

Activity

Underline the missing-subject fragment in each selection. Then rewrite that part of the selection needed to correct the fragment. Use one of the two methods of correction described above.

1. I tried on an old suit hanging in our basement closet. And discovered, to my surprise, that it was too tight to button.

    _____

    _____

    _____

2. When Mary had a sore throat, friends told her to gargle with salt water. Or suck on an ice cube. The worst advice she got was to avoid swallowing.

_____

_____

_____

3. One of my grade-school teachers embarrassed us with her sarcasm. Also, seated us in rows from the brightest student to the dumbest. I can imagine the pain the student in the last seat must have felt.

_____

_____

_____

A Review: How to Check for Sentence Fragments

1 Read your paper aloud from the *last* sentence to the *first*. You will be better able to see and hear whether each word group you read is a complete thought.

2 Ask yourself of any word group you think is a fragment: Does this contain a subject and a verb and express a complete thought?

3 More specifically, be on the lookout for the most common fragments:

- Dependent-word fragments (starting with words like *after, because, since, when,* and *before*)

- *-ing* and *to* fragments (*-ing* or *to* at or near the start of a word group)

- Added-detail fragments (starting with words like *for example, such as, also,* and *especially*)

- Missing-subject fragments (a verb is present but not the subject)

■ Review Test 1

Turn each of the following word groups into a complete sentence. Use the space provided.

Example   With sweaty palms

With sweaty palms, I walked in for the job interview.

Even when it rains

The football teams practice even when it rains.

1. When the alarm sounded

_____

2. In order to save some money

_____

3. Was late for the game

_____

4. To pass the course

_____

5. Peter, who is very impatient

_____

6. During the holiday season

_____

7. The store where I worked

_____

8. Before the movie started

_____

9. Down in the basement

_____

10. Feeling very confident

_____

■ **Review Test 2**

Each word group in the student paragraph below is numbered. In the space provided, write C if a word group is a *complete sentence;* write F if it is a *fragment.* You will find seven fragments in the paragraph.

A Disastrous First Date

1. _____
2. _____
3. _____
4. _____
5. _____
6. _____
7. _____
8. _____
9. _____
10. _____
11. _____
12. _____
13. _____
14. _____
15. _____
16. _____
17. _____
18. _____
19. _____
20. _____

¹My first date with Donna was a disaster. ²I decided to take her to a small Italian restaurant. ³That my friends told me had reasonable prices. ⁴I looked over the menu and realized I could not pronounce the names of the dishes. ⁵Such as "veal piccante," and "fettucini Alfredo." ⁶Then, I noticed a burning smell. ⁷The candle on the table was starting to blacken. ⁸And scorch the back of my menu. ⁹Trying to be casual, I quickly poured half my glass of water on the menu. ¹⁰When the waiter returned to our table. ¹¹He asked me if I wanted to order some wine. ¹²I ordered a bottle of Blue Nun. ¹³The only wine that I had heard of and could pronounce. ¹⁴The waiter brought the wine, poured a small amount into my glass, and waited. ¹⁵I said, "You don't have to stand there. We can pour the wine ourselves." ¹⁶After the waiter put down the wine bottle and left. ¹⁷Donna told me I was supposed to taste the wine. ¹⁸Feeling like a complete fool. ¹⁹I managed to get through the dinner. ²⁰However, for weeks afterward, I felt like jumping out a tenth-story window.

On separate paper, correct the fragments you have found. Attach each fragment to the sentence that comes before or after it, or make whatever other change is needed to turn the fragment into a sentence.

■ **Review Test 3**

Underline the two fragments in each selection. Then rewrite the selection in the space provided, making the changes needed to correct the fragments.

Example The people at the diner save money. By watering down the coffee. Also, using the cheapest grade of hamburger. Few people go there anymore.

The people at the diner save money by watering down the coffee. Also,

they use the cheapest grade of hamburger.

1. Gathering speed with enormous force. The plane was suddenly in the air. Then it began to climb sharply. And several minutes later leveled off.

_____

_____

_____

2. Before my neighbors went on vacation. They asked me to watch their house. I agreed to check the premises once a day. Also, to take in their mail.

_____

_____

_____

3. Running untouched into the end zone. The halfback raised his arms in triumph. Then he slammed the football to the ground. And did a little victory dance.

_____

_____

_____

4. It's hard to keep up with bills. Such as the telephone, gas, and electricity. After you finally mail the checks. New ones seem to arrive a day or two later.

_____

_____

_____

5. While a woman ordered twenty pounds of cold cuts. Customers at the deli counter waited impatiently. The woman explained that she was in charge of a school picnic. And apologized for taking up so much time.

_____

_____

_____

■ Review Test 4

Write quickly for five minutes about what you like to do in your leisure time. Don't worry about spelling, punctuation, finding exact words, or organizing your thoughts. Just focus on writing as many words as you can without stopping.

After you have finished, go back and make whatever changes are needed to correct any sentence fragments in your writing.

RUN-ONS

Introductory Project

A run-on occurs when two sentences are run together with no adequate sign given to mark the break between them. Shown below are four run-on sentences and four correctly marked sentences. See if you can complete the statement that explains how each run-on is corrected.

1. He is the meanest little kid on his block he eats only the heads of animal crackers. *Run-on*

   He is the meanest little kid on his block. He eats only the heads of animal crackers. *Correct*

   The run-on has been corrected by using a _____ and a capital letter to separate the two complete thoughts.

2. Fred Grencher likes to gossip about other people, he doesn't like them to gossip about him. *Run-on*

   Fred Grencher likes to gossip about other people, but he doesn't like them to gossip about him. *Correct*

   The run-on has been corrected by using a joining word, _____, to connect the two complete thoughts.

3. The chain on my bike likes to chew up my pants, it leaves grease marks on my ankle as well. *Run-on*

   The chain on my bike likes to chew up my pants; it leaves grease marks on my ankle as well. *Correct*

   The run-on has been corrected by using a _____ to connect the two closely related thoughts.

4. The window shade snapped up like a gunshot, her cat leaped four feet off the floor. *Run-on*

   When the window shade snapped up like a gunshot, her cat leaped four feet off the floor. *Correct*

   The run-on has been corrected by using the subordinating word _____ to connect the two closely related thoughts.

Answers are on page 562.

WHAT ARE RUN-ONS?

A *run-on* is two complete thoughts that are run together with no adequate sign given to mark the break between them.* Some run-ons have no punctuation at all to mark the break between the thoughts. Such run-ons are known as *fused sentences:* they are fused or joined together as if they were only one thought.

Fused Sentence

My grades are very good this semester my social life rates only a C.

Fused Sentence

Our father was a madman in his youth he would do anything on a dare.

In other run-ons, known as *comma splices,* a comma is used to connect or ''splice'' together the two complete thoughts. However, a comma alone is *not enough* to connect two complete thoughts. Some stronger connection than a comma alone is needed.

Comma Splice

My grades are very good this semester, my social life rates only a C.

Comma Splice

Our father was a madman in his youth, he would do anything on a dare.

Comma splices are the most common kind of run-on mistake. Students sense that some kind of connection is needed between two thoughts, and so put a comma at the dividing point. But the comma alone is not sufficient, and a stronger, clearer mark between the two thoughts is needed.

---

* *Note:* Some instructors refer to each complete thought in a run-on as an *independent clause.* A *clause* is simply a group of words having a subject and a verb. A clause may be *independent* (expressing a complete thought and able to stand alone) or *dependent* (not expressing a complete thought and not able to stand alone). A run-on is two independent clauses that are run together with no adequate sign given to mark the break between them.

CORRECTING RUN-ONS

Here are four common methods of correcting a run-on:

1   Use a period and a capital letter to break the two complete thoughts into separate sentences:

My grades are very good this semester. My social life rates only a C.
Our father was a madman in his youth. He would do anything on a dare.

2   Use a comma plus a joining word (*and, but, for, or, nor, so, yet*) to connect the two complete thoughts:

My grades are very good this semester, but my social life rates only a C.
Our father was a madman in his youth, for he would do anything on a dare.

3   Use a semicolon to connect the two complete thoughts:

My grades are very good this semester; my social life rates only a C.
Our father was a madman in his youth; he would do anything on a dare.

4   Use subordination:

Although my grades are very good this semester, my social life rates only a C.
Because my father was a madman in his youth, he would do anything on a dare.

The following pages will give you practice in all four methods of correcting a run-on. The use of subordination will be explained further on page 523, in a section of the book that deals with sentence variety.

Method 1: Period and a Capital Letter

One way of correcting a run-on is to use a period and a capital letter at the break between the two complete thoughts. Use this method especially if the thoughts are not closely related or if another method would make the sentence too long.

Activity 1

Locate the split in each of the following run-ons. Each is a *fused sentence* — that is, each consists of two sentences that are fused or joined together with no punctuation at all between them. Reading each sentence aloud will help you "hear" where a major break or split in the thought occurs. At such a point, your voice will probably drop and pause.

Correct the run-on sentence by putting a period at the end of the first thought and a capital letter at the start of the next thought.

Example   Martha Grencher shuffled around the apartment in her slippers. Her husband couldn't stand their slapping sound on the floor.

1. The goose down jacket was not well-made little feathers leaked out of the seams.

2. Phil cringed at the sound of the dentist's drill it buzzed like a fifty-pound mosquito.

3. Last summer no one swam in the lake a little boy had dropped his pet piranhas into the water.

4. A horse's teeth never stop growing they will eventually grow outside the horse's mouth.

5. Sue's doctor told her he was an astrology nut she did not feel good about learning that.

6. Ice water is the best remedy for a burn using butter is like adding fat to a flame.

7. In the apartment the air was so dry that her skin felt parched the heat was up to eighty degrees.

8. My parents bought me an ant farm it's going to be hard to find tractors that small.

9. Lobsters are cannibalistic this is one reason they are hard to raise in captivity.

10. Julia placed an egg timer next to the phone she did not want to talk more than three minutes on her long-distance calls.

Locate the split in each of the following run-ons. Some of the run-ons are fused sentences, and some of them are *comma splices* — run-ons spliced or joined together only with a comma. Correct each run-on by putting a period at the end of the first thought and a capital letter at the start of the next thought.

1. A bird got into the house through the chimney we had to catch it before our cat did.

2. Some so-called health foods are not so healthy, many are made with oils that raise cholesterol levels.

3. We sat only ten feet from the magician, we still couldn't see where all the birds came from.

4. Rich needs only five hours of sleep each night his wife needs at least seven.

5. Our image of dentistry will soon change dentists will use lasers instead of drills.

6. Gale entered her apartment and jumped with fright someone was leaving through her bedroom window.

7. There were several unusual hair styles at the party one woman had bright green braids.

8. Jon saves all of his magazines, once a month, he takes them to a nearby nursing home.

9. The doctor seemed to be in a rush, I still took time to ask all the questions that were on my mind.

10. When I was little, my brother tried to feed me flies, he told me they were raisins.

A Warning: Words That Can Lead to Run-Ons: People often write run-on sentences when the second complete thought begins with one of the following words:

I	we	there	now
you	they	this	then
he, she, it		that	next

Remember to be on the alert for run-on sentences whenever you use one of these words in writing a paper.

Activity

Write a second sentence to go with each of the sentences that follow. Start the second sentence with the word given in italics. Your sentences can be serious or playful.

Example   *She*   Jackie works for the phone company. _She climbs telephone poles in all kinds of weather._

It   1.   The alarm clock is unreliable. _____

_____

He   2.   My uncle has a peculiar habit. _____

_____

Then   3.   Lola studied for the math test for two hours. _____

_____

It   4.   I could not understand why the car would not start. _____

_____

There   5.   We saw all kinds of litter on the highway. _____

_____

Method 2: Comma and a Joining Word

Another way of correcting a run-on sentence is to use a comma plus a joining word to connect the two complete thoughts. Joining words (also called *conjunctions*) include *and, but, for, or, nor, so,* and *yet.* Here is what the four most common joining words mean:

and   in addition to, along with

His feet hurt from the long hike, and his stomach was growling.

(*And* means "in addition": His feet hurt from the long hike; *in addition,* his stomach was growling.)

but   however, except, on the other hand, just the opposite

I remembered to get the cocoa, but I forgot the marshmallows.

(*But* means "however": I remembered to get the cocoa; *however,* I forgot the marshmallows.)

for   because, the reason why, the cause for something

She was afraid of not doing well in the course, for she had always had bad luck with English before.

(*For* means "because" or "the reason why": She was afraid of not doing well in the course; *the reason why* was that she had always had bad luck with English before.)

Note: If you are not comfortable using *for,* you may want to use *because* instead of *for* in the activities that follow. If you do use *because,* omit the comma before it.

so   as a result, therefore

The windshield wiper was broken, so she was in trouble when the rain started.

(*So* means "as a result": The windshield wiper was broken; *as a result,* she was in trouble when the rain started.)

Activity 1

Insert the joining word (*and, but, for, so*) that logically connects the two thoughts in each sentence.

1. The couple wanted desperately to buy the house, _____ they did not qualify for a mortgage.
2. A lot of men today get their hair styled, _____ they use perfume and other cosmetics as well.
3. Clyde asked his wife if she had any bandages, _____ he had just sliced his finger with a paring knife.
4. He failed the vision part of his driver's test, _____ he did not get his driver's license that day.
5. The restaurant was beautiful, _____ the food was overpriced.

Activity 2

Add a complete and closely related thought to go with each of the following statements. Use a comma plus the italicized joining word when you write the second thought.

> ***Example*** *for* Lola spent the day walking barefoot, *for the heel of one of her shoes had come off.*

but 1. She wanted to go to the party _____

_____

and 2. Tony washed his car in the morning _____

_____

so 3. The day was dark and rainy _____

_____

for 4. I'm not going to eat in the school cafeteria anymore _____

_____

but 5. I asked my brother to get off the telephone _____

_____

Method 3: Semicolon

A third method of correcting a run-on sentence is to use a semicolon to mark the break between two thoughts. A *semicolon* (;) is made up of a period above a comma and is sometimes called a *strong comma*. The semicolon signals more of a pause than a comma alone but not quite the full pause of a period.

Semicolon Alone: Here are some earlier sentences that were connected with a comma plus a joining word. Notice that a semicolon, unlike the comma alone, can be used to connect the two complete thoughts in each sentence:

A lot of men today get their hair styled; they use perfume and other cosmetics as well.

She was afraid of not doing well in the course; she had always had bad luck with English before.

The restaurant was beautiful; the food was overpriced.

Use of the semicolon can add to sentence variety. For some people, however, the semicolon is a confusing mark of punctuation. Keep in mind that if you are not comfortable using it, you can and should use one of the first two methods of correcting a run-on sentence.

Activity

Insert a semicolon where the break occurs between the two complete thoughts in each of the following sentences.

Example   I missed the bus by seconds; there would not be another for half an hour.

1. I spend eight hours a day in a windowless office it's a relief to get out in the open air after work.
2. The audience howled with laughter the comedian enjoyed a moment of triumph.
3. It rained all week parts of the highway were flooded.
4. Tony never goes to a certain gas station anymore he found out that the service manager overcharged him for a valve job.
5. The washer shook and banged with its unbalanced load then it began to walk across the floor.

Semicolon with a Transitional Word: A semicolon is sometimes used with a transitional word and a comma to join two complete thoughts:

We were short of money; therefore, we decided not to eat out that weekend.

The roots of a geranium have to be crowded into a small pot; otherwise, the plants may not flower.

I had a paper to write; however, my brain had stopped working for the night.

On the opposite page is a list of common transitional words (also known as *adverbial conjunctions*). Brief meanings are given for most of the words.

Transitional Word	Meaning
however	but
nevertheless	however
on the other hand	however
instead	as a substitute
meanwhile	in the intervening time
otherwise	under other conditions
indeed	in fact
in addition	
also	in addition
moreover	in addition
furthermore	in addition
as a result	
thus	as a result
consequently	as a result
therefore	as a result

Activity 1

Choose a logical transitional word from the list in the box and write it in the space provided. Put a semicolon *before* the connector and a comma *after* it.

Example Exams are over _____*; however,*_____ I still feel tense and nervous.

1. I did not understand her point _____ I asked her to repeat it.

2. With his thumbnail, Tony tried to split open the cellophane covering on the new record album _____ the cellophane refused to tear.

3. Post offices are closed for today's holiday _____ no mail will be delivered.

4. They decided not to go to the movie _____ they went to play miniature golf.

5. I had to skip lunch _____ I would be late for class.

Activity 2

Punctuate each sentence by using a semicolon and a comma.

Example   My brother's asthma was worsening; as a result, he quit the soccer team.

1. Bill ate an entire pizza for supper in addition he had a big chunk of pound cake for dessert.
2. The man leaned against the building in obvious pain however no one stopped to help him.
3. Our instructor was absent therefore the test was postponed.
4. I had no time to type up the paper instead I printed it out neatly in black ink.
5. Lola loves the velvety texture of cherry Jell-O moreover she loves to squish it between her teeth.

Method 4: Subordination

A fourth method of joining related thoughts is to use subordination. *Subordination* is a way of showing that one thought in a sentence is not as important as another thought.

Here are three earlier sentences that have been recast so that one idea is subordinated to (made less important than) the other idea:

When the window shade snapped up like a gunshot, her cat leaped four feet off the floor.

Because it rained all week, parts of the highway were flooded.

Although my grades are very good this year, my social life rates only a C.

Notice that when we subordinate, we use dependent words like *when, because,* and *although.* Here is a brief list of common dependent words:

Common Dependent Words		
after	before	unless
although	even though	until
as	if	when
because	since	while

Subordination is explained further on page 523.

Activity

Choose a logical dependent word from the box above and write it in the space provided.

Example _____Because_____ I had so much to do, I never even turned on the TV last night.

1. _____ we emerged from the darkened theater, it took several minutes for our eyes to adjust to the light.

2. _____ "All Natural" was printed in large letters on the yogurt carton, the fine print listing of the ingredients told a different story.

3. I can't study for the test this weekend _____ my boss wants me to work overtime.

4. _____ the vampire movie was over, my children were afraid to go to bed.

5. _____ you have a driver's license and two major credit cards, that store will not accept your check.

A Review: How to Check for Run-On Sentences

1 To see if a sentence is a run-on, read it aloud and listen for a break marking two complete thoughts. Your voice will probably drop and pause at the break.

2 To check an entire paper, read it aloud from the _last_ sentence to the _first_. Doing so will help you hear and see each complete thought.

3 Be on the lookout for words that can lead to run-on sentences:

I	he, she, it	they	this	next
you	we	there	that	then

4 Correct run-on sentences by using one of the following methods:

- Period and capital letter
- Comma and joining word (_and, but, for, or, nor, so, yet_)
- Semicolon
- Subordination

■ Review Test 1

Some of the run-ons that follow are fused sentences, having no punctuation between the two complete thoughts; others are comma splices, having only a comma between the two complete thoughts. Correct the run-ons by using one of the following three methods:

- ■ Period and a capital letter
- ■ Comma and a joining word
- ■ Semicolon

Do not use the same method of correction for every sentence.

Example   Three people did the job, *but* I could have done it alone.

1. The impatient driver tried to get a jump on the green light he kept edging his car into the intersection.

2. The course on the history of UFOs sounded interesting, it turned out to be very dull.

3. That clothing store is a strange place to visit you keep walking up to dummies that look like real people.

4. Everything on the menu of the Pancake House sounded delicious they wanted to order the entire menu.

5. Bill pressed a cold washcloth against his eyes, it helped relieve his headache.

6. Craig used to be a fast-food junkie now he eats only vegetables and sunflower seeds.

7. I knew my term paper was not very good, I placed it in a shiny plastic cover to make it look better.

8. Lola enjoys watching a talk show, Tony prefers watching a late movie.

9. My boss does not know what he is doing half the time then he tries to tell me what to do.

10. In the next minute, 100 people will die, over 240 babies will be born.

■ Review Test 2

Correct the run-on in each sentence by using subordination. Choose from among the following dependent words:

after	before	unless
although	even though	until
as	if	when
because	since	while

Example My eyes have been watering all day, I can tell the pollen count is high.

Because my eyes have been watering all day, I can tell the pollen count

is high.

1. There are a number of suits and jackets on sale, they all have very noticeable flaws.

   _____

   _____

2. Rust has eaten a hole in the muffler, my car sounds like a motorcycle.

   _____

   _____

3. I finished my household chores, I decided to do some shopping.

   _____

   _____

4. The power went off for an hour during the night, all the clocks in the house must be reset.

   _____

   _____

5. Electric cars eliminate auto pollution, the limited power of the car's battery is a serious problem.

   _____

   _____

■ **Review Test 3**

Write quickly for five minutes about what you did this past weekend. Don't worry about spelling, punctuation, finding exact words, or organizing your thoughts. Just focus on writing as many words as you can without stopping.

After you have finished, go back and make whatever changes are needed to correct any run-ons in your writing.

STANDARD
ENGLISH
VERBS

Introductory Project

Underline what you think is the correct form of the verb in each of the sentences below:

As a boy, he (enjoy, enjoyed) watching nature shows on television.
He still (enjoy, enjoys) watching such shows today as an adult.

When my car was new, it always (start, started) in the morning.
Now it (start, starts) only sometimes.

A couple of years ago, when Alice (cook, cooked) dinner, you needed an antacid tablet.
Now, when she (cook, cooks), neighbors invite themselves over to eat with us.

On the basis of the above examples, see if you can complete the following statements:

1. The first example in each pair refers to a (past, present) action, and the regular verb has an _____ ending.
2. The second example in each pair refers to a (past, present) action, and the regular verb has an _____ ending.

Answers are on page 562.

Many people have grown up in communities where nonstandard verb forms are used in everyday life. Such forms include *I thinks, he talk, it done, we has, you was,* and *she don't.* Community dialects have richness and power but are a drawback in college and the world at large, where standard English verb forms must be used. Standard English helps ensure clear communication among English-speaking people everywhere, and it is especially important in the world of work.

This chapter compares the community dialect and the standard English forms of one regular verb and three common irregular verbs.

REGULAR VERBS: DIALECT AND STANDARD FORMS

The chart below compares the community dialect (nonstandard) and the standard English forms of the regular verb *smile.*

SMILE

Community Dialect (Do not use in your writing)		***Standard English*** (Use for clear communication)	
Present tense			
I smiles	we smiles	I smile	we smile
you smiles	you smiles	you smile	you smile
he, she, it smile	they smiles	he, she, it smiles	they smile
Past tense			
I smile	we smile	I smiled	we smiled
you smile	you smile	you smiled	you smiled
he, she, it smile	they smile	he, she, it smiled	they smiled

One of the most common nonstandard forms results from dropping the endings of regular verbs. For example, people might say "David never *smile* anymore" instead of "David never *smiles* anymore." Or they will say "Before he lost his job, David *smile* a lot" instead of "Before he lost his job, David *smiled* a lot." To avoid such nonstandard usage, memorize the forms shown above for the regular verb *smile.* Then use the activities that follow to help make the inclusion of verb endings a writing habit.

Present Tense Endings

The verb ending -s or -es is needed with a regular verb in the present tense when the subject is *he, she, it,* or any *one person* or *thing.* Consider the following examples of present tense endings.

He	He yell*s.*
She	She throw*s* things.
It	It really anger*s* me.
One person	Their son storm*s* out of the house.
One person	Their frightened daughter crouch*es* behind the bed.
One thing	At night the house shake*s.*

Activity 1

All but one of the ten sentences that follow need -s or -es verb endings. Cross out the nonstandard verb forms and write the standard forms in the spaces provided. Mark the one sentence that needs no change with a C for *correct.* One example is given for you.

Example *wants* Pat always ~~want~~ the teacher's attention.

_____ 1. That newspaper print nothing but bad news.

_____ 2. The gourmet ice cream bar sell for almost two dollars.

_____ 3. Pat gossip about me all the time.

_____ 4. Whole-wheat bread taste better to me than rye bread.

_____ 5. Bob weaken his lungs by smoking so much.

_____ 6. The sick baby scream whenever her mother puts her down.

_____ 7. You make me angry sometimes.

_____ 8. Clyde drive twenty-five miles to work each day.

_____ 9. She live in a rough section of town.

_____ 10. Martha relax by drinking a glass of wine every night.

Activity 2

Rewrite the short selection below, adding present -s or -es verb endings wherever needed.

> The man lounge on his bed and watch a spider as it crawl across the ceiling. It come closer and closer to a point directly above his head. It reach the point and stop. If it drop now, it will fall right into his mouth. For a while he attempt to ignore the spider. Then he move nervously off the bed.

_____

_____

_____

_____

_____

_____

_____

Past Tense Endings

The verb ending -d or -ed is needed with a regular verb in the past tense.

A midwife delivered my baby.
The visitor puzzled over the campus map.
The children watched cartoons all morning.

Activity 1

All but one of the ten sentences that follow need -d or -ed verb endings. Cross out the nonstandard verb forms and write the standard forms in the spaces provided. Mark the one sentence that needs no change with a C.

Example   *failed*_____   This morning I ~~fail~~ a chemistry quiz.

_____   1. The customer twist his ankle on the diner's slippery steps.

_____   2. The Vietnamese student struggle with the new language.

_____   3. The sick little boy start to cry again.

_____   4. The tired mother turned on the TV for him.

——————   5. I miss quite a few days of class early in the semester.

——————   6. The weather forecaster promise blue skies, but rain began early this morning.

——————   7. Sam attempt to put out the candle flame with his finger.

——————   8. However, he end up burning himself.

——————   9. Carlo thread the film through the reels of the projector.

—————— 10. As Alice was about to finish work last night, a man came into the diner and order two dozen hamburgers.

Activity 2

Rewrite the short selection below, adding past tense -*d* or -*ed* verb endings wherever needed.

I smoke for two years and during that time suffer no real side effects. Then my body attack me. I start to have trouble falling asleep, and I awaken early every morning. My stomach digest food very slowly, so that at lunchtime I seem to be still full with breakfast. My lips and mouth turn dry and I swallow water constantly. Also, mucus fill my lungs and I cough a lot. I decide to stop smoking when my wife insist I take out more life insurance for our family.

———————————————————————————————

———————————————————————————————

———————————————————————————————

———————————————————————————————

———————————————————————————————

———————————————————————————————

———————————————————————————————

———————————————————————————————

———————————————————————————————

———————————————————————————————

———————————————————————————————

THREE COMMON IRREGULAR VERBS:
DIALECT AND STANDARD FORMS

The following charts compare the community dialect and the standard English forms of the common irregular verbs *be, have,* and *do.* (For more on irregular verbs, see pages 382–389.)

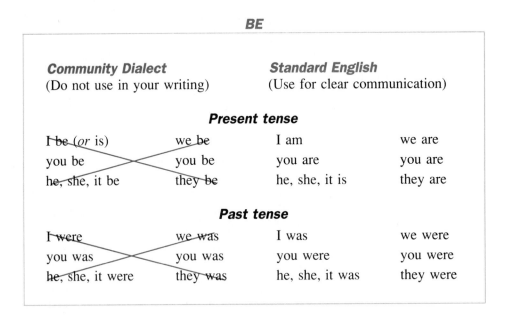

BE

Community Dialect (Do not use in your writing)		Standard English (Use for clear communication)	
Present tense			
I be (*or* is)	we be	I am	we are
you be	you be	you are	you are
he, she, it be	they be	he, she, it is	they are
Past tense			
I were	we was	I was	we were
you was	you was	you were	you were
he, she, it were	they was	he, she, it was	they were

HAVE

Community Dialect (Do not use in your writing)		Standard English (Use for clear communication)	
Present tense			
I has	we has	I have	we have
you has	you has	you have	you have
he, she, it have	they has	he, she, it has	they have
Past tense			
I has	we has	I had	we had
you has	you has	you had	you had
he, she, it have	they has	he, she, it had	they had

DO

	Community Dialect (Do not use in your writing)		Standard English (Use for clear communication)

Present tense

Community Dialect		Standard English	
I does	we does	I do	we do
you does	you does	you do	you do
he, she, it do	they does	he, she, it does	they do

Past tense

Community Dialect		Standard English	
I done	we done	I did	we did
you done	you done	you did	you did
he, she, it done	they done	he, she, it did	they did

Note: Many people have trouble with one negative form of *do*. They will say, for example, ''He don't agree'' instead of ''He doesn't agree,'' or they will say ''The door don't work'' instead of ''The door doesn't work.'' Be careful to avoid the common mistake of using *don't* instead of *doesn't*.

Activity 1

Underline the standard form of *be, have,* or *do.*

1. When Walt (have, has) his own house, he will install built-in stereo speakers in every room.
2. The children (is, are) ready to go home.
3. Whenever we (do, does) the laundry, our clothes are spotted with blobs of undissolved detergent.
4. Ed and Arlene (was, were) ready to leave for the movies when the baby began to wail.
5. Our art class (done, did) the mural on the wall of the cafeteria.
6. If Maryanne (have, has) the time, she will help us set up the projector and tape the wires to the floor.
7. Curtis (be, is) the best Ping-Pong player in the college.
8. That mechanic always (do, does) a good job when he fixes my car.
9. The mice in our attic (have, has) chewed several holes in our ceiling.
10. The science instructor said that the state of California (be, is) ready for a major earthquake any day.

Activity 2

Fill in each blank with the standard form of *be, have,* or *do.*

1. My car _____ a real personality.

2. It acts as if it _____ human.

3. On cold mornings, it _____ not want to start.

4. Like me, the car _____ a problem dealing with freezing weather.

5. I don't want to get out of bed, and my car _____ not like leaving the garage.

6. Also, we _____ the same feeling about rainstorms.

7. I hate driving to school in a downpour and so _____ the car.

8. When the car _____ stopped at a light, it stalls.

9. The habits my car _____ may be annoying.

10. But they _____ understandable.

■ Review Test 1

Underline the standard verb form.

1. Martha (argue, argues) just to hear herself talk.
2. Those shoppers (do, does) not seem to know their way around the market; they keep retracing their steps.
3. The cheap ballpoint pen (leak, leaked) all over the lining of my pocketbook.
4. Pat (bag, bagged) the dirty laundry and threw it into the car.
5. If you (has, have) any trouble with the assignment, give me a call.
6. Whenever the hairdresser (do, does) my hair, she cuts one side shorter than the other.
7. Lola often (watch, watches) TV after her parents have gone to bed.
8. Two of the players (was, were) suspended from the league for ten games for using drugs.
9. Jeannie (has, have) only one eye; she lost the other years ago after falling on some broken glass.
10. I remember how my wet mittens (use, used) to steam on the hot school radiator.

■ Review Test 2

Cross out the two nonstandard verb forms in each sentence below. Then write the standard English verbs in the spaces provided.

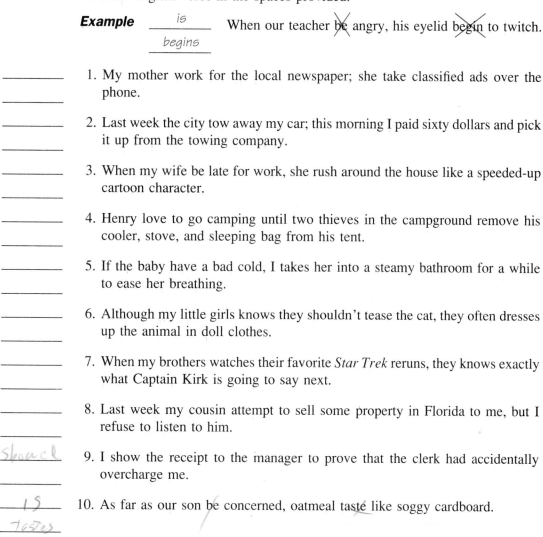

Example _____is_____ When our teacher ~~be~~ angry, his eyelid ~~begin~~ to twitch.
_____begins_____

_____ 1. My mother work for the local newspaper; she take classified ads over the
_____ phone.

_____ 2. Last week the city tow away my car; this morning I paid sixty dollars and pick
_____ it up from the towing company.

_____ 3. When my wife be late for work, she rush around the house like a speeded-up
_____ cartoon character.

_____ 4. Henry love to go camping until two thieves in the campground remove his
_____ cooler, stove, and sleeping bag from his tent.

_____ 5. If the baby have a bad cold, I takes her into a steamy bathroom for a while
_____ to ease her breathing.

_____ 6. Although my little girls knows they shouldn't tease the cat, they often dresses
_____ up the animal in doll clothes.

_____ 7. When my brothers watches their favorite *Star Trek* reruns, they knows exactly
_____ what Captain Kirk is going to say next.

_____ 8. Last week my cousin attempt to sell some property in Florida to me, but I
_____ refuse to listen to him.

___showed___ 9. I show the receipt to the manager to prove that the clerk had accidentally
_____ overcharge me.

___is___ 10. As far as our son be concerned, oatmeal taste like soggy cardboard.
___tastes___

IRREGULAR VERBS

Introductory Project

You may already have a sense of which common English verbs are regular and which are not. To test yourself, fill in the past tense and past participle of the verbs below. Five are regular verbs and so take *-d* or *-ed* in the past tense and past participle. Five are irregular verbs and will probably not sound right when you try to add *-d* or *-ed*. Write I for irregular in front of these verbs. Also, see if you can write in their irregular verb forms. (The item at the top is an example.)

Present	Past	Past Participle
shout	*shouted*	*shouted*
1. crawl	_____	_____
2. bring	_____	_____
3. use	_____	_____
4. do	_____	_____
5. give	_____	_____
6. laugh	_____	_____
7. go	_____	_____
8. scare	_____	_____
9. dress	_____	_____
10. see	_____	_____

Answers are on page 562.

A BRIEF REVIEW OF REGULAR VERBS

Every verb has four principal parts: present, past, past participle, and present participle. These parts can be used to build all the verb tenses (the times shown by a verb).

The past and past participle of a regular verb are formed by adding *-d* or *-ed* to the present. The *past participle* is the form of the verb used with the helping verbs *have, has,* or *had* (or some form of *be* with passive verbs). The *present participle* is formed by adding *-ing* to the present. Here are the principal forms of some regular verbs:

Present	*Past*	*Past Participle*	*Present Participle*
crash	crashed	crashed	crashing
shiver	shivered	shivered	shivering
kiss	kissed	kissed	kissing
apologize	apologized	apologized	apologizing
tease	teased	teased	teasing

Most verbs in English are regular.

LIST OF IRREGULAR VERBS

Irregular verbs have irregular forms in the past tense and past participle. For example, the past tense of the irregular verb *know* is *knew;* the past participle is *known.*

Almost everyone has some degree of trouble with irregular verbs. When you are unsure about the form of a verb, you can check the list of irregular verbs on the following pages. (The present participle is not shown on this list because it is formed simply by adding *-ing* to the base form of the verb.) Or you can check a dictionary, which gives the principal parts of irregular verbs.

Present	Past	Past Participle
arise	arose	arisen
awake	awoke *or* awaked	awoken *or* awaked
be (am, are, is)	was (were)	been
become	became	become
begin	began	begun
bend	bent	bent
bite	bit	bitten
blow	blew	blown
break	broke	broken
bring	brought	brought
build	built	built
burst	burst	burst
buy	bought	bought
catch	caught	caught
choose	chose	chosen
come	came	come
cost	cost	cost
cut	cut	cut
do (does)	did	done
draw	drew	drawn
drink	drank	drunk
drive	drove	driven
eat	ate	eaten
fall	fell	fallen
feed	fed	fed
feel	felt	felt
fight	fought	fought
find	found	found
fly	flew	flown
freeze	froze	frozen
get	got	got *or* gotten
give	gave	given
go (goes)	went	gone
grow	grew	grown

Present	*Past*	*Past Participle*
have (has)	had	had
hear	heard	heard
hide	hid	hidden
hold	held	held
hurt	hurt	hurt
keep	kept	kept
know	knew	known
lay	laid	laid
lead	led	led
leave	left	left
lend	lent	lent
let	let	let
lie	lay	lain
lose	lost	lost
make	made	made
meet	met	met
pay	paid	paid
ride	rode	ridden
ring	rang	rung
run	ran	run
say	said	said
see	saw	seen
sell	sold	sold
send	sent	sent
shake	shook	shaken
shrink	shrank	shrunk
shut	shut	shut
sing	sang	sung
sit	sat	sat
sleep	slept	slept
speak	spoke	spoken
spend	spent	spent
stand	stood	stood
steal	stole	stolen

Present	Past	Past Participle
stick	stuck	stuck
sting	stung	stung
swear	swore	sworn
swim	swam	swum
take	took	taken
teach	taught	taught
tear	tore	torn
tell	told	told
think	thought	thought
wake	woke *or* waked	woken *or* waked
wear	wore	worn
win	won	won
write	wrote	written

Activity 1

Cross out the incorrect verb form in each of the following sentences. Then write the correct form of the verb in the space provided.

Example *drew* The little boy drawed on the marble table with permanent ink.

_____ 1. Tomatoes were once thought to be poisonous, and they were growed only as ornamental shrubs.

_____ 2. Juan has rode the bus to school for two years while saving for a car.

_____ 3. My cats have tore little holes in all my good wool sweaters.

_____ 4. The pipes in the bathroom freezed last winter, and they burst when they thawed.

_____ 5. Every time my telephone has rang today, there has been bad news on the line.

_____ 6. Only seven people have ever knowed the formula for Coca-Cola.

_____ 7. Amy blowed up animal-shaped balloons for her son's birthday party.

_____ 8. I shaked the bottle angrily until the catsup began to flow.

_____ 9. While waiting for the doctor to arrive, I sitted in a plastic chair for over two hours.

_____ 10. The pile of bones on the plate showed how much chicken the family had ate.

Activity 2

For each of the italicized verbs, fill in the three missing forms in the following order:

a Present tense, which takes an *-s* ending when the subject is *he, she, it,* or any *one person* or *thing* (see page 375)

b Past tense

c Past participle — the form that goes with the helping verb *have, has,* or *had*

Example My uncle likes to *give* away certain things. He (a) _____*gives*_____ old, threadbare clothes to the Salvation Army. Last year he (b) _____*gave*_____ me a worthless television set in which the picture tube was burned out. He has (c) _*given*_____ away stuff that a junk dealer would reject.

1. I like to *freeze* Hershey bars. A Hershey bar (a) _____ in half an hour. Once I (b) _____ a bottle of Pepsi. I put it in the freezer to chill and then forgot about it. Later I opened the freezer and discovered that it had (c) _____ and exploded.

2. I *know* the girl in the lavender bikini. She (a) _____ me, too. I (b) _____ her brother before I met her. I have (c) _____ him since boyhood.

3. An acquaintance of mine is a shoplifter, although he knows it's wrong to *steal*. He (a) _____ candy bars from supermarkets. Last month he (b) _____ a Sony Walkman and was caught by a detective. He has (c) _____ pants and shirts by wearing several layers of clothes out of a store.

4. I *go* to parties a lot. Often Camille (a) _____ with me. She (b) _____ with me just last week. I have (c) _____ to parties every Friday for the past month.

5. My brother likes to *throw* things. Sometimes he (a) _____ socks in his bureau drawer. In high school he (b) _____ footballs while quarterbacking the team. And he has (c) _____ Frisbees in our backyard for as long as I can remember.

6. I *see* her every weekend. She (*a*) _____ her other friends during the week. We first (*b*) _____ each other on a cold Saturday night last winter, when we went for supper at an Indian restaurant. Since then we have (*c*) _____ each other every weekend except when my car was broken down.

7. I often *lie* down for a few minutes after a hard day's work. Sometimes my cat (*a*) _____ down near me. Yesterday was Saturday, so I (*b*) _____ in bed all morning. I probably would have (*c*) _____ in bed all afternoon, but I wanted to get some planting done in my vegetable garden.

8. I *do* not understand the assignment. It simply (*a*) _____ not make sense to me. I was surprised to learn that Shirley (*b*) _____ understand it. In fact, she had already (*c*) _____ the assignment.

9. I often find it hard to *begin* writing a paper. The assignment that I must do (*a*) _____ to worry me while I'm watching television, but I seldom turn off the set. Once I waited until the late movie had ended before I (*b*) _____ to write. If I had (*c*) _____ earlier, I would have gotten a decent night's sleep.

10. Martha likes to *eat*. She (*a*) _____ as continuously as some people smoke. Once she (*b*) _____ a large pack of cookies in half an hour. Even if she has (*c*) _____ a heavy meal, she often starts munching snacks right afterward.

■ **Review Test 1**

Underline the correct verb in the parentheses.

1. I (shaked, shook) the bottle of medicine before I took a teaspoon of it.

2. Chico came into the gym and (began, begun) to practice on the parallel bars.

3. Over half the class has (taken, took) this course on a pass-fail basis.

4. Even though my father (teached, taught) me how to play baseball, I never enjoyed any part of the game.

5. Because I had (lended, lent) him the money, I had a natural concern about what he did with it.

6. The drugstore clerk (gave, gived) him the wrong change.

7. Lola (brang, brought) a sweatshirt with her, for she knew the mountains got cold at night.

8. My sister (was, be) at school when a stranger came asking for her at our home.

9. The mechanic (did, done) an expensive valve job on my engine without getting my permission.

10. The basketball team has (broke, broken) the school record for most losses in one year.

11. Someone (leaved, left) his books in the classroom.

12. That jacket was (tore, torn) during the football game.

13. If I hadn't (threw, thrown) away the receipt, I could have gotten my money back.

14. I would have (become, became) very angry if you had not intervened.

15. As the flowerpot (fell, falled) from the windowsill, the little boy yelled, "Bombs away!"

■ Review Test 2

Write short sentences that use the form requested for the following irregular verbs.

Example   Past of *grow*   *I grew eight inches in one year.* _____

1. Past of *know* _____

2. Past of *take* _____

3. Past participle of *give* _____

4. Past participle of *write* _____

5. Past of *bring* _____

6. Past participle of *speak* _____

7. Present of *begin* _____

8. Past of *go* _____

9. Past participle of *see* _____

10. Past of *drive* _____

SUBJECT-VERB AGREEMENT

Introductory Project

As you read each pair of sentences below, place a check mark beside the sentence that you think uses the underlined word correctly.

There <u>was</u> too many people talking at once. _____

There <u>were</u> too many people talking at once. _____

The onions in that spaghetti sauce <u>gives</u> me heartburn. _____

The onions in that spaghetti sauce <u>give</u> me heartburn. _____

The mayor and her husband <u>attends</u> our church. _____

The mayor and her husband <u>attend</u> our church. _____

Everything <u>seem</u> to slow me down when I'm in a hurry. _____

Everything <u>seems</u> to slow me down when I'm in a hurry. _____

Answers are on page 562.

A verb must agree with its subject in number. A *singular subject* (one person or thing) takes a singular verb. A *plural subject* (more than one person or thing) takes a plural verb. Mistakes in subject-verb agreement are sometimes made in the following situations (each situation is explained on the following pages):

1 When words come between the subject and the verb
2 When a verb comes before the subject
3 With compound subjects
4 With indefinite pronouns

WORDS BETWEEN SUBJECT AND VERB

Words that come between the subject and the verb do not change subject-verb agreement. In the sentence

The mean <u>cockroaches</u> behind my stove <u>get</u> high on Raid.

the subject (<u>cockroaches</u>) is plural and so the verb (<u>get</u>) is plural. The words *behind my stove* that come between the subject and verb do not affect subject-verb agreement.

To help find the subject of certain sentences, you should cross out prepositional phrases (see page 335):

<u>Nell</u>, ~~with her three dogs close behind~~, <u>runs</u> around the park every day.

The <u>seams</u> ~~in my new coat~~ <u>have split</u> after only two wearings.

Activity

Underline the subject and lightly cross out any words that come between the subject and the verb. Then double-underline the verb choice in parentheses that you believe is correct.

1. The decisions of the judge (seems, seem) questionable.

2. A hamburger with a double order of french fries (is, are) my usual lunch.

3. One of my son's worst habits (is, are) leaving an assortment of dirty plates on the kitchen counter every morning.

4. The rust spots on the back of Emily's car (needs, need) to be cleaned with a special polish.

5. The collection of medicine bottles in my bathroom (overflows, overflow) the cabinet shelves.

VERB BEFORE SUBJECT

A verb agrees with its subject even when the verb comes *before* the subject. Words that may precede the subject include *there, here,* and, in questions, *who, which, what,* and *where.*

On Bill's doorstep <u>were</u> two <u>police officers</u>.
There <u>are</u> many pizza <u>places</u> in our town.
Here <u>is</u> your <u>receipt</u>.
Where <u>are</u> <u>they</u> <u>going</u> to sleep?

If you are unsure about the subject, ask *who* or *what* of the verb. With the first example above, you might ask, "*Who* were on the doorstep?" The answer, *police officers,* is the subject.

Activity

Write the correct form of the verb in the space provided.

was, were   1.   There _____ not enough glasses for all the guests at the party.

is, are   2.   Here _____ the tickets for tonight's ball game.

do, does   3.   Where _____ you go when you want to be alone?

is, are   4.   There _____ too many people in the room for me to feel comfortable.

was, were   5.   Stuffed into the mailbox _____ ten pieces of junk mail and three ripped magazines.

COMPOUND SUBJECTS

Subjects joined by *and* generally take a plural verb.

<u>Maple syrup</u> and <u>sweet butter</u> <u>taste</u> delicious on pancakes.
<u>Fear</u> and <u>ignorance</u> <u>have</u> a lot to do with hatred.

When subjects are joined by *either . . . or, neither . . . nor, not only . . . but also,* the verb agrees with the subject closer to the verb.

Either the <u>Oak Ridge Boys</u> or <u>Randy Travis</u> <u>deserves</u> the award for the best country album of the year.

The nearer subject, *Randy Travis,* is singular, and so the verb is singular.

Activity

Write the correct form of the verb in the space provided.

is, are   1. An egg and a banana _____ required for the recipe.

was, were   2. Owning a car and having money in my pocket _____ the chief ambitions of my adolescence.

visits, visit   3. My aunt and uncle from Ireland _____ us every other summer.

was, were   4. Before they saw a marriage therapist, Peter and Jenny _____ planning to get divorced.

acts, act   5. Not only the landlady but also her children _____ unfriendly to us.

INDEFINITE PRONOUNS

The following words, known as *indefinite pronouns,* always take singular verbs:

(*-one* words)	(*-body* words)	(*-thing* words)	
one	nobody	nothing	each
anyone	anybody	anything	either
everyone	everybody	everything	neither
someone	somebody	something	

Note: *Both* always takes a plural verb.

Activity

Write the correct form of the verb in the space provided.

pitches, pitch   1. If each of us _____ in, we can finish this job in an hour.

was, were   2. Everybody in the theater _____ getting up and leaving before the movie ended.

provides, provide   3. Neither of the restaurants _____ facilities for the handicapped.

likes, like   4. No one in our family _____ housecleaning, but we all take a turn at it.

steals,   5. Someone in our neighborhood _____ vegetables from people's
steal   gardens.

■ Review Test

Underline the correct verb in parentheses.

1. The lettuce in most of the stores in our area now (costs, cost) almost one dollar a head.

2. Nobody in the class of fifty students (understands, understand) how to solve the equation on the blackboard.

3. The packages in the shopping bag (was, were) a wonderful mystery to the children.

4. My exercise class of five students (meets, meet) every Thursday afternoon.

5. Anyone who (steals, steal) my purse won't find much inside it.

6. Business contacts and financial backing (is, are) all that I need to establish my career as a dress designer.

7. Each of those breakfast cereals (contains, contain) a high proportion of sugar.

8. The serious look in that young girl's eyes (worries, worry) me.

9. All of the cars on my block (has, have) to be moved one day a month for street cleaning.

10. The job is not for people who (stumbles, stumble) over tough decisions.

CONSISTENT
VERB
TENSE

Introductory Project

See if you can find and underline the two mistakes in verb tense in the following selection.

Tony's eyes burned and itched all day long. When he looked at them in a mirror, he also discovers there were red blotches on his neck. He spoke to his mother about the symptoms, and she said that maybe he was allergic to something. Then he remembers he had been cuddling the kitten that Lola had just bought the day before. "Good grief. I must be allergic to cats," he said to himself.

Answers are on page 562.

KEEPING TENSES CONSISTENT

Do not shift verb tenses unnecessarily. If you begin writing a paper in the present tense, don't shift suddenly to the past. If you begin in the past, don't shift without reason to the present. Notice the inconsistent verb tenses in the following selection:

> The shoplifter *walked* quickly toward the front of the store. When a clerk *shouts* at him, he *started* to run.

The verbs must be consistently in the present tense:

> The shoplifter *walks* quickly toward the front of the store. When a clerk *shouts* at him, he *starts* to run.

Or the verbs must be consistently in the past tense:

> The shoplifter *walked* quickly toward the front of the store. When a clerk *shouted* at him, he *started* to run.

Activity 1

In each selection one verb must be changed so that it agrees in tense with the other verbs. Cross out the incorrect verb and write the correct form in the space provided.

Example   *carried*   Ted wanted to be someplace else when the dentist ~~carries~~ in a long needle.

_____   1. I played my stereo and watched television before I decide to do some homework.

_____   2. The hitchhiker stopped me as I walks from the turnpike rest station and said, "Are you on your way to San Jose?"

_____   3. Some students attend all their classes in school. They listen carefully during lectures but they don't take notes. As a result, they often failed tests.

_____   4. His parents stayed together for his sake; only after he graduates from college were they divorced.

_____   5. In the movie, artillery shells exploded on the hide of the reptile monster. It just grinned, tosses off the shells, and kept eating people.

———— 6. Several months a year, monarch butterflies come to live in a spot along the California coast. Thousands and thousands of them hang from the trees and fluttered through the air in large groups.

———— 7. After waking up each morning, Harry stays in bed for a while. First he stretches and yawned loudly, and then he plans his day.

———— 8. The salespeople at Biggs' Department Store are very helpful. When people asked for a product the store doesn't carry or is out of, the salesperson recommends another store.

———— 9. Part-time workers at the company are the first to be laid off. They are also paid less, and they received no union representation.

———— 10. Smashed cars, ambulances, and police cars blocked traffic on one side of the highway. On the other side, traffic slows down as drivers looked to see what happened.

Activity 2

Change verbs where needed in the following selection so that they are consistently in the past tense. Cross out each incorrect verb and write the correct form above it, as shown in the example. You will need to make nine corrections.

Late one rainy night, Sheila woke to the sound of steady dripping. When she got out of bed to investigate, a drop of cold water ~~splashes~~ *splashed* onto her arm. She looks up just in time to see another drop form on the ceiling, hang suspended for a moment, and fall to the carpet. Stumbling to the kitchen, Sheila reaches deep into one of the cabinets and lifts out a large roasting pan. As she did so, pot lids and baking tins clattered out and crash onto the counter. Sheila ignored them, stumbled back to the bedroom, and places the pan on the floor under the drip. But a minute after sliding her icy feet under the covers, Sheila realized she is in trouble. The sound of each drop hitting the metal pan echoed like a gunshot in the quiet room. Sheila feels like crying, but she finally thought of a solution. She got out of bed and returns a minute later with a thick bath towel. She lined the pan with the towel and crawls back into bed.

■ Review Test

Change verbs where needed in the following selection so that they are consistently in the past tense. Cross out each incorrect verb and then write the correct form in the space provided. You will need to make ten corrections in all.

Balancing the green plastic bag full of trash, Craig yanked the front door open. As he stepped onto the front porch, he notices that a light snow was already falling. He remembers that when he called to rent the cabin, he was told that it was not too early to expect snow in this mountain community. He glances up at the sky and then walks briskly to the end of the driveway. There he deposited the overflowing bag into one of the large trash cans. Shivering from the cold, he turned around and starts back toward the house, but then he pauses suddenly. At the southwest corner of the cabin, standing on its hind legs, was an enormous black bear. For a long terrible second, Craig was positive the bear was staring right at him. Looking for a promising direction to run, Craig turns around and saw a small bear cub scampering away from behind another garbage can. Before Craig had time to react, the large bear went down on all fours, sprints past the house, and started after the cub. Craig breathed a sigh of relief, races into the cabin, and locks the door behind him.

1. _____   6. _____

2. _____   7. _____

3. _____   8. _____

4. _____   9. _____

5. _____   10. _____

PRONOUN AGREEMENT, REFERENCE, AND POINT OF VIEW

Introductory Project

Read each pair of sentences below. Then put a check mark beside the sentence that you think uses the underlined word or words correctly.

Someone in my neighborhood lets their dog run loose. _____

Someone in my neighborhood lets his or her dog run loose. _____

After Tony reviewed his notes with Bob, he passed the exam with ease. _____

After reviewing his notes with Bob, Tony passed the exam with ease. _____

I dislike waitressing, for you can never count on a fair tip. _____

I dislike waitressing, for I can never count on a fair tip. _____

Answers are on page 562.

Pronouns are words that take the place of nouns (persons, places, or things). In fact, the word *pronoun* means "for a noun." Pronouns are shortcuts that keep you from unnecessarily repeating words in writing. Here are some examples of pronouns:

Shirley had not finished *her* paper. (*Her* is a pronoun that takes the place of *Shirley's*.)

Tony swung so heavily on the tree branch that *it* snapped. (*It* replaces *branch*.)

When the three little pigs saw the wolf, *they* pulled out cans of Mace. (*They* is a pronoun that takes the place of *pigs*.)

This section presents rules that will help you avoid three common mistakes people make with pronouns. The rules are as follows:

1 A pronoun must agree in number with the word or words it replaces.
2 A pronoun must refer clearly to the word it replaces.
3 Pronouns should not shift unnecessarily in point of view.

PRONOUN AGREEMENT

A pronoun must agree in number with the word or words it replaces. If the word a pronoun refers to is singular, the pronoun must be singular; if that word is plural, the pronoun must be plural. (Note that the word a pronoun refers to is also known as the *antecedent*.)

Barbara agreed to lend me (her) Willie Nelson albums.

People walking the trail must watch (their) step because of snakes.

In the first example, the pronoun *her* refers to the singular word *Barbara;* in the second example, the pronoun *their* refers to the plural word *People*.

Activity

Write the appropriate pronoun (*their, they, them, it*) in the blank space in each of the following sentences.

Example   I lifted the pot of hot potatoes carefully, but _____*it*_____ slipped out of my hand.

1. The value that people receive for _____ dollars these days is rapidly diminishing.

2. Fred never misses his daily workout; he believes _____ keeps him healthy.

3. Sometimes, in marriage, partners expect too much from _____ mates.

4. For some students, college is often their first experience with an undisciplined learning situation, and _____ are not always ready to accept the responsibility.

5. Our new neighbors moved in three months ago, but I have yet to meet _____ .

Indefinite Pronouns

The following words, known as *indefinite pronouns*, are always singular.

(*-one* words)	(*-body* words)	
one	nobody	each
anyone	anybody	either
everyone	everybody	neither
someone	somebody	

If a pronoun in a sentence refers to one of these above singular words, the pronoun should be singular.

Each father felt that (his) child should have won the contest.

One of the women could not find (her) purse.

Everyone must be in (his) seat before the instructor takes the roll.

In each example, the circled pronoun is singular because it refers to one of the special singular words.

Note: The last example is correct if everyone in the class is a man. If everyone in the class was a woman, the pronoun would be *her*. If the class had both women and men, the pronoun form would be *his or her:*

Everyone must be in his or her seat before the instructor takes the roll.

Some writers follow the traditional practice of using *his* to refer to both women and men. Many now use *his or her* to avoid an implied sexual bias. To avoid using *his* or the somewhat awkward *his or her,* a sentence can often be rewritten in the plural:

Students must be in their seats before the instructor takes the roll.

Activity

Underline the correct pronoun.

1. Someone has blocked the parking-lot exit with (his, their) car.
2. Everyone in the women's group has volunteered some of (her, their) time for the voting drive.
3. Neither of the men arrested as terrorists would reveal information about (his, their) group.
4. Not one of the women coaches will be returning to (her, their) job next year.
5. Each of the President's advisers offered (his or her, their) opinion about the rail strike.

PRONOUN REFERENCE

A sentence may be confusing and unclear if a pronoun appears to refer to more than one word, or if the pronoun does not refer to any specific word. Look at this sentence:

Joe almost dropped out of high school, for he felt *they* emphasized discipline too much.

Who emphasized discipline too much? There is no specific word that *they* refers to. Be clear:

Joe almost dropped out of high school, for he felt *the teachers* emphasized discipline too much.

Here are sentences with other kinds of faulty pronoun references. Read the explanations of why they are faulty and look carefully at how they are corrected.

Faulty	*Clear*
June told Margie that *she* lacked self-confidence.	June told Margie, ''You lack self-confidence.''
(*Who* lacked self-confidence: June or Margie? Be clear.)	(Quotation marks, which can sometimes be used to correct an unclear reference, are explained on pages 458–465.)
Nancy's mother is a hairdresser, but Nancy is not interested in *it*.	Nancy's mother is a hairdresser, but Nancy is not interested in becoming one.
(There is no specific word that *it* refers to. It would not make sense to say, ''Nancy is not interested in hairdresser.'')	
Ron blamed the police officer for the ticket, *which* was foolish.	Foolishly, Ron blamed the police officer for the ticket.
(Does *which* mean that the officer's giving the ticket was foolish, or that Ron's blaming the officer was foolish? Be clear.)	

Activity

Rewrite each of the following sentences to make clear the vague pronoun reference. Add, change, or omit words as necessary.

Example   Our cat was friends with our hamster until he bit him.

Until the cat bit the hamster, the two were friends.

1. Maria's mother let her wear her new earrings to school.

   _____

2. When I asked why I failed my driver's test, he said I drove too slowly.

   _____

3. Dad ordered my brother to paint the garage because he didn't want to do it.

   _____

4. Herb dropped his psychology courses because he thought they assigned too much reading.

   _____

5. I love Parmesan cheese on veal, but it does not always digest well.

   _____

PRONOUN POINT OF VIEW

Pronouns should not shift their point of view unnecessarily. When writing a paper, be consistent in your use of first-, second-, or third-person pronouns.

Type of Pronoun	Singular	Plural
First-person pronouns	I (my, mine, me)	we (our, us)
Second-person pronouns	you (your)	you (your)
Third-person pronouns	he (his, him) she (her) it (its)	they (their, them)

Note: Any person, place, or thing, as well as any indefinite pronoun like *one, anyone, someone,* and so on (page 401), is a third-person word.

For instance, if you start writing in the third person *she,* don't jump suddenly to the second person *you.* Or if you are writing in the first person *I,* don't shift unexpectedly to *one.* Look at the examples.

Inconsistent	Consistent
I enjoy movies like *The Return of the Vampire* that frighten *you.* (The most common mistake people make is to let a *you* slip into their writing after they start with another pronoun.)	I enjoy movies like *The Return of the Vampire* that frighten me.
As soon as a person walks into Helen's apartment, *you* can tell that Helen owns a cat. (Again, the *you* is a shift in point of view.)	As soon as a person walks into Helen's apartment, *he or she* can tell that Helen owns a cat. (See also the note on *his or her* references on pages 401–402.)

Activity

Cross out inconsistent pronouns in the following sentences, and write the correct form of the pronoun above each crossed-out word.

Example My dreams are always the kind that haunt ~~you~~ *me* the next day.

1. Whenever we take our children on a trip, you have to remember to bring snacks, tissues, and toys.

2. In our society, we often need a diploma before you are hired for a job.

3. A worker can take a break only after a relief person comes to take your place.

4. If a student organizes time carefully, you can accomplish a great deal of work.

5. Although I know you should watch your cholesterol intake, I can never resist an ear of corn dripping with melted butter.

■ Review Test 1

Cross out the pronoun error in each sentence and write the correction in the space provided at the left. Then circle the letter that correctly describes the type of error that was made.

Examples

his (or her)

Each player took ~~their~~ position on the court.
Mistake in: a. pronoun reference (b.) pronoun agreement

the store

I was angry when ~~they~~ wouldn't give me cash back when I returned the sweater I had bought.
Mistake in: (a.) pronoun reference   b. pronoun point of view

I

I love Jello because ~~you~~ can eat about five bowls of it and still not feel full.
Mistake in: a. pronoun agreement   (b.) pronoun point of view

1. Dan asked Mr. Sanchez if he could stay an extra hour at work today.
    Mistake in: a. pronoun reference   b. pronoun agreement

2. Both the front door and the back door of the abandoned house had fallen off its hinges.
    Mistake in: a. pronoun agreement   b. pronoun point of view

3. I hate going to the supermarket because you always have trouble finding a parking space there.
    Mistake in: a. pronoun agreement   b. pronoun point of view

4. Norm was angry when they raised the state tax on cigarettes again.
    Mistake in: a. pronoun agreement   b. pronoun reference

5. Every one of those musicians who played for two hours in the rain truly earned their money last night.
    Mistake in: a. pronoun agreement   b. pronoun reference

_____   6. As I entered the house, you could hear someone giggling in the hallway.
Mistake in: a. pronoun reference   b. pronoun point of view

_____   7. Each of the beauty queens is asked a thought-provoking question and then judged on their answer.
Mistake in: a. pronoun agreement   b. pronoun reference

_____   8. Sometimes I take the alternate route, but it costs you five dollars in tolls.
Mistake in: a. pronoun agreement   b. pronoun point of view

_____   9. At the dental office, I asked him if it was really necessary to take x-rays of my mouth again.
Mistake in: a. pronoun agreement   b. pronoun reference

_____   10. My favorite subject is abnormal psychology because the case studies make you seem so normal by comparison.
Mistake in: a. pronoun agreement   b. pronoun point of view

■ Review Test 2

Underline the correct word in parentheses.

1. As we sat in class waiting for the test results, (you, we) could feel the tension.

2. Hoping to be first in line when (they, the ushers) opened the doors, we arrived two hours early for the concert.

3. If a person really wants to appreciate good coffee, (he or she, you, they) should drink it black.

4. I am hooked on science fiction stories because they allow (you, me) to escape to other worlds.

5. Lois often visits the reading center in school, for she finds that (they, the tutors) give her helpful instruction.

6. Nobody seems to know how to add or subtract without (his or her, their) pocket calculator anymore.

7. Cindy is the kind of woman who will always do (their, her) best.

8. Each of my brothers has had (his, their) apartment broken into.

9. If someone is going to write a composition, (he or she, you, they) should prepare at least one rough draft.

10. I've been taking cold medicine, and now (it, the cold) is better.

PRONOUN TYPES

This chapter describes some common types of pronouns: subject and object pronouns, possessive pronouns, and demonstrative pronouns.

SUBJECT AND OBJECT PRONOUNS

Pronouns change their form depending on the purpose they serve in a sentence. In the box that follows is a list of subject and object pronouns.

Subject Pronouns	Object Pronouns
I	me
you	you (no change)
he	him
she	her
it	it (no change)
we	us
they	them

Subject Pronouns

Subject pronouns are subjects of verbs.

> *She* is wearing blue nail polish on her toes. (*She* is the subject of the verb *is wearing*.)
>
> *They* ran up three flights of steps. (*They* is the subject of the verb *ran*.)
>
> *We* children should have some privacy too. (*We* is the subject of the verb *should have*.)

Rules for using subject pronouns, and several kinds of mistakes people sometimes make with subject pronouns, are explained below.

1   Use a subject pronoun in spots where you have a compound (more than one) subject.

Incorrect	*Correct*
Sally and *me* are exactly the same size.	Sally and *I* are exactly the same size.
Her and *me* share our wardrobes with each other.	*She* and *I* share our wardrobes with each other.

Hint: If you are not sure what pronoun to use, try each pronoun by itself in the sentence. The correct pronoun will be the one that sounds right. For example, "Her shares her wardrobe" does not sound right; "she shares her wardrobe" does.

2   Use a subject pronoun after forms of the verb *be*. Forms of *be* include *am, are, is, was, were, has been*, and *have been*.

> It was *I* who called you a minute ago and then hung up.
>
> It may be *they* entering the diner.
>
> It was *he* who put the white tablecloth into the washing machine with a red sock.

The sentences above may sound strange and stilted to you because they are seldom used in conversation. When we speak with one another, forms such as "It was me," "It may be them," and "It is her" are widely accepted. In formal writing, however, the grammatically correct forms are still preferred.

Hint: To avoid having to use the subject pronoun form after *be*, you can simply reword a sentence. Here is how the preceding examples could be reworded:

> I was the one who called you a minute ago and then hung up.
>
> They may be the ones entering the diner.
>
> He put the white tablecloth into the washing machine with a red sock.

3 Use subject pronouns after *than* or *as*. The subject pronoun is used because a verb is understood after the pronoun.

> Mark can hold his breath longer than *I* (can). (The verb *can* is understood after *I*.)
>
> Her thirteen-year-old daughter is as tall as *she* (is). (The verb *is* is understood after *she*.)
>
> You drive much better than *he* (drives). (The verb *drives* is understood after *he*.)

Hint: Avoid mistakes by mentally adding the ''missing'' verb at the end of the sentence.

Object Pronouns

Object pronouns (*me, him, her, us, them*) are the objects of verbs or prepositions. (*Prepositions* are connecting words like *for, at, about, to, before, by, with,* and *of.* See also page 335.)

> Lee pushed *me.* (*Me* is the object of the verb *pushed.*)
>
> We dragged *them* all the way home. (*Them* is the object of the verb *dragged.*)
>
> She wrote all about *us* in her diary. (*Us* is the object of the preposition *about.*)
>
> Vera passed a note to *him* as she walked to the pencil sharpener. (*Him* is the object of the preposition *to.*)

People are sometimes uncertain about what pronoun to use when two objects follow the verb.

Incorrect	*Correct*
I argued with his sister and *he.*	I argued with his sister and *him.*
The cashier cheated Rick and *I.*	The cashier cheated Rick and *me.*

Hint: If you are not sure what pronoun to use, try each pronoun by itself in the sentence. The correct pronoun will be the one that sounds right. For example, "I argued with he" does not sound right; "I argued with him" does.

Activity

Underline the correct subject or object pronoun in each of the following sentences. Then show whether your answer is a subject or an object pronoun by circling the S or O in the margin. The first one is done for you as an example.

(S) O   1. Kenny and (<u>she</u>, her) kept dancing even after the band stopped playing.

S O   2. The letters Mom writes to Estelle and (I, me) are always typewritten in red.

S O   3. No one has more nerve than (he, him).

S O   4. Their relay team won because they practiced more than (we, us).

S O   5. (We, Us) choir members get to perform for the governor.

S O   6. The rest of (they, them) came to the wedding by train.

S O   7. (She, Her) and Sammy got divorced and then remarried.

S O   8. My sister keeps track of all the favors she does for my brother and (I, me).

S O   9. Tony and (he, him) look a lot alike, but they're not even related.

S O   10. Our neighbors asked Maria and (I, me) to help with their parents' surprise party.

POSSESSIVE PRONOUNS

Possessive pronouns show ownership or possession.

Using a small branch, Stu wrote *his* initials in the wet cement.
The furniture is *mine,* but the car is hers.

Here is a list of possessive pronouns:

my, mine	our, ours
your, yours	your, yours
his	their, theirs
her, hers	
its	

Note: A possessive pronoun *never* uses an apostrophe. (See also page 453.)

Incorrect	*Correct*
That earring is *hers'*.	That earring is *hers*.
The orange cat is *theirs'*.	The orange cat is *theirs*.

Activity

Cross out the incorrect pronoun form in each of the sentences below. Write the correct form in the space at the left.

Example ___*hers*___ Those gloves are ~~hers'~~.

_____ 1. A porcupine has no quills on its' belly.

_____ 2. The stereo set is theirs'.

_____ 3. You can easily tell which team is ours' by when we cheer.

_____ 4. The car with the pink car seats is hers'.

_____ 5. Grandma's silverware and dishes will be yours' when you get married.

DEMONSTRATIVE PRONOUNS

Demonstrative pronouns point to or single out a person or thing. There are four demonstrative pronouns:

this	these
that	those

Generally speaking, *this* and *these* refer to things close at hand; *that* and *those* refer to things farther away. The four pronouns are commonly used in the role of demonstrative adjectives as well.

This milk has gone sour.

My son insists on saving all *these* hot rod magazines.

I almost tripped on *that* roller skate at the bottom of the steps.

Those plants in the corner don't get enough light.

Note: Do not use *them, this here, that there, these here,* or *those there* to point out. Use only *this, that, these,* or *those*.

Activity

Cross out the incorrect form of the demonstrative pronoun and write the correct form in the space provided.

Example _Those_ ~~Those them~~ tires look worn.

_____ 1. This here child has a high fever.

_____ 2. These here pants I'm wearing are so tight I can hardly breathe.

_____ 3. Them kids have been playing in the alley all morning.

_____ 4. That there umpire won't stand for any temper tantrums.

_____ 5. I save them old baby clothes for my daughter's dolls.

■ Review Test

Underline the correct word in the parentheses.

1. If I left dinner up to (he, him), we'd have Cheerios every night.

2. Julia's words may have come from the script, but the smile is all (hers', hers).

3. My boyfriend offered to drive his mother and (I, me) to the mall to shop for his birthday present.

4. (Them, those) little marks on the floor are scratches, not crumbs.

5. I took a picture of my brother and (I, me) looking into the hallway mirror.

6. When Lin and (she, her) drove back from the airport, they talked so much that they missed their exit.

7. (That there, That) orange juice box says "Fresh," but the juice is made from concentrate.

8. Eliot swears that he dreamt about (she, her) and a speeding car the night before Rose was injured in a car accident.

9. The waitress brought our food to the people at the next table, and gave (theirs, theirs') to us.

10. Since it was so hot out, Lana and (he, him) felt they had a good excuse to study at the beach.

ADJECTIVES
AND
ADVERBS

ADJECTIVES

What Are Adjectives?

Adjectives describe nouns (names of persons, places, or things) or pronouns.

Ernie is a *rich* man. (The adjective *rich* describes the noun *man*.)

He is also *generous*. (The adjective *generous* describes the pronoun *he*.)

Our *gray* cat sleeps a lot. (The adjective *gray* describes the noun *cat*.)

She is *old*. (The adjective *old* describes the pronoun *she*.)

Adjectives usually come before the word they describe (as in *rich man* and *gray cat*). But they also come after forms of the verb *be* (*is, are, was, were*, and so on). They also follow verbs such as *look, appear, seem, become, sound, taste,* and *smell.*

That speaker was *boring*. (The adjective *boring* describes the speaker.)

The Petersons are *homeless*. (The adjective *homeless* describes the Petersons.)

The soup looked *good*. (The adjective *good* describes the soup.)

But it tasted *salty*. (The adjective *salty* describes the pronoun *it*.)

Using Adjectives to Compare

For all one-syllable adjectives and some two-syllable adjectives, add *-er* when comparing two things and *-est* when comparing three or more things.

> My sister's handwriting is *neater* than mine, but Mother's is the *neatest*.
>
> Canned juice is sometimes *cheaper* than fresh juice, but frozen juice is often the *cheapest*.

For some two-syllable adjectives and all longer adjectives, add *more* when comparing two things and *most* when comparing three or more things.

> Typing something is *more efficient* than writing it out by hand, but the *most efficient* way to write is on a computer.
>
> Jeans are generally *more comfortable* than slacks, but sweat pants are the *most comfortable* of all.

You can usually tell when to use *more* and *most* by the sound of a word. For example, you can probably tell by its sound that ''carefuller'' would be too awkward to say and that *more careful* is thus correct. In addition, there are many words for which both *-er* or *-est* and *more* or *most* are equally correct. For instance, either ''a more fair rule'' or ''a fairer rule'' is correct.

To form negative comparisons, use *less* and *least*.

> When kids called me ''Dum-dum,'' I tried to look *less* hurt than I felt.
>
> They say men gossip *less* than women do, but I don't believe it.
>
> Suzanne is the most self-centered, *least* thoughtful person I know.

Points to Remember about Comparing

Point 1: Use only one form of comparison at a time. In other words, do not use both an *-er* ending and *more* or both an *-est* ending and *most*:

Incorrect	Correct
My Southern accent is always *more stronger* after I visit my family in Georgia.	My Southern accent is always *stronger* after I visit my family in Georgia.
My *most luckiest* day was the day I met my wife.	My *luckiest* day was the day I met my wife.

Point 2: Learn the irregular forms of the words shown below.

	Comparative (for Comparing Two Things)	*Superlative (for Comparing Three or More Things)*
bad	worse	worst
good, well	better	best
little (in amount)	less	least
much, many	more	most

Do not use both *more* and an irregular comparative or *most* and an irregular superlative.

Incorrect	*Correct*
It is *more better* to stay healthy than to have to get healthy.	It is *better* to stay healthy than to have to get healthy.
Yesterday I went on the *most best* date of my life — and all we did was go on a picnic.	Yesterday I went on the *best* date of my life — and all we did was go on a picnic.

Activity

Add to each sentence the correct form of the word in the margin.

bad

Examples The _____worst_____ scare I ever had was when I thought my son was on an airplane that crashed.

wonderful

The day of my divorce was even ___more wonderful___ than the day of my wedding.

good

1. The _____ way to diet is gradually.

popular

2. Vanilla ice cream is even _____ than chocolate ice cream.

bad

3. One of the _____ things you can do to people is ignore them.

light

4. A pound of feathers is no _____ than a pound of stones.

little

5. The _____ expensive way to accumulate a wardrobe is by buying used clothing whenever possible.

ADVERBS

What Are Adverbs?

Adverbs describe verbs, adjectives, or other adverbs. They usually end in *-ly*.

The referee *suddenly* stopped the fight. (The adverb *suddenly* describes the verb *stopped*.)

Her yellow rosebushes are *absolutely* beautiful. (The adverb *absolutely* describes the adjective *beautiful*.)

The auctioneer spoke so *terribly* fast that I couldn't understand him. (The adverb *terribly* describes the adverb *fast*.)

A Common Mistake with Adverbs and Adjectives

People often mistakenly use an adjective instead of an adverb after a verb.

Incorrect	*Correct*
I jog *slow*.	I jog *slowly*.
The nervous witness spoke *quiet*.	The nervous witness spoke *quietly*.
The first night I quit smoking, I wanted a cigarette *bad*.	The first night I quit smoking, I wanted a cigarette *badly*.

Activity

Underline the adjective or adverb needed. (Remember that adjectives describe nouns, and adverbs describe verbs or other adverbs.)

1. During a quiet moment in class, my stomach rumbled (loud, loudly).

2. I'm a (slow, slowly) reader, so I have to put aside more time to study than some of my friends.

3. Thinking no one was looking, the young man (quick, quickly) emptied his car's ashtray onto the parking lot.

4. The kitchen cockroaches wait (patient, patiently) in the shadows; at night they'll have the place to themselves.

5. I hang up the phone (immediate, immediately) whenever the speaker is a recorded message.

Well and *Good*

Two words that are often confused are *well* and *good*. *Good* is an adjective; it describes nouns. *Well* is usually an adverb; it describes verbs. *Well* (rather than *good*) is also used when referring to a person's health.

Activity

Write *well* or *good* in each of the sentences that follow.

1. I could tell by the broad grin on Ginny's face that the news was _____ .

2. They say he sang so _____ that even the wind stopped to listen.

3. The food at the salad bar must not have been too fresh because I didn't feel _____ after dinner.

4. When I want to do a really _____ job of washing the floor, I do it on my hands and knees.

5. The best way to get along _____ with our boss is to stay out of his way.

■ Review Test

Underline the correct word in the parentheses.

1. In Egypt, silver was once (more valued, most valued) than gold.

2. After seeing Mark get sick, I didn't feel too (good, well) myself.

3. The (littler, less) coffee I drink, the better I feel.

4. Light walls make a room look (more large, larger) than dark walls do.

5. One of the (unfortunatest, most unfortunate) men I know is a millionaire.

6. The moths' (continuous, continuously) thumping against the screen got on my nerves.

7. The Amish manage (good, well) without radios, telephones, or television.

8. A purple crocus had burst (silent, silently) through the snow outside our window.

9. It is (good, better) to teach people to fish than to give them fish.

10. Today a rocket can reach the moon more (quick, quickly) than it took a stagecoach to travel from one end of England to the other.

MISPLACED
MODIFIERS

Introductory Project

Because of misplaced words, each of the sentences below has more than one possible meaning. In each case, see if you can explain the intended meaning and the unintended meaning. Also, circle the words that you think create the confusion because they are misplaced.

1. The sign in the restaurant window reads, "Wanted: Young Man— To Open Oysters with References."

   Intended meaning: _____

   Unintended meaning: _____

2. Clyde and Charlotte decided to have two children on their wedding day.

   Intended meaning: _____

   Unintended meaning: _____

3. The students no longer like the math instructor who failed the test.

   Intended meaning: _____

   Unintended meaning: _____

Answers are on page 562.

WHAT MISPLACED MODIFIERS ARE
AND HOW TO CORRECT THEM

Modifiers are descriptive words. *Misplaced modifiers* are words that, because of awkward placement, do not describe the words the writer intended them to describe. Misplaced modifiers often obscure the meaning of a sentence. To avoid them, place words as close as possible to what they describe.

Misplaced Words	*Correctly Placed Words*
Tony bought an old car from a crooked dealer *with a faulty transmission.* (The dealer had a faulty transmission?)	Tony bought an old car with a faulty transmission from a crooked dealer. (The words describing the old car are now placed next to "car.")
I *nearly* earned a hundred dollars last week. (You just missed earning a hundred dollars, but in fact earned nothing?)	I earned nearly a hundred dollars last week. (The meaning — that you earned a little under a hundred dollars — is now clear.)
Bill yelled at the howling dog *in his underwear.* (The *dog* wore underwear?)	Bill, in his underwear, yelled at the howling dog. (The words describing Bill are placed next to him.)

Activity

Underline the misplaced word or words in each sentence. Then rewrite the sentence, placing related words together and thereby making the meaning clear.

Examples   The suburbs <u>nearly</u> had five inches of rain.
The suburbs had nearly five inches of rain.

We could see the football stadium <u>driving across the bridge</u>.
Driving across the bridge, we could see the football stadium.

1. I saw mountains of uncollected trash walking along the city streets.

   _____

2. I almost had a dozen job interviews after I sent out my résumé.

   _____

3. Bill swatted the wasp that stung him with a newspaper.

_____

4. Joanne decided to live with her grandparents when she attended college to save money.

_____

5. Charlotte returned the hamburger to the supermarket that was spoiled.

_____

6. Roger visited the old house still weak with the flu.

_____

7. The phone almost rang fifteen times last night.

_____

8. My uncle saw a kangaroo at the window under the influence of whiskey.

_____

9. We decided to send our daughter to college on the day she was born.

_____

10. Fred always opens the bills that arrive in the mailbox with a sigh.

_____

■ **Review Test**

Write M for *misplaced* or C for *correct* in front of each sentence.

_____ 1. Rita found it difficult to mount the horse wearing tight jeans.

_____ 2. Rita, wearing tight jeans, found it difficult to mount the horse.

_____ 3. I noticed a crack in the window walking into the delicatessen.

_____ 4. Walking into the delicatessen, I noticed a crack in the window.

_____ 5. A well-worn track shoe was found on the locker bench with holes in it.

_____ 6. A well-worn track shoe with holes in it was found on the locker bench.

_____ 7. I almost caught a hundred lightning bugs.

_____ 8. I caught almost a hundred lightning bugs.

_____ 9. In a secondhand store, Willie found a television set that had been stolen from me last month.

_____ 10. Willie found a television set in a secondhand store that had been stolen from me last month.

_____ 11. Willie found, in a secondhand store, a television set that had been stolen from me last month.

_____ 12. There were four cars parked outside the café with Minnesota license plates.

_____ 13. There were four cars with Minnesota license plates parked outside the café.

_____ 14. The President was quoted on the *NBC Evening News* as saying that the recession was about to end.

_____ 15. The President was quoted as saying that the recession was about to end on the *NBC Evening News*.

DANGLING
MODIFIERS

WHAT DANGLING MODIFIERS ARE
AND HOW TO CORRECT THEM

A modifier that opens a sentence must be followed immediately by the word it is meant to describe. Otherwise, the modifier is said to be *dangling*, and the sentence takes on an unintended meaning. For example, in the sentence

While smoking a pipe, my dog sat with me by the crackling fire.

the unintended meaning is that the *dog* was smoking the pipe. What the writer meant, of course, was that *he,* the writer, was smoking the pipe. He should have said,

While smoking a pipe, *I* sat with my dog by the crackling fire.

The dangling modifier could also be corrected by placing the subject within the opening word group:

While *I* was smoking my pipe, my dog sat with me by the crackling fire.

Here are other sentences with dangling modifiers. Read the explanations of why they are dangling and look carefully at how they are corrected.

Dangling	*Correct*
Swimming at the lake, a rock cut Sue's foot. (*Who* was swimming at the lake? The answer is not *rock* but *Sue.* The subject *Sue* must be added.)	Swimming at the lake, Sue cut her foot on a rock. *Or:* When Sue was swimming at the lake, she cut her foot on a rock.
While eating my sandwich, five mosquitoes bit me. (*Who* is eating the sandwich? The answer is not *five mosquitoes,* as it unintentionally seems to be, but *I.* The subject *I* must be added.)	While *I* was eating my sandwich, five mosquitoes bit me. *Or:* While eating my sandwich, *I* was bitten by five mosquitoes.
Getting out of bed, the tile floor was so cold that Maria shivered all over. (*Who* got out of bed? The answer is not *tile floor* but *Maria.* The subject *Maria* must be added.)	Getting out of bed, *Maria* found the tile floor so cold that she shivered all over. *Or:* When *Maria* got out of bed, the tile floor was so cold that she shivered all over.

Dangling	*Correct*
To join the team, a C average or better is necessary.	To join the team, *you* must have a C average or better.
(*Who* is to join the team? The answer is not *C average* but *you*. The subject *you* must be added.)	*Or:* For *you* to join the team, a C average or better is necessary.

The preceding examples make clear the two ways of correcting a dangling modifier. Decide on a logical subject and do one of the following:

1 Place the subject *within* the opening word group:

When Sue was swimming at the lake, she cut her foot on a rock.

Note: In some cases an appropriate subordinating word such as *When* must be added, and the verb may have to be changed slightly as well.

2 Place the subject right *after* the opening word group:

Swimming at the lake, Sue cut her foot on a rock.

Activity

Ask *Who?* of the opening words in each sentence. The subject that answers the question should be nearby in the sentence. If it is not, provide the logical subject by using either method of correction described above.

Example While sleeping at the campsite, a Frisbee hit Bill on the head.

While Bill was sleeping at the campsite, a Frisbee hit him

on the head.

or _While sleeping at the campsite, Bill was hit on the head by_

a Frisbee.

1. Watching the horror movie, goose bumps covered my spine.

_____

_____

2. After putting on a corduroy shirt, the room didn't seem as cold.

_____

_____

3. Flunking out of school, my parents demanded that I get a job.

_____

_____

4. Covered with food stains, my mother decided to wash the tablecloth.

_____

_____

5. Joining several college clubs, Mike's social life became more active.

_____

_____

6. While visiting the Jungle Park Safari, a baboon scrambled onto the hood of their car.

_____

_____

7. Under attack by beetles, Charlotte sprayed her roses with insecticide.

_____

_____

8. Standing at the ocean's edge, the wind coated my glasses with a salty film.

_____

_____

9. Braking the car suddenly, my shopping bags tumbled off the seat.

_____

_____

10. Using binoculars, the hawk was clearly seen following its prey.

_____

_____

✓ ■ **Review Test**

Place a D for *dangling* or a C for *correct* in front of each sentence. Remember that the opening words are a dangling modifier if they have no logical subject to modify.

_____ 1. Advertising in the paper, Frank's car was quickly sold.

_____ 2. By advertising in the paper, Frank quickly sold his car.

_____ 3. After painting the downstairs, the house needed airing to clear out the fumes.

_____ 4. After we painted the downstairs, the house needed airing to clear out the fumes.

_____ 5. Frustrated by piles of homework, Wanda was tempted to watch television.

_____ 6. Frustrated by piles of homework, Wanda's temptation was to watch television.

_____ 7. After I waited patiently in the bank line, the teller told me I had filled out the wrong form.

_____ 8. After waiting patiently in the bank line, the teller told me I had filled out the wrong form.

_____ 9. When dieting, desserts are especially tempting.

_____ 10. When dieting, I find desserts especially tempting.

_____ 11. Looking through the telescope, I saw a brightly lit object come into view.

_____ 12. As I was looking through the telescope, a brightly lit object came into view.

_____ 13. Looking through the telescope, a brightly lit object came into my view.

_____ 14. Weighing thousands of pounds, no one knows how the enormous stones were brought to Stonehenge.

_____ 15. No one knows how the enormous stones, weighing thousands of pounds, were brought to Stonehenge.

FAULTY PARALLELISM

Introductory Project

Read aloud each pair of sentences below. Put a check mark beside the sentence that reads most smoothly and clearly and sounds most natural.

I made resolutions to study more, to lose weight, and watching less TV. _____

I made resolutions to study more, to lose weight, and to watch less TV. _____

A consumer group rates my car as noisy, expensive, and not having much safety. _____

A consumer group rates my car as noisy, expensive, and unsafe.

_____

Lola likes wearing soft sweaters, eating exotic foods, and to bathe in Calgon bath oil. _____

Lola likes wearing soft sweaters, eating exotic foods, and bathing in Calgon bath oil. _____

Single life offers more freedom of choice; more security is offered by marriage. _____

Single life offers more freedom of choice; marriage offers more security. _____

Answers are on page 563.

PARALLELISM EXPLAINED

Words in a pair or a series should have a parallel structure. By balancing the items in a pair or a series so that they have the same kind of structure, you will make a sentence clearer and easier to read. Notice how the parallel sentences that follow read more smoothly than the nonparallel ones.

Nonparallel (Not Balanced)	*Parallel (Balanced)*
I made resolutions to lose weight, to study more, and *watching* less TV.	I made resolutions to lose weight, to study more, and to watch less TV. (A balanced series of *to* verbs: *to lose, to study, to watch*)
A consumer group rates my car as noisy, expensive, and *not having much safety*.	A consumer group rates my car as noisy, expensive, and unsafe. (A balanced series of descriptive words: *noisy, expensive, unsafe*)
Lola likes wearing soft sweaters, eating exotic foods, and *to bathe* in Calgon bath oil.	Lola likes wearing soft sweaters, eating exotic foods, and bathing in Calgon bath oil. (A balanced series of *-ing* words: *wearing, eating, bathing*)
The single life offers more freedom of choice; *more security is offered by marriage*.	The single life offers more freedom of choice; marriage offers more security. (Balanced verbs and word order: *single life offers . . . ; marriage offers . . .*)

You need not worry about balanced sentences when writing first drafts. But when you rewrite, you should try to put matching words and ideas into matching structures. Such parallelism will improve your writing style.

Activity

The unbalanced part of each of the following sentences is *italicized*. Rewrite the unbalanced part so that it matches the rest of the sentence. The first one is done for you as an example.

1. Woody Allen's films are clever, well-acted, and *have a lot of humor*.
   <u> humorous </u>

2. Filling out an income tax form is worse than wrestling a bear or *to walk* on hot coals. _____

3. The study-skills course taught me how to take more effective notes, to read a textbook chapter, and *preparing* for exams. _____

4. Lola plans to become a model, a lawyer, or *to go into nursing.* _____

5. Martha Grencher likes *to water* her garden, walking her fox terrier, and arguing with her husband. _____

6. Filled with talent and *ambitious,* Charlie plugged away at his sales job.
_____

7. When I saw my roommate with my girlfriend, I felt worried, angry, and *embarrassment* as well. _____

8. Cindy's cat likes sleeping in the dryer, lying in the bathtub, and *to chase* squirrels. _____

9. The bacon was fatty, *grease was on the potatoes,* and the eggs were cold.
_____

10. People in the lobby munched popcorn, sipped sodas, and *were shuffling* their feet impatiently. _____

■ **Review Test 1**

On separate paper, write five sentences of your own that use parallel structure.

■ **Review Test 2**

Draw a line under the unbalanced part of each sentence. Then rewrite the unbalanced part so that it matches the other item or items in the sentence. The first one is done for you as an example.

1. Our professor warned us that he would give surprise tests, <u>the assignment of</u> term papers, and allow no makeup exams.

   <u>*assign term papers*</u>

2. Pesky mosquitoes, humidity that is high, and sweltering heat make summer an unpleasant time for me.
_____

3. I want a job that pays high wages, provides a complete benefits package, and offering opportunities for promotion.
_____

4. My teenage daughter enjoys shopping for new clothes, to try different cosmetics, and reading beauty magazines.

   _____

5. My car needed the brakes replaced, the front wheels aligned, and recharging of the battery.

   _____

6. I had to correct my paper for fragments, misplaced modifiers, and there were apostrophe mistakes.

   _____

7. They did not want a black-and-white TV set, but a color set could not be afforded.

   _____

8. The neighborhood group asked the town council to repair the potholes and that a traffic light be installed.

   _____

9. Having a headache, my stomach being upset, and a bad case of sunburn did not put me in a good mood for the evening.

   _____

10. The Gray Panthers is an organization that not only aids older citizens but also providing information for their families.

    _____

PAPER FORMAT

When you hand in a paper for any of your courses, probably the first thing you will be judged on is its format. It is important, then, that you do certain things to make your papers look attractive, neat, and easy to read.

Here are guidelines to follow in preparing a paper for an instructor:

1 Use full-sized theme or typewriter paper, 8½ by 11 inches.

2 Leave wide margins (1 to 1½ inches) on all four sides of each page. In particular, do not crowd the right-hand or bottom margin. The white space makes your paper more readable; also, the instructor has room for comments.

3 If you write by hand,

   a Use a blue or black pen (*not* a pencil).

   b Be careful not to overlap letters or to make decorative loops on letters. On narrow-ruled paper, write only on every other line.

   c Make all your letters distinct. Pay special attention to *a, e, i, o,* and *u* — five letters that people sometimes write illegibly.

   d Keep your capital letters clearly distinct from small letters. You may even want to print all the capital letters.

4 Center the title of your paper on the first line of page 1. Do *not* put quotation marks around the title or underline the title or put a period after the title. Capitalize all the major words in a title, including the first word. Small connecting words within a title like *of, for, the, in,* and *to* are not capitalized. Skip a line between the title and the first line of your text.

5 Indent the first line of each paragraph about five spaces (half an inch) from the left-hand margin.

6 Make commas, periods, and other punctuation marks firm and clear. Leave a slight space after each period. When you type, leave a double space after a period.

7 Whenever possible, avoid breaking (hyphenating) words at the end of lines. If you must break a word, break only between syllables (see page 482). Do not break words of one syllable.

8 Write your name, the date, and the course number where your instructor asks for them.

Also keep in mind these important points about the *title* and *first sentence* of your paper:

9 The title should simply be several words that tell what the paper is about. It should usually *not* be a complete sentence. For example, if you are writing a paper about one of the most frustrating jobs you have ever had, the title could be just ''A Frustrating Job.''

10 Do not rely on the title to help explain the first sentence of your paper. The first sentence must be independent of the title. For instance, if the title of your paper is ''A Frustrating Job,'' the first sentence should *not* be ''It was working as a baby-sitter.'' Rather, the first sentence might be ''Working as a baby-sitter was the most frustrating job I ever had.''

Activity 1

Identify the mistakes in format in the following lines from a student theme. Explain the mistakes in the spaces provided. One mistake is described for you as an example.

	"an unpleasant dining companion"
	My little brother is often an unpleasant dining companion. Last
	night was typical. For one thing, his appearance was disgusting
	His shoes were not tied, and his shirt was unbuttoned and han-
	ging out of his pants, which he had forgotten to zip up. Traces
	of his afternoon snack of grape juice and chocolate cookies were

1. Hyphenate only between syllables _____

2. _____

3. _____

4. _____

5. _____

6. _____

Activity 2

As already stated, a title should tell in several words (but *not* a complete sentence) what a paper is about. Often a title can be based on the topic sentence—the sentence that expresses the main idea of the paper. Following are five topic sentences from student papers. Write a suitable and specific title for each paper, basing the title on the topic sentence. (Note the example.)

Example   *Compromise in a Relationship*

Learning how to compromise is essential to a good relationship.

1. Title: _____
   Some houseplants are dangerous to children and pets.

2. Title: _____
   A number of fears haunted me when I was a child.

3. Title: _____
   To insulate a house properly, several important steps should be taken.

4. Title: _____
   My husband is compulsively neat.

5. Title: _____
   There are a number of drawbacks to having a roommate.

Activity 3

As has already been stated, you must *not* rely on the title to help explain your first sentence. In four of the five sentences that follow, the writer has, inappropriately, used the title to help explain the first sentence.

Rewrite the four sentences so that they stand independent of the title. Write *Correct* under the one sentence that is independent of the title.

Example   Title: My Career Plans
First sentence: They have changed in the last six months.

Rewritten: *My career plans have changed in the last six months.*

1. Title: Contending with Dogs
   First sentence: This is the main problem in my work as a mail carrier.

   Rewritten: _____

   _____

2. Title: Study Skills
    First sentence: They are necessary if a person is to do well in college.

    Rewritten: _____

    _____

3. Title: Summer Vacation
    First sentence: Contrary to popular belief, a summer vacation can be the most miserable experience of the year.

    Rewritten: _____

    _____

4. Title: My Wife and the Sunday Newspaper
    First sentence: My wife has a peculiar way of reading it.

    Rewritten: _____

    _____

5. Title: Traffic Circles
    First sentence: They are one of the chief hazards today's driver must confront.

    Rewritten: _____

    _____

■ **Review Test**

In the space provided on the opposite page, rewrite the following sentences from a student paper. Correct the mistakes in format.

		"disciplining our children"
		My husband and I are becoming experts in disciplining our child-
		ren. We have certain rules that we insist upon, and if there are
		any violations, we are swift to act. When our son simply doesn't
		do what he is told to do, he must write that particular action
		twenty times. For example, if he doesn't brush his teeth, he
		writes, "I must brush my teeth." If a child gets home after the

CAPITAL LETTERS

Introductory Project

You probably know a good deal about the uses of capital letters. Answering the questions below will help you check your knowledge.

1. Write the full name of a person you know: _____

2. In what city and state were you born? _____

3. What is your present street address? _____

4. Name a country where you would like to travel: _____

5. Name a school that you attended: _____

6. Give the name of a store where you buy food: _____

7. Name a company where someone you know works: _____

8. What day of the week gives you the best chance to relax? _____

9. What holiday is your favorite? _____

10. What brand of toothpaste do you use? _____

11. Give the brand name of a candy or gum you like: _____

12. Name a song or a television show you enjoy: _____

13. Give the title of a magazine you read: _____

Items 14–16: Three capital letters are needed in the lines below. Underline the words that you think should be capitalized. Then write them, capitalized, in the spaces provided.

the masked man reared his silvery-white horse, waved good-bye, and rode out of town. My heart thrilled when i heard someone say, "that was the Lone Ranger. You don't see his kind much, anymore."

14. _____ 15. _____ 16. _____

Answers are on page 563.

436

MAIN USES OF CAPITAL LETTERS

Capital letters are used with:

1 The first word in a sentence or direct quotation
2 Names of persons and the word *I*
3 Names of particular places
4 Names of days of the week, months, and holidays
5 Names of commercial products
6 Names of organizations such as religious and political groups, associations, companies, unions, and clubs
7 Titles of books, magazines, newspapers, articles, stories, poems, films, television shows, songs, papers that you write, and the like

Each use is illustrated on the pages that follow.

First Word in a Sentence or Direct Quotation

The panhandler touched me and asked, "Do you have any change?"
↑   ↑
(Capitalize the first word in the sentence.)   (Capitalize the first word in the direct quotation.)

"If you want a ride," said Brenda, "get ready now. Otherwise, I'm going alone."

(*If* and *Otherwise* are capitalized because they are the first words of sentences within a direct quotation. But *get* is not capitalized because it is part of the first sentence within the quotation.)

Names of Persons and the Word *I*

Last night I ran into Tony Curry and Lola Morrison.

Names of Particular Places

Charlotte graduated from Fargone High School in Orlando, Florida. She then moved with her parents to Bakersfield, California, and worked for a time there at Alexander's Gift House. Eventually she married and moved with her husband to the Naval Reserve Center in Atlantic County, New Jersey. She takes courses two nights a week at Stockton State College. On weekends she and her family often visit the nearby Wharton State Park and go canoeing on the Mullica River. She does volunteer work at Atlantic City Hospital in connection with the First Christian Church. In addition, she works during the summer as a hostess at Convention Hall and the Holiday Inn.

But: Use small letters if the specific name of a place is not given.

Charlotte sometimes remembers her unhappy days in high school and at the gift shop where she worked after graduation. She did not imagine then that she would one day be going to college and doing volunteer work for a church and a hospital in the community where she and her husband live.

Names of Days of the Week, Months, and Holidays

I was angry at myself for forgetting that Sunday was Mother's Day.

During July and August, Fred's company works a four-day week, and he has Mondays off.

Bill still has a scar on his ankle from a cherry bomb that exploded near him on a Fourth of July and a scar on his arm where he stabbed himself with a fishhook on a Labor Day weekend.

But: Use small letters for the seasons — summer, fall, winter, spring.

Names of Commercial Products

Clyde uses Scope mouthwash, Certs mints, and Dentyne gum to drive away the taste of the Marlboro cigarettes and White Owl cigars that he always smokes.

My sister likes to play Monopoly and Sorry; I like chess and poker; my brother likes Scrabble, baseball, and table tennis.

But: Use small letters for the *type* of product (mouthwash, mints, gum, cigarettes, and so on).

Names of Organizations Such as Religious and Political Groups, Associations, Companies, Unions, and Clubs

Fred Grencher was a Lutheran for many years but converted to Catholicism when he married. Both he and his wife, Martha, are members of the Democratic Party. Both belong to the American Automobile Association. Martha works part time as a refrigerator salesperson at Sears. Fred is a mail carrier and belongs to the Postal Clerks' Union.

Tony met Lola when he was a Boy Scout and she was a Campfire Girl; she asked him to light her fire.

Titles of Books, Magazines, Newspapers, Articles, Stories, Poems, Films, Television Shows, Songs, Papers That You Write, and the Like

On Sunday Lola read the first chapter of *I Know Why the Caged Bird Sings,* a book required for her writing course. She looked through her parents' copy of *The New York Times.* She then read an article titled ''Thinking about a Change in Your Career'' and a poem titled ''Some Moments Alone'' in *Cosmopolitan* magazine. At the same time she played an old Beatles album, *Abbey Road.* In the evening she watched *60 Minutes* on television and a movie, *Sudden Impact,* starring Clint Eastwood. Then from 11 P.M. to midnight she worked on a paper titled ''Uses of Leisure Time in Today's Culture'' for her sociology class.

Activity

Cross out the words that need capitals in the following sentences. Then write the capitalized forms of the words in the spaces provided. The number of spaces tells you how many corrections to make in each case.

Example I brush with crest toothpaste but get cavities all the time. _____*Crest*_____

1. A spokesperson for general motors announced that the prices of all chevrolets will rise next year.

   _____   _____   _____

2. Steve graduated from Maplewood high school in june 1988.

_____   _____   _____

3. The mild-mannered reporter named clark kent said to the Wolfman, ''you'd better think twice before you mess with me, Buddy.''

_____   _____   _____

4. While watching television, Bill drank four pepsis, ate an entire package of ritz crackers, and finished up a bag of oreo cookies.

_____   _____   _____

5. A greyhound bus almost ran over Tony as he was riding his yamaha to a friend's home in florida.

_____   _____   _____

6. Before I lent my polaroid camera to Janet, I warned her, ''be sure to return it by friday.''

_____   _____   _____

7. Before christmas George took his entire paycheck, went to sears, and bought a twenty-inch zenith color television.

_____   _____   _____

8. On their first trip to New York City, Fred and Martha visited the empire State Building and Times square. They also saw the New York mets play at Shea Stadium.

_____   _____   _____

9. Clyde was listening to Tina Turner's recording of ''Proud mary,'' Charlotte was reading an article in *Reader's digest* titled ''let's Stop Peddling Sex,'' and their son was watching *sesame Street*.

_____   _____   _____   _____

10. When a sign for a howard johnson's rest stop appeared on the turnpike, anita said, ''let's stop here and stretch our legs for a bit.''

_____   _____   _____   _____

OTHER USES OF CAPITAL LETTERS

Capital letters are also used with:

1 Names that show family relationships
2 Titles of persons when used with their names
3 Specific school courses
4 Languages
5 Geographic locations
6 Historical periods and events
7 Races, nations, and nationalities
8 Opening and closing of a letter

Each use is illustrated on the pages that follow.

Names That Show Family Relationships

I got Mother to baby-sit for me.

I went with Grandfather to the church service.

Uncle Carl and Aunt Lucy always enclose five dollars with birthday cards.

But: Do not capitalize words like *mother, father, grandmother, aunt,* and so on, when they are preceded by a possessive word (*my, your, his, her, our, their*).

I got my mother to baby-sit for me.

I went with my grandfather to the church service.

My uncle and aunt always enclose five dollars with birthday cards.

Titles of Persons When Used with Their Names

I wrote to Senator Grabbel and Congresswoman Punchie.

Professor Snorrel sent me to Chairperson Ruck, who sent me to Dean Rappers.

He drove to Dr. Helen Thompson's office after the cat bit him.

But: Use small letters when titles appear by themselves, without specific names.

I wrote to my senator and congresswoman.

The professor sent me to the chairperson, who sent me to the dean.

He drove to the doctor's office after the cat bit him.

Specific School Courses

I got an A in both Accounting I and Small Business Management, but I got a C in Human Behavior.

But: Use small letters for general subject areas.

I enjoyed my business courses but not my psychology or language courses.

Languages

She knows German and Spanish, but she speaks mostly American slang.

Geographic Locations

I grew up in the Midwest. I worked in the East for a number of years and then moved to the West Coast.

But: Use small letters in directions.

A new high school is being built at the south end of town.

Because I have a compass in my car, I know that I won't be going east or west when I want to go north.

Historical Periods and Events

Hector did well answering an essay question about the Second World War, but he lost points on a question about the Great Depression.

Races, Nations, Nationalities

The research study centered on African Americans and Hispanics.

They have German knives and Danish glassware in the kitchen, an Indian wood carving in the bedroom, Mexican sculptures in the study, and an Oriental rug in the living room.

Opening and Closing of a Letter

Dear Sir:

Dear Madam:

Sincerely yours,

Truly yours,

Note: Capitalize only the first word in a closing.

Activity

Cross out the words that need capitals in the following sentences. Then write the capitalized forms of the words in the spaces provided. The number of spaces tells you how many corrections to make in each case.

1. Although my grandfather spoke german and polish, my mother never learned either language.

   _____   _____

2. The chain letter began, ''dear friend — You must mail twenty copies of this letter if you want good luck.''

   _____   _____

3. Tomorrow in our history class, dr. connalley will start lecturing on the civil war.

   _____   _____   _____   _____

4. aunt Sarah and uncle Hal, who are mormons, took us to their church services when we visited them in the midwest.

   _____   _____   _____   _____

5. My sister has signed up for a course titled eastern religions; she'll be studying buddhism and hinduism.

   _____   _____   _____   _____

UNNECESSARY USE OF CAPITALS

Many errors in capitalization are caused by using capitals where they are not needed.

Activity

Cross out the incorrectly capitalized words in the following sentences. Then write the correct forms of the words in the spaces provided. The number of spaces tells you how many corrections to make in each sentence.

1. Although the Commercials say that Things go better with Coke, I prefer Root Beer.

   _____   _____   _____   _____

2. The old man told the Cabdriver, ''I want to go out to the Airport and don't try to cheat me.''

   _____   _____

3. A front-page Newspaper story about the crash of a commercial Jet has made me nervous about my Overseas trip.

   _____   _____   _____

4. During a Terrible Blizzard in 1888, People froze to Death on the streets of New York.

   _____   _____   _____   _____

5. I asked the Bank Officer at Citibank, ''How do I get an identification Card to use the automatic teller machines?''

   _____   _____   _____

■ Review Test 1

Cross out the words that need capitals in the following sentences. Then write the capitalized forms of the words in the spaces provided. The number of spaces tells you how many corrections to make in each sentence.

1. wanda and i agreed to meet on saturday before the football game.

   _____   _____   _____

2. Between Long island and the atlantic Ocean lies a long, thin sandbar called fire island.

   _____   _____   _____   _____

3. When I'm in the supermarket check-out line, it seems as if every magazine on display has an article called "how You Can Lose Twenty pounds in two weeks."

_____   _____   _____   _____

4. At the bookstore, each student received a free sample pack of bayer aspirin, arrid deodorant, and prell shampoo.

_____   _____   _____

5. "can't you be quiet?" I pleaded. "do you always have to talk while I'm watching *general hospital* on television?"

_____   _____   _____   _____

6. On father's day, the children drove home and took their parents out to dinner at the ramada inn.

_____   _____   _____   _____

7. I will work at the holly Day School on mondays and fridays for the rest of september.

_____   _____   _____   _____

8. glendale bank, where my sister Kathy works, is paying for her night course titled business accounting I.

_____   _____   _____   _____

9. I subscribe to one newspaper, the *daily planet,* and two magazines, *people* and *glamour.*

_____   _____   _____   _____

10. On thanksgiving my brother said, "let's hurry and eat so i can go watch the football game on our new sony TV."

_____   _____   _____   _____

■ Review Test 2

On separate paper,

1. Write seven sentences demonstrating the seven main uses of capital letters.
2. Write eight sentences demonstrating the eight additional uses of capital letters.

NUMBERS
AND
ABBREVIATIONS

NUMBERS

1 Spell out numbers that can be expressed in one or two words. Otherwise, use numerals — the numbers themselves.

> During the past five years, over twenty-five barracuda have been caught in the lake.
>
> The parking fine was ten dollars.
>
> In my grandmother's attic are eighty-four pairs of old shoes.

But

> Each year about 250 baby trout are added to the lake.
>
> My costs after contesting a parking fine in court were $135.
>
> Grandmother has 382 back copies of *Reader's Digest* in her attic.

2 Be consistent when you use a series of numbers. If some numbers in a sentence or paragraph require more than two words, then use numerals throughout the selection:

> During his election campaign, State Senator Mel Grabble went to 3 county fairs, 16 parades, 45 cookouts, and 112 club dinners, and delivered the same speech 176 times.

3 Use numerals for dates, times, addresses, percentages, and parts of a book.

> The letter was dated April 3, 1872.
>
> My appointment was at 6:15. (*But:* Spell out numbers before *o'clock*. For example: The doctor didn't see me until seven o'clock.)
>
> He lives at 212 West 19th Street.
>
> About 20 percent of our class has dropped out of school.
>
> Turn to page 179 in Chapter 8 and answer questions 1 – 10.

Activity

Cross out the mistakes in numbers and write the corrections in the spaces provided.

1. Rich was born on February fifteenth, nineteen seventy.

    _____

    _____

2. When the 2 children failed to return from school, over 50 people volunteered to search for them.

    _____

    _____

3. At 1 o'clock in the afternoon last Thursday, an earthquake destroyed at least 20 buildings in the town.

    _____

    _____

ABBREVIATIONS

While abbreviations are a helpful time-saver in note-taking, you should avoid most abbreviations in formal writing. Listed below are some of the few abbreviations that can acceptably be used in compositions. Note that a period is used after most abbreviations.

1 Mr., Mrs., Ms., Jr., Sr., Dr. when used with proper names:

   Mr. Tibble   Dr. Stein   Ms. O'Reilly

2 Time references:

   A.M. or a.m.   P.M. or p.m.   B.C. or A.D.

3 First or middle name in a signature:

   R. Anthony Curry   Otis T. Redding   J. Alfred Prufrock

4 Organizations, technical words, and trade names known primarily by their initials:

   FBI   UN   CBS   FM   STP

Activity

Cross out the words that should not be abbreviated and correct them in the spaces provided.

1. On a Sat. morning I will never forget, Dec. 5, 1992, at ten min. after eight, I came downstairs and discovered that I had been robbed.

    _____   _____   _____

2. For six years I lived at First Ave. and Gordon St., right next to Shore Memorial Hosp., in San Fran., Calif.

    _____   _____   _____   _____   _____

3. Before her biol. and Eng. exams, Linda was so nervous that her doc. gave her a tranq.

    _____   _____   _____   _____

■ Review Test

Cross out the mistakes in numbers and abbreviations and correct them in the spaces provided.

1. At three-fifteen p.m., an angry caller said a bomb was planted in a bus stat. locker.

    _____   _____

2. Page eighty-two is missing from my chem. book.

    _____   _____

3. Martha has over 200 copies of *People* mag.; she thinks they may be worth money someday.

    _____   _____

4. When I was eight yrs. old, I owned three cats, two dogs, and 4 rabbits.

    _____   _____

5. Approx. half the striking workers returned to work on Jan. third, nineteen eighty-five.

    _____   _____   _____   _____

APOSTROPHE

Introductory Project

1. Larry's motorcycle
   my sister's boyfriend
   Grandmother's shotgun
   the men's room
   Dionne Warwick's new album

   What is the purpose of the *'s* in the examples above?

   _____

2. They didn't mind when their dog bit people, but now they're leashing him because he's eating all their garden vegetables.

   What is the purpose of the apostrophe in *didn't, they're,* and *he's?*

   _____

3. I used to believe that vampires lived in the old coal bin of my cellar. The vampire's whole body recoiled when he saw the crucifix.

   Fred ate two baked potatoes.
   One baked potato's center was still hard.

   In each of the sentence pairs above, why is the *'s* used in the second

   sentence but not in the first? _____

   _____

Answers are on page 563.

The two main uses of the apostrophe are:

1 To show the omission of one or more letters in a contraction
2 To show ownership or possession

Each use is explained on the pages that follow.

APOSTROPHE IN CONTRACTIONS

A contraction is formed when two words are combined to make one word. An apostrophe is used to show where letters are omitted in forming the contraction. Here are two contractions:

have + not = haven't (the *o* in *not* has been omitted)
I + will = I'll (the *wi* in *will* has been omitted)

The following are some other common contractions:

I + am = I'm	it + is = it's
I + have = I've	it + has = it's
I + had = I'd	is + not = isn't
who + is = who's	could + not = couldn't
do + not = don't	I + would = I'd
did + not = didn't	they + are = they're

Note: Will + not has an unusual contraction: won't.

Activity 1

Combine the following words into contractions. One is done for you.

1. we + are = ___we're___   6. you + have = _____

2. are + not = _____   7. has + not = _____

3. you + are = _____   8. who + is = _____

4. they + have = _____   9. does + not = _____

5. would + not = _____   10. there + is = _____

Activity 2

Write the contractions for the words in parentheses. One is done for you.

1. (Are not) _____ you coming with us to the concert?

2. (I am) _____ going to take the car if (it is) _____ all right with you.

3. (There is) _____ an extra bed upstairs if (you would) _____ like to stay here for the night.

4. (I will) _____ give you the name of the personnel director, but there (is not) _____ much chance that (he will) _____ speak to you.

5. Linda (should not) _____ complain about the cost of food if (she is) _____ not willing to grow her own by planting a backyard garden.

Note: Even though contractions are common in everyday speech and in written dialog, usually it is best to avoid them in formal writing.

APOSTROPHE TO SHOW OWNERSHIP OR POSSESSION

To show ownership or possession, we can use such words as *belongs to, possessed by, owned by,* or (most commonly) *of.*

the jacket that *belongs to* Tony
the grades *possessed by* James
the gas station *owned by* our cousin
the footprints *of* the animal

But the apostrophe plus *s* (if the word is not a plural ending in *-s*) is often the quickest and easiest way to show possession. Thus we can say:

Tony's jacket
James's grades
our cousin's gas station
the animal's footprints

Points to Remember

The *'s* goes with the owner or possessor (in the examples given, *Tony, cousin, the animal*). What follows is the person or thing possessed (in the examples given, *the jacket, gas station, footprints*).

When *'s* is handwritten, there should always be a break between the word and the *'s*.

Tony's   not   Tony's

Yes   No

A singular word ending in *-s* (such as *James* on the preceding page) also shows possession by adding an apostrophe plus *s* (*James's*).

Activity 1

Rewrite the italicized part of each of the sentences below, using the *'s* to show possession. Remember that the *'s* goes with the owner or possessor.

Examples   *The toys belonging to the children* filled an entire room.

              The children's toys

1. *The roller skates owned by Pat* have been stolen.

   _____

2. *The visit of my cousin* lasted longer than I wanted it to.

   _____

3. *The fenders belonging to the car* are badly rusted.

   _____

4. *The prescription of a doctor* is needed for the pills.

   _____

5. *The jeep owned by Doris* was recalled because of an engine defect.

   _____

6. Is this *the hat of somebody*?

   _____

7. The broken saddle produced a sore on *the back of the horse*.

   _____

8. *The two dogs belonging to my neighbor* ripped open the trash bags.

_____

9. *The energy level possessed by the little boy* is much higher than hers.

_____

10. *The foundation of the house* is crumbling.

_____

Activity 2

Add *'s* to each of the following words to make them the possessors or owners of something. Then write sentences using the words. Your sentences can be serious or playful. One is done for you.

1. dog _____*dog's*_____   *That dog's bite is worse than his bark.*

2. instructor _____   _____

   _____   _____

3. Lola _____   _____

   _____   _____

4. store _____   _____

   _____   _____

5. mother _____   _____

   _____   _____

Apostrophe versus Possessive Pronouns

Do not use an apostrophe with possessive pronouns. They already show ownership. Possessive pronouns include *his, hers, its, yours, ours,* and *theirs.*

Correct	*Incorrect*
The bookstore lost its lease.	The bookstore lost its' lease.
The racing bikes were theirs.	The racing bikes were theirs'.
The change is yours.	The change is yours'.
His problems are ours, too.	His' problems are ours', too.
His skin is more sunburned than hers.	His' skin is more sunburned than hers.'

Apostrophe versus Simple Plurals

When you want to make a word plural, just add an -s at the end of the word. Do *not* add an apostrophe. For example, the plural of the word *movie* is *movies,* not *movie's* or *movies'*. Look at this sentence:

Lola adores Tony's broad shoulders, rippling muscles, and warm eyes.

The words *shoulders, muscles,* and *eyes* are simple plurals, meaning *more than one shoulder*, *more than one muscle*, *more than one eye*. The plural is shown by adding -s only. On the other hand, the *'s* after *Tony* shows possession — that Tony owns the shoulders, muscles, and eyes.

Activity

In the space provided under each sentence, add the one apostrophe needed and explain why the other word or words ending in *s* are simple plurals.

Example Karens tomato plants are almost six feet tall.

Karens: *Karen's, meaning "the plants belonging to Karen"*

plants: *simple plural meaning "more than one plant"*

1. My fathers influence on his brothers has been enormous.

   fathers: _____

   brothers: _____

2. Phils job — slaughtering pigs — was enough to make him a vegetarian.

   Phils: _____

   pigs: _____

3. As Tinas skill at studying increased, her grades improved.

   Tinas: _____

   grades: _____

4. When I walked into my doctors office, there were six people waiting who also had appointments.

   doctors: _____

   appointments: _____

5. I asked the record clerk for several blank cassette tapes and Whitney Houstons new album.

tapes: _____

Houstons: _____

6. After six weeks without rain, the nearby streams started drying up, and the lakes water level fell sharply.

weeks: _____

streams: _____

lakes: _____

7. Everyone wanted to enroll in Dr. Lerners class, but all the sections were closed.

Lerners: _____

sections: _____

8. When the brakes failed on Phils truck, he narrowly avoided hitting several parked cars and two trees.

Phils: _____

cars: _____

trees: _____

9. My familys favorite breakfast is bacon, eggs, and home-fried potatoes.

familys: _____

eggs: _____

potatoes: _____

10. We like Floridas winters, but we prefer to spend the summers in other states.

Floridas: _____

winters: _____

summers: _____

states: _____

Apostrophe with Plural Words Ending in -s

Plurals that end in -s show possession simply by adding the apostrophe, rather than an apostrophe plus s.

My *parents'* station wagon is ten years old.

The many *students'* complaints were ignored by the high school principal.

All the *Boy Scouts'* tents were damaged by the hail storm.

Activity

In each sentence, cross out the one plural word that needs an apostrophe. Then write the word correctly, with the apostrophe, in the space provided.

Example _____soldiers'_____ All the ~~soldiers~~ rifles were cleaned for inspection.

1. My parents car was stolen last night.

   _____

2. The transit workers strike has just ended.

   _____

3. Two of our neighbors homes are up for sale.

   _____

4. The door to the ladies room is locked.

   _____

5. When students gripes about the cafeteria were ignored, many started to bring their own lunches.

   _____

■ Review Test 1

In each sentence, cross out the two words that need apostrophes. Then write the words correctly in the spaces provided.

1. The contestants face fell when she learned she had won a years supply of Ajax cleanser.

   _____   _____

2. Weve been trying for weeks to see that movie, but theres always a long line.

   _____   _____

3. Freds car wouldnt start until the baby-faced mechanic replaced its spark plugs and points.

   _____   _____

4. The citys budget director has trouble balancing his own familys books.

   _____   _____

5. Taking Dianes elderly parents to church every week is one example of Toms generous behavior.

   _____   _____

6. Heres a checklist of points to follow when youre writing your class reports.

   _____   _____

7. Lola shops in the mens store for jeans and the childrens department for belts.

   _____   _____

8. The cats babies are under my chair again; I cant find a way to keep her from bringing them near me.

   _____   _____

9. Because of a family feud, Julie wasnt invited to a barbecue at her cousins house.

   _____   _____

10. Phyllis grade was the highest in the class, and Lewis grade was the lowest.

    _____   _____

■ Review Test 2

Make the following words possessive and then use at least five of them in a not-so-serious paragraph that tells a story. In addition, use at least three contractions in the paragraph.

mugger	restaurant	Tony	student
New York	sister	children	vampire
duck	Jay Leno	boss	Eddie Murphy
customer	bartender	police car	yesterday
instructor	someone	mob	Chicago

QUOTATION MARKS

Introductory Project

Read the following scene and underline all the words enclosed within quotation marks. Your instructor may also have you dramatize the scene, with one person reading the narration and two persons acting the two speaking parts of the young man and the old woman. The two speakers should imagine the scene as part of a stage play and try to make their words seem as real and true-to-life as possible.

An old woman in a Rolls-Royce was preparing to back into a parking space. Suddenly a small sports car appeared and pulled into the space. "That's what you can do when you're young and fast," the young man in the car yelled to the old woman. As he strolled away, laughing, he heard a terrible crunching sound. "What's that noise?" he said. Turning around, he saw the old woman backing repeatedly into and crushing his small car. "You can't do that, old lady!" he yelled.

"What do you mean, I can't?" she chuckled, as metal grated against metal. "This is what you can do when you're old and rich."

1. On the basis of the above selection, what is the purpose of quotation marks?

   _____

2. Do commas and periods that come after a quotation go inside or outside the quotation marks?

   _____

Answers are on page 563.

458

The two main uses of quotation marks are:

1 To set off the exact words of a speaker or a writer
2 To set off the titles of short works

Each use is explained on the pages that follow.

QUOTATION MARKS TO SET OFF
EXACT WORDS OF A SPEAKER OR A WRITER

Use quotation marks when you want to show the exact words of a speaker or a writer.

> ''Say something tender to me,'' whispered Lola to Tony.
> (Quotation marks set off the exact words that Lola spoke to Tony.)
>
> Mark Twain once wrote, ''The more I know about human beings, the more I like my dog.''
> (Quotation marks set off the exact words that Mark Twain wrote.)
>
> ''The only dumb question,'' the instructor said, ''is the one you don't ask.''
> (Two pairs of quotation marks are used to enclose the instructor's exact words.)
>
> Sharon complained, ''I worked so hard on this paper. I spent two days getting information in the library and two days writing it. Guess what grade I got on it.''
> (Note that the end quotation marks do not come until the end of Sharon's speech. Place quotation marks before the first quoted word of a speech and after the last quoted word. As long as no interruption occurs in the speech, do not use quotation marks for each new sentence.)

Punctuation Hint: In the four examples above, notice that a comma sets off the quoted part from the rest of the sentence. Also observe that commas and periods at the end of a quotation always go *inside* quotation marks.

Complete the following statements that explain how capital letters, commas, and periods are used in quotations. Refer to the four examples as guides.

1. Every quotation begins with a _____ letter.
2. When a quotation is split (as in the sentence above about dumb questions), the second part does not begin with a capital letter unless it is a _____ sentence.

3. _____ are used to separate the quoted part of a sentence from the rest of the sentence.

4. Commas and periods that come at the end of a quotation should go _____ the quotation marks.

The answers are *capital, new, Commas,* and *inside.*

Activity 1

Place quotation marks around the exact words of a speaker or writer in the sentences that follow.

1. Take some vitamin C for your cold, Lola told Tony.
2. How are you doing in school? my uncle always asks me.
3. An epitaph on a tombstone in Georgia reads, I told you I was sick!
4. Dave said, Let's walk faster. I think the game has already started.
5. Lincoln wrote, To'sin by silence when they should protest makes cowards of men.
6. Thelma said, My brother is so lazy that, if opportunity knocked, he'd resent the noise.
7. It's extremely dangerous to mix alcohol and pills, Dr. Wilson reminded us. The combination could kill you.
8. Ice-cold drinks! shouted the vendor selling lukewarm drinks.
9. Be careful not to touch the fence, the guard warned. It's electrified.
10. Just because I'm deaf, Lynn said, many people treat me as if I were stupid.

Activity 2

1. Write a sentence in which you quote a favorite expression of someone you know. Identify the relationship of the person to you.

   Example   *One of my father's favorite expressions is, "Don't sweat the small stuff."*

   _____

   _____

   _____

2. Write a quotation that contains the words *Tony asked Lola.* Write a second quotation that includes the words *Lola replied.*

_____

_____

_____

_____

3. Copy a sentence or two that interests you from a book or magazine. Identify the title and author of the work.

Example   In Night Shift, Stephen King writes, "I don't like to sleep

with one leg sticking out. Because if a cool hand ever

reached out from under the bed and grasped my ankle,

I might scream."

_____

_____

_____

_____

Indirect Quotations

An indirect quotation is a rewording of someone else's comments, rather than a word-for-word direct quotation. The word *that* often signals an indirect quotation. Quotation marks are *not* used with indirect quotations.

Direct Quotation	*Indirect Quotation*
Fred said, ''The distributor cap on my car is cracked.''	Fred said that the distributor cap on his car was cracked.
(Fred's exact spoken words are given, so quotation marks are used.)	(We learn Fred's words *in*directly, so no quotation marks are used.)
Sally's note to Jay read, ''I'll be working late. Don't wait up for me.''	Sally left a note for Jay saying she would be working late and he should not wait up for her.
(The exact words that Sally wrote in the note are given, so quotation marks are used.)	(We learn Sally's words indirectly, so no quotation marks are used.)

Activity

Rewrite the following sentences, changing words as necessary to convert the sentences into direct quotations. The first one is done for you as an example.

1. Fred asked Martha if he could turn on the football game.

   _____

2. Martha said that he could listen to the game on the radio.

   _____

3. Fred replied he was tired of being told what to do.

   _____

4. Martha said that as long as she was bigger and stronger, she would make the rules.

   _____

5. Fred said that the day would come when the tables would be turned.

   _____

QUOTATION MARKS TO SET OFF TITLES OF SHORT WORKS

Titles of short works are usually set off by quotation marks, while titles of long works are underlined. Use quotation marks to set off the titles of such short works as articles in books, newspapers, or magazines; chapters in a book; short stories; poems; and songs.

On the other hand, you should underline the titles of books, newspapers, magazines, plays, movies, record albums, and television shows.

Quotation Marks	*Underlines*
the article ''The Mystique of Lawyers''	in the book <u>Verdicts on Lawyers</u>
the article ''Getting a Fix on Repairs''	in the newspaper <u>The New York Times</u>
the article ''Animal Facts and Fallacies''	in the magazine <u>Reader's Digest</u>
the chapter ''Why Do Men Marry?''	in the book <u>Passages</u>
the story ''The Night the Bed Fell''	in the book <u>A Thurber Carnival</u>

Quotation Marks	*Underlines*
the poem "A Prayer for My Daughter"	in the book <u>Poems of W. B. Yeats</u>
the song "Beat It"	in the album <u>Thriller</u>
	the television show <u>Cheers</u>
	the movie <u>Gone with the Wind</u>

Note: In printed works, titles of books, newspapers, and so on are set off by italics — slanted type that looks *like this* — instead of being underlined.

Activity

Use quotation marks or underlines as needed.

1. I Was a Playboy Bunny is the first chapter of Gloria Steinem's book Outrageous Acts and Everyday Rebellions.

2. No advertising is permitted in Consumer Reports, a nonprofit consumer magazine.

3. I cut out an article from Newsweek called The Bad News about Our High Schools to use in my sociology report.

4. Tony's favorite television show is Star Trek, and his favorite movie is The Night of the Living Dead.

5. Our instructor gave us a week to buy the textbook titled Personal Finance and to read the first chapter, Work and Income.

6. Every holiday season, our family watches the movie A Christmas Carol on television.

7. Fred bought the Ladies' Home Journal because he wanted to read the cover article titled Secrets Men Never Tell You.

8. Edgar Allan Poe's short story The Murders in the Rue Morgue and his poem The Raven are in a paperback titled Great Tales and Poems of Edgar Allan Poe.

9. When Elaine got her TV Guide, she read an article titled The New Comedians and then thumbed through the listings to see who would be the guests that week on the Tonight Show.

10. The night before his exam, he discovered with horror that the chapter Becoming Mature was missing from Childhood and Adolescence, the psychology text that he had bought secondhand.

OTHER USES OF QUOTATION MARKS

1   Quotation marks are used to set off special words or phrases from the rest of a sentence:

Many people spell the words "a lot" as *one* word, "alot," instead of correctly spelling them as two words.

I have trouble telling the difference between "their" and "there."

Note: In printed works, *italics* are often used to set off special words or phrases. That is usually done in this book, for example.

2   Quotation marks are also used to mark off a quotation within a quotation:

The instructor said, "Know the chapter titled 'Status Symbols' in *Adolescent Development* if you expect to pass the test."

Lola said, "One of my favorite Mae West lines is 'I used to be Snow White, but I drifted.' "

Note: A quotation within a quotation is indicated by *single* quotation marks.

■ Review Test 1

Insert quotation marks where needed in the sentences that follow.

1. Don't you ever wash your car? Lola asked Tony.
2. When the washer tilted and began to buzz, Martha shouted, Let's get rid of that blasted machine!
3. Take all you want, read the sign above the cafeteria salad bar, but please eat all you take.
4. After scrawling formulas all over the board with lightning speed, my math instructor was fond of asking, Any questions now?
5. Move that heap! the truck driver yelled. I'm trying to make a living here.
6. I did a summary of an article titled Aspirin and Heart Attacks in the latest issue of Time.

7. Writer's block is something that happens to everyone at times, the instructor explained. You simply have to keep writing to break out of it.

8. A passenger in the car ahead of Clyde threw food wrappers and empty cups out the window. That man, said Clyde to his son, is a human pig.

9. If you are working during the day, said the counselor, the best way to start college is with a night course or two.

10. I told the dentist that I wanted Novocain. Don't be a sissy, he said. A little pain won't hurt you. I told him that a little pain wouldn't bother him, but it would bother me.

■ Review Test 2

Go through the comics section of a newspaper to find a comic strip that amuses you. Be sure to choose a strip where two or more characters are speaking to each other. Write a full description that will enable people who have not read the comic strip to visualize it clearly and appreciate its humor. Describe the setting and action in each panel, and enclose the words of the speakers in quotation marks.

COMMA

Introductory Project

Commas often (though not always) signal a minor break, or pause, in a sentence. Each of the six pairs of sentences below illustrates one of the six main uses of the comma. Read each pair of sentences aloud and place a comma wherever you feel a slight pause occurs.

1. a. Frank's interests are Maria television and sports.
   b. My mother put her feet up sipped some iced tea and opened the newspaper.

2. a. Although the Lone Ranger used lots of silver bullets he never ran out of ammunition.
   b. To open the cap of the aspirin bottle you must first press down on it.

3. a. Kitty Litter and Dredge Rivers Hollywood's leading romantic stars have made several movies together.
   b. Sarah who is my next-door neighbor just entered the hospital with an intestinal infection.

4. a. The wedding was scheduled for four o'clock but the bride changed her mind at two.
   b. Verna took three coffee breaks before lunch and then she went on a two-hour lunch break.

5. a. Lola's mother asked her ''What time do you expect to get home?''
   b. ''Don't bend over to pat the dog'' I warned ''or he'll bite you.''

6. a. Roy ate seventeen hamburgers on July 29 1992 and lived to tell about it.
   b. Roy lives at 817 Cresson Street Detroit Michigan.

Answers are on page 563.

466

SIX MAIN USES OF THE COMMA

Commas are used mainly as follows:

1 To separate items in a series
2 To set off introductory material
3 Before and after words that interrupt the flow of thought in a sentence
4 Between two complete thoughts connected by *and, but, for, or, nor, so, yet*
5 To set off a direct quotation from the rest of a sentence
6 For certain everyday material

Each use is explained on the pages that follow.

You may find it helpful to remember that the comma often marks a slight pause, or break, in a sentence. Read aloud the sentence examples given for each rule, and listen for the minor pauses, or breaks, that are signaled by commas.

Comma between Items in a Series

Use commas to separate items in a series.

> Do you drink tea with milk, lemon, or honey?
>
> Today the dishwasher stopped working, the garbage bag split, and the refrigerator turned into an icebox.
>
> The television talk shows enraged him so much he did not know whether to laugh, cry, or throw up.
>
> Jan awoke from a restless, nightmare-filled sleep.

Notes

a The final comma in a series is optional, but it is often used.
b A comma is used between two descriptive words in a series only if *and* inserted between the words sounds natural. You could say:

> Jan awoke from a restless *and* nightmare-filled sleep.

But notice in the following sentence that the descriptive words do not sound natural when *and* is inserted between them. In such cases, no comma is used.

> Wanda drove a bright blue Corvette. (A bright *and* blue Corvette doesn't sound right, so no comma is used.)

Activity

Place commas between items in a series.

1. Superman believes in truth justice and the American way.
2. My father taught me to swim by talking to me in a calm manner holding my hand firmly and throwing me into the pool.
3. Paul added white wine mushrooms salt pepper and oregano to his spaghetti sauce.
4. Baggy threadbare jeans feel more comfortable than pajamas to me.
5. Mark grabbed a tiny towel bolted out of the bathroom and ran toward the ringing phone.

Comma after Introductory Material

Use a comma to set off introductory material.

After punching the alarm clock with his fist, Bill turned over and went back to sleep.

Looking up in the sky, I saw a man who was flying faster than a speeding bullet.

Holding a baited trap, Clyde cautiously approached the gigantic mousehole.

In addition, he held a broom in his hand.

Also, he wore a football helmet in case a creature should leap out at his head.

Notes:

a If the introductory material is brief, the comma is sometimes omitted. In the activities here, you should use the comma.

b A comma is also used to set off extra material at the end of a sentence. Here are two earlier sentences where this comma rule applies:

A sudden breeze shot through the windows, driving the stuffiness out of the room.

I love to cook and eat Italian food, especially spaghetti and lasagna.

COMMA **469**

Activity

Place commas after introductory material.

1. When the president entered the room became hushed.
2. Feeling brave and silly at the same time Tony volunteered to go on stage and help the magician.
3. While I was eating my tuna sandwich the cats circled my chair like hungry sharks.
4. Because my parents died when I was young I have learned to look after myself. Even though I am now independent I still carry a special loneliness within me.
5. At first putting extra hot pepper flakes on the pizza seemed like a good idea. However I felt otherwise when flames seemed about to shoot out of my mouth.

Comma around Words Interrupting the Flow of Thought

Use commas before and after words or phrases that interrupt the flow of thought in a sentence.

My brother, a sports nut, owns over five thousand baseball cards.

That game show, at long last, has been canceled.

The children used the old Buick, rusted from disuse, as a backyard clubhouse.

Usually you can ''hear'' words that interrupt the flow of thought in a sentence. However, if you are not sure that certain words are interrupters, remove them from the sentence. If it still makes sense without the words, you know that the words are interrupters and the information they give is nonessential. Such nonessential information is set off with commas. In the sentence

Dody Thompson, who lives next door, won the javelin-throwing competition.

the words *who lives next door* are extra information, not needed to identify the subject of the sentence, *Dody Thompson*. Put commas around such nonessential information. On the other hand, in the sentence

The woman who lives next door won the javelin-throwing competition.

the words *who lives next door* supply essential information — information needed for us to identify the woman being spoken of. If the words were removed from the sentence, we would no longer know who won the competition. Commas are *not* used around such essential information.

Here is another example:

Wilson Hall, which the tornado destroyed, was ninety years old.

Here the words *which the tornado destroyed* are extra information, not needed to identify the subject of the sentence, *Wilson Hall.* Commas go around such non-essential information. On the other hand, in the sentence

The building which the tornado destroyed was ninety years old.

the words *which the tornado destroyed* are needed to identify the building. Commas are *not* used around such essential information.

As noted above, however, most of the time you will be able to "hear" words that interrupt the flow of thought in a sentence and will not have to think about whether the words are essential or nonessential.

Activity

Use commas to set off interrupting words.

1. On Friday my day off I went to get a haircut.
2. Dracula who had a way with women is Tony's favorite movie hero. He feels that the Wolfman on the other hand showed no class in handling women.
3. Many people forget that Franklin Roosevelt one of our most effective presidents was also handicapped.
4. Mowing the grass especially when it is six inches high is my least favorite job.
5. A jar of chicken noodle soup which was all there was in the refrigerator did not make for a very satisfying meal.

Comma between Complete Thoughts

Use a comma between two complete thoughts connected by *and, but, for, or, nor, so, yet.*

The wedding was scheduled for four o'clock, but the bride changed her mind at two.

We could always tell when our instructor felt disorganized, for his shirt would not be tucked in.

Rich has to work on Monday nights, so he tapes the TV football game on his video recorder.

Notes

a The comma is optional when the complete thoughts are short.

Grace's skin tans and Mark's skin freckles.
Her soda turned watery for the ice melted quickly.
The day was overcast so they didn't go swimming.

b Be careful not to use a comma in sentences having *one* subject and a *double* verb. The comma is used only in sentences made up of two complete thoughts (two subjects and two verbs). In the following sentence, there is only one subject (*Bill*) with a double verb (*will go* and *forget*). Therefore, no comma is needed:

Bill will go partying tonight and forget all about tomorrow's exam.

Likewise, the following sentence has only one subject (*Rita*) and a double verb (*was* and *will work*); therefore, no comma is needed:

Rita was a waitress at the Holiday Inn last summer and probably will work there this summer.

Activity

Place a comma before a joining word that connects two complete thoughts (two subjects and two verbs). Remember, do *not* place a comma within sentences that have only one subject and a double verb.

1. The oranges in the refrigerator were covered with blue mold and the potatoes in the cupboard felt like sponges.
2. All the slacks in the shop were on sale but not a single pair was my size.
3. Martha often window-shops in the malls for hours and comes home without buying anything.
4. Tony left the dentist's office with his mouth still numb from Novocain and he talked with a lisp for two hours.
5. I covered the walls with three coats of white paint but the purple color underneath still showed through.
6. The car squealed down the entrance ramp and sped recklessly out onto the freeway.

7. The dancers in the go-go bar moved like wound-up Barbie dolls and the men in the audience sat as motionless as stones.

8. The aliens in the science fiction film visited our planet in peace but we greeted them with violence.

9. I felt like shouting at the gang of boys but didn't dare open my mouth.

10. Lenny claims he wants to succeed in college but he has missed classes all semester.

Comma with Direct Quotations

Use a comma to set off a direct quotation from the rest of a sentence.

His father shouted, ''Why don't you go out and get a job?''

''Our modern world has lost a sense of the sacredness of life,'' the speaker said.

''No,'' said Celia to Jerry. ''I won't go to the roller derby with you.''

''Can anyone remember,'' wrote Emerson, ''when the times were not hard and money not scarce?''

Note: Commas and periods at the end of a quotation go inside quotation marks. See also page 459.

Activity

Use commas to set off quotations from the rest of the sentence.

1. The man yelled ''Call an ambulance, somebody!''

2. My partner on the dance floor said ''Don't be so stiff. You look as if you'd swallowed an umbrella.''

3. The question on the anatomy test read ''What human organ grows faster than any other, never stops growing, and always remains the same size?''

4. The student behind me whispered ''The skin.''

5. ''My stomach hurts'' Bruce said ''and I don't know whether it was the hamburger or the math test.''

Comma with Everyday Material

Use a comma with certain everyday material.

Persons Spoken To

Tina, go to bed if you're not feeling well.
Cindy, where did you put my shoes?
Are you coming with us, Bob?

Dates

March 4, 1992, is when Martha buried her third husband.

Addresses

Tony's grandparents live at 183 Roxborough Avenue, Cleveland, Ohio 44112.

Note: No comma is used to mark off the zip code.

Openings and Closings of Letters

Dear Santa,
Dear Larry,
Sincerely yours,
Truly yours,

Note: In formal letters, a colon is used after the opening: Dear Sir: *or* Dear Madam:

Numbers

The dishonest dealer turned the used car's odometer from 98,170 miles to 39,170 miles.

Activity

Place commas where needed.

1. I expected you to set a better example for the others Mike.
2. Janet with your help I passed the test.
3. The movie stars Kitty Litter and Dredge Rivers were married on September 12 1991 and lived at 3865 Sunset Boulevard Los Angeles California for one month.
4. They received 75000 congratulatory fan letters and were given picture contracts worth $3000000 in the first week of their marriage.
5. Kitty left Dredge on October 12 1991 and ran off with their marriage counselor.

■ Review Test 1

Insert commas where needed. In the space provided below each sentence, summarize briefly the rule that explains the use of the comma or commas.

1. The best features of my new apartment are its large kitchen its bay windows and its low rent.

   _____

2. Because we got in line at dawn we were among the first to get tickets for the concert.

   _____

3. "When will someone invent a telephone" Lola asked "that will let you know who's calling before you answer it?"

   _____

4. Without opening his eyes Simon stumbled out of bed and opened the door for the whining dog.

   _____

5. I think David that you had better ask someone else for your $2500 loan.

   _____

6. Hot dogs are the most common cause of choking deaths in children for a bite-size piece can easily plug up a toddler's throat.

   _____

7. Tax forms though shortened and revised every year never seem to get any simpler.

   _____

8. Sandra may decide to go to college full-time or she may enroll in a couple of evening courses.

_____

9. I remember how with the terrible cruelty of children we used to make fun of the retarded girl who lived on our street.

_____

10. Although that old man on the corner looks like a Skid Row bum he is said to have a Swiss bank account.

_____

■ Review Test 2

Insert commas where needed.

1. My dog who is afraid of the dark sleeps with a night-light.
2. ''I wish there were some pill'' said Chuck ''that would give you the equivalent of eight hours' sleep in four hours.''
3. The hot dogs at the ball park tasted delicious but they reacted later like delayed time bombs.
4. Janice attended class for four hours worked at the hospital for three hours and studied at home for two hours.
5. The old man as he gasped for air tried to assure the hospital clerk that he had an insurance card somewhere.
6. George and Ida sat down to watch the football game with crackers sharp cheese salty pretzels and two frosty bottles of beer.
7. Although I knew exactly what was happening the solar eclipse gave me a strong feeling of anxiety.
8. The company agreed to raise a senior bus driver's salary to $28000 by January 1 1995.
9. Even though King Kong was holding her at the very top of the Empire State Building Fay Wray kept yelling at him ''Let me go!''
10. Navel oranges which Margery as a little girl called belly-button oranges are her favorite fruit.

■ Review Test 3

On separate paper, write six sentences, with each sentence demonstrating one of the six main comma rules.

OTHER PUNCTUATION MARKS

Introductory Project

Each of the sentences below needs one of the following punctuation marks:

$$; \quad — \quad \text{-} \quad (\ ) \quad :$$

See if you can insert the correct mark in each sentence. Each mark is used once.

1. The following holiday plants are poisonous and should be kept away from children and pets holly, mistletoe, and poinsettias.
2. The freeze dried remains of Annie's canary were in a clear bottle on her bookcase.
3. William Shakespeare 1564 – 1616 married a woman eight years his senior when he was eighteen.
4. Grooming in space is more difficult than on Earth no matter how much astronauts comb their hair, for instance, it still tends to float loosely around their heads.
5. I opened the front door, and our cat walked in proudly with a live bunny hanging from his mouth.

Answers are on page 564.

COLON (:)

Use the colon at the end of a complete statement to introduce a list, a long quotation, or an explanation.

List

The following were my worst jobs: truck loader in an apple plant, assembler in a battery factory, and attendant in a state mental hospital.

Long Quotation

Thoreau explains in *Walden:* "I went to the woods because I wished to live deliberately, to front only the essential facts of life, and see if I could not learn what it had to teach, and not, when I came to die, discover that I had not lived."

Explanation

There are two softball leagues in our town: the fast-pitch league and the lob-pitch league.

Activity

Place colons where needed.

1. Foods that are high in cholesterol include the following eggs, butter, milk, cheese, shrimp, and well-marbled meats.
2. All the signs of the flu were present hot and cold spells, heavy drainage from the sinuses, a bad cough, and an ache through the entire body.
3. In his book *Illiterate America*, Jonathan Kozol has written "Twenty-five million American adults cannot read the poison warnings on a can of pesticide, a letter from their child's teacher, or the front page of a daily paper. An additional 35 million read only at a level which is less than equal to the full survival needs of our society. Together, these 60 million people represent more than one-third of the entire adult population."

SEMICOLON (;)

The main use of the semicolon is to mark a break between two complete thoughts, as explained on page 365. Another use of the semicolon is to mark off items in a series when the items themselves contain commas. Here are some examples:

> Winning prizes at the national flower show were Roberta Collins, Alabama, azaleas; Sally Hunt, Kentucky, roses; and James Weber, California, Shasta daisies.

> The following books must be read for the course: *The Color Purple,* by Alice Walker; *In Our Time,* by Ernest Hemingway; and *Man's Search for Meaning,* by Viktor Frankl.

Activity

Place semicolons where needed.

1. The specials at the restaurant today are eggplant Parmesan, for $3.95 black beans and rice, for $2.95 and chicken potpie, for $4.95.
2. The top of the hill offered an awesome view of the military cemetery thousands of headstones were ranged in perfect rows.
3. Lola's favorite old movies are *To Catch a Thief,* starring Cary Grant and Grace Kelly *Animal Crackers,* a Marx Brothers comedy and *The Wizard of Oz,* with Judy Garland.

DASH (—)

A dash signals a degree of pause longer than a comma but not as complete as a period. Use a dash to set off words for dramatic effect:

> I didn't go out with him a second time — once was more than enough.

> Some of you — I won't mention you by name — cheated on the test.

> It was so windy that the VW passed him on the highway — overhead.

Notes

a   The dash is formed on the typewriter by striking the hyphen twice (- -). In handwriting, the dash is as long as two letters would be.

b   Be careful not to overuse dashes.

Activity

Place dashes where needed.

1. Riding my bike, I get plenty of exercise especially when dogs chase me.
2. I'm advising you in fact, I'm telling you not to bother me again.
3. The package finally arrived badly damaged.

HYPHEN (-)

1 Use a hyphen with two or more words that act as a single unit describing a noun.

> The fast-talking salesman was so good that he went into politics. (*Fast* and *talking* combine to describe the salesman.)
>
> I both admire and envy her well-rounded personality.
>
> When the dude removed his blue-tinted shades, Lonnell saw the spaced-out look in his eyes.

2 Use a hyphen to divide a word at the end of a line of writing or typing. When you need to divide a word at the end of a line, divide it between syllables. Use your dictionary to be sure of correct syllable divisions (see also page 482).

> When Tom lifted up the hood of his Toyota, he realized that one of the radi-ator hoses had broken.

Notes

a Do not divide words of one syllable.
b Do not divide a word if you can avoid doing so.

Activity

Place hyphens where needed.

1. High flying jets and gear grinding trucks are constant sources of noise pollution in our neighborhood, and consequently we are going to move.
2. When Linda turned on the porch light, ten legged creatures scurried every where over the crumb filled floor.
3. Fred had ninety two dollars in his pocket when he left for the supermarket, and he had twenty two dollars when he got back.

PARENTHESES ( )

Parentheses are used to set off extra or incidental information from the rest of a sentence:

> The section of that book on the medical dangers of abortion (pages 35 to 72) is outdated.

> Yesterday at Hamburger House (my favorite place to eat), the guy who makes french fries asked me to go out with him.

Note: Do not use parentheses too often in your writing.

Activity

Add parentheses where needed.

1. Certain sections of the novel especially Chapter 5 made my heart race with suspense.
2. Did you hear that George Linda's first husband just got remarried?
3. Sigmund Freud 1856 – 1939 was the founder of psychoanalysis.

■ Review Test

At the appropriate spot, place the punctuation mark shown in the margin.

; 1. Mary's savings have dwindled to nothing she's been borrowing from me to pay her rent.

— 2. There's the idiot I'd know him anywhere who dumped trash on our front lawn.

- 3. Today's two career couples spend more money on eating out than their parents did.

: 4. Ben Franklin said "If a man empties his purse into his head, no man can take it away from him. An investment in knowledge always pays the best interest."

( ) 5. One-fifth of our textbook pages 401 – 498 consists of footnotes and a bibliography.

USING
THE
DICTIONARY

The dictionary is a valuable tool. To take advantage of it, you need to understand the main kinds of information that a dictionary gives about a word. Look at the information provided for the word *murder* in the following entry from the *American Heritage Dictionary of the English Language:**

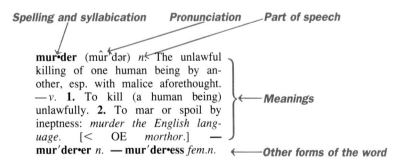

Spelling and syllabication   *Pronunciation*   *Part of speech*

mur•der (mûr′dər) *n.* The unlawful killing of one human being by another, esp. with malice aforethought. —*v.* **1.** To kill (a human being) unlawfully. **2.** To mar or spoil by ineptness: *murder the English language.* [< OE *morthor.*] — **mur′der•er** *n.* — **mur′der•ess** *fem.n.*

Meanings

Other forms of the word

SPELLING

The first bit of information, in the boldface (heavy type) entry itself, is the spelling of *murder*. You probably already know the spelling of *murder,* but if you didn't, you could find it by pronouncing the syllables in the word carefully and then looking it up in the dictionary.

Use your dictionary to correct the spelling of the following words:

compatable _____	insite _____
althogh _____	troble _____
aksident _____	untill _____
embelish _____	easyer _____
systimatise _____	prepostrous _____
shedule _____	comotion _____
attenshun _____	Vasaline _____
wierd _____	fatel _____
hurryed _____	busines _____
alright _____	jenocide _____
fony _____	poluted _____
kriterion _____	perpose _____
hetirosexual _____	chalange _____

SYLLABICATION

The second bit of information that the dictionary gives, also in the boldface entry, is the syllabication of *murder*. Note that a dot separates the syllables.

Use your dictionary to mark the syllable divisions in the following words. Also indicate how many syllables are in each word.

j i t t e r	(_____ syllables)
m o t i v a t e	(_____ syllables)
o r a n g u t a n	(_____ syllables)
i n c o n t r o v e r t i b l e	(_____ syllables)

Noting syllable divisions will enable you to *hyphenate* a word: divide it at the end of one line of writing and complete it at the beginning of the next line. You can correctly hyphenate a word only at a syllable division, and you may have to check your dictionary to make sure of a word's syllable divisions.

PRONUNCIATION

The third bit of information in the dictionary entry is the pronunciation of *murder:* (mûr'dər). You already know how to pronounce *murder,* but if you didn't, the information within the parentheses would serve as your guide. Use your dictionary to complete the following exercises that relate to pronunciation.

Vowel Sounds

You will probably use the pronunciation key in your dictionary mainly as a guide to pronouncing different vowel sounds. (Vowels are the letters *a, e, i, o,* and *u.*) Here is part of the pronunciation key in the *American Heritage Dictionary:*

ă bat / ā way / ĕ ebb / ē equal / ĭ if

The key tells you, for example, that the sound of the short *a* is pronounced like the *a* in *bat,* the sound of the long *a* is like the *a* in *way,* and the sound of the short *e* is like the *e* in *ebb.*

Now look at the pronunciation key in your dictionary. The key is probably located in the front of the dictionary or at the bottom of every page. What common word in the key tells you how to pronounce each of the following sounds?

ī _____   ŭ _____

ŏ _____   o͞o _____

ō _____   o͞o _____

(Note that the long vowel always has the sound of its own name.)

The Schwa (ə)

The symbol ə looks like an upside-down *e.* It is called a *schwa,* and it stands for the unaccented sound in such words as *ago, item, easily, gallop,* and *circus.* More approximately, it stands for the sound *uh* — like the *uh* that speakers sometimes make when they hesitate in their speech. Perhaps it would help to remember that *uh,* as well as ə, could be used to represent the schwa sound.

Here are some of the many words in which the sound appears: *imitation* (*im-uh-tā'shuhn* or *im-ə-tā'shən*); *elevate* (*el'uh-vāt* or *el'ə-vāt*); *horizon* (*huh-rī'zuhn* or *hə-rī'zən*). Open your dictionary to any page, and you will almost surely be able to find words that make use of the schwa in the pronunciation in parentheses after the main entry.

In the spaces below, write three words that make use of the schwa, and their pronunciations.

1. _____   (_____)

2. _____   (_____)

3. _____   (_____)

Accent Marks

Some words contain both a primary accent, shown by a heavy stroke ('), and a secondary accent, shown by a lighter stroke ('). For example, in the word *contro-versy* (kon'trə vûr'se), the stress, or accent, goes chiefly on the first syllable (kon'), and to a lesser extent, on the third syllable (vûr').

Use your dictionary to add accent marks to the following words:

preclude (pri klo͞od)
atrophy (at rə fē)
inveigle (in vā gəl)
ubiquitous (yo͞o bik wi təs)
prognosticate (prog nos ti kāt)

Full Pronunciation

Now use your dictionary to write the full pronunciation (the information given in parentheses) for each of the following words.

1. inveigh _____

2. diatribe _____

3. raconteur _____

4. panacea _____

5. esophagus _____

6. Cesarean _____

7. clandestine _____

8. vicarious _____

9. quiescent _____

10. parsimony _____

11. penchant _____

12. antipathy _____

13. capricious _____

14. schizophrenia _____

15. euphemism _____

16. internecine _____

17. amalgamate _____

18. quixotic _____

19. laissez-faire _____

20. antidisestablishmentarianism (This word is probably not in a paperback dictionary, but if you can say *establish* and if you break the rest of the word into individual syllables, you should be able to pronounce it.)

_____

Now practice pronouncing each word. Use the pronunciation key in your dictionary as an aid to sounding out each syllable. Do *not* try to pronounce a word all at once; instead, work on mastering *one syllable at a time.* When you can pronounce each of the syllables in a word successfully, then say them in sequence, add the accent, and pronounce the entire word.

PARTS OF SPEECH

The next bit of information that the dictionary gives about *murder* is *n.* This abbreviation means that the meanings of *murder* as a noun will follow.

Use your dictionary if necessary to fill in the meanings of the following abbreviations:

v. = _____   sing. = _____

adj. = _____   pl. = _____

PRINCIPAL PARTS OF IRREGULAR VERBS

Murder is a regular verb and forms its principal parts by adding *-ed, -ed,* and *-ing* to the stem of the verb. When a verb is irregular, the dictionary lists its principal parts. For example, with *give* the present tense comes first (the entry itself, *give*). Next comes the past tense (*gave*), and then the past participle (*given*) — the form of the verb used with such helping words as *have, had,* and *was.* Then comes the present participle (*giving*) — the *-ing* form of the verb.

Look up the parts of the following irregular verbs and write them in the spaces provided. The first one is done for you.

Present	*Past*	*Past Participle*	*Present Participle*
tear	tore	torn	tearing
go			
know			
steal			

PLURAL FORMS OF IRREGULAR NOUNS

The dictionary supplies the plural forms of all irregular nouns. (Regular nouns like *murder* form the plural by adding *-s* or *-es*.) Give the plurals of the following nouns. If two forms are shown, write down both.

analysis _____

dictionary _____

criterion _____

activity _____

thesis _____

MEANINGS

When a word has more than one meaning, the meanings are numbered in the dictionary, as with the verb *murder*. In many dictionaries, the most common meanings of a word are presented first. The introductory pages of your dictionary will explain the order in which meanings are presented.

Use the sentence context to try to explain the meaning of the underlined word in each of the following sentences. Write your definition in the space provided. Then look up and record the dictionary meaning of the word. Be sure you pick out the meaning that fits the word as it is used in the sentence.

1. I spend an <u>inordinate</u> amount of time watching television.

Your definition: _____

Dictionary definition: _____

2. I appreciated her <u>candid</u> remark that my pants were so baggy they made me look like a clown.

    Your definition: _____

    Dictionary definition: _____

3. The FBI <u>squelched</u> the terrorists' plan to plant a bomb in the White House.

    Your definition: _____

    Dictionary definition: _____

4. One of the <u>cardinal</u> rules in our house was, ''Respect other people's privacy.''

    Your definition: _____

    Dictionary definition: _____

5. A special <u>governor</u> prevents the school bus from traveling more than fifty-five miles an hour.

    Your definition: _____

    Dictionary definition: _____

ETYMOLOGY

Etymology refers to the origin and historical development of a word. Such information is usually enclosed in brackets and is more likely to be present in a hardbound desk dictionary than in a paperback one. Good desk dictionaries include the following:

American Heritage Dictionary of the English Language
The Random House College Dictionary
Webster's New Collegiate Dictionary
Webster's New World Dictionary

A good desk dictionary will tell you, for example, that the word *berserk* derives from the name of a tribe of Scandinavian warriors who would work themselves into a frenzy during battle. The word is now a general term to describe someone whose actions are frenzied or crazed.

See if your dictionary says anything about the origins of the following words:

bikini _____

sandwich _____

tantalize _____

breakfast _____

USAGE LABELS

As a general rule, use only standard English words in your writing. If a word is not standard English, your dictionary will probably give it a usage label like one of the following: *informal, nonstandard, slang, vulgar, obsolete, archaic, rare.*

Look up the following words and record how your dictionary labels them. Remember that a recent hardbound desk dictionary will always be the best source of information about usage.

flunk _____

tough (meaning ''unfortunate, too bad'') _____

creep (meaning ''an annoying person'') _____

ain't _____

scam _____

SYNONYMS

A *synonym* is a word that is close in meaning to another word. Using synonyms helps you avoid unnecessary repetition of the same word in a paper. A paperback dictionary is not likely to give you synonyms for words, but a good desk dictionary will. (You might also want to own a *thesaurus,* a book that lists synonyms and antonyms. An *antonym* is a word approximately opposite in meaning to another word.)

Consult a desk dictionary that gives synonyms for the following words, and write the synonyms in the spaces provided.

heavy _____

escape _____

necessary _____

IMPROVING SPELLING

Poor spelling often results from bad habits developed in early school years. With work, such habits can be corrected. If you can write your name without misspelling it, there is no reason why you can't do the same with almost any word in the English language. Following are six steps you can take to improve your spelling.

STEP 1: USE THE DICTIONARY

Get into the habit of using the dictionary. When you write a paper, allow yourself time to look up the spelling of all those words you are unsure about. Do not overlook the value of this step just because it is such a simple one. By using the dictionary, you can probably make yourself a 95 percent better speller.

STEP 2: KEEP A PERSONAL SPELLING LIST

Keep a list of words you misspell and study the words regularly. Use the chart on page 567 as a starter. When you accumulate additional words, put them on the back page of a frequently used notebook or on a separate sheet of paper titled ''Personal Spelling List.''

To master the words on your list, do the following:

1 Write down any hint that will help you remember the spelling of a word. For example, you might want to note that *occasion* is spelled with two *c*'s, or that *all right* is two words, not one word.

2 Study a word by looking at it, saying it, and spelling it. You may also want to write out the word one or more times, or "air-write" it with your finger in large, exaggerated motions.

3 When you have trouble spelling a long word, try to break the word down into syllables and see whether you can spell the syllables. For example, *inadvertent* can be spelled easily if you can hear and spell in turn its four syllables: *in ad ver tent.* Or the word *consternation* can be spelled easily if you hear and spell in turn its four syllables: *con ster na tion.* Remember, then: try to see, hear, and spell long words in terms of their syllables.

4 Keep in mind that review and repeated self-testing are the keys to effective learning. When you are learning a series of words, go back after studying each new word and review all the preceding ones.

STEP 3: MASTER COMMONLY CONFUSED WORDS

Master the meanings and spellings of the commonly confused words on pages 502–511. Your instructor may assign twenty words for you to study at a time and give you a series of quizzes until you have mastered the words.

STEP 4: UNDERSTAND BASIC SPELLING RULES

Explained briefly here are three rules that may improve your spelling. While exceptions sometimes occur, the rules hold true most of the time.

1 Changing *y* to *i*. When a word ends in a consonant plus *y*, change *y* to *i* when you add an ending.

try	+ ed	= tried	easy	+ er	= easier
defy	+ es	= defies	carry	+ ed	= carried
ready	+ ness	= readiness	penny	+ less	= penniless

2 Final silent *e*. Drop a final *e* before an ending that starts with a vowel (the vowels are *a, e, i, o,* and *u*).

create + ive = creative   believe + able = believable

nerve + ous = nervous   share + ing = sharing

Keep the final *e* before an ending that starts with a consonant.

extreme + ly = extremely   life + less = lifeless

hope   + ful = hopeful   excite + ment = excitement

3 Doubling a final consonant. Double the final consonant of a word when all three of the following are true:

  a The word is one syllable or is accented on the last syllable.

  b The word ends in a single consonant preceded by a single vowel.

  c The ending you are adding starts with a vowel.

shop   + er = shopper   thin   + est = thinnest

equip + ed = equipped   submit + ed = submitted

swim   + ing = swimming   drag   + ed = dragged

Activity

Combine the following words and endings by applying the three rules above.

1. worry   + ed   = _____
2. write   + ing = _____
3. marry   + es   = _____
4. run   + ing = _____
5. terrify + ed   = _____
6. dry   + es   = _____
7. forget   + ing = _____
8. care   + ful = _____
9. control + ed   = _____
10. debate + able = _____

STEP 5: STUDY A BASIC WORD LIST

Study the spellings of the words in the following list. They are five hundred of the words most often used in English. Your instructor may assign twenty-five or fifty words for you to study at a time and give you a series of quizzes until you have mastered the list.

Five Hundred Basic Words

ability	another	begin	chair
absent	answer	being	change
accept	anxious	believe	charity
accident	appetite	between	cheap
ache	apply	bicycle	cheat
across	approach	black	cheek
address	approve	blue	chicken
advertise	argue	board	chief
advice	around	borrow	children
after	arrange	bottle	choose
again	attempt	bottom	church
against	attention	brake	cigarette
agree	August	breast	citizen
all right	automobile	breathe	city
almost	autumn	brilliant	close
a lot	avenue	brother	clothing
already	awful	building	coffee
also	awkward	bulletin	collect
although	back	bureau	college
always	balance	business	color
amateur	bargain	came	come
American	beautiful	can't	comfortable **100**
among	because	careful **75**	company
amount	become **50**	careless	condition
angry **25**	been	cereal	conversation
animal	before	certain	copy

daily	eight	general	into
danger	either	get	iron
daughter	empty	good	itself
daybreak	English	grammar	January
dear	enough	great **175**	July
death	entrance	grocery	June
December	evening	grow	just
decide	everything	guess	kindergarten
deed	examine	half	kitchen
dentist	except	hammer	knock
deposit	exercise	hand	knowledge
describe	exit	handkerchief	labor
did	expect **150**	happy	laid
died	fact	having	language
different	factory	head	last
dinner	family	heard	laugh
direction	far	heavy	learn
discover	February	high	led
disease	few	himself	left
distance	fifteen	hoarse	leisure
doctor **125**	fight	holiday	length
does	flower	home	lesson **225**
dollar	forehead	hospital	letter
don't	foreign	house	life
doubt	forty	however	light
down	forward	hundred	listen
dozen	found	hungry	little
during	fourteen	husband	loaf
each	Friday	instead	loneliness
early	friend	intelligence **200**	long
earth	from	interest	lose
easy	gallon	interfere	made
education	garden	interrupt	making

many	not	pillow	repeat
March	nothing	place	resource
marry	November **275**	plain	restaurant
match	now	please	ribbon
matter	number	pocket	ridiculous
may	ocean	policeman	right **350**
measure	o'clock	possible	said
medicine	October	post office	same
men	offer	potato	sandwich
middle	often	power	Saturday
might	old	prescription	say
million	omit	president	school
minute	once	pretty	scissors
mistake **250**	one	probably	season
Monday	only	promise	see
money	operate	psychology	sentence
month	opinion	public **325**	September
more	opportunity	pursue	service
morning	optimist	put	seventeen
mother	original	quart	several
mountain	ought	quarter	shoes
mouth	ounce	quick	should
much	overcoat	quiet	sight
must	pain	quit	since
nail	paper	quite	sister
near	part	quiz	sixteenth
needle	peace	raise	sleep
neither	pear **300**	read	smoke
never	pencil	ready	soap
newspaper	penny	really	soldier
nickel	people	reason	something **375**
niece	perfect	receive	sometimes
night	period	recognize	soul
ninety	person	refer	soup
noise	picture	religion	south
none	piece	remember	stamp

state	they	upon	well
still	thing	used	went
stockings	thirteen	usual	were
straight	this	valley	what
street	though	value	whether **475**
strong	thousand	variety	which
student	thread	vegetable	while
studying	three	very	white
such	through	view	whole
suffer	Thursday	villain **450**	whose
sugar	ticket	visitor	wife
suit	time	voice	window
summer	tired	vote	winter
Sunday	today	wage	without
supper	together **425**	wagon	woman
sure	tomorrow	waist	wonder
sweet	tongue	wait	won't
take	tonight	wake	work
teach	touch	walk	world
tear **400**	toward	warm	worth
telegram	travel	warning	would
telephone	trouble	Washington	writing
tenant	trousers	watch	written
tenth	truly	water	wrong
than	twelve	wear	year
Thanksgiving	uncle	weather	yesterday
that	under	Wednesday	yet
theater	understand	week	young
them	United States	weigh	your
there	until	welcome	you're **500**

Note: Two spelling mistakes that students often make are to write *a lot* as one word (*alot*) and to write *all right* as one word (*alright*). Do not write either *a lot* or *all right* as one word.

STEP 6: USE ELECTRONIC AIDS

There are three electronic aids that may help your spelling. First, many *electronic typewriters* on the market today will beep automatically when you misspell a word. They include built-in dictionaries that will then give you the correct spelling. Smith-Corona, for example, has a series of portable typewriters with an "Auto-Spell" feature that start at around $150 at discount stores.

Second, a *computer with a spell-checker* will identify incorrect words and suggest correct spellings. If you know how to write on the computer, you will have no trouble learning how to use the spell-check feature.

Third, *electronic spell-checkers* are pocket-size devices that look much like the pocket calculators you may carry to your math class. They are the latest example of how technology can help the learning process. Electronic spellers can be found in the typewriter or computer section of any discount store, at prices in the $100 range. The checker includes a tiny keyboard. You type out the word the way you think it is spelled, and the checker quickly provides you with the correct spelling of related words. Some of these checkers even *pronounce* the word aloud for you.

VOCABULARY DEVELOPMENT

A good vocabulary is a vital part of effective communication. A command of many words will make you a better writer, speaker, listener, and reader. Studies have shown that students with a strong vocabulary, and students who work to improve a limited vocabulary, are more successful in school. And one research study found that *a good vocabulary, more than any other factor, was common to people enjoying successful careers.* This section will describe three ways of developing your word power: (1) regular reading, (2) vocabulary wordsheets, and (3) vocabulary study books. You should keep in mind from the start, however, that none of the approaches will help unless you truly decide that vocabulary development is an important goal. Only when you have this attitude can you begin doing the sustained work needed to improve your word power.

REGULAR READING

Through reading a good deal, you will learn words by encountering them a number of times in a variety of sentences. Repeated exposure to a word in context will eventually make it a part of your working language.

You should develop the habit of reading a daily newspaper and one or more weekly magazines like *Time, Newsweek,* or even *People,* as well as monthly magazines suited to your interests. In addition, you should try to read some books for pleasure. This may be especially difficult at times when you also have textbook reading to do. Try, however, to redirect a regular half hour to one hour of your recreational time to reading books, rather than watching television, listening to music, or the like. Doing so, you may eventually reap the rewards of an improved vocabulary *and* the discovery that reading can be truly enjoyable. If you would like some recommendations, ask your instructor for a copy of the ''List of Interesting Books'' in the Instructor's Manual of *English Skills.*

VOCABULARY WORDSHEETS

Vocabulary wordsheets are another means of vocabulary development. You should first mark off words in your reading that you want to learn. After you have accumulated a number of words, sit down with a dictionary and look up basic information about each of them. Put this information on a wordsheet like the one shown below. Be sure also to write down a sentence in which each word appears. A word is always best learned not in a vacuum but in the context of surrounding words.

Study each word as follows. First, make sure you can correctly pronounce the word and its derivations. (Pages 483–485 explain the dictionary pronunciation key that will help you pronounce each word properly.) Second, study the main meanings of the word until you can say them without looking at them. Finally, spend a moment looking at the example of the word in context. Follow the same process with the second word. Then, after testing yourself on the first and the second words, go on to the third word. Remember to continue to test yourself on all the words you have studied after you learn each new word. Repeated self-testing is a key to effective learning.

Activity

Locate four words in your reading that you would like to master. Enter them in the spaces on the vocabulary wordsheet below and fill in all the needed information. Your instructor may then check your wordsheet and perhaps give you a quick oral quiz on selected words.

You may receive a standing assignment to add five words a week to a wordsheet and to study the words. Note that you can create your own wordsheets using loose-leaf paper, or your instructor may give you copies of the wordsheet that appears below.

Vocabulary Wordsheet

1. Word: _____*formidable*_____ Pronunciation: *(fôr′ mi də bəl)*

   Meanings: _____*1. feared or dreaded*_____

   _____*2. extremely difficult*_____

   Other forms of the word: *formidably   formidability*

   Use of the word in context: *Several formidable obstacles stand between Matt and his goal.*

2. Word: _____ Pronunciation: _____

   Meanings: _____

   _____

   _____

   Other forms of the word: _____

   Use of the word in context: _____

   _____

3. Word: _____ Pronunciation: _____

   Meanings: _____

   _____

   _____

   Other forms of the word: _____

   Use of the word in context: _____

   _____

4. Word: _____ Pronunciation: _____

   Meanings: _____

   _____

   _____

   Other forms of the word: _____

   Use of the word in context: _____

   _____

5. Word: _____ Pronunciation: _____

   Meanings: _____

   _____

   _____

   Other forms of the word: _____

   Use of the word in context: _____

   _____

VOCABULARY STUDY BOOKS

A third way to increase your word power is to use vocabulary study books. One well-known series of books that may be available in the learning-skills center at your school is the EDL *Word Clues*. These books help you learn a word by asking you to look at the context — or the words around the unfamiliar word — to unlock its meaning. This method is called *using context clues,* or *word clues.*

Many vocabulary books and programs are available. The best are those that present words in one or more contexts and then provide several reinforcement activities for each word. These books will help you increase your vocabulary if you have the determination required to work with them on a regular basis.

COMMONLY
CONFUSED
WORDS

Answers are on page 564.

HOMONYMS

The commonly confused words on the following pages have the same sounds but different meanings and spellings; such words are known as *homonyms*. Complete the activity for each set of words, and check off and study the words that give you trouble.

all ready   completely prepared
already   previously; before

> We were *all ready* to start the play, but the audience was still being seated.
> I have *already* called the police.

Fill in the blanks: I am _____ for the economics examination because I have _____ studied the chapter three times.

brake   stop; the stopping device in a vehicle
break   come apart

> His car bumper has a sticker reading, "I *brake* for animals."
> "I am going to *break* up with Bill if he keeps seeing other women," said Rita.

Fill in the blanks: When my car's emergency _____ slipped, the car rolled back and demolished my neighbor's rose garden, causing a _____ in our good relations with each other.

coarse   rough
course   part of a meal; a school subject; direction; certainly (as in *of course*)

> By the time the waitress served the customers the second *course* of the meal, she was aware of their *coarse* eating habits.

Fill in the blanks: Ted felt that the health instructor's humor was too _____ for his taste and was glad when he finished the _____ .

hear   perceive with the ear
here   in this place

>   "The salespeople act as though they don't see or *hear* me, even though I've been standing *here* for fifteen minutes," the woman complained.

Fill in the blanks: "Did you _____ about the distinguished visitor who just came into town and is staying _____ at this very hotel?"

hole   an empty spot
whole   entire

>   "I can't believe I ate the *whole* pizza," moaned Ralph. "I think it's going to make a *hole* in my stomach lining."

Fill in the blanks: The _____ time I was at the party I tried to conceal the _____ I had in my trousers.

its   belonging to it
it's   the shortened form for "it is" or "it has"

>   The car blew *its* transmission (the transmission belonging to it, the car).
>   *It's* (it has) been raining all week and *it's* (it is) raining now.

Fill in the blanks: _____ hot and unsanitary in the restaurant kitchen I work in, and I don't think the restaurant deserves _____ good reputation.

knew   past form of *know*
new   not old

>   "I had *new* wallpaper put up," said Sarah.
>   "I *knew* there was some reason the place looked better," said Bill.

Fill in the blanks: Lola _____ that getting her hair cut would give her face a _____ look.

know   to understand
no   a negative

"I don't *know* why my dog Fang likes to attack certain people," said Martha.
"There's *no* one thing the people have in common."

Fill in the blanks: I _____ of _____ way of telling
whether that politician is honest or not.

pair   a set of two
pear   a fruit

"What a great *pair* of legs Tony has," said Lola to Vonnie. Tony didn't hear
her, for he was feeling very sick after munching on a green *pear*.

Fill in the blanks: In his lunch box was a _____ of _____.

passed   went by; succeeded in; handed to
past   a time before the present; beyond, as in "We worked past closing time."

Someone *passed* him a wine bottle; it was the way he chose to forget his
unhappy *past*.

Fill in the blanks: I walked _____ the instructor's office but was
afraid to ask her whether or not I had _____ the test.

peace   calm
piece   a part

Nations often risk world *peace* by fighting over a *piece* of land.

Fill in the blanks: Martha did not have any _____ until she gave
her pet dog a _____ of her meat loaf.

plain   simple; a flat area
plane   aircraft

The *plain,* unassuming young man on the *plane* suddenly jumped up with a grenade in his hand and announced, ''We're all going to Tibet.''

Fill in the blanks: The game-show contestant opened the small box wrapped in

_____ brown paper and found inside the keys to his own jet

_____ .

principal   main; a person in charge of a school
principle   a law, standard, or rule

Note: It might help to remember that the *e* in *principle* is also in *rule* — the meaning of *principle.*

Pete's high school *principal* had one *principal* problem: Pete. This was because there were only two *principles* in Pete's life: rest and relaxation.

Fill in the blanks: The _____ reason she dropped out of school

was that she believed in the _____ of complete freedom of choice.

right   correct; opposite of *left*
write   what you do in English

If you have the *right* course card, I'll *write* your name on the class roster.

Fill in the blanks: Eddie thinks I'm weird since I _____ with both

my _____ and left hands.

than (thăn)   used in comparisons
then (thĕn)   at that time

Note: It might help to remember that the*n* is also a tim*e* signal.

> When we were kids, my friend Elaine had prettier clothes *than* I did. I really envied her *then*.

Fill in the blanks: Marge thought she was better _____ the rest of us, but _____ she got the lowest grade on the history test.

their   belonging to them
there   at that place; a neutral word used with verbs like *is, are, was, were, have,* and *had*
they're   the shortest form of ''they are''

> Two people own that van over *there* (at that place). *They're* (they are) going to move out of *their* apartment (the apartment belonging to them) and into the van, in order to save money.

Fill in the blanks: _____ not going to invite us to _____ table because _____ is no room for us to sit down.

threw   past form of *throw*
through   from one side to the other; finished

> The fans *threw* so much litter on the field that the teams could not go *through* with the game.

Fill in the blanks: When Mr. Jefferson was _____ screaming about the violence on television, he _____ the newspaper at his dog.

to   a verb part, as in *to smile;* toward, as in "I'm going *to* heaven"

too   overly, as in "The pizza was *too* hot"; also, as in "The coffee was hot, *too.*"

two   the number 2

> Tony drove *to* the park *to* be alone with Lola. (The first *to* means "toward"; the second *to* is a verb part that goes with *be*.)
>
> Tony's shirt is *too* tight; his pants are tight, *too.* (The first *too* means "overly"; the second *too* means "also.")
>
> You need *two* hands (2 hands) to handle a Whopper.

Fill in the blanks: _____ times tonight, you have been _____ ready _____ make assumptions without asking questions first.

wear   to have on
where in what place

> Fred wanted to *wear* his light pants on the hot day, but he didn't know *where* he had put them.

Fill in the blanks: _____ exactly on my leg should I _____ this elastic bandage?

weather atmospheric conditions
whether if it happens that; in case; if

> Some people go on vacations *whether* or not the *weather* is good.

Fill in the blanks: I always ask Bill _____ or not we're going to have a storm, for he can feel rainy _____ approaching in his bad knee.

whose belonging to whom
who's   the shortened form for "who is" and "who has"

> *Who's* the teacher *whose* students are complaining?

Fill in the blanks: _____ the guy _____ car I saw you in?

your   belonging to you
you're   the shortened form of "you are"

    You're (meaning "you are") not going to the fair unless *your* brother (the brother belonging to you) goes with you.

Fill in the blanks: _____ going to have to put aside individual differences and play together for the sake of _____ team.

OTHER WORDS FREQUENTLY CONFUSED

Following is a list of other words that people frequently confuse. Complete the activities for each set of words, and check off and study the words that give you trouble.

a, an   Both *a* and *an* are used before other words to mean, approximately, "one."

Generally you should use *an* before words starting with a vowel (*a, e, i, o, u*):

    an ache   an experiment   an elephant   an idiot   an ox

Generally you should use *a* before words starting with a consonant (all other letters):

    a Coke   a brain   a cheat   a television   a gambler

Fill in the blanks: The girls had _____ argument over _____ former boyfriend.

accept (ăk sĕpt')   receive; agree to
except (ĕk sĕpt')   exclude; but

    "I would *accept* your loan," said Bill to the bartender, "*except* that I'm not ready to pay 25 percent interest."

Fill in the blanks: _____ that she can't _____ any criticism, Lori is a good friend.

advice (ăd vīs′)   a noun meaning "an opinion"
advise (ăd vīz′)   a verb meaning "to counsel, to give advice"

I *advise* you to take the *advice* of your friends and stop working so hard.

Fill in the blanks: I _____ you to listen carefully to any _____ you get from your boss.

affect (uh fĕkt′)   a verb meaning "to influence"
effect (ĭ fĕkt′)   a verb meaning "to bring about something"; a noun meaning "result"

The full *effects* of marijuana and alcohol on the body are only partly known; however, both drugs clearly *affect* the brain in various ways.

Fill in the blanks: The new tax laws go into _____ next month, and they are going to _____ your income tax deductions.

among   implies three or more
between   implies only two

We had to choose from *among* 125 shades of paint but *between* only 2 fabrics.

Fill in the blanks: The layoff notices distributed _____ the unhappy workers gave them a choice _____ working for another month at full pay and leaving immediately with two weeks' pay.

beside   along the side of
besides   in addition to

I was lucky I wasn't standing *beside* the car when it was hit.
Besides being unattractive, these uniforms are impractical.

Fill in the blanks: _____ the alarm system hooked up to the door, our neighbors keep a gun _____ their beds.

desert (dĕz'ərt)   a stretch of dry land; (di zûrt') to abandon one's post or duty
dessert (dĭ zûrt') last part of a meal

Sweltering in the *desert*, I was tormented by the thought of an icy *dessert*.

Fill in the blanks: After their meal, they carried their _____ into
the living room so that they would not miss the start of the old _____
movie about Lawrence of Arabia.

fewer   used with things that can be counted
less   refers to amount, value, or degree

There were *fewer* than seven people in all my classes today.
I seem to feel *less* tired when I exercise regularly.

Fill in the blanks: With _____ people driving large cars, we are
importing _____ oil than we used to.

loose (loos)   not fastened; not tight-fitting
lose (looz)   misplaced; fail to win

Phil's belt is so *loose* that he always looks ready to *lose* his pants.

Fill in the blanks: At least once a week our neighbors _____ their
dog; it's because they let him run _____.

quiet (kwī'ĭt)   peaceful
quite (kwīt)   entirely; really; rather

After a busy day, the children are now *quiet,* and their parents are *quite* tired.

Fill in the blanks: The _____ halls of the church become
_____ lively during square-dance evenings.

though (thō)   despite the fact that
thought (thôt)   past form of *think*

Even *though* she worked, she *thought* she would have time to go to school.

Fill in the blanks: Susan _____ she would like the job, but even
_____ the pay was good, she hated the traveling involved.

■ Review Test 1

Underline the correct word in the parentheses. If necessary, look back at the explanations of the words instead of trying to guess.

1. Please take my (advice, advise) and (where, wear) something warm and practical, rather (than, then) something fashionable and flimsy.

2. Glen felt that if he could (loose, lose) twenty pounds, the (affect, effect) on his social life might be dramatic.

3. (Their, There, They're) going to show seven horror films at (their, there, they're) Halloween night festival; I hope you'll be (their, there, they're).

4. (Your, You're) going to have to do (a, an) better job on (your, you're) final exam if you expect to pass the (coarse, course).

5. Those (to, too, two) issues are (to, too, two) hot for any politician (to, too, two) handle.

6. Even (though, thought) the (brakes, breaks) on my car were worn, I did not have (quiet, quite) enough money to get them replaced (right, write) away.

7. (Accept, Except) for the fact that my neighbor receives most of his mail in (plain, plane) brown wrappers, he is (know, no) stranger (than, then) anyone else in this rooming house.

8. Because the Randalls are so neat and fussy, (its, it's) hard (to, too, two) feel comfortable when (your, you're) in (their, there, they're) house.

9. (Whose, Who's) the culprit who left the paint can on the table? The paint has ruined a (knew, new) tablecloth, and (its, it's) soaked (threw, through) the linen and (affected, effected) the varnish stain on the table.

10. I would have been angry at the car that (passed, past) me at ninety miles an hour on the highway, (accept, except) that I (knew, new) it would not get (passed, past) the speed trap (to, too, two) miles down the road.

■ **Review Test 2**

On separate paper, write short sentences using the ten words shown below.

their	principal
its	except
you're	past
too	through
then	who's

EFFECTIVE
WORD
CHOICE

Introductory Project

Put a check beside the sentence in each pair that you feel makes more effective use of words.

1. I flipped out when Faye broke our date. _____

   I got very angry when Faye broke our date. _____

2. Doctors as dedicated as Dr. Curtin are few and far between. _____

   Doctors as dedicated as Dr. Curtin are rare. _____

3. Yesterday I ascertained that Elena and Judd broke up. _____

   Yesterday I found out that Elena and Judd broke up. _____

4. Judging by the looks of things, it seems to me that it will probably rain very soon. _____

   It looks as though it will rain soon. _____

Now see if you can circle the correct number in each case:

Pair (1, 2, 3, 4) contains a sentence with slang.

Pair (1, 2, 3, 4) contains a sentence with a cliché.

Pair (1, 2, 3, 4) contains a sentence with a pretentious word.

Pair (1, 2, 3, 4) contains a wordy sentence.

Answers are on page 564.

Choose your words carefully when you write. Always take the time to think about your word choices rather than simply using the first word that comes to mind. You want to develop the habit of selecting words that are appropriate and exact for your purposes. One way you can show sensitivity to language is by avoiding slang, clichés, pretentious words, and wordiness.

SLANG

We often use slang expressions when we talk because they are so vivid and colorful. However, slang is usually out of place in formal writing. Here are some examples of slang expressions:

My girlfriend *got straight* with me by saying she wanted to see other men.

Rick spent all Saturday *messing around* with his stereo.

My boss keeps *riding* me about coming to work on time.

The tires on the Corvette make the car look like *something else*.

The crowd was *psyched up* when the game began.

Slang expressions have a number of drawbacks: they go out of date quickly, they become tiresome if used excessively in writing, and they may communicate clearly to some readers but not to others. Also, the use of slang can be a way of evading the specific details that are often needed to make one's meaning clear in writing. For example, in ''The tires on the Corvette make the car look like something else,'' the writer has not provided the specific details about the tires necessary for us to understand the statement clearly. In general, then, you should avoid slang in your writing. If you are in doubt about whether an expression is slang, it may help to check a recently published hardbound dictionary.

Activity

Rewrite the following sentences, replacing the italicized slang words with more formal ones.

Example   The movie was a *real bomb*, so we *cut out* early.
The movie was terrible, so we left early.

1. My boss *came down on me* for *goofing off* on the job.

_____

_____

2. The car was a *steal* for the money until the owner *jacked up* the price.

_____

_____

3. If the instructor stops *hassling* me, I am going to *get my act together* in the course.

_____

_____

CLICHÉS

A cliché is an expression that has been worn out through constant use. Some typical clichés are listed below:

Clichés

all work and no play	saw the light
at a loss for words	short but sweet
better late than never	sigh of relief
drop in the bucket	singing the blues
easier said than done	taking a big chance
had a hard time of it	time and time again
in the nick of time	too close for comfort
in this day and age	too little, too late
it dawned on me	took a turn for the worse
it goes without saying	under the weather
last but not least	where he (she) is
make ends meet	coming from
on top of the world	word to the wise
sad but true	work like a dog

Clichés are common in speech but make your writing seem tired and stale. Also, clichés—like slang—are often a way of evading the specific details that you must work to provide in your writing. You should, then, avoid clichés and try to express your meaning in fresh, original ways.

Activity

Underline the cliché in each of the following sentences. Then substitute specific, fresh words for the trite expression.

Example   I passed the test by the skin of my teeth.

I barely passed the test.

1. Anyone turning in a paper late is throwing caution to the winds.

    _____

2. Judy doesn't make any bones about her ambition.

    _____

3. I met with my instructor to try to iron out the problems in my paper.

    _____

PRETENTIOUS WORDS

Some people feel they can improve their writing by using fancy and elevated words rather than simple and natural words. They use artificial and stilted language that more often obscures their meaning than communicates it clearly.

Here are some unnatural-sounding sentences:

I comprehended her statement.

While partaking of our morning meal, we engaged in an animated conversation.

I am a stranger to excessive financial sums.

Law enforcement officers directed traffic when the lights malfunctioned.

The same thoughts can be expressed more clearly and effectively by using plain, natural language, as below:

I understood what she said.

While eating breakfast, we had a lively talk.

I have never had much money.

Police officers directed traffic when the lights stopped working.

Activity

Cross out the artificial words in each sentence. Then substitute clear, simple language for the artificial words.

Example   The manager ~~reproached~~ me for my ~~tardiness~~.
The manager criticized me for being late.

1. One of Tina's objectives in life is to accomplish a large family.

   _____

   _____

2. Upon entering our residence, we detected smoke in the atmosphere.

   _____

   _____

3. I am not apprehensive about the test, which encompasses five chapters of the book.

   _____

   _____

WORDINESS

Wordiness — using more words than necessary to express a meaning — is often a sign of lazy or careless writing. Your readers may resent the extra time and energy they must spend when you have not done the work needed to make your writing direct and concise.

Here are examples of wordy sentences:

Anne is of the opinion that the death penalty should be allowed.

I would like to say that my subject in this paper will be the kind of generous person that my father was.

Omitting needless words improves the sentences:

Anne supports the death penalty.

My father was a generous person.

In the box on the following page is a list of some wordy expressions that could be reduced to single words.

Wordy Form	Short Form
a large number of	many
a period of a week	a week
arrive at an agreement	agree
at an earlier point in time	before
at the present time	now
big in size	big
owing to the fact that	because
during the time that	while
five in number	five
for the reason that	because
good benefit	benefit
in every instance	always
in my own opinion	I think
in the event that	if
in the near future	soon
in this day and age	today
is able to	can
large in size	large
plan ahead for the future	plan
postponed until later	postponed
red in color	red
return back	return

Activity

Rewrite the following sentences, omitting needless words.

1. After a lot of careful thinking, I have arrived at the conclusion that drunken drivers should receive jail terms.

   _____

   _____

2. The movie that I went to last night, which was fairly interesting, I must say, was enjoyed by me and my girlfriend.

   _____

   _____

3. Owing to inclement weather conditions of wind and rain, we have decided not to proceed with the athletic competition about to take place on the baseball diamond.

   _____

   _____

4. Without any question, there should be a law making it a requirement for parents of young children to buckle the children into car seats for safety.

   _____

   _____

5. Beyond a doubt, the only two things you can rely or depend on would be the sure facts that death comes to everyone and that the government will tax your yearly income.

   _____

   _____

■ Review Test 1

Certain words are italicized in the following sentences. In the space provided, identify the words as *slang* (S), *clichés* (C), or *pretentious words* (PW). Then rewrite the sentences, replacing the words with more effective diction.

_____ 1. We're *psyched* for tonight's concert, which is going to be *totally awesome*.

   _____

   _____

_____ 2. Getting good grades in college courses is sometimes *easier said than done*.

   _____

   _____

_____ 3. I *availed myself* of the chance to *participate* in the computer course.

   _____

   _____

_____ 4. The victims of the car accident were shaken but *none the worse for wear*.

   _____

   _____

_____ 5. My roommate *pulled an all-nighter* and almost *conked out* during the exam.

   _____

   _____

■ **Review Test 2**

Rewrite the following sentences, omitting needless words.

1. Workers who are on a part-time basis are attractive to a business because they do not have to be paid as much as full-time workers for a business.

   _____

   _____

2. During the time that I was sick and out of school, I missed a total of three math tests.

   _____

   _____

3. The game, which was scheduled for later today, has been canceled by the officials because of the rainy weather.

   _____

   _____

4. At this point in time, I am quite undecided and unsure about just which classes I will take during this coming semester.

   _____

   _____

5. An inconsiderate person located in the apartment next to mine keeps her radio on too loud a good deal of the time, with the result being that it is disturbing to everyone in the neighboring apartments.

   _____

   _____

SENTENCE VARIETY

One aspect of effective writing is to vary the kinds of sentences you write. If every sentence follows the same pattern, writing may become monotonous to read. This chapter explains four ways you can create variety and interest in your writing style. The first two ways involve coordination and subordination — important techniques for achieving different kinds of emphasis in writing.

The following are four methods you can use to make your sentences more varied and more sophisticated:

1   Add a second complete thought (coordination)
2   Add a dependent thought (subordination)
3   Begin with a special opening word or phrase
4   Place adjectives or verbs in a series

Each method will be discussed in turn.

ADD A SECOND COMPLETE THOUGHT

When you add a second complete thought to a simple sentence, the result is a compound (or double) sentence. The two complete statements in a compound sentence are usually connected by a comma plus a joining, or coordinating, word (*and, but, for, or, nor, so, yet*).

A compound sentence is used when you want to give equal weight to two closely related ideas. The technique of showing that ideas have equal importance is called *coordination.*

Following are some compound sentences. Each contains two ideas that the writer regards as equal in importance.

Bill has stopped smoking cigarettes, but he is now addicted to chewing gum.

I repeatedly failed the math quizzes, so I decided to drop the course.

Stan turned all the lights off, and then he locked the office door.

Activity

Combine the following pairs of simple sentences into compound sentences. Use a comma and a logical joining word (*and, but, for, so*) to connect each pair.

Note: If you are not sure what *and, but, for,* and *so* mean, review pages 363–364.

Example   ■ The record kept skipping.
■ There was dust on the needle.

The record kept skipping, for there was dust on the needle.

1. ■ The line at the deli counter was long.
    ■ Jake took a numbered ticket anyway.

    _____

2. ■ Vandals smashed the car's headlights.
    ■ They slashed the tires as well.

    _____

3. ■ I married at age seventeen.
    ■ I never got a chance to live on my own.

    _____

4. ■ Mold grew on my leather boots.
    ■ The closet was warm and humid.

    _____

    _____

5. ■ My father has a high cholesterol count.
    ■ He continues to eat red meat almost every day.

    _____

    _____

ADD A DEPENDENT THOUGHT

When you add a dependent thought to a simple sentence, the result is a complex sentence.* A dependent thought begins with a word or phrase like one of the following:

Dependent Words

after	if, even if	when, whenever
although, though	in order that	where, wherever
as	since	whether
because	that, so that	which, whichever
before	unless	while
even though	until	who, whoever
how	what, whatever	whose

A complex sentence is used when you want to emphasize one idea over another within the sentence. Look at the following complex sentence:

Although I lowered the thermostat, my heating bill remained high.

The idea that the writer wants to emphasize here — *my heating bill remained high* — is expressed as a complete thought. The less important idea — *Although I lowered my thermostat* — is subordinated to this complete thought. The technique of giving one idea less emphasis than another is called *subordination*.

Following are other examples of complex sentences. In each case, the part starting with the dependent word is the less emphasized part of the sentence.

Even though I was tired, I stayed up to watch the horror movie.

Before I take a bath, I check for spiders in the tub.

When Ivy feels nervous, she pulls on her earlobe.

---

* The two parts of a complex sentence are sometimes called an *independent clause* and a *dependent clause*. A *clause* is simply a word group that contains a subject and a verb. An independent clause expresses a complete thought and can stand alone. A dependent clause does not express a complete thought in itself and "depends on" the independent clause to complete its meaning. Dependent clauses always begin with a dependent, or subordinating, word.

Activity

Use logical subordinating words to combine the following pairs of simple sentences into sentences that contain a dependent thought. Place a comma after a dependent statement when it starts the sentence.

Example ▪ Our team lost.
▪ We were not invited to the tournament.

Because our team lost, we were not invited to the _____

tournament. _____

1. ▪ I receive my degree in June.
▪ I will begin applying for jobs.

_____

_____

2. ▪ Lola doesn't enjoy cooking.
▪ She often eats at fast-food restaurants.

_____

_____

3. ▪ I sent several letters of complaint.
▪ The electric company never corrected my bill.

_____

_____

4. ▪ Neil's car went into a skid.
▪ He took his foot off the gas pedal.

_____

_____

5. ▪ The final exam covered sixteen chapters.
▪ The students complained.

_____

_____

BEGIN WITH A SPECIAL OPENING WORD OR PHRASE

Among the special openers that can be used to start sentences are *-ed* words, *-ing* words, *-ly* word, *to* word groups, and prepositional phrases. Here are examples of all five kinds of openers:

-ed *word*	Tired from a long day of work, Sharon fell asleep on the sofa.
-ing *word*	Using a thick towel, Mel dried his hair quickly.
-ly *word*	Reluctantly, I agreed to rewrite the paper.
to *word group*	To get to the church on time, you must leave now.
Prepositional phrase	With Fred's help, Martha planted the evergreen shrubs.

Activity

Combine the simple sentences into one sentence by using the opener shown in the margin and omitting repeated words. Use a comma to set off the opener from the rest of the sentence.

Example   *-ing word:*   ■ The toaster refused to pop up.
■ It buzzed like an angry hornet.

Buzzing like an angry hornet, the toaster refused

to pop up.

-ed *word*

1. ■ Bill was annoyed by the poor TV reception.
   ■ He decided to get a new antenna.

_____

_____

-ing *word*

2. ■ The star player glided down the court.
   ■ He dribbled the basketball like a pro.

_____

_____

-ly *word*

3. ■ Food will run short on our crowded planet.
   ■ It is inevitable.

_____

_____

to *word group*

4. ■ Bill rented a limousine for the night.
   ■ He wanted to make a good impression.

_____

_____

prepositional phrase

5. ■ Lisa answered the telephone.
   ■ She did this at 4 A.M.

_____

_____

-ed *word*

6. ■ Nate dreaded the coming holidays.
   ■ He was depressed by his recent divorce.

_____

_____

-ing *word*

7. ■ The people pressed against the doors of the theater.
   ■ They pushed and shoved each other.

_____

_____

-ly *word*

8. ■ I waited in the packed emergency room.
   ■ I was impatient.

_____

_____

to *word group*

9. ■ The little boy likes to annoy his parents.
   ■ He pretends he can't hear them.

_____

_____

Prepositional phrase

10. ■ People must wear white-soled shoes.
    ■ They must do this in the gym.

_____

_____

PLACE ADJECTIVES OR VERBS IN A SERIES

Various parts of a sentence may be placed in a series. Among these parts are adjectives (descriptive words) and verbs. Here are examples of both in a series.

Adjectives   The *black, smeary* newsprint rubbed off on my *new butcher-block* table.

Verbs   The quarterback *fumbled* the ball, *recovered* it, and *sighed* with relief.

Activity

Combine the simple sentences into one sentence by using adjectives or verbs in a series and by omitting repeated words. In most cases, use a comma between the adjectives or verbs in a series.

Example   ■ Before Christmas, I made fruitcakes.
■ I decorated the house.
■ I wrapped dozens of toys.

Before Christmas, I made fruitcakes, decorated the house, and wrapped dozens of toys.

1.   ■ My lumpy mattress was giving me a cramp in my neck.
■ It was causing pains in my back.
■ It was making me lose sleep.

_____

_____

2.   ■ Lights appeared in the fog.
■ The lights were flashing.
■ The lights were red.
■ The fog was gray.
■ The fog was soupy.

_____

_____

3. ■ Before going to bed, I locked all the doors.
   ■ I activated the burglar alarm.
   ■ I slipped a kitchen knife under my mattress.

   _____

   _____

4. ■ Lola picked sweater hairs off her coat.
   ■ The hairs were fuzzy.
   ■ The hairs were white.
   ■ The coat was brown.
   ■ The coat was suede.

   _____

   _____

5. ■ The contact lens fell onto the floor.
   ■ The contact lens was thin.
   ■ The contact lens was slippery.
   ■ The floor was dirty.
   ■ The floor was tiled.

   _____

   _____

■ Review Test 1

On separate paper, use coordination or subordination to combine the following groups of simple sentences into one or more longer sentences. Omit repeated words. Since various combinations are possible, you might want to jot down several combinations in each case. Then read them aloud to find the combination that sounds best.

Keep in mind that very often, the relationship among ideas in a sentence will be clearer when subordinating rather than coordinating words are used.

Example ■ I don't like to ask for favors.
   ■ I must borrow money from my brother-in-law.
   ■ I know he won't turn me down.
   ■ I still feel guilty about it.

   I don't like to ask for favors, but I must borrow money from my

   brother-in-law. Although I know he won't turn me down, I still

   feel guilty about it.

Comma Hints

a   Use a comma at the end of a word group that starts with a subordinating word (as in ''Although I know he won't turn me down, . . .'').

b   Use a comma between independent word groups connected by *and, but, for, or, nor, so, yet* (as in ''I don't like to ask for favors, but . . .'').

1. ■ My grandmother is eighty-six.
   ■ She drives to Florida alone every year.
   ■ She believes in being self-reliant.

2. ■ His name was called.
   ■ Luis walked into the examining room.
   ■ He was nervous.
   ■ He was determined to ask the doctor for a straight answer.

3. ■ They left twenty minutes early for class.
   ■ They were late anyway.
   ■ The car overheated.

4. ■ John failed the midterm exam.
   ■ He studied harder for the final.
   ■ He passed it.

5. ■ A volcano erupts.
   ■ It sends tons of ash into the air.
   ■ This creates flaming orange sunsets.

6. ■ Tony got home from the shopping mall.
   ■ He discovered his rented tuxedo did not fit.
   ■ The jacket sleeves covered his hands.
   ■ The pants cuffs hung over his shoes.

7. ■ The boys waited for the bus.
   ■ The wind shook the flimsy shelter.
   ■ They shivered with cold.
   ■ They were wearing thin jackets.

8. ■ The engine almost caught.
   ■ Then it died.
   ■ I realized no help would come.
   ■ I was on a lonely road.
   ■ It was very late.

9. ■ Miriam wanted white wall-to-wall carpeting.
   ■ She knew it was a bad buy.
   ■ It would look beautiful.
   ■ It would be very hard to clean.

10. ■ Gary was leaving the store.
    ■ The shoplifting alarm went off.
    ■ He had not stolen anything.
    ■ The clerk had forgotten to remove the magnetic tag.
    ■ The tag was on a shirt Gary had bought.

■ Review Test 2

On separate paper, write two sentences of your own that begin with (1) *-ed* words, (2) *-ing* words, (3) *-ly* words, (4) *to* word groups, and (5) prepositional phrases. Also write two sentences of your own that contain (6) a series of adjectives and (7) a series of verbs.

COMBINED MASTERY TESTS

SENTENCE FRAGMENTS AND RUN-ONS

■ Combined Mastery Test 1

The word groups below are numbered 1 through 20. In the space provided for each, write C if a word group is a complete sentence, write F if it is a sentence fragment, and write R-O if it is a run-on. Then correct the errors.

1. _____
2. _____
3. _____
4. _____
5. _____
6. _____
7. _____
8. _____
9. _____
10. _____
11. _____
12. _____
13. _____
14. _____
15. _____
16. _____
17. _____
18. _____
19. _____
20. _____

[1]I had a frightening dream last night, I dreamed that I was walking high up on an old railroad trestle. [2]It looked like the one I used to walk on recklessly. [3]When I was about ten years old. [4]At that height, my palms were sweating, just as they did when I was a boy. [5]I could see the ground out of the corners of my eyes, I felt a swooning, sickening sensation. [6]Suddenly, I realized there were rats below. [7]Thousands upon thousands of rats. [8]They knew I was up on the trestle, they were laughing. [9]Because they were sure they would get me. [10]Their teeth glinted in the moonlight, their red eyes were like thousands of small reflectors. [11]That almost blinded my sight. [12]Sensing there was something even more hideous behind me. [13]I kept moving forward. [14]Then I realized that I was coming to a gap in the trestle. [15]There was no way I could stop or go back I would have to cross over that empty gap. [16]I leaped out in despair. [17]Knowing I would never make it. [18]And felt myself falling helplessly down to the swarm of rejoicing rats. [19]I woke up bathed in sweat. [20]Half expecting to find a rat in my bed.

Score Number correct _____ × 5 = _____ percent

SENTENCE FRAGMENTS AND RUN-ONS

■ Combined Mastery Test 2

The word groups below are numbered 1 through 20. In the space provided for each, write C if a word group is a complete sentence, write F if it is a sentence fragment, and write R-O if it is a run-on. Then correct the errors.

1. _____

2. _____

3. _____

4. _____

5. _____

6. _____

7. _____

8. _____

9. _____

10. _____

11. _____

12. _____

13. _____

14. _____

15. _____

16. _____

17. _____

18. _____

19. _____

20. _____

[1]My sister asked my parents and me to give up television for two weeks. [2]As an experiment for her psychology class. [3]We were too embarrassed to refuse, we reluctantly agreed. [4]The project began on a Monday morning. [5]To help us resist temptation. [6]My sister unplugged the living room set. [7]That evening the four of us sat around the dinner table much longer than usual, we found new things to talk about. [8]Later we played board games for several hours, we all went to bed pleased with ourselves. [9]Everything went well until Thursday evening of that first week. [10]My sister went out after dinner. [11]Explaining that she would be back about ten o'clock. [12]The rest of us then decided to turn on the television. [13]Just to watch the network news. [14]We planned to unplug the set before my sister got home. [15]And pretend nothing had happened. [16]We were settled down comfortably in our respective chairs, unfortunately, my sister walked in at that point and burst out laughing. [17]"Ah ha! I caught you," she cried. [18]She explained that part of the experiment was to see if we would stick to the agreement. [19]Especially during her absence. [20]She had predicted we would weaken, it turned out she was right.

Score Number correct _____ × 5 = _____ percent

VERBS

■ Combined Mastery Test 1

Each sentence contains a mistake involving (1) standard English or irregular verb forms, (2) subject-verb agreement, or (3) consistent verb tense. Circle the letter that identifies the mistake. Then cross out the incorrect verb and write the correct form in the space provided.

_____ 1. One of my apartment neighbors always keep the radio on loudly all night.
Mistake in: a. Subject-verb agreement   b. Verb tense

_____ 2. The more the instructor explained the material and the more he wroted on the board, the more confused I got.
Mistake in: a. Irregular verb form   (b) Verb tense

_____ 3. I grabbed the last carton of skim milk on the supermarket shelf but when I checks the date on it, I realized it was not fresh.
Mistake in: a. Subject-verb agreement   (b) Verb tense

_____ 4. This morning my parents argued loudly, but later they apologized to each other and embrace.
Mistake in: a. Subject-verb agreement   (b) Verb tense

_____ 5. When the bell rang, Mike takes another bite of his sandwich and then prepared for class.
Mistake in: a. Irregular verb form   (b) Verb tense

_____ 6. Someone called Marion at the office to tell her that her son had been bit by a stray dog.
Mistake in: a. Irregular verb form   b. Verb tense

_____ 7. Because I had throwed away the sales slip, I couldn't return the microwave.
Mistake in: a. Irregular verb form   b. Verb tense

_____ 8. My dog and cat usually ignores each other, but once in a while they fight.
Mistake in: a. Subject-verb agreement   b. Verb tense

_____ 9. From the back of our neighborhood bakery comes some of the best smells in the world.
Mistake in: a. Subject-verb agreement   b. Verb tense

_____ 10. The cost of new soles and heels are more than those old shoes are worth.
Mistake in: a. Subject-verb agreement   b. Verb tense

Score   Number correct _____ × 5 = _____ percent

VERBS

■ Combined Mastery Test 2

Each sentence contains a mistake involving (1) standard English or irregular verb forms, (2) subject-verb agreement, or (3) consistent verb tense. Circle the letter that identifies the mistake. Then cross out the incorrect verb and write the correct form in the space provided.

_____ 1. My friend's bitter words had stinged me deeply.
   Mistake in: a. Irregular verb form   b. Verb tense

_____ 2. After she poured the ammonia in the bucket, Karen reels backward because the strong fumes made her eyes tear.
   Mistake in: a. Subject-verb agreement   b. Verb tense

_____ 3. Flying around in space is various pieces of debris from old space satellites.
   Mistake in: a. Subject-verb agreement   b. Verb tense

_____ 4. Eileen watched suspiciously as a strange car drived back and forth in front of her house.
   Mistake in: a. Irregular verb form   b. Verb tense

_____ 5. Both crying and laughing helps us get rid of tension.
   Mistake in: a. Subject-verb agreement   b. Verb tense

_____ 6. All my clothes were dirty, so I stayed up late and washes a load for tomorrow.
   Mistake in: a. Subject-verb agreement   b. Verb tense

_____ 7. McDonald's has selled enough hamburgers to reach to the moon.
   Mistake in: a. Irregular verb form   b. Verb tense

_____ 8. When Fred peeled back the bedroom wallpaper, he discovered another layer of wallpaper and uses a steamer to get that layer off.
   Mistake in: a. Subject-verb agreement   b. Verb tense

_____ 9. Rosie searched for the fifty-dollar bill she had hid somewhere in her dresser.
   Mistake in: a. Irregular verb form   b. Verb tense

_____ 10. The realistic yellow tulips on the gravestone is made of a weather-resistant fabric.
   Mistake in: a. Subject-verb agreement   b. Verb tense

Score   Number correct _____ × 5 = _____ percent

CAPITAL LETTERS AND PUNCTUATION

■ Combined Mastery Test 1

Each of the following sentences contains an error in capitalization or punctuation. Refer to the box below and write, in the space provided, the letter identifying the error. Then correct the error.

a.	missing capital	c.	missing quotation marks
b.	missing apostrophe	d.	missing comma

_____ 1. Maggie's aerobics class has been canceled this week so she's decided to go running instead.

_____ 2. ''One of the striking differences between a cat and a lie, wrote Mark Twain, ''is that a cat has only nine lives.''

_____ 3. My uncles checks are printed to look like Monopoly money.

_____ 4. Did you know someone is turning the old school on ninth Street into a restaurant named Home Economics?

_____ 5. My parents always ask me where Im going and when I'll be home.

_____ 6. She doesn't talk about it much, but my aunt has been a member of alcoholics Anonymous for ten years.

_____ 7. The sweating straining horses neared the finish line.

_____ 8. Whenever he gave us the keys to the car, my father would say, Watch out for the other guy.''

_____ 9. If you're going to stay up late be sure to turn down the heat before going to bed.

_____ 10. I decided to have a glass of apple juice rather than order a pepsi.

Score Number correct _____ × 10 = _____ percent

CAPITAL LETTERS AND PUNCTUATION

■ Combined Mastery Test 2

Each of the following sentences contains an error in capitalization or punctuation. Refer to the box below and write, in the space provided, the letter identifying the error. Then correct the error.

a. missing capital	c. missing quotation marks
b. missing apostrophe	d. missing comma

_____ 1. Even though I hadn't saved the receipt I was able to return the blender to Sears.

_____ 2. "The diners food is always reliable," said Stan. "It's consistently bad."

_____ 3. Some people are surprised to hear that manhattan is an island.

_____ 4. "To love oneself, said Oscar Wilde, "is the beginning of a life-long romance."

_____ 5. The airplane was delayed for more than three hours and the passengers were getting impatient.

_____ 6. Leslie said to the woman behind her in the theater, "will you stop talking, please?"

_____ 7. Charles corns are as good predictors of the weather as the TV weather report.

_____ 8. "Before you can reach your goals," says my grandfather, you have to believe you can reach them."

_____ 9. There is little evidence that king Arthur, the legendary knight, really existed.

_____ 10. My cousin learned to cook when he was an Army cook during World war II.

Score Number correct _____ × 10 = _____ percent

WORD USE

■ Combined Mastery Test 1

Each of the following sentences contains a mistake identified in the left-hand margin. Underline the mistake and then correct it in the space provided.

Slang
1. Because Lisa has a lot of pull at work, she always has first choice of vacation time.

Wordiness
2. Truthfully, I've been wishing that the final could be postponed to a much later date sometime next week.

Cliché
3. Meg hoped her friends would be green with envy when they saw her new boyfriend.

Pretentious language
4. Harold utilizes old coffee cans to water his house plants.

Adverb error
5. The sled started slow and then picked up speed as the icy hill became steeper.

Error in comparison
6. When the weather is dry, my arthritis feels more better.

Confused word
7. If you neglect your friends, their likely to become former friends.

Confused word
8. She's the neighbor who's dog is courting my dog.

Confused word
9. If you don't put cans, jars, and newspapers on the curb for recycling, the township won't pick up you're garbage.

Confused word
10. "Its the most economical car you can buy," the announcer said.

Score Number correct _____ × 10 = _____ percent

WORD USE

■ Combined Mastery Test 2

Each of the following sentences contains a mistake identified in the left-hand margin. Underline the mistake and then correct it in the space provided.

Slang   1. After coming in to work late all last week, Sheila was canned.

_____

Wordiness   2. At this point in time, I'm not really sure what my major will be.

_____

Cliché   3. Jan and Stan knew they could depend on their son in their hour of need.

_____

Pretentious language   4. I plan to do a lot of comparison shopping before procuring a new dryer.

_____

Adverb error   5. The children sat very quiet as their mother read the next chapter of *Charlie and the Chocolate Factory*.

_____

Error in comparison   6. The respectfuller you treat people, the more they are likely to deserve your respect.

_____

Confused word   7. Insert the disk in the computer before you turn on its' power switch.

_____

Confused word   8. "My advise to you," said my grandmother, "is to focus on your strengths, not your fears."

_____

Confused word   9. The principle advantage to the school cafeteria is that it's three blocks from a Wendy's.

_____

Confused word   10. My parents mean well, but there goals for me aren't my goals.

_____

> *Score*   Number correct _____ × 10 = _____ percent

EDITING
TESTS

EDITING AND PROOFREADING
FOR SENTENCE-SKILLS MISTAKES

The twelve tests in this chapter will give you practice in editing and proofreading for sentence-skills mistakes. People often find it hard to edit and proofread a paper carefully. They have put so much work into their writing, or so little, that it's almost painful for them to look at the paper one more time. You may simply have to *force* yourself to edit and proofread. Remember that eliminating sentence-skills mistakes will improve an average paper and help ensure a strong grade on a good paper. Further, as you get into the habit of checking your papers, you will also get into the habit of using the sentence skills consistently. They are a basic part of clear and effective writing.

■ Editing Test 1

Identify the five mistakes in paper format in the student paper that follows. From the box below, choose the letters that describe the five mistakes and write those letters in the spaces provided.

---

a. The title should not be underlined.
b. The title should not be set off in quotation marks.
c. There should not be a period at the end of the title.
d. All the major words in the title should be capitalized.
e. The title should be just several words and not a complete sentence.
f. The first line of the paper should stand independent of the title.
g. A line should be skipped between the title and the first line of the paper.
h. The first line of the paper should be indented.
i. The right-hand margin should not be crowded.
j. Hyphenation should occur only between syllables.

"my candy apple adventure"

	It was the best event of my day. I loved the sweetness that
	filled my mouth as I bit into the sugary coating. With my second
	bite, I munched contentedly on the apple underneath. Its
	crunchy tartness was the perfect balance to the smooth sweet-
	ness of the outside. Then the apple had a magical effect on me.
	Suddenly I remembered when I was seven years old, walking
	through the county fair grounds, holding my father's hand. We
	stopped at a refreshment stand, and he bought us each a
	candy apple. I had never had one before, and I asked him what it
	was. "This is a very special fruit," he said. "If you ever feel sad,
	all you have to do is eat a candy apple, and it will bring you
	sweetness." Now, years later, his words came back to me, and
	as I ate my candy apple, I felt the world turn sweet once more.

1. _____   2. _____   3. _____   4. _____   5. _____

■ Editing Test 2

Identify the sentence-skills mistakes at the underlined spots in the selection that follows. From the box below, choose the letter that describes each mistake and write it in the space provided. The same mistake may appear more than once.

a. sentence fragment	d. apostrophe mistake
b. run-on	e. faulty parallelism
c. mistake in subject-verb agreement	

Looking Out for Yourself

It's sad but true that "If you don't look out for yourself, no one else will." For example, some people have a false idea about the power of a college degree, they think <u>1</u> that once they possesses the degree, the world will be waiting on their doorstep. In fact, <u>2</u> nobody is likely to be on their doorstep unless, through advance planning, they has <u>3</u> prepared themselves for a career. The kind in which good job opportunities exist. Even <u>4</u> after a person has landed a job, however, a healthy amount of self-interest is needed. People who hide in corners or with hesitation to let others know about their skills doesn't <u>5</u> <u>6</u> get promotions or raises. Its important to take credit for a job well done, whether it <u>7</u> involves writing a report, organized the office filing system, or calming down an angry <u>8</u> customer. Also, people should feel free to ask the boss for a raise. If they work hard <u>9</u> and really deserve it. Those who look out for themselves get the rewards, people who <u>10</u> depend on others to help them along get left behind.

1. _____	3. _____	5. _____	7. _____	9. _____
2. _____	4. _____	6. _____	8. _____	10. _____

■ **Editing Test 3**

Identify the sentence-skills mistakes at the underlined spots in the selection that follows. From the box below, choose the letter that describes each mistake and write it in the space provided. The same mistake may appear more than once.

a. sentence fragment	e. missing commas around an interrupter
b. run-on	
c. mistake in verb tense	f. mistake with quotation marks
d. irregular verb mistake	g. apostrophe mistake

Deceptive Appearances

Appearances can be deceptive. While looking through a library window yesterday, I saw a neatly groomed woman walk by. Her clothes were skillfully <u>tailored her</u> makeup

 ₁

was perfect. <u>Then thinking no one was looking she</u> crumpled a piece of paper in her hand.

 ₂

<u>And tossed it into a nearby hedge.</u> Suddenly she no longer <u>looks</u> attractive to me. On

 ₃ ₄

another occasion, I started talking to a person in my psychology class named Eric. Eric

seemed to be a great person. He always got the class laughing with his <u>jokes, on</u> the days

 ₅

when Eric was absent, I think even the professor missed his lively personality. Eric asked

me <u>''if I wanted to get a Coke in the cafeteria,''</u> and I felt happy he had <u>chose</u> me to be a

 ₆ ₇

friend. <u>While we were sitting in the cafeteria.</u> Eric took out an envelope with several kinds

 ₈

of pills inside. ''Want one?'' he asked. ''They're uppers.'' I didn't want <u>one, I</u> felt dis-

 ₉

appointed. <u>Erics</u> terrific personality was the product of the pills he took.

 ₁₀

1. _____   3. _____   5. _____   7. _____   9. _____

2. _____   4. _____   6. _____   8. _____   10. _____

■ Editing Test 4

Identify the sentence-skill mistakes at the underlined spots in the selection that follows. From the box below, choose the letter that describes each mistake and write it in the space provided. The same mistake may appear more than once.

a. sentence fragment	e. apostrophe mistake
b. run-on	f. dangling modifier
c. irregular verb mistake	g. missing quotation marks
d. missing comma after introductory words	

A Horrifying Moment

The most horrifying moment in my life occurred in the dark hallway. <u>Which led to my apartment house.</u> Though the hallway light was <u>out I</u> managed to find my apartment ₁ ₂ door. However, I could not find the keyhole with my door key. I then pulled a book of matches from my pocket. <u>Trying to strike a match,</u> the entire book of matches <u>bursted</u> ₃ ₄ into flames. I flicked the matches away but not before my coat sleeve <u>catched</u> fire. ₅ Within seconds, my arm was like a torch. <u>Struggling to unsnap the buttons of my coat,</u> ₆ flames began to sear my skin. I was quickly going into shock. <u>And began screaming in</u> ₇ <u>pain.</u> A <u>neighbors</u> door opened and a voice cried out, <u>My God!</u> I was pulled through an ₈ ₉ apartment and put under a bathroom shower, which extinguished the flames. I suffered third degree burns on my <u>arm, I</u> felt lucky to escape with my life. ₁₀

1. _____   3. _____   5. _____   7. _____   9. _____

2. _____   4. _____   6. _____   8. _____   10. _____

■ Editing Test 5

Identify the sentence-skills mistakes at the underlined spots in the selection that follows. From the box below, choose the letter that describes each mistake and write it in the space provided. The same mistake may appear more than once.

a. sentence fragment	e. faulty parallelism
b. run-on	f. apostrophe mistake
c. missing capital letter	g. missing quotation mark
d. mistake in subject-verb agreement	h. missing comma after introductory words

Why I Didn't Go to Church

I almost never attended church in my boyhood years. There was an unwritten code that the guys on the corner <u>was</u> not to be seen in <u>churches'</u>. Although there <u>was</u> many
$\qquad$ 1 $\qquad\qquad$ 2 $\qquad\qquad$ 3
days when I wanted to attend a church, I felt I had no choice but to stay away. If the guys had heard I had gone to church, they would have said things like, <u>''hey, angel,</u>
$\qquad\qquad$ 4
when are you going to <u>fly?</u> With my group of friends, <u>its</u> amazing that I developed any
$\qquad$ 5 $\qquad\qquad$ 6
religious feeling at all. Another reason for not going to church was my father. When he was around the <u>house he</u> told my mother, ''Mike's not going to church. No boy of mine
$\qquad$ 7
is a sissy.'' My mother and sister went to <u>church, I</u> sat with my father and read the
$\qquad$ 8
Sunday paper or <u>watching television.</u> I did not start going to church until years later.
$\qquad$ 9
<u>When I no longer hung around with the guys on the corner or let my father have power</u>
$\qquad$ 10
over me.

1. _____ 3. _____ 5. _____ 7. _____ 9. _____

2. _____ 4. _____ 6. _____ 8. _____ 10. _____

■ Editing Test 6

Identify the sentence-skills mistakes at the underlined spots in the selection that follows. From the box below, choose the letter that describes each mistake and write it in the space provided. The same mistake may appear more than once.

a. sentence fragment	f. missing comma between two complete thoughts
b. run-on	
c. nonparallel structure	g. missing comma after introductory words
d. missing apostrophe	
e. missing quotation mark	h. misspelled word

Anxiety and the Telephone

Not many of us would want to do without our <u>telephones but</u> there are times when
                                               1
the phone is a source of anxiety. For example, you might be walking up to your front
door. <u>When you hear the phone ring.</u> You struggle to find your key, to unlock the door,
                 2
and <u>getting</u> to the phone quickly. You know the phone will stop ringing the instant you
      3
pick up the <u>receiver, then</u> you wonder if you missed the call that would have made you a
             4
<u>millionare</u> or introduced you to the love of your life. Another time, you may have called
 5
in sick to work with a phony excuse. All day long, <u>youre</u> afraid to leave the house in case
                                      6
the boss calls back. <u>And asks himself why you were feeling well enough to go out.</u> In addi-
                                     7
tion, you worry that you might unthinkingly pick up the phone and say in a cheerful voice,
'' <u>Hello,</u> completely <u>forgeting</u> to use your fake cough. In cases like <u>these having</u> a telephone
   8   9   10
is more of a curse than a blessing.

1. _____	3. _____	5. _____	7. _____	9. _____
2. _____	4. _____	6. _____	8. _____	10. _____

■ Editing Test 7

See if you can locate and correct the ten sentence-skills mistakes in the following passage. The mistakes are listed in the box below. As you locate each mistake, write the number of the sentence containing that mistake. Use the spaces provided.

5 sentence fragments

_____ _____ _____ _____ _____

5 run-ons

_____ _____ _____ _____ _____

Family Stories

¹When I was little, my parents invented some strange stories to explain everyday events to me, my father, for example, told me that trolls lived in our house. ²When objects such as scissors or pens were missing. ³My father would look at me and say, ''The trolls took them.'' ⁴For years, I kept a flashlight next to my bed. ⁵Hoping to catch the trolls in the act as they carried away our possessions. ⁶Another story I still remember is my mother's explanation of pussy willows. ⁷After the fuzzy gray buds emerged in our backyard one spring. ⁸I asked Mom what they were. ⁹Pussy willows, she explained, were cats who had already lived nine lives, in this tenth life, only the tips of the cats' tails were visible to people. ¹⁰All the tails looked alike. ¹¹So that none of the cats would be jealous of the others. ¹²It was also my mother who created the legend of the birthday fairy, this fairy always knew which presents I wanted. ¹³Because my mother called up on a special invisible telephone. ¹⁴Children couldn't see these phones, every parent had a direct line to the fairy. ¹⁵My parents' stories left a great impression on me, I still feel a surge of pleasure when I think of them.

■ Editing Test 8

See if you can locate and correct the ten sentence-skills mistakes in the following passage. The mistakes are listed in the box below. As you locate each mistake, write the number of the sentence containing that mistake. Use the spaces provided.

---

1 sentence fragment _____

1 run-on _____

1 nonstandard verb _____

1 missing comma around

   an interrupter _____

2 missing commas between items

   in a series _____   _____

2 apostrophe mistakes _____

   _____

1 capital letter mistake _____

1 homonym mistake _____

Search for the Perfect Pajama Bottoms

¹I met a strange character in my job as an Orderly at a state mental hospital. ²Jacks illness was such that it led to a strange behavior. ³Jack, a middle-aged man spent most of the day obsessed with pajama bottoms. ⁴He seemed to think that someone might have a better pair of pajama bottoms than his. ⁵He would sneak into other patients rooms. ⁶And steal there pajama bottoms. ⁷Other times, he would undress a distracted patient to try on his pajamas. ⁸He would try them on find they were not perfect and discard them. ⁹Some days, I'd come upon three or four patients with no pajama bottoms on, I'd know Jack was on the prowl again. ¹⁰He were tireless in his search for the perfect pair of pajama bottoms. ¹¹It was clearly a search that would never end in success.

■ Editing Test 9

See if you can locate and correct the ten sentence-skills mistakes in the following passages. The mistakes are listed in the box below. As you locate each mistake, write the number of the sentence containing that mistake. Use the spaces provided.

2 sentence fragments 8_____ _10_____	1 missing comma after introductory words _7_____
1 run-on 8, 9 15	2 apostrophe mistakes 14_____
1 irregular verb mistake 3_____	_____
1 missing comma between items in a series 3, 5_____	1 nonparallel structure 3 1_____
	1 missing quotation mark ____4

Fred's Funeral

[1]Sometimes when Fred feels undervalued and depression, he likes to imagine his own funeral. [2]He pictures all the people who will be there. [3]He hears their hushed words sees their tears, and feels their grief. [4]He glows with a warm sadness as the minister begins a eulogy by saying, "Fred Grencher was no ordinary man. . . .'' [5]As the minister talks on Freds eyes grow moist. [6]He laments his own passing and feels altogether appreciated and wonderful.

Feeding Time

[7]Recently I was at the cathouse in the zoo. [8]Right before feeding time. [9]The tigers and lions were lying about on benches and little stands. [10]Basking in the late-afternoon sun. [11]They seemed tame and harmless. [12]But when the meat was brung in, a remarkable change occurred. [13]All the cats got up and moved toward the food. [14]I was suddenly aware of the rippling muscles' of their bodies and their large claws and teeth. [15]They seemed three times bigger, I could feel their power.

■ Editing Test 10

See if you can locate and correct the ten sentence-skills mistakes in the following passage. The mistakes are listed in the box below. As you locate each mistake, write the number of the sentence containing that mistake. Use the spaces provided.

---

1 run-on _____

1 mistake in subject-verb agreement _____

1 missing comma after introductory words _____

2 missing commas around an interrupter _____ _____

1 missing comma between items in a series _____

2 apostrophe mistakes _____ _____

2 missing quotation marks _____ _____

Walking Billboards

¹Many Americans have turned into driving, walking billboards. ²As much as we all claim to hate commercials on television we dont seem to have any qualms about turning ourselves into commercials. ³Our car bumpers for example advertise lake resorts underground caverns, and amusement parks. ⁴Also, we wear clothes marked with other peoples initials and slogans. ⁵Our fascination with the names of designers show up on the backs of our sneakers, the breast pockets of our shirts, and the right rear pockets of our blue jeans. ⁶And we wear T-shirts filled with all kinds of advertising messages. ⁷For instance, people are willing to wear shirts that read, ''Dillon Construction,'' ''Nike,'' or even I Got Crabs at Ed's Seafood Palace. ⁸In conclusion, we say we hate commercials, we actually pay people for the right to advertise their products.

■ **Editing Test 11**

See if you can locate and correct the ten sentence-skills mistakes in the following passage. The mistakes are listed in the box below. As you locate each mistake, write the number of the sentence containing that mistake. Use the spaces provided.

3 sentence fragments _____	1 mistake in pronoun point of
_____ _____	view _____
2 run-ons _____ _____	1 dangling modifier _____
1 irregular verb mistake _____	1 missing comma between
1 nonparallel structure _____	two complete thoughts _____

Too Many Cooks

¹The problem in my college dining hall was the succession of incompetent cooks who were put in charge. ²During the time I worked there, I watched several cooks come and go. ³The first of these was Irving. ⁴He was skinny and greasy like the undercooked bacon he served for breakfast. ⁵Irving drank, by late afternoon he begun to sway as he cooked. ⁶Once, he looked at the brightly colored photograph on the orange juice machine. ⁷And asked why the TV was on. ⁸Having fired Irving, Lonnie was hired. ⁹Lonnie had a soft, round face that resembled the Pillsbury Doughboy's but he had the size and temperament of a large bear. ¹⁰He'd wave one paw and growl if you entered the freezers without his permission. ¹¹He also had poor eyesight. ¹²This problem caused him to substitute flour for sugar and using pork for beef on a regular basis. ¹³After Lonnie was fired, Enzo arrived. ¹⁴Because he had come from Italy only a year or two previously. ¹⁵He spoke little English. ¹⁶In addition, Enzo had trouble with seasoning and spices. ¹⁷His vegetables were too salty, giant bay leaves turned up in everything. ¹⁸Including the scrambled eggs. ¹⁹The cooks I worked for in the college dining hall would have made Julia Child go into shock.

■ Editing Test 12

See if you can locate and correct the ten sentence-skills mistakes in the following passage. The mistakes are listed in the box below. As you locate each mistake, write the number of the sentence containing that mistake. Use the spaces provided.

2 sentence fragments _____ _____	1 missing comma between two complete thoughts _____
2 run-ons _____ _____	1 missing comma between items in a series _____
1 mistake in pronoun point of view _____	1 missing quotation mark _____
1 apostrophe mistake _____	1 misspelled word _____

My Ideal Date

[1]Here are the ingredients for my ideal date, first of all, I would want to look as stunning as possible. [2]I would be dressed in a black velvet jumpsuit. [3]That would fit me like a layer of paint. [4]My acessories would include a pair of red satin spike heels a diamond hair clip, and a full-length black mink coat. [5]My boyfriend, Tony, would wear a sharply tailored black tuxedo, a white silk shirt, and a red bow tie. [6]The tux would emphasize Tony's broad shoulders and narrow waist, and you would see his chest muscles under the smooth shirt fabric. [7]Tony would pull up to my house in a long, shiny limousine, then the driver would take us to the most exclusive and glittery nightclub in Manhattan. [8]All eyes would be on us as we entered and photographers would rush up to take our picture for *People* magazine. [9]As we danced on the lighted floor of the club, everyone would step aside to watch us perform our moves. [10]After several bottles of champagne, Tony and I would head for the top floor of the World Trade Center. [11]As we gazed out over the lights' of the city, Tony would hand me a small velvet box containing a fifty-carat ruby engagement ring. [12]And ask me to marry him. [13]I would thank Tony for a lovely evening and tell him gently, ''Tony, I don't plan to marry until I'm thirty.

SENTENCE-SKILLS ACHIEVEMENT TEST

PART 1

This test will help you measure your improvement in important sentence skills. Certain parts of the following word groups are underlined. Write X in the answer space if you think a mistake appears at the underlined part. Write C in the answer space if you think the underlined part is correct.

A series of headings ("Sentence Fragments," "Run-Ons," and so on) will give you clues for the mistakes to look for. However, you do not have to understand the label to find a mistake. What you are checking for is your own sense of effective written English.

Sentence Fragments

_____ 1. After a careless driver hit my motorcycle, I decided to buy a car. At least I would have more protection against other careless drivers.

_____ 2. I was never a good student in high school. Because I spent all my time socializing with my group of friends. Good grades were not something that my group really valued.

_____ 3. The elderly couple in the supermarket were not a pleasant sight. Arguing with each other. People pretended not to notice them.

_____ 4. Using a magnifying glass, the little girls burned holes in the dry leaf. They then set some tissue paper on fire.

_____ 5. My brother and I seldom have fights about what to watch on television. Except with baseball games. I get bored watching this sport.

_____ 6. My roommate and I ate, talked, danced, and sang at a fish-fry party the other night. Also, we played cards until 3 A.M. As a result, we both slept until noon the next day.

Run-Ons

———— 7. She decided to quit her high-pressured <u>job, she</u> didn't want to develop heart trouble.

———— 8. His car's wheels were not balanced <u>properly, for</u> the car began to shake when he drove over forty miles an hour.

———— 9. I got through the interview without breaking out in a sweat <u>mustache, I</u> also managed to keep my voice under control.

———— 10. The craze for convenience in our country has gone too <u>far. There</u> are drive-in banks, restaurants, and even churches.

———— 11. My most valued possession is my stoneware <u>cooker, I</u> can make entire meals in it at a low cost.

———— 12. The shopping carts outside the supermarket seemed welded <u>together, Rita</u> could not separate one from another.

Standard English Verbs

———— 13. I am going to borrow my father's car if he <u>agree</u> to it.

———— 14. For recreation he sets up hundreds of dominoes, and then he <u>knocks</u> them over.

———— 15. He <u>stopped</u> taking a nap after supper because he then had trouble sleeping at night.

———— 16. There was no bread for sandwiches, so he <u>decided</u> to drive to the store.

Irregular Verbs

———— 17. I learned that Dennis had <u>began</u> to see someone else while he was still going out with me.

———— 18. That woman has never <u>ran</u> for political office before.

———— 19. I <u>knowed</u> the answer to the question, but I was too nervous to think of it when the instructor called on me.

———— 20. They had <u>ate</u> the gallon of natural vanilla ice cream in just one night.

Subject-Verb Agreement

———— 21. Her watchband <u>have</u> to be fixed.

———— 22. There <u>is</u> two minutes left in the football game.

———— 23. He believes films that feature violence <u>is</u> a disgrace to our society.

———— 24. The plastic slipcovers that she bought <u>have</u> begun to crack.

Consistent Verb Tense

_____ 25. Myra wanted to watch the late movie, but she was so tired she <u>falls</u> asleep before it started.

_____ 26. When the mailman arrived, I <u>hoped</u> the latest issue of *People* magazine would be in his bag.

_____ 27. Juan ran down the hall without looking and <u>trips</u> over the toy truck sitting on the floor.

_____ 28. Debbie enjoys riding her bike in the newly built park, which <u>features</u> a special path for bikers and runners.

Pronoun Agreement, Reference, and Point of View

_____ 29. At the Saturday afternoon movie we went to, children were making so much noise that <u>you</u> could not relax.

_____ 30. We did not return to the amusement park, for <u>we</u> had to pay too much for the rides and meals.

_____ 31. Drivers should check the oil level in <u>their</u> cars every three months.

_____ 32. At the hospital, I saw mothers with tears in their eyes wandering down the hall, hoping that <u>her</u> child's operation was a success.

_____ 33. Sharon's mother was overjoyed when <u>Sharon</u> became pregnant.

_____ 34. You must observe all the rules of the game, even if you do not always agree with <u>it</u>.

Pronoun Types

_____ 35. Nancy and <u>her</u> often go to dating bars.

_____ 36. No one in the class is better at computers than <u>he</u>.

Adjectives and Adverbs

_____ 37. The little girl spoke so <u>quiet</u> I could hardly hear her.

_____ 38. Lola looks <u>more better</u> than Gina in a leather coat.

Misplaced Modifiers

_____ 39. I saw sharks <u>scuba-diving</u>.

_____ 40. <u>With a mile-wide grin,</u> Betty turned in her winning raffle ticket.

_____ 41. I bought a beautiful blouse in a local store <u>with long sleeves and French cuffs.</u>

_____ 42. I first spotted the turtle <u>playing tag on the back lawn.</u>

Dangling Modifiers

_____ 43. When seven <u>years old</u>, Jeff's father taught him to play ball.

_____ 44. <u>Running across the field</u>, I caught the Frisbee.

_____ 45. <u>Turning on the ignition</u>, the car backfired.

_____ 46. <u>Looking at my watch</u>, a taxi nearly ran me over.

Faulty Parallelism

_____ 47. Much of my boyhood was devoted to getting into rock fights, crossing railway trestles, and <u>the hunt for rats in drainage tunnels</u>.

_____ 48. I put my books in my locker, changed into my gym clothes, and <u>hurried to the playing field</u>.

_____ 49. Ruth begins every day with warm-up exercises, a half-hour run, and <u>taking a hot shower</u>.

_____ 50. In the evening I plan to write a paper, <u>to watch a movie</u>, and to read two chapters in my biology text.

Capital Letters

_____ 51. When the can of <u>drano</u> didn't unclog the sink, Hal called a plumber.

_____ 52. I asked Cindy, ''<u>what</u> time will you be leaving?''

_____ 53. I have to get an allergy shot once a <u>Week</u>.

_____ 54. Mother ordered the raincoat at the catalog store on <u>Monday</u>, and it arrived four days later.

Apostrophe

_____ 55. I asked the record clerk if the store had Stevie <u>Wonders</u> latest album.

_____ 56. He's failing the course because he doesn't have any confidence in his <u>ability to do the work</u>.

_____ 57. Clyde was incensed at the dentist who charged him fifty dollars to fix his <u>son's</u> tooth.

_____ 58. I <u>cant</u> believe that she's not coming to the dance.

Quotation Marks

_____ 59. "Don't forget to water the grass, my sister said.

_____ 60. Martha said to Fred at bedtime, "Why is it that men's pajamas always have such baggy bottoms?" "You look like a circus clown in that flannel outfit."

_____ 61. The red sign on the door read, "Warning — open only in case of an emergency."

_____ 62. "I can't stand that commercial," said Sue. "Do you mind if I turn off the television?"

Comma

_____ 63. Hard-luck Sam needs a loan, a good-paying job, and someone to show an interest in him.

_____ 64. Even though I was tired I agreed to go shopping with my parents.

_____ 65. Power, not love or money, is what most politicians want.

_____ 66. The heel on one of Lola's shoes came off, so she spent the day walking barefoot.

_____ 67. "Thank goodness I'm almost done" I said aloud with every stroke of the broom.

_____ 68. I hated to ask Anita who is a very stingy person to lend me the money.

Commonly Confused Words

_____ 69. To succeed in the job, you must learn how to control your temper.

_____ 70. Fortunately, I was not driving very fast when my car lost its' brakes.

_____ 71. Put your packages on the table over their.

_____ 72. There are too many steps in the math formula for me to understand it.

_____ 73. The counseling center can advise you on how to prepare for an interview.

_____ 74. Who's white Eldorado is that in front of the house?

Effective Word Use

_____ 75. The teacher called to discuss Ron's social maladjustment difficulties.

_____ 76. I thought the course would be a piece of cake, but a ten-page paper was required.

_____ 77. When my last class ended, I felt as free as a bird.

_____ 78. Spike gave away his television owing to the fact that it distracted him from studying.

PART 2 (OPTIONAL)

Do the following at your instructor's request. This second part of the test will provide more detailed information about skills you need to know. On separate paper, number and correct all the items you have marked with an X. For example, suppose you had marked the word groups below with an X. (Note that these examples are not taken from the test.)

4. If football games disappeared entirely from television. I would not even miss them. Other people in my family would perish.

7. The kitten suddenly saw her reflection in the mirror, she jumped back in surprise.

15. The tree in my cousins front yard always sheds its leaves two weeks before others on the street.

29. When we go out to a restaurant we always order something we would not cook for ourselves.

Here is how you should write your corrections on a separate sheet of paper.

4. television, I
7. mirror, and
15. cousin's
29. restaurant, we

There are over forty corrections to make in all.

APPENDIX

ANSWERS
AND
CHARTS

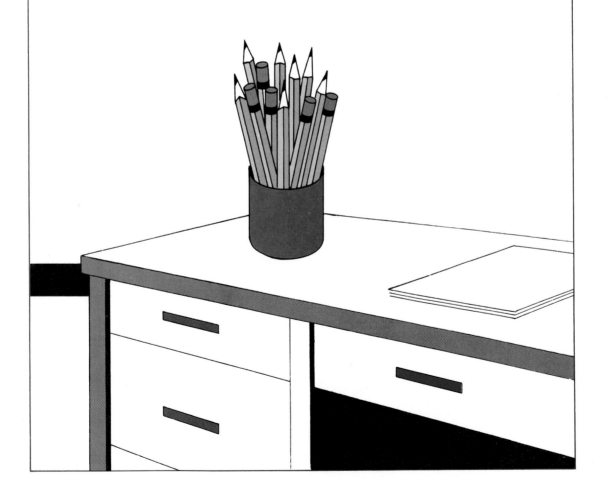

PREVIEW

This Appendix provides answers for the Sentence-Skills Diagnostic Test on pages 327–332 and for the Introductory Projects in Part Five. It also gives three useful charts: an assignment chart and a spelling list to be filled in by the student, and a general form for planning a paragraph.

ANSWERS TO SENTENCE-SKILLS DIAGNOSTIC TEST AND INTRODUCTORY PROJECTS

SENTENCE-SKILLS DIAGNOSTIC TEST (pages 327–332)

Sentence Fragments

1. X
2. C
3. X
4. X
5. C
6. X

Run-Ons

7. C
8. X
9. X
10. X
11. C
12. X

Standard English Verbs

13. C
14. C
15. X
16. X

Irregular Verbs

17. X
18. C
19. C
20. X

Subject-Verb Agreement

21. X
22. X
23. C
24. X

Consistent Verb Tense

25. X
26. C
27. C
28. X

Pronoun Agreement, Reference, and Point of View

29. X
30. C
31. X
32. C
33. X
34. C

Pronoun Types

35. X
36. C

Adjectives and Adverbs

37. X
38. X

Misplaced Modifiers

39. X
40. C
41. X
42. X

Dangling Modifiers

43. C
44. X
45. C
46. X

Faulty Parallelism

47. X
48. C
49. X
50. C

Capital Letters

51. X
52. X
53. C
54. X

Apostrophe

55. C
56. X
57. X
58. C

Quotation Marks

59. C
60. X
61. X
62. C

Comma

63. X
64. X
65. C
66. X
67. C
68. X

Commonly Confused Words

69. X
70. X
71. C
72. X
73. X
74. C

Effective Word Use

75. X
76. X
77. X
78. X

INTRODUCTORY PROJECTS

Sentence Fragments (page 341)

1. verb
2. subject
3. subject . . . verb
4. express a complete thought

Run-Ons (page 358)

1. period
2. *but*
3. semicolon
4. *When*

Standard English Verbs (page 373)

enjoyed . . . enjoys; started . . . starts;
cooked . . . cooks
1. past . . . *-ed*
2. present . . . *-s*

Irregular Verbs (page 382)

1. crawled, crawled (regular)
2. brought, brought (irregular)
3. used, used (regular)
4. did, done (irregular)
5. gave, given (irregular)
6. laughed, laughed (regular)
7. went, gone (irregular)
8. scared, scared (regular)
9. dressed, dressed (regular)
10. saw, seen (irregular)

Subject-Verb Agreement (page 390)

The second sentence in each pair is correct.

Consistent Verb Tense (page 395)

discovered . . . remembered

Pronoun Agreement, Reference, and Point of View (page 399)

The second sentence in each pair is correct.

Misplaced Modifiers (page 418)

1. Intended: A young man with references is wanted to open oysters.
   Unintended: The oysters have references.
2. Intended: On their wedding day, Clyde and Charlotte decided they would have two childen.
   Unintended: Clyde and Charlotte decided to have two children who would magically appear on the day of their wedding.
3. Intended: The students who failed the test no longer like the math instructor.
   Unintended: The math instructor failed the test.

Dangling Modifiers (page 422)

1. Intended: My dog sat with me as I smoked a pipe.
   Unintended: My dog smoked a pipe.
2. Intended: He looked at a leather-skirted woman.
   Unintended: His sports car looked at a leather-skirted woman.
3. Intended: A beef pie baked for several hours.
   Unintended: Grandmother baked for several hours.

Faulty Parallelism (page 427)

The second sentence in each pair reads more smoothly and clearly.

Capital Letters (page 436)

All the answers to questions 1 to 13 should be in capital letters.

14. The   15. I   16. "That . . ."

Apostrophe (page 449)

1. The purpose of the 's is to show possession (that Larry owned the motorcycle, the boyfriend belonged to the sister, Grandmother owned the shotgun, and so on).
2. The purpose of the apostrophe is to show the omission of one or more letters in the contractions — two words shortened to form one word.
3. The 's shows possession in each of the second sentences: the body of the vampire; the center of the baked potato. In each first sentence, the s is used to form simple plurals: more than one vampire; more than one potato.

Quotation Marks (page 458)

1. The purpose of quotation marks is to set off the exact words of a speaker. (The words that the young man actually spoke aloud are set off with quotation marks, as are the words that the old woman spoke aloud.)
2. Commas and periods go inside quotation marks.

Comma (page 466)

1. a. Frank's interests are Maria, television, and sports.
   b. My mother put her feet up, sipped some iced tea, and opened the newspaper.
2. a. Although the Lone Ranger used lots of silver bullets, he never ran out of ammunition.
   b. To open the cap of the aspirin bottle, you must first press down on it.
3. a. Kitty Litter and Dredge Rivers, Hollywood's leading romantic stars, have made several movies together.
   b. Sarah, who is my next-door neighbor, just entered the hospital with an intestinal infection.
4. a. The wedding was scheduled for four o'clock, but the bride changed her mind at two.
   b. Verna took three coffee breaks before lunch, and then she went on a two-hour lunch break.
5. a. Lola's mother asked her, "What time do you expect to get home?"
   b. "Don't bend over to pat the dog," I warned, "or he'll bite you."
6. a. Roy ate seventeen hamburgers on July 29, 1992, and lived to tell about it.
   b. Roy lives at 817 Cresson Street, Detroit, Michigan.

Other Punctuation Marks (page 476)

1. pets: holly
2. freeze-dried
3. Shakespeare (1564–1616)
4. Earth; no
5. proudly—with

Commonly Confused Words (page 501)

Your mind and body . . . . *There* is a lot of evidence . . . .
then it will . . . said *to* have . . . *It's* not clear

Effective Word Choice (page 513)

1. ''Flipped out'' is slang.
2. ''Few and far between'' is a cliché.
3. ''Ascertained'' is a pretentious word.
4. The first sentence here is wordy.

CHARTS

ASSIGNMENT CHART

Use this chart to record daily or weekly assignments in your composition class. You might want to print writing assignments and their due dates in capital letters so that they stand out clearly.

Date Given	Assignment	Date Due

Date Given	Assignment	Date Due

SPELLING LIST

Enter here the words that you misspelled in your papers (note the examples). If you add to and study this list regularly, you will not repeat the same mistakes in your writing.

Incorrect Spelling	Correct Spelling	Points to Remember
alright	all right	two words
ocasion	occasion	two "c"s

FORM FOR PLANNING A PARAGRAPH

To write an effective paragraph, first prepare an outline. Often (though not always) you may be able to use a form like the one below.

Topic sentence: _____

_____

_____

Support (1): _____

Details:

Support (2): _____

Details:

Support (3): _____

Details:

INDEX